T0135266

Advances in Intelligent Systems and Computing

Volume 950

The series "Advances in Intelligent Systems and Computing" contains publications on theory, applications, and design methods of Intelligent Systems and Intelligent Computing. Virtually all disciplines such as engineering, natural sciences, computer and information science, ICT, economics, business, e-commerce, environment, healthcare, life science are covered. The list of topics spans all the areas of modern intelligent systems and computing such as: computational intelligence, soft computing including neural networks, fuzzy systems, evolutionary computing and the fusion of these paradigms, social intelligence, ambient intelligence, computational neuroscience, artificial life, virtual worlds and society, cognitive science and systems, Perception and Vision, DNA and immune based systems, self-organizing and adaptive systems, e-Learning and teaching, human-centered and human-centric computing, recommender systems, intelligent control, robotics and mechatronics including human-machine teaming, knowledge-based paradigms, learning paradigms, machine ethics, intelligent data analysis, knowledge management, intelligent agents, intelligent decision making and support, intelligent network security, trust management, interactive entertainment, Web intelligence and multimedia.

The publications within "Advances in Intelligent Systems and Computing" are primarily proceedings of important conferences, symposia and congresses. They cover significant recent developments in the field, both of a foundational and applicable character. An important characteristic feature of the series is the short publication time and world-wide distribution. This permits a rapid and broad dissemination of research results.

** Indexing: The books of this series are submitted to ISI Proceedings, EI-Compendex, DBLP, SCOPUS, Google Scholar and Springerlink **

More information about this series at http://www.springer.com/series/11156

Francisco Martínez Álvarez ·
Alicia Troncoso Lora · José António Sáez Muñoz ·
Héctor Quintián · Emilio Corchado
Editors

14th International Conference on Soft Computing Models in Industrial and Environmental Applications (SOCO 2019)

Seville, Spain, May 13–15, 2019, Proceedings

 Springer

Editors
Francisco Martínez Álvarez
Data Science and Big Data Lab
Pablo de Olavide University
Seville, Spain

Alicia Troncoso Lora
Data Science and Big Data Lab
Pablo de Olavide University
Seville, Spain

José António Sáez Muñoz
University of Salamanca
Salamanca, Spain

Héctor Quintián
Department of Industrial Engineering
University of A Coruña
A Coruña, Spain

Emilio Corchado
University of Salamanca
Salamanca, Spain

ISSN 2194-5357   ISSN 2194-5365 (electronic)
Advances in Intelligent Systems and Computing
ISBN 978-3-030-20054-1   ISBN 978-3-030-20055-8 (eBook)
https://doi.org/10.1007/978-3-030-20055-8

This Springer imprint is published by the registered company Springer Nature Switzerland AG
The registered company address is: Gewerbestrasse 11, 6330 Cham, Switzerland

Preface

This volume of Advances in Intelligent and Soft Computing contains accepted papers presented at SOCO 2019 conference held in the beautiful and historic city of Seville (Spain), in May 2019.

Soft computing represents a collection or set of computational techniques in machine learning, computer science and some engineering disciplines, which investigate, simulate and analyze very complex issues and phenomena.

After a thorough peer review process, the 14th SOCO 2019 International Program Committee selected 57 papers which are published in these conference proceedings and represented an acceptance rate of 45%. In this relevant edition, a special emphasis was put on the organization of special sessions. Four special sessions were organized related to relevant topics as: Soft Computing Methods in Manufacturing and Management Systems; Soft Computing Applications in the Field of Industrial and Environmental Enterprises; Optimization, Modeling and Control by Soft Computing Techniques and Soft Computing in Aerospace, Mechanical and Civil Engineering: New methods and Industrial Applications.

The selection of papers was extremely rigorous in order to maintain the high quality of the conference, and we would like to thank the members of the Program Committees for their hard work in the reviewing process. This is a crucial process to the creation of a high-standard conference, and the SOCO conference would not exist without their help.

SOCO 2019 enjoyed outstanding keynote speeches by distinguished guest speakers: Prof. Dieu Tien Bui (University of South-Eastern Norway, Norway), Prof. Juan Manuel Corchado (University of Salamanca, Spain) and Prof. Julien Jacques (University of Lyon, France).

SOCO 2019 has teamed up with Neurocomputing (Elsevier) and Logic Journal of the IGPL (Oxford Academic) for a suite of special issues including selected papers from SOCO 2019. Furthermore, papers from a particular special session will be also considered for publication in special issues in Cybernetics and Systems: An International Journal (Taylor & Francis) and Expert Systems (Wiley).

Particular thanks go as well to the conference main sponsors, Startup Ole and IEEE SMC Spanish Chapter, who jointly contributed in an active and constructive manner to the success of this initiative.

We would like to thank all the special session organizers, contributing authors, as well as the members of the Program Committees and the Local Organizing Committee for their hard and highly valuable work. Their work has helped to contribute to the success of the SOCO 2019 event.

May 2019

Francisco Martínez Álvarez
Alicia Troncoso Lora
José António Sáez Muñoz
Héctor Quintián
Emilio Corchado

Organization

General Chairs

Francisco Martínez Álvarez	Pablo de Olavide University, Spain
Alicia Troncoso Lora	Pablo de Olavide University, Spain
Emilio Corchado	University of Salamanca, Spain

International Advisory Committee

Ashraf Saad	Armstrong Atlantic State University, USA
Amy Neustein	Linguistic Technology Systems, USA
Ajith Abraham	Machine Intelligence Research Labs (MIR Labs), Europe
Jon G. Hall	The Open University, UK
Paulo Novais	Universidade do Minho, Portugal
Amparo Alonso Betanzos (President)	Spanish Association for Artificial Intelligence (AEPIA), Spain
Michael Gabbay	King's College London, UK
Aditya Ghose	University of Wollongong, Australia
Saeid Nahavandi	Deakin University, Australia
Henri Pierreval	LIMOS UMR CNRS 6158 IFMA, France

Program Committee Chairs

Emilio Corchado	University of Salamanca, Spain
Francisco Martínez Álvarez	Pablo de Olavide University, Spain
Alicia Troncoso Lora	Pablo de Olavide University, Spain
Héctor Quintián	University of A Coruña, Spain

Program Committee

Albeto Herreros López	University of Valladolid, Spain
Alfredo Jimenez	KEDGE Business School, Spain
Andreea Vescan	Babes-Bolyai University, Romania
Angel Arroyo	University of Burgos, Spain
Anna Bartkowiak	University of Wroclaw, Poland
Anna Burduk	Wrocław University of Technology, Poland
Anton Koval	Zhytomyr State Technological University, Ukraine
Antonio Bahamonde	University of Oviedo, Spain
Camelia Chira	Babes-Bolyai University, Romania
Camelia Serban	Babes-Bolyai University, Romania
Camelia-M. Pintea	Technical University of Cluj-Napoca, North University Center at Baia Mare, Romania
Carlos Cambra	University of Burgos, Spain
Carlos Pereira	ISEC, Portugal
Carmen Benavides	University of León, Spain
Castejon Limas	University of León, Spain
Damian Krenczyk	Silesian University of Technology, Poland
Daniela Perdukova	Technical University of Kosice, Slovakia
David Alvarez Leon	University of León, Spain
David Griol	University Carlos III de Madrid, Spain
Dragan Simic	University of Novi Sad, Serbia
Eduardo Solteiro Pires	UTAD, Portugal
Eleni Mangina	University of A Coruña, Ireland
Eloy Irigoyen	University of the Basque Country, Spain
Enrique De La Cal Marín	University of Oviedo, Spain
Enrique Dominguez	University of Malaga, Spain
Enrique Onieva	University of Deusto, Spain
Esteban García-Cuesta	Universidad Europea de Madrid, Spain
Esteban Jove	University of A Coruña, Spain
Eva Volna	University of Ostrava, Czechia
Fernando Sanchez Lasheras	University of Oviedo, Spain
Florentino Fdez-Riverola	University of Vigo, Spain
Francisco Martínez Álvarez	Pablo de Olavide University, Spain
Georgios Ch. Sirakoulis	Democritus University of Thrace, Greece
Grzegorz Ćwikła	Silesian University of Technology, Poland
Héctor Quintián	University of A Coruña, Spain
Henri Pierreval	LIMOS-IFMA, France
Humberto Bustince	UPNA, Spain
Isaias Garcia	University of León, Spain
Iwona Pisz	Opole University, Poland
Jaime A. Rincon	Universitat Politècnica de València, Spain

Javier Sanchis Saez	Universitat Politècnica de València, Spain
Jesús D. Santos	University of Oviedo, Spain
Jiri Pospichal	University of Ss. Cyril and Methodius, Slovakia
Jorge García-Gutiérrez	University of Seville, Spain
Jose Gamez	University of Castilla-La Mancha, Spain
José Valente de Oliveira	Universidade do Algarve, Portugal
Jose Alfredo Ferreira Costa	Universidade Federal do Rio Grande do Norte, Brazil
José F. Torres	Pablo de Olavide University, Spain
Jose Luis Calvo-Rolle	University of A Coruña, Spain
José Luis Casteleiro-Roca	University of A Coruña, Spain
Jose M. Molina	University Carlos III de Madrid, Spain
Jose Manuel Gonzalez-Cava	University of La Laguna, Spain
Jose Manuel Lopez-Guede	University of the Basque Country, Spain
José Ramón Villar	University of Oviedo, Spain
Juan Gomez Romero	University of Granada, Spain
Juan Mendez	University of La Laguna, Spain
Julio César Puche Regaliza	University of Burgos, Spain
Krzysztof Kalinowski	Silesian University of Technology, Poland
Leocadio G. Casado	University of Almeria, Spain
Lidia Sánchez-González	University of León, Spain
Luis Magdalena	Universidad Politécnica de Madrid, Spain
Luis Alfonso Fernández Serantes	FH-Joanneum University of Applied Sciences, Spain
Luis Paulo Reis	University of Porto - FEUP/LIACC, Portugal
M. Chadli	University of Picardie Jules Verne, France
Maciej Grzenda	Warsaw University of Technology, Poland
Manuel Mejia-Lavalle	Cenidet, Mexico
Marcin Iwanowski	Warsaw University of Technology, Poland
Marcin Paprzycki	IBS PAN and WSM, Poland
Maria Tomas Rodriguez	The City University of London, UK
Maria Luisa Sanchez	University of Oviedo, Spain
Marius Balas	Aurel Vlaicu University of Arad, Romania
Matilde Santos	Universidad Complutense de Madrid, Spain
Mehmet Emin Aydin	University of the West of England, UK
Michal Wozniak	Wroclaw University of Technology, Poland
Mitiche Lahcene	University of Djelfa, Algeria
Oscar Castillo	Tijuana Institute of Technology, Mexico
Paul Eric Dossou	ICAM, France
Paulo Moura Oliveira	UTAD, Portugal
Paulo Novais	University of Minho, Portugal
Petr Dolezel	University of Pardubice, Czechia
Przemyslaw Korytkowski	West Pomeranian University of Technology, Szczecin, Poland
Reggie Davidrajuh	University of Stavanger, Norway

Richard Duro	University of A Coruña, Spain
Robert Burduk	Wroclaw University of Technology, Poland
Sebastian Saniuk	University of Zielona Gora, Poland
Sebastián Ventura	University of Cordoba, Spain
Stefano Pizzuti	Energy New technologies and sustainable Economic development Agency (ENEA), Italy
Sung-Bae Cho	Yonsei University, South Korea
Tzung-Pei Hong	National University of Kaohsiung, Taiwan
Urko Zurutuza	Mondragon University, Spain
Valeriu Manuel Ionescu	University of Pitesti, Romania
Vicente Matellan	University of Leon, Spain
Wei-Chiang Hong	Jiangsu Normal University, China
Wilfried Elmenreich	Alpen-Adria-Universität Klagenfurt, Austria
Zita Vale	GECAD - ISEP/IPP, Portugal

Special Sessions

Soft Computing Methods in Manufacturing and Management Systems

Program Committee

Damian Krenczyk (Organizer)	Silesian University of Technology, Poland
Bożena Skołud (Organizer)	Silesian University of Technology, Poland
Anna Burduk (Organizer)	Wrocław University of Technology, Poland
Krzysztof Kalinowski (Organizer)	Silesian University of Technology, Poland
Arkadiusz Gola	Lublin University of Technology, Poland
Bozena Skolud	Silesian University of Technology, Poland
Cezary Grabowik	Silesian University of Technology, Poland
Franjo Jovic	University of Osijek, Croatia
Grzegorz Ćwikła	Silesian University of Technology, Poland
Hongze Ma	Kone, Finland
Ivan Kuric	University of Zilina, Slovakia
Iwona Pisz	Opole University, Poland
Laszlo Dudas	University of Miskolc, Hungary
Paul Eric Dossou	ICAM, France
Reggie Davidrajuh	University of Stavanger, Norway
Sebastian Saniuk	University of Zielona Gora, Poland
Wojciech Bozejko	Wroclaw University of Technology, Poland

Soft Computing Applications in the Field of Industrial and Environmental Enterprises

Program Committee

Álvaro Herrero (Organizer)	University of Burgos, Spain
Alfredo Jimenez (Organizer)	KEDGE Business School, Spain
Alberto Rivas	University of Salamanca, Spain
Alfonso González Briones	University of Salamanca, Spain
Angel Arroyo	University of Burgos, Spain
Camelia Chira	Babes-Bolyai University, Romania
David Griol	Universidad Carlos III de Madrid, Spain
Dragan Simic	University of Novi Sad, Serbia
Jose Luis Calvo-Rolle	University of A Coruña, Spain
José Luis Casteleiro	University of A Coruña, Spain
José Ramón Villar	University of Oviedo, Spain
Julio César Puche Regaliza	University of Burgos, Spain
Manuel Grana	University of the Basque Country, Spain
Mercedes Rodríguez Sastre	Rey Juan Carlos I University, Spain
Montserrat Jimenez	Rey Juan Carlos I University, Spain
Pablo Chamoso	University of Salamanca, Spain
Pedro Antonio Gutierrez	University of Cordoba, Spain
Roberto Casado-Vara	University of Salamanca, Spain
Sung-Bae Cho	Yonsei University, South Korea

Optimization, Modeling and Control by Soft Computing Techniques

Program Committee

Eloy Irigoyen Gordo (Organizer)	University of the Basque Country, Spain
Matilde Santos (Organizer)	Universidad Complutense de Madrid, Spain
Jose Luis Calvo-Rolle (Organizer)	University of A Coruña, Spain
Mikel Larrea (Organizer)	University of the Basque Country, Spain
Agustin Jimenez	Polytechnic University of Madrid, Spain
Antonio Robles Alvarez	University of Oviedo, Spain
Antonio Sala	Universitat Politècnica de València, Spain
Antonio Javier Barragán	University of Huelva, Spain
Basil Al-Hadithi	Universidad Politécnica de Madrid, Spain
Fernando Castaño Romero	Polytechnic University of Madrid, Spain
Gerardo Beruvides	Hitachi Europe, Germany
Isabel García-Morales	University of Málaga, Spain
Javier Muguerza	University of the Basque Country, Spain

Jesus Fernandez-Lozano	University of Málaga, Spain
Jesus Lozano	University of Extremadura, Spain
Jose Manuel Lopez-Guede	University of the Basque Country, Spain
Jose-Luis Diez	Universitat Politècnica de València, Spain
Joseba Quevedo	Universitat Politècnica de Catalunya, Spain
Juan Albino Mendez Perez	Universidad de La Laguna, Spain
Juan Albino Mendez Perez	University of La Laguna, Spain
Luciano Alonso	University of Cantabria, Spain
Luis Magdalena	Polytechnic University of Madrid, Spain
Raquel Martinez	University of the Basque Country, Spain
Vicente Gomez-Garay	University of the Basque Country, Spain
Xavier Blasco	Universitat Politècnica de València, Spain

Soft Computing in Aerospace, Mechanical and Civil Engineering: New Methods and Industrial Applications

Program Committee

Soledad Le Clainche (Organizer)	Universidad Politécnica de Madrid, Spain
José M. Pérez (Organizer)	Universidad Politécnica de Madrid, Spain
Ricardo Vinuesa (Organizer)	KTH Royal Institute of Technology, Sweden
Esteban Ferrer	ETSIAE-Universidad Politecnica de Madrid (UPM), Spain
Francisco Martínez Álvarez	Pablo de Olavide University, Spain
Héctor Quintián	University of A Coruña, Spain
José F. Torres	Pablo de Olavide University, Spain
Jose Luis Calvo Rolle	University of A Coruña, Spain
Josem. Perez	Universidad Politécnica de Madrid, Spain

SOCO 2019 Organizing Committee

Francisco Martínez Álvarez	Pablo de Olavide University, Spain
Alicia Troncoso Lora	Pablo de Olavide University, Spain
José F. Torres Maldonado	Pablo de Olavide University, Spain
David Gutiérrez-Avilés	Pablo de Olavide University, Spain
Rubén Pérez Chacón	Pablo de Olavide University, Spain
Ricardo L. Talavera Llames	Pablo de Olavide University, Spain
Federico Divina	Pablo de Olavide University, Spain
Gualberto Asencio Cortés	Pablo de Olavide University, Spain
Miguel García Torres	Pablo de Olavide University, Spain
Cristina Rubio Escudero	University of Seville, Spain
María Martínez Ballesteros	University of Seville, Spain

Álvaro Herrero	University of Burgos, Spain
José Antonio Sáez-Muñoz	University of Salamanca, Spain
Héctor Quintián	University of A Coruña, Spain
Emilio Corchado	University of Salamanca, Spain

Contents

Contents

Machine Learning

Indexes to Find the Optimal Number of Clusters in a Hierarchical Clustering

José David Martín-Fernández[(✉)], José María Luna-Romera, Beatriz Pontes,
and José C. Riquelme-Santos

University of Seville, 41012 Seville, Spain
jfernandez94@us.es
http://grupo.us.es/minerva/

Abstract. Clustering analysis is one of the most commonly used techniques for uncovering patterns in data mining. Most clustering methods require establishing the number of clusters beforehand. However, due to the size of the data currently used, predicting that value is at a high computational cost task in most cases. In this article, we present a clustering technique that avoids this requirement, using hierarchical clustering. There are many examples of this procedure in the literature, most of them focusing on the dissociative or descending subtype, while in this article we cover the agglomerative or ascending subtype. Being more expensive in computational and temporal cost, it nevertheless allows us to obtain very valuable information, regarding elements membership to clusters and their groupings, that is to say, their dendrogram. Finally, several sets of data have been used, varying their dimensionality. For each of them, we provide the calculations of internal validation indexes to test the algorithm developed, studying which of them provides better results to obtain the best possible clustering.

Keywords: Machine Learning · Hierarchical clustering ·
Internal validation indexes

1 Introduction

In recent years, the size of the information available for various types of studies has grown considerably. Areas like medicine [1], social networks [2], energy [3] or electronic consumption [4] are just a few examples of this, with an increasing amount of data. This information needs to be processed to get some useful knowledge.

Among the different possible solutions to data analysis we focus on Machine Learning techniques, allowing us to extract the main features and a model covering the main information in a dataset. One of the most used model is called clustering, which determines the number of instances of a certain grouping within the data under study. Within the existing grouping variants, hierarchical clustering provides us with very interesting additional information. We can see the

© Springer Nature Switzerland AG 2020
F. Martínez Álvarez et al. (Eds.): SOCO 2019, AISC 950, pp. 3–13, 2020.
https://doi.org/10.1007/978-3-030-20055-8_1

evolution of the clusters in each step of the algorithm, thus studying the grouping of X elements within the data. There exists two subtypes within hierarchical clustering: dissociative or descending, starting from a group with all the elements, and ending in a cluster for each instance in the dataset; or agglomerative or ascending, starting with as many clusters as exists elements in the dataset and ending with a single agglomerative cluster with all of them.

Nowadays, there exists several frameworks to work with Machine Learning techniques to obtain knowledge. One of the most known is Apache Hadoop [5], that is built around the programming model based on the Google paradigm MapReduce [6]. Moreover, one of the most widely used open source projects is Apache Spark [7]. In the Google paradigm, it is read and written from the hard disk on many occasions, which reduces produces a detriment in the speed of data processing. Spark, the number of write/read cycles on the disk, so that intermediate calculations are logically and quickly stored in RAM. To do this, Spark uses a data structure called *"Resilient Distributed Datasets"* (RDD), that are specially designed to parallelize cache calculations with high data volume. In addition, this system contains the scalable library for Machine Learning (*MLlib*), with a series of such as algorithms classification, regression, recommendation systems and clustering techniques, will be of great help to achieve our goal [8].

The purpose of this article is to present a new agglomerative clustering technique implemented in Apache Spark. We have tested our algorithm using diverse datasets, that were created by means of a random database generator. Furthermore, we have applied different internal clustering validation indexes (CVIs) [9] in order to test our clustering results and compare the CVIs performance between the agglomerative hierarchical clustering implemented (AHC) and that provided by Spark as a dissociative modality, Bisecting K-Means (BK-Means) [10].

The rest of the document is organized as follows. Section 2 reviews different types of clustering techniques, as well as the CVIs used during our experimentation. Section 3 describes the algorithm and its implementation. Section 4 presents the experiments carried out and, finally, Sect. 5 summarizes the main conclusions of this work.

2   Related Work

In this section are reviewed the main grouping methods, as well as the internal validation indexes that have been used in our experimentation.

2.1   Clustering Methods

There are several types of clustering algorithms, which could be classified into the following categories depending on the method we use [11]:

- *Grouping by partitions:* Given a set of n elements, the partition method builds K groups, where each partition represents a cluster and $K \leq n$. It's based on

the principle of distance between the individuals, so given an initial K, a first solution could be obtained. Then the process consists in iterating over the dataset, moving objects between groups and trying to improve the previous solution.
- *Density-based methods:* In this approach it is possible to obtain a clustering whose groups are made up of high-density areas and separated from each other by low-density areas.
- *Grid-based methods:* It consist in dividing the elements into a finite cell space which is part of a grid structure. It is applied independently to the size of the data, and the difference is given by the number of cells in each dimension of the generated space.
- *Hierarchical clustering:* It groups data to form a set, or to separate some already existing sets to give origin to other two. Thus, the distance is minimized or the similarity between them is maximized. It is possible to choose different measures to quantify both distance and similarity in this type of grouping.

Within the family of hierarchical algorithms there are two versions or strategies that can be used:

- *Dissociative or descending:* It starts with a cluster that includes all the objects, from which successive divisions are made, forming smaller groups until as many groups as there are elements in the dataset are obtained.
- *Agglomerative or ascending:* It works the opposite way to the descending version. It starts with as many clusters as there are elements in our dataset. At each step, more and more clusters of instances are formed until you end up with a single cluster made up of all available data.

From the first group of algorithms, we can find numerous examples in the literature [12–14]. However, in this work we present a version of the agglomerative option. The main problem of this implementation lies in the computation time needed when treating with large amount of data. Hence, there are few examples in the literature of implementations of this type of strategies [15, 16].

2.2 Validation Indexes

The validation of the results obtained by clustering algorithms is a fundamental part of the clustering process. CVI have been typically used to evaluate the partition obtained. Most popular CVIs are *Dunn* [17] and *Silhouette* [18]. Furthermore, indexes presented in [9], have been used in this work, being some of them a simplification of *Dunn* and *Silhouette*.

Dunn and Silhouette. We describe in the following the two most used validation indexes in the literature, which have been implemented in this work for our experiments:

– *Dunn*: Measure widely applied in literature, but open to the possibility of choosing between several variants for calculation. The index is defined by:

$$Dunn = \frac{Min(Inter\text{-}cluster)}{Max(Intra\text{-}cluster)},$$ (1)

where *Inter-cluster* is computed as the distance between all the points of a certain cluster M to all the points of a cluster N, and divided by the product of the number of elements in both clusters. *Intra-cluster* represents the distance between all the elements that are part of a cluster, divided by the number of instances within that set.

– *Silhouette*: Silhouette distance is calculated for each point i of a cluster:

$$Silhouette_i = \frac{(b_i - a_i)}{Max(b_i, a_i)},$$ (2)

where b_i is the shortest distance between point i to the rest of points of any cluster in which i is not a part of; and a_i is the average distance between point i and the rest of the points of the clusters to which it belongs. Silhouette value is in the interval $-1 \le Silhouette_i \le 1$, being its optimal value equals to 1.

Other Indexes. As aforementioned, we have used three other validation indexes from [9] in order to check and compare the goodness of our clustering algorithm. *Davis-Bouldin* is included [19], which uses data object quantities and features inherent to the dataset to set the compactness and separation of the clusters; *BD-Silhouette* is a simplification of the traditional *Silhouette* index based on using intra-cluster and inter-cluster distances to the centroid of each cluster, rather than every element within them; and *BD-Dunn*, which also simplifies the calculations of the internal validation index *Dunn* is used the centroid of each group of data.

3   Our Proposal

In this section we present our approach for AHC. Our technique starts from as many clusters as instances and, in an ascending way, groups them until it reaches a single cluster. In the next, we show the pseudocode of our strategy is shown in Algorithm 1:

The algorithm receives as parameters: a RDD of objects of type *"Distance"* [20], (class created internally to represent the distance between any two elements of the database); the number of clusters to be obtained; the strategy for computing the distances; and the total number of instances in the dataset.

Line 10 refers to the calculation of the Cartesian product, which is necessary to find the distances between the points or clusters of each iteration with respect to the other elements of the RDD of objects of type *"Distance"*. In addition, the step performed on line 11 is configurable according to the designated strategy for calculating the distance between elements in the database, being the implemented options *"min"*, *"max"* and *"avg"*. They refer to the minimum, maximum or mean distance between the distances of the remaining points from each of the points that make up the pair found during line 2 of the algorithm, respectively. In the literature, each of these strategies establishes a different hierarchical clustering typology. Being "minimum or simple link grouping" (*simple linkage*), "maximum or complete link grouping" (*complete linkage*) and "average or average link grouping" (*average linkage*), respectively [21].

Algorithm 1. Agglomerative Hierarchical Clustering (AHC)

Input: RDD with objects *Distance*, number of clusters, strategy for the distance between elements and number of elements of the DataSet.
Output: Hierarchical clustering model.
1: **for** $a \leftarrow 0$ to $(elementsDataSetNumber - numClusters)$ **do**
2:     Find the next cluster as the pair of elements with the shortest distance between them within the RDD of *Distance* (they can be two DataSet points; a DataSet object and a cluster; or two clusters).
3:     Save the elements of the pair that make up the new cluster.
4:     Update the hierarchical clustering model with the cluster found on line 2 and its elements.
5:     **if** a < (elementsDataSetNumber - numClusters - 1) **then**
6:         Delete from the RDD of *Distance* the match found on line 2.
7:         Search the RDD for *Distance* for all the relationships between the first point or cluster found on line 2 and the rest of the elements.
8:         Search the RDD for *Distance* for all the relationships between the second point or cluster found on line 2 and the rest of the elements.
9:         Delete from the RDD of *Distance* all items found on lines 7 and 8.
10:        Calculate the Cartesian product from the elements found in lines 7 and 8.
11:        Add to the RDD of *Distance* the distances from the cluster found on line 2 to the other elements calculated on line 10.
12:    **end if**
13:    Every 5 iterations make a backup copy of the RDD of *Distance*.
14: **end for**

3.1 Implementation

For the creation of this hierarchical grouping, several variants can be made depending on the basis for storing the information, using *"Resilient Distributed Dataset"* (RDD) or *"DataFrames"*. Both objects are provided by Apache Spark. In addition, some functions from MLlib were used, which allows us to delegate some calculations of our algorithm.

Following Spark recommendations, *collect()* and *coalesce()* [22] methods were used in order to accelerate the process. With the *collect()* method, it is possible to obtain data stored in memory during previous calculations, so that it is not necessary to wait for the executions *lazy* of the Spark framework. Through the use of the *coalesce()* method, it is possible to reduce considerably the partitions in which the data are parallelized during the execution of our algorithm. Spark divides the stored data into four times the number of working nodes being used.

Spark offers other alternatives to solve this issue, such as using the *count()* method, which count the number of elements of a given RDD, thus forcing the system to use the data stored in memory. Finally, this alternative was replaced by the former one in our implementation due to the results in terms of execution time between both during the performance tests of the algorithm.

In addition, through experimentation, *checkpoints* each several iterations helps to achieve better performance of the algorithm. Therefore, it is necessary to remember the data that are within the RDD of the distances between all elements. After several performance tests, 5 iterations were inferred as the optimum number for these backups. As for the *coalesce()* method, tests were carried out with several multiples of 4, following Spark's recommendations that establish to use them by multiplying by the number of CPUs used during the execution. The best configuration was to use 8 partitions to distribute the data since the equipment where the tests have been performed has 4 CPUs (it is described in Sect. 4.1). Following Spark's recommendations, it is one of the configurations that usually work best.

4 Experimentation

In this section we present the experimental setup and results obtained using our AHC approach and the comparison with respect to the use of the dissociative clustering algorithm BK-Means.

4.1 Working Environment and Datasets

Our goal is to check the goodness of the different grouping obtained by our algorithm, using several datasets and evaluating the results by means of multiple CVIs. The experiments were executed in: IntelliJ IDEA development environment; the Apache Spark framework using the Scala language; and the Machine Learning library provided by the MLlib framework; a computer with an Intel Core i7-7700HQ CPU with 4 cores of 2.8 GHz, 16 GB of RAM, an SSD of 256 GB and a HDD of 1 TB.

A total of 60 datasets have been used in our experimentation, which were generated by using the database generator in [9]. This tool allowed us to configure the desired number of clusters, dimensions, and the number of points for each cluster.

For the experimentation, three different configurations for the number of clusters (K) have been used: 3, 5 and 7; 20 different configurations for the dimensionality of the data: from 1 to 20; and 100 points for each of cluster. In order to achieve the 20 different dimensions expressed above, a dataset has been taken as the basis for each different K with 20 total dimensions, from which the different dimensions from 1 to 20 have been selected.

4.2   Experimental Results

In order to study which CVI offers the best results using our hierarchical algorithm as the basis for clustering, we must first define how to measure this goodness. In our experiments, the modality *"avg"* distance of the hierarchical clustering algorithm has been chosen, explained in Sect. 3. As for the configuration
of the BK-Means algorithm, all the default values for parameters have been set.
Since the number of clusters (K) for each of the datasets is known beforehand,
we would check in how many datasets each of the indexes matches K. After
executing the algorithm for each of the datasets, different cluster numbers on
the resulting model for each CVI has been tested. Specifically, K values from
3 to 9 have been used $(3 < K < 9)$. This interval has been chosen in order to
guarantee that all K values can be found, since, the minimum would be 3 and
the maximum would be 7 in the data used during the experimentation.

The results have been grouped according to two criteria: by the number of
clusters, and by the number of dimensions in each of the databases. Following the
first of the criteria, each index could obtain a maximum of 20 hits in each of the
numbers of clusters. Whereas for the second criterion, each index could obtain a
maximum of 3 hits for each of the dimensions available in our databases. In this
sense, each of the indexes will have two different hits: one grouping all the cluster
numbers for each dimension, being able to obtain a maximum of 20; and another
grouping all dimensions for each cluster number, being able to obtain a maximum
of 3. The results can be summarized in the following Tables 1, 2 and 3:

Table 1. Summary of hits of each index grouped by the number of clusters in each
dataset.

Index	AHC (K3)	AHC (K5)	AHC (K7)	BK-Means (K3)	BK-Means (K5)	BK-Means (K7)	AHC (Total)	BK-Means (Total)
Silhouette	**20**	18	14	**20**	18	7	**52**	**45**
Dunn	19	18	14	17	18	7	51	42
Silhouette-BD	17	15	9	13	14	3	41	30
Dunn-BD	19	18	13	17	18	7	50	42
Davis-Bouldin	**20**	18	14	18	18	7	**52**	43

Table 2. Summary of hits of each index grouped by the number of dimension in each
dataset in AHC execution.

Index	1	2	3	4	5	6	7	8	9	10	11	12	13	14	15	16	17	18	19	20	Total
Silhouette	1	1	2	2	2	2	3	3	3	3	3	3	3	3	3	3	3	3	3	3	14
Dunn	1	1	1	1	2	2	2	3	3	3	3	3	3	3	3	3	3	3	3	3	14
Silhouette-BD	0	0	1	2	1	2	3	2	3	2	3	3	2	2	2	2	2	3	3	3	7
Dunn-BD	0	1	2	2	2	2	3	3	2	3	3	3	3	3	3	3	3	3	3	3	13
Davis-Bouldin	1	1	2	2	2	2	3	3	3	3	3	3	3	3	3	3	3	3	3	3	14

Table 3. Summary of hits of each index grouped by the number of dimension in each dataset in BK-Means execution.

Index	1	2	3	4	5	6	7	8	9	10	11	12	13	14	15	16	17	18	19	20	Total
Silhouette	1	1	2	2	2	2	3	2	3	3	3	3	3	2	3	2	2	2	2	2	**7**
Dunn	0	0	1	2	2	2	3	2	3	3	3	3	3	2	3	2	2	2	2	2	**7**
Silhouette-BD	0	0	0	1	0	0	1	2	2	3	3	2	2	2	2	2	2	2	2	2	2
Dunn-BD	0	0	1	2	2	2	3	2	3	3	3	3	3	2	3	2	2	2	2	2	**7**
Davis-Bouldin	1	0	1	2	2	2	3	2	3	3	3	3	3	2	3	2	2	2	2	2	**7**

Calculating the total success percentage of each of the CVIs studied would be as simple as dividing the "Total" columns of the previous tables by the maximum number of clusters and dimensions of those groupings, 60 and 20, respectively. Figure 1 summarizes the global results for each CVI, showing the percentage of total hits, and grouping the tests carried out by the number of clusters, and by the number of dimensions for each dataset.

The best indexes taking into account the grouping of databases by the number of clusters have been *Silhouette* and *Davis-Bouldin* in the case of using our AHC algorithm, both obtaining a success rate of 87%. Studying the case of the BK-Means algorithm it can be observed how the best index has been the only *Silhouette*, with a 75% success rate. If we study the percentages by the number of dimensions, we find the same previous winners plus *Dunn* in the case of using our AHC algorithm, all obtaining a 70% success rate on the number of dimensions. However, in the case of using the BK-Means algorithm, the *Dunn-BD* index would have to be added to the previous winners, all with a 35% success rate.

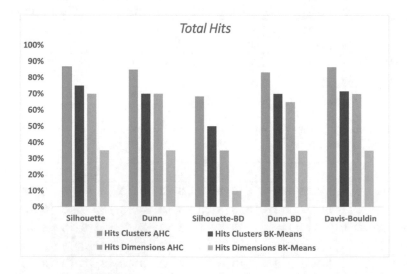

Fig. 1. Percentage of total hits for each of the indexes.

A much smaller percentage than in the case of the AHC algorithm. So it can be concluded that the best indexes to validate our hierarchical clustering algorithm are both *Silhouette* and *Davis-Bouldin*, as they are the ones that are most accurate in all the conditions studied in the case of using our AHC algorithm and the only *Silhouette* in the case of the BK-Means algorithm.

We have found that the greater the number of clusters, by the more difficult the finding the optimum number of clusters. Specifically, the worst results were obtained for $K = 7$. On the other hand, from the point of view of dimensions, all indexes have had problems when the databases had a low number of dimensions. More specifically from 1 to 6 in the most cases, concluding that the greater the number of dimensions, the better all the indexes studied in this article behave with our hierarchical clustering algorithm.

With respect to computation time, the following Table compare the different algorithms used Table 4:

Table 4. Average computation time for each type of hierarchical clustering.

Hierarchical type	Clustering (300p)	Clustering (500p)	Clustering (700p)
BK-Means	1.95 s	1.86 s	1.68 s
AHC	43.12 s	105.94 s	156.85 s

As it can be seen, the executions have been grouped according to the number of points of each dataset studied with 300, 500 and 700 points. In all the variants studied, the BK-Means algorithm has obtained better computation times than the proposed AHC algorithm, as expected.

5 Conclusions

This article presents a new approach for AHC, together with a performance comparison involving several clustering validation indexes. On the one hand, the effectiveness of the generated model by this clustering algorithm has been verified. On the other hand, we have established *Silhouette* and *Davis-Bouldin* as the best validation indexes for our algorithm. In addition, better results than the BK-Means algorithm developed by Apache Spark have been achieved in all indexes.

Moreover, all the indexes under study have problems in finding the optimum number of clusters when the number of clusters is high, and the dimensions of the points are reduced. Therefore, with the developed algorithm in environments may be used a high number of dimensions, since knowing a priori the optimal number of clusters in a database is not an easy task.

For future work, the main objective is to improve the implementation of this hierarchical clustering algorithm; by increasing the number of instances that can be introduced, and also trying to reduce the computation time with respect to the dissociative version (BK-Means).

All the code generated, as well as the used databases during the study can be found at the following link: https://github.com/Joseda13/LinkageClustering.

References

1. Krumholz, H.M.: Big data and new knowledge in medicine: the thinking, training, and tools needed for a learning health system. Health Aff. **33**(7), 1163–1170 (2014)
2. Su, Z., Xu, Q., et al.: Big data in mobile social networks: a QoE-oriented framework. IEEE Netw. **30**(1), 52–57 (2016)
3. Pérez-Chacón, R., Luna-Romera, J.M., et al.: Big data analytics for discovering electricity consumption patterns in smart cities. Energies **11**(3), 683 (2018)
4. Guo, H., Liu, Z., et al.: Big earth data: a new challenge and opportunity for digital earth's development. Int. J. Digit. Earth **10**(1), 1–12 (2017)
5. Dean, J., Ghemawat, S.: MapReduce. Commun. ACM **51**(1), 107 (2008)
6. Ghemawat, S., Gobioff, H., et al.: The Google file system. In: Proceedings of the Nineteenth ACM Symposium on Operating Systems Principles - SOSP 2003, vol. 37, no. 5, p. 29. ACM Press, New York (2003)
7. Apache Spark: Lightning-Fast C. C. https://spark.apache.org/
8. Apache Spark: Clustering Documentation (2019). https://spark.apache.org/docs/2.2.0/ml-clustering.html
9. Luna-Romera, J.M., García-Gutiérrez, J., et al.: An approach to validity indices for clustering techniques in Big Data. Progr. Artif. Intell. **7**, 1–14 (2017)
10. Apache Spark: Clustering - Bisecting k-means. https://spark.apache.org/docs/2.2.0/mllib-clustering.html#bisecting-k-means
11. Fahad, A., Alshatri, N., et al.: A survey of clustering algorithms for big data: taxonomy and empirical analysis. IEEE Trans. Emerg. Top. Comput. **2**(3), 267–279 (2014)
12. Sharma, A., López, Y., et al.: Divisive hierarchical maximum likelihood clustering. BMC Bioinform. **18**(S16), 546 (2017)
13. Kim, E., Oh, W., et al.: Divisive hierarchical clustering towards identifying clinically significant pre-diabetes subpopulations. In: AMIA – Annual Symposium Proceedings. AMIA Symposium, vol. 2014, pp. 1815–1824 (2014)
14. Patnaik, A.K., Bhuyan, P.K., et al.: Divisive Analysis (DIANA) of hierarchical clustering and GPS data for level of service criteria of urban streets. Alexandria Eng. J. **55**(1), 407–418 (2016)
15. Loewenstein, Y., Portugaly, E.: Efficient algorithms for accurate hierarchical clustering of huge datasets: tackling the entire protein space. Bioinformatics **24**(13), i41–i49 (2008)
16. Uchiyama, I.: Hierarchical clustering algorithm for comprehensive orthologous-domain classification in multiple genomes. Nucleic Acids Res. **34**(2), 647–658 (2006)
17. Dunn, J.C.: Well-separated clusters and optimal fuzzy partitions. J. Cybern. **4**(1), 95–104 (1974)
18. Rousseeuw, P.J.: Silhouettes: a graphical aid to the interpretation and validation of cluster analysis. J. Comput. Appl. Math. **20**, 53–65 (1987)
19. Davies, D.L., Bouldin, D.W.: A cluster separation measure. IEEE Trans. Pattern Anal. Mach. Intell. **PAMI-1**(2), 224–227 (1979)

20. Martín-Fernández, J.D., Luna-Romera, J.M.: Distance class (2018). https://github.com/Joseda13/linkage/blob/master/src/main/scala/es/us/linkage/Distance.scala
21. Hastie, T., Tibshirani, R., et al.: The Elements of Statistical Learning. Springer Series in Statistics. Springer, New York (2009)
22. Spark, A.: ScalaDoc - RDD. https://spark.apache.org/docs/latest/api/scala/index.html#org.apache.spark.rdd.RDD

Analysis and Application of Normalization Methods with Supervised Feature Weighting to Improve K-means Accuracy

Iratxe Niño-Adan[1]([⊠]), Itziar Landa-Torres[2], Eva Portillo[3], and Diana Manjarres[1]

[1] Tecnalia Research and Innovation, 48160 Derio, Spain
iratxe.nino@tecnalia.com
[2] Petronor Innovación S.L., 48550 Muskiz, Spain
[3] Department of Automatic Control and System Engineering,
School of Engineering, University of the Basque Country, UPV/EHU,
48013 Bilbao, Spain

Abstract. Normalization methods are widely employed for transforming the variables or features of a given dataset. In this paper three classical feature normalization methods, Standardization (St), Min-Max (MM) and Median Absolute Deviation (MAD), are studied in different synthetic datasets from UCI repository. An exhaustive analysis of the transformed features' ranges and their influence on the Euclidean distance is performed, concluding that knowledge about the group structure gathered by each feature is needed to select the best normalization method for a given dataset. In order to effectively collect the features' importance and adjust their contribution, this paper proposes a two-stage methodology for normalization and supervised feature weighting based on a Pearson correlation coefficient and on a Random Forest Feature Importance estimation method. Simulations on five different datasets reveal that our two-stage proposed methodology, in terms of accuracy, outperforms or at least maintains the K-means performance obtained if only normalization is applied.

Keywords: Normalization · Standardization ·
Weighted Euclidean Distance · Pearson correlation · Random Forest ·
K-means

1 Introduction

K-means algorithm is one of the most popular clustering algorithms [1]. It is an iterative partitioning algorithm, i.e., given a set S of n samples described by m variables, the algorithm splits the n samples into k disjoint clusters. These iteratively calculated clusters aim to minimize the Euclidean distance between

© Springer Nature Switzerland AG 2020
F. Martínez Álvarez et al. (Eds.): SOCO 2019, AISC 950, pp. 14–24, 2020.
https://doi.org/10.1007/978-3-030-20055-8_2

each sample and the centroid of its cluster, defined as the mean of the samples in the cluster. As widely known, the Euclidean distance, is sensitive to the variables' scale. Thus, variables with high scale and variance can dominate the Euclidean distance value and influence the performance of the algorithm. Consequently, different approaches like normalization or feature weighting have been proposed in order to minimize such dominances. Some of the most employed techniques are presented in what follows.

On the one hand, normalization and standardization methods are widely used as a preprocessing step. It is believed to be a necessary process when the variables present different scales [2], since it is assumed that these methods equalize the contribution of each variable [3] on the Euclidean distance. Albeit the amount of existing methods [4,5], as authors in [6] remark, in many cases the traditional Standarization or Z-score is used without taking into account other choices. In fact, the precision reached by different clustering methods for several datasets and their normalized versions after applying seven different normalization methods is analysed in [7] and it is concluded that Min-Max normalization obtains better results in terms of Adjusted Rand Index [8].

However, for adequately defining the prepossessing methodology, not only the selection of the normalization method should depend on the properties of the dataset, but also the choice of transforming or not the data with a normalization method should be carefully analysed. If a group structure exists in the data, i.e., if some variables are more discriminant for partitioning the samples according to the real clusters, the normalization can mask such structure [9], or even emphasize the contribution of less discriminant variables in the Euclidean distance. Then, as concluded in [10], it would be interesting to mix standardization and variable selection methods in order to obtain the representative variables of the group structure.

On the other hand, also related to minimizing the dominance of certain variables, the Weighted Euclidean distance adds a weight to each component, aiming at adjusting their contribution on the calculation of the distance proportionally to their capacity to distinguish the group structure. Recent proposals for calculating these weights can be divided into unsupervised and supervised approaches. Unsupervised methods are commonly utilised to update the weights at each step of the algorithm. Some proposals focus on updating the weights considering the inter and intra-cluster dispersion, such as [11] or [12]. The authors in [11] propose an adaptive weighted K-means that adjusts the current weight value using the distance between samples clustered from the previous iteration. Similarly, in [12] the authors propose a method to calculate the weights using the dispersion of the current partition of the data. However, this work uses a prefixed constant parameter β when estimating the weights that has to be selected in advance. The work presented in [13] includes the weight entropy on the objective function to identify the important variables. Regarding optimization based approaches, different methods can be found, ranging from more traditional techniques like Polack-Ribière optimization procedure [14] to evolutionary algorithms like Particle Swarm Optimization [15].

Other contributions employ the knowledge of the real cluster distribution to estimate the weights, following a supervised approach. For instance: similarly to [15], authors in [16] use a partial optimization in which the weights are elevated to a parameter β that requires an exhaustive search based on the accuracy achieved. Another approach can be found in [17] where a weighted Naive-Bayes algorithm classifier is presented. The authors calculate the pondered mutual information to estimate the correlation between the features and also between each feature and the class label to employ them as weights. Another proposal can be found in [18] where an hybrid algorithm that combines a genetic algorithm (GA) with a Support Vector Machine (SVM) classifier is presented. In [19] Information Gain is employed to extract the weights for a Feature Weighted K-Nearest Neighbor algorithm in order to predict stock markets indices. Authors in [20] present a hybrid method that by means of a feature weighting K-means obtains complex-value features for a Complex-Valued Artifitial Neural Network (CVANN) to improve the Parkinson disease diagnosis. Finally, authors in [21] apply a weighted K-means to detect intrusive activity in a network. First, a Random Forest classification algorithm is utilised to build intrusion patterns and to calculate the importance of the features. Then, these are employed by the weighted K-means to distinguish the anomalous network behaviour.

Due to the lack of information about the real cluster configuration, the unsupervised weighting methods can only assign the weights to optimize the inter and intra-cluster distance, which not necessarily entails that the obtained clusters correspond to the real ones. Similarly, the normalization methods applied without information about the importance of each variable can obscure the underlying group structure. In the case of the supervised feature weighting, although the labels are available, the selection of the method that best represents the variables' importance [22] and their degree of influence in the Euclidean distance not to disturb the contribution of the rest of them is a non trivial task.

In this work, an analysis of the normalization methods and supervised feature weighting factors applied to different synthetic datasets has been conducted. The hypothesis raised are described in Sect. 2. A two-stage methodology, including normalization and supervised feature weighting methods are presented in Sect. 3. Finally, Sect. 4 depicts the results obtained for five different datasets and the achieved conclusions are summarized in Sect. 5.

2   Hypothesis and Foundations

The hypothesis and foundations followed about the possible impact of transforming the features are described below.

1. As explained in Sect. 1, normalization methods are believed to equalize the contribution of each variable. However, each method emphasizes differently the contribution of the features. Given a certain dataset comprised of 2 dimensional samples and based on the premise that the clusters C_1, C_2 are represented by their centroids $c_1 = [\overline{x_1}, \overline{y_1}]$, $c_2 = [\overline{x_2}, \overline{y_2}]$ respectively, when a

new sample $[x_a, y_a]$ is considered, it is assigned to its closest cluster. Assuming $[x_a, y_a] - c_1 = [d_1, d_2]$, $[x_a, y_a] - c_2 = [d_3, d_4]$, $d_1 = d_4$, if $d_2 < d_3$, the sample is assigned to cluster C_1. However, if each feature is normalized by S_1, S_2 respectively, the sample is allocated in C_1 if and only if $\frac{d_1^2 - d_3^2}{S_1^2} < \frac{d_4^2 - d_2^2}{S_2^2}$, yielding to different clustering results depending on these values. This new clustering distribution will affect the accuracy obtained by K-means.

2. Normalizing the features of the dataset by the same method results on different transformation of the features. That means so, depending on the presence of outliers or the distribution of the variable, each variable can be more or less compressed, which means after all that the normalization process itself weights the features, i.e., applying the same method to different features can result on an excessive compression of the internal values of a representative feature. Then, depending on the aforementioned values S_1, S_2 the contribution of each feature in the results can vary substantially. For instance, if $S_1 \gg S_2$, then $\frac{d_1^2 - d_3^2}{S_1^2} \ll \frac{d_4^2 - d_2^2}{S_2^2}$ resulting on a excessive compression of the first feature, that alters the distance to the centroids and consequently the Euclidean distance calculation which affects the accuracy.

3. The presence of outliers affects the centroids of a certain distribution. For example, assuming the presence of one outlier $q = r + p$ due to the first feature of the dataset, being p the second highest value on the feature and assuming that both correspond to a cluster C_1, then, if the outlier is replaced by p ($q^* = p$), the obtained centroid is $\overline{c_1^*} = \overline{c_1} - (r/|C_1|)$, reducing the value of d_3, perhaps sufficiently to make $d_2 > d_3^*$ and to erroneously cluster the sample $[x_a, y_a]$ in C_2, and worsen the accuracy of the algorithm.

All the explained transformations over the features can lead to features contributing more than others, and without a priori knowledge of the feature relevance, no conclusions of which method is better can be taken.

3   Proposed Two-Stage Methodology for Normalization and Feature Weighting

In this work the sequential use of normalization and feature weighting is proposed for improving the accuracy of the K-means algorithm.

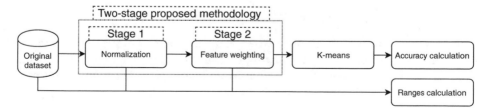

Fig. 1. Proposed two-stage methodology for normalization and feature weighting.

As shown in Fig. 1, the original dataset is firstly normalized (stage 1) and then, a feature weighting strategy (stage 2) is specially tailored in order to further modify the features so as to enhance the clustering outcome. The proposed two-stage transformation methodology is designed to intelligently adapt the features by means of analysing the effect of the transformations on the Euclidean distance and by adding the information of the relevance of the features. Then the K-means algorithm is applied and the accuracy is computed. Besides, the range of the features of each dataset is calculated in order to analyse their contribution on the Euclidean distance calculation.

3.1 First Stage: Normalization Methods

From the wide range of existing normalization methods, three of the most popular ones have been selected to conduct this work:

- Standardization (St) [5,7]: it is usually recommended for features that follow a normal distribution, despite being sensitive to the presence of outliers.
- Min-Max normalization (MM) [5,7]: it transforms the feature to unit range. As standardization, this method is also affected by the presence of outliers.
- Median Absolute Deviation (MAD) normalization [2,5]: it is a robust version of the St due to the utilization of central values, avoiding the outliers influence.

Note that the above mentioned normalization methods are a linear transformation of the original features X_j, with $j \in \{1, ..., m\}$ resulting in a transformed feature value $Z_j = \frac{X_j - p_j}{d_j}$ where p_j and d_j are statistics of position and dispersion of X_j, respectively. Table 1 gathers the statistics employed by each normalization method.

Table 1. Values for d_j and p_j used by each normalization method.

Normalization	d_j	p_j
St	$std(X_j)$	$mean(X_j)$
MM	$range(X_j)$	$min(X_j)$
MAD	$MAD(X_j)$	$median(X_j)$

3.2 Second Stage: Feature Weighting Strategy

The proposed feature weighting strategy aims at emphasizing the effect of the normalization by means of multiplying each feature by weights obtained from the real labels. In this work two different feature weighting strategies are selected to be applied after the normalization method:

- Random Forest for feature importance estimation (RF): this algorithm is considered due to its capability for obtaining the correlations between features and labels, even for nonlinear relationships among them. After applying the RF algorithm [23], the Mean Decrease Gini [24] is employed to calculate the feature weights. This way, a vector of weights W in which each position j establishes the relative importance of each feature respect to the real label is obtained in such a way that the sum of the weights is one.
- Pearson correlation coefficient [22]: in this case categorical labels can not be utilised. Then, the centroids are calculated employing the vector of real labels L, and each label $L_i, i \in \{1, ..., n\}$ is replaced by the centroid to which the sample belongs to. Hence, a new matrix of labels $L_{i,j}^*, i \in \{1, ..., n\}, j \in \{1, ..., m\}$ is created. Then a new vector r is calculated; it comprises the Pearson correlation coefficient r_j between each feature X_j of the dataset and the new labels $L_j^*, j = \{1, ..., m\}$. Finally, as the addition of weights must be one, the formed vector is transformed as follows:

$$ w_j = \frac{|r_j|}{\sum_{j=1}^{m} |r_j|} \tag{1} $$

The estimated weights are then multiplied by the normalized dataset ($P_j = w_j \cdot Z_j$), giving rise to the final input dataset P.

4   Results

In order to validate the proposed methodology, the following five datasets from the UCI repository have been chosen (see Table 2).

Table 2. Description of the employed datasets from the UCI repository [25].

Dataset	Samples	Features	N clusters	Samples by cluster
Iris	150	4	3	[50, 50, 50]
Wine	178	13	3	[59, 71, 48]
Wholesale customers (Channels)	440	6	2	[298, 142]
HTRU2 (Pulsars)	17898	8	2	[16259, 1639]
(Breast) diagnostic	569	30	2	[212, 357]

Table 3 shows the best accuracy values obtained by K-means over 100 Monte Carlo. The accuracy is computed for each Monte Carlo simulation as the percentage of the total number of well clustered samples among the total number of samples in the datasets. The maximum accuracy value obtained is presented. The applied methods are: original dataset, the selected normalization approaches (St, MM, MAD) described in Sect. 3.1 and the proposed two-stage methodology with all the possible combinations. Results show that the proposed methodology improves or at least equals the accuracy obtained.

Table 3. Best accuracy obtained over 100 Monte Carlo for the selected datasets.

Methods	Iris	Wine	Channels	Pulsars	Breast
Original dataset	89.33	70.22	59.32	76.70	85.41
St	83.33	97.19	77.05	93.66	91.04
MM	88.67	95.51	77.50	91.86	92.79
MAD	81.33	94.38	76.36	90.97	86.99
St+Pearson	88	96.07	77.50	96.25	92.79
MM+Pearson	90.67	94.94	**80**	**96.95**	**93.67**
MAD+Pearson	85.33	**97.19**	75.91	92.79	86.82
St+RF	96	96.07	75.91	96.07	91.56
MM+RF	96	95.51	75.68	96.41	93.32
MAD+RF	**96.67**	96.07	76.14	95.09	89.46

In order to show the contribution of each feature on the Euclidean distance, Tables 4, 5, 6 and 7 depict the features' ranges for the original dataset and after the different transformations. Although the range is not totally representative of the feature contribution on the Euclidean distance (i.e. the presence of outliers and the imbalanced datasets can also affect the K-means clustering results), the ranges put into relation with the corresponding accuracy results can be used to infer the features' influence in relative terms, i.e. the range can be considered as an indicator of the feature influence. Therefore, the conclusions obtained by the feature ranges analysis serve as a clustering performance indicator in this study. Note that the weights w_j achieved by each method are included in Tables 4, 5, 6 and 7 as an indicator of the feature importance. For the sake of brevity, the study does not include the results for the 30 features of the Breast dataset.

Table 4. Features' ranges for all the transformation methods for Iris dataset.

j	Original	St	MM	MAD	Pearson			RF			Weights	
					St	MM	MAD	St	MM	MAD	P	RF
1	3.6	4.347	1	5.143	1.022	0.235	1.209	0.395	0.091	0.468	0.235	0.091
2	2.4	5.535	1	9.6	1.035	0.187	1.796	0.136	0.025	0.235	0.187	0.025
3	5.9	3.344	1	4.72	0.969	0.29	1.368	**1.54**	**0.46**	**2.173**	**0.290**	**0.460**
4	2.4	3.145	1	3.429	0.906	0.288	0.987	1.334	0.424	1.454	0.288	0.424

Each row in Tables 4, 5, 6 and 7 shows the range for each feature for all the transformation methods described. As explained before, normalizing a dataset by the same method results in a different transformation of its features, compressing or expanding them in different degrees. As observed in Table 4, originally the second and fourth features presented the same range, consequently both

Table 5. Features' ranges for all the transformation methods for Wine dataset.

j	Original	St	MM	MAD	Pearson			RF			Weights	
					St	MM	MAD	St	MM	MAD	P	RF
1	3.8	4.681	1	5.588	0.439	0.094	0.524	0.518	0.111	0.619	0.094	0.111
2	5.06	4.529	1	9.731	0.297	0.066	0.638	0.191	0.042	0.411	0.066	0.042
3	1.87	6.816	1	11.688	0.298	0.044	0.511	0.077	0.011	0.131	0.044	0.011
4	19.4	5.809	1	9.463	0.377	0.065	0.614	0.133	0.023	0.217	0.065	0.023
5	92	6.441	1	9.2	0.273	0.042	0.391	0.226	0.035	0.323	0.042	0.035
6	2.9	4.634	1	5.743	0.401	0.087	0.497	0.178	0.039	0.221	0.087	0.039
7	4.74	4.745	1	5.677	0.487	0.103	0.583	0.764	0.161	0.914	**0.103**	0.161
8	0.853	4.259	1	6.235	0.251	0.059	0.367	0.057	0.013	0.084	0.059	0.013
9	3.17	5.538	1	8.342	0.338	0.061	0.509	0.145	0.026	0.219	0.061	0.026
10	11.72	5.055	1	7.762	0.463	0.092	0.711	0.734	0.145	1.128	0.092	0.145
11	1.23	5.381	1	7.455	0.474	0.088	0.657	0.483	0.09	0.669	0.088	0.090
12	2.73	3.845	1	5.25	0.383	0.1	0.523	0.422	0.11	0.576	0.100	0.110
13	1402	4.452	1	6.923	0.45	0.101	0.699	0.864	0.194	1.343	0.101	**0.194**

Table 6. Features' ranges for all the transformation methods for Channels.

j	Original	St	MM	MAD	Pearson			RF			Weights	
					St	MM	MAD	St	MM	MAD	P	RF
1	112148	8.867	1	18.946	0.703	0.079	1.503	0.471	0.053	1.006	0.079	0.053
2	73443	9.951	1	29.855	2.15	0.216	6.449	1.839	0.185	5.516	0.216	0.185
3	92777	9.763	1	29.991	2.787	0.285	8.561	2.358	0.241	7.243	0.285	0.241
4	60844	12.533	1	56.103	1.187	0.095	5.315	0.971	0.077	4.345	0.095	0.077
5	40824	8.562	1	57.057	2.553	**0.298**	**17.015**	**3.337**	**0.39**	**22.239**	**0.298**	**0.390**
6	47940	16.999	1	75.2	0.446	0.026	1.975	0.908	0.053	4.018	0.026	0.053

Table 7. Features' ranges for all the transformation methods for Pulsars dataset.

j	Original	St	MM	MAD	Pearson			RF			Weights	
					St	MM	MAD	St	MM	MAD	P	RF
1	186.805	7.282	1	14.456	1.201	0.165	2.385	1.369	0.188	2.718	0.165	0.188
2	74.007	10.815	1	17.254	0.964	0.089	1.538	0.495	0.046	0.79	0.089	0.046
3	9.946	9.347	1	45.896	1.813	**0.194**	8.904	**3.55**	**0.38**	17.431	**0.194**	**0.380**
4	69.894	11.332	1	147.737	1.97	0.174	25.69	1.922	0.17	25.06	0.174	0.170
5	223.179	7.572	1	189.441	0.744	0.098	18.612	0.41	0.054	10.259	0.098	0.054
6	103.272	5.304	1	20.005	0.639	0.12	2.41	0.43	0.081	1.621	0.120	0.081
7	37.679	8.362	1	15.544	0.801	0.096	1.489	0.331	0.04	0.614	0.096	0.040
8	1192.978	11.2	1	23.158	0.711	0.064	1.471	0.471	0.042	0.974	0.064	0.042

contribute similarly to the Euclidean distance. However, after the application of St normalization, the second feature presents a range 2.3 times higher than the original, while the fourth feature has only expanded 1.3 times its range. In the case of MAD normalization, the transformed ranges present new ranges 4 and 2.43 times higher than the original ones, respectively.

Furthermore,   comparing   the   transformation caused by different normalization methods over the same dataset, it can be observed that they distinctly influence the same feature. For instance, in Table 5 it is shown how the second feature – that originally presented a range of 5.06 – after applying St normalization its range is 4.529, lower than the original. However, MAD normalization obtains a range of 9.731, higher than the original. Thus, St compresses the second feature while MAD normalization expands it. Conversely, Table 5 depicts that employing the St normalization for the seventh feature the range is almost equal to the original, i.e., no transformation has been done over such feature. Nevertheless, MAD expands that range up to 5.677.

On the basis of the remarks made, due to the statistics employed by each normalization method (see Table 1), each feature is differently transformed. Thereupon, without having any knowledge about the representativeness of the group structure gathered in each feature (obtained by the real labels), no conclusion about the suitability of normalizing the dataset or the method which will obtain better results can be drawn. In fact, as it can be noticed from the accuracy obtained in Table 3 for the Iris dataset, normalizing can deteriorate the accuracy obtained by the raw dataset. Likewise, the accuracy depicted in the Table 3 for all the datasets shows that all normalization methods attain lower values than the proposed two-stage methodology.

Regarding the feature importance estimation, Tables 4, 5, 6 and 7 depict the weights obtained by the two described methods in Sect. 3. It can be easily shown that the importance, and hence, the calculated weights vary considerably depending on the method selected; in the case of Iris dataset, the weights obtained by Random Forest (which is more discriminant between the maximum and minimum values), in combination with the MAD normalization, represent the best approach. However, for the Wine dataset, MAD normalization combined with the weights calculated by Pearson reaches the highest accuracy. In contrast, the datasets formed by two clusters, i.e. Channels, Pulsars and Breast, obtain the best results when the weights are obtained by Pearson, as depicted in Tables 6 and 7. Furthermore, regarding Breast dataset, the approach that most fairly reflects these weights in the ranges of the features is the MM normalization combined with the weights of Pearson. Also, the results obtained by Wine, Channels, Pulsars and Breast validate the use of the real centroids in the Pearson correlation formula to calculate the weights that each feature should present in the Euclidean distance calculation for the K-means clustering.

In the light of the results, due to the inclusion of feature importance knowledge on the degree of contribution of each feature, it can be concluded that the two-stage methodology performs better than maintaining the original dataset or just normalizing in a single step basis.

5   Conclusions

This work has presented a thoughtful analysis of the transformation that normalization methods introduce to the features and their effect on the Euclidean distance. Three different normalization methods, i.e. Standardization (St), Min-Max (MM) and Median/Median Absolute Deviation (MAD), and two supervised feature weighting approaches based on a Pearson correlation coefficient and on a Random Forest Feature Importance calculation method have been studied in different synthetic datasets from UCI repository. It has been observed that the joint application of a normalization and a feature weighting approach clearly outperforms the K-means performance obtained if only normalization is applied. The proposed methodology is capable of gathering the importance of each feature and adjusting its contribution on the Euclidean distance in an effective way. Similarly, a deep analysis of the ranges of the transformed features is presented in order to infer knowledge about the group structure gathered by each feature and hence, being able to propose the best normalization method for a given dataset.

Acknowledgement. This work has been supported in part by the ELKARTEK program (SeNDANEU KK-2018/00032), the HAZITEK program (DATALYSE ZL-2018/00765) of the Basque Government and a TECNALIA Research and Innovation PhD Scholarship.

References

1. Blömer, J., Lammersen, C., Schmidt, M., Sohler, C.: Theoretical analysis of the k-means algorithm–a survey. In: Algorithm Engineering, pp. 81–116. Springer, Heidelberg (2016)
2. Daszykowski, M., Kaczmarek, K., Vander Heyden, Y., Walczak, B.: Robust statistics in data analysis—a review: basic concepts. Chemometr. Intell. Lab. Syst. **85**(2), 203–219 (2007)
3. Aksoy, S., Haralick, R.M.: Feature normalization and likelihood-based similarity measures for image retrieval. Pattern Recogn. Lett. **22**(5), 563–582 (2001)
4. Jain, A., Nandakumar, K., Ross, A.: Score normalization in multimodal biometric systems. Pattern Recogn. **38**(12), 2270–2285 (2005)
5. Pan, J., Zhuang, Y., Fong, S.: The impact of data normalization on stock market prediction: using SVM and technical indicators. In: International Conference on Soft Computing in Data Science, pp. 72–88. Springer, Heidelberg (2016)
6. Milligan, G.W.: Clustering validation: results and implications for applied analyses. In: Clustering and Classification, pp. 341–375. World Scientific, Singapore (1996)
7. Milligan, G.W., Cooper, M.C.: A study of standardization of variables in cluster analysis. J. Classif. **5**(2), 181–204 (1988)
8. Hubert, L., Arabie, P.: Comparing partitions. J. Classif. **2**(1), 193–218 (1985)
9. Dillon, W.R., Mulani, N., Frederick, D.G.: On the use of component scores in the presence of group structure. J. Consum. Res. **16**(1), 106–112 (1989)
10. Steinley, D.: Standardizing variables in k-means clustering. In: Classification, Clustering, and Data Mining Applications, pp. 53–60. Springer, Heidelberg (2004)

11. Tsai, C.Y., Chiu, C.C.: Developing a feature weight self-adjustment mechanism for a k-means clustering algorithm. Comput. Stat. Data Anal. **52**(10), 4658–4672 (2008)
12. Huang, J.Z., Ng, M.K., Rong, H., Li, Z.: Automated variable weighting in k-means type clustering. IEEE Trans. Pattern Anal. Mach. Intell. **27**(5), 657–668 (2005)
13. Jing, L., Ng, M.K., Huang, J.Z.: An entropy weighting k-means algorithm for subspace clustering of high-dimensional sparse data. IEEE Trans. Knowl. Data Eng. **19**(8), 1026–1041 (2007)
14. Makarenkov, V., Legendre, P.: Optimal variable weighting for ultrametric and additive trees and k-means partitioning: methods and software. J. Classif. **18**(2), 245–271 (2001)
15. Lu, Y., Wang, S., Li, S., Zhou, C.: Particle swarm optimizer for variable weighting in clustering high-dimensional data. Mach. Learn. **82**(1), 43–70 (2011)
16. Chan, E.Y., Ching, W.K., Ng, M.K., Huang, J.Z.: An optimization algorithm for clustering using weighted dissimilarity measures. Pattern Recogn. **37**(5), 943–952 (2004)
17. Jiang, L., Zhang, L., Li, C., Wu, J.: A correlation-based feature weighting filter for Naive Bayes. IEEE Trans. Knowl. Data Eng. (2018)
18. Phan, A.V., Le Nguyen, M., Bui, L.T.: Feature weighting and SVM parameters optimization based on genetic algorithms for classification problems. Appl. Intell. **46**(2), 455–469 (2017)
19. Chen, Y., Hao, Y.: A feature weighted support vector machine and k-nearest neighbor algorithm for stock market indices prediction. Expert Syst. Appl. **80**, 340–355 (2017)
20. Gürüler, H.: A novel diagnosis system for Parkinson's disease using complex-valued artificial neural network with k-means clustering feature weighting method. Neural Comput. Appl. **28**(7), 1657–1666 (2017)
21. Elbasiony, R.M., Sallam, E.A., Eltobely, T.E., Fahmy, M.M.: A hybrid network intrusion detection framework based on random forests and weighted k-means. Ain Shams Eng. J. **4**(4), 753–762 (2013)
22. Wei, P., Lu, Z., Song, J.: Variable importance analysis: a comprehensive review. Reliab. Eng. Syst. Saf. **142**, 399–432 (2015)
23. Breiman, L.: Random forests. Mach. Learn. **45**(1), 5–32 (2001)
24. Louppe, G., Wehenkel, L., Sutera, A., Geurts, P.: Understanding variable importances in forests of randomized trees. In: Advances in Neural Information Processing Systems, pp. 431–439 (2013)
25. UCI Machine Learning Repository. http://archive.ics.uci.edu/ml

Classifying Excavator Operations with Fusion Network of Multi-modal Deep Learning Models

Jin-Young Kim and Sung-Bae Cho[✉]

Department of Computer Science, Yonsei University, Seoul, South Korea
{seago0828, sbcho}@yonsei.ac.kr

Abstract. Prognostics and health management (PHM) aims to offer comprehensive solutions for managing equipment health. Classifying the excavator operations plays an important role in measuring the lifetime, which is one of the tasks in PHM because the effect on the lifetime depends on the operations performed by the excavator. Several researchers have struggled with classifying the operations with either sensor or video data, but most of them have difficulties with the use of single modal data only, the surrounding environment, and the exclusive feature extraction for the data in different domains. In this paper, we propose a fusion network that classifies the excavator operations with multi-modal deep learning models. Trained are multiple classifiers with specific type of data, where feature extractors are reused to place at the front of the fusion network. The proposed fusion network combines a video-based model and a sensor-based model based on deep learning. To evaluate the performance of the proposed method, experiments are conducted with the data collected from real construction workplace. The proposed method yields the accuracy of 98.48% which is higher than conventional methods, and the multi-modal deep learning models can complement each other in terms of precision, recall, and F1-score.

Keywords: Excavator · Classification · Deep learning · Multi-modal data · Autoencoder · Feature extraction

1 Introduction

Remaining useful life (RUL) prediction is an attempt to predict the time period over which a device can perform normal operation at a certain level [1]. This is an important step that allows us to keep track of the expected lifetime of the equipment, or the time of replacement or repair, without interruption. Because we can expect the burden on the equipment depending on the mode of operation the equipment is performing, we can predict the RUL if we look at the operation history of the equipment. However, the process of manually identifying and classifying work histories (e.g., sensor record, survey record, etc.) is costly. Therefore, it is necessary to study the operation mode, and classify the operation mode accurately.

We can classify the operation mode of the equipment through the sensor log of the equipment or the video that shows the operation of the equipment. As direct viewing and classifying of videos can be highly accurate, the method of automatic classification

© Springer Nature Switzerland AG 2020
F. Martínez Álvarez et al. (Eds.): SOCO 2019, AISC 950, pp. 25–34, 2020.
https://doi.org/10.1007/978-3-030-20055-8_3

of video has been studied extensively. However, as shown in Fig. 1, the performance varies greatly depending on the surrounding environment (weather, time, etc.) [2–4]. On the other hand, classifying the operation mode using sensor values is not significantly affected by the surrounding environment. However, as shown in Fig. 2, classification model using sensor values can hardly distinguish when the instrument performs similar operations. To show that it is difficult to classify the operation mode using the sensor data of the excavation used in this paper, the average and variance of the sensor values for each operation are illustrated in Fig. 3. Even though the statistics of the sensor data do not differ significantly from one operation to another, it is difficult to classify the operation of the equipment by using single-modal data. If we use the video and sensor values simultaneously, we can construct a model with higher performance and robustness.

Fig. 1. Examples of video data. From left to right, images show the operations of leveling, excavation, rock excavation, and run.

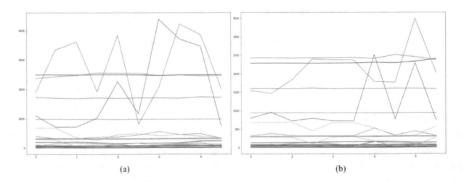

Fig. 2. A sensor graph used in this paper. The sensor values when the equipment excavates the ground (left) and rock (right). X- and y-axes represent time and sensor values, respectively.

In this paper, we propose a method to classify the operation mode by using both data at the same time to overcome the limitation of using one kind of data. Figure 4

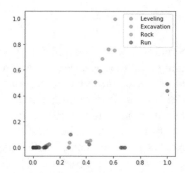

Fig. 3. Statistics of sensor values. X- and y-axes represent the average and standard deviation for each operation, respectively.

shows some examples of operation mode of excavator. The method is based on deep learning models to effectively extract the characteristics of each data [5–8]. The classification model for each data is constructed and learned first, and the output of the last hidden layer of each model is set as extracted features, which are concatenated into the input of the final classification model to classify the operation mode. The main contributions of this paper are as follows.

- We construct a model that extracts the characteristics of multi-modal data (i.e., video and sensor data) which are different kinds of data, design a fusion network that integrates the features, and evaluate the performance with several experiments.
- We collect the data used in the real workplace and use it in experiments to show the practicality of the proposed model.

The paper is organized as follows. In Sect. 2, we present relevant works that solve the classification problem with one kind of data. We propose a method in Sect. 3 that classifies the operation mode with multi-modal data. Section 4 shows the experimental results to evaluate the proposed method. The conclusions and discussion of this paper are covered in Sect. 5.

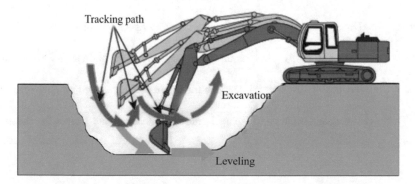

Fig. 4. Examples of operation mode of excavator. Excavation is the digging operation and leveling is the operation to level the ground.

2 Related Work

In order to classify the operation mode of equipment by video data or sensor information, a method of processing each data is needed. Many researchers have carried out to perform tasks with video or sensor information, as summarized in Table 1.

In the sensor-based model, various feature extraction techniques are used. Dao et al. performed multi-sensor classification using a sparse representation framework [9]. Chavez-Garcia and Aycard performed tasks using localization and mapping, detecting and tracking the moving objects [10]. Cao et al. classified the excavation equipment by machine learning method after preprocessing sensor signals with mel-frequency cepstral coefficient (MFCC) technique [11, 12]. Choi and Cho constructed modular Bayesian networks and an optimal stimulus decision module for predicting emotion with various sensor information [13]. Kim and Cho studied a method to predict energy demand by processing power demand [14].

In the study on video-based model, many attempts have been made to effectively classify and use various features in video [2–4]. Donahue et al. performed frame-by-frame feature extraction based on the convolutional neural network and sequential information processing through recurrent neural network [15]. Zha et al. extracted features of each image frame and classified them with support vector machine (SVM) [16]. Ye et al. used a method of extracting features by applying not only video frames but also optical flow [17]. Wu et al. proposed a method to separate motion and background in video to classify it [18].

Table 1. Summary of the related work.

Category	Author	Description
Sensor-based model	Dao et al. [9]	Propose sparse representation framework
	Chavez-Garcia and Aycard [10]	Use localization, mapping, detecting and tracking the moving objects
	Cao et al. [11]	Present extreme learning model to classify the excavator
	Choi and Cho [13]	Construct modular Bayesian networks to predict emotions in physical space with various sensor information
Video-based model	Donahue et al. [15]	Feature extraction by using convolutional neural network
	Zha et al. [16]	Extract feature of each frame and classify video using SVM
	Ye et al. [17]	Use video frame and extract feature through optical flow technique
	Wu et al. [18]	Separate motion and background features to classify video

3   Proposed Method

The overall structure of the proposed method is shown in Fig. 5. It consists of a video-based model, a sensor-based model, and a fusion network that combines the two models. In order to extract the characteristics of video frames, we use an encoder, which is pre-trained with an autoencoder, in a video-based model. The features extracted for each frame enter the input of the time series model and are converted into the characteristics of the video data. The sensor-based model extracts the features by considering the various sensor information with the convolutional neural network, and then outputs the sensor data characteristics to the time series model. The fusion network concatenates the characteristics of each data obtained from the two models and uses them as inputs to classify the operation mode of the equipment.

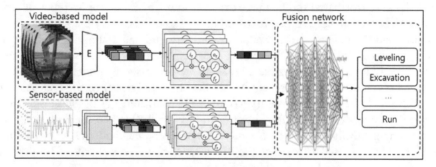

Fig. 5. The overall structure of the proposed model. It consists of a video-based model, a sensor-based model, and a fusion network. The video-based model extracts the frame feature with the encoder of the autoencoder and processes the sequence information to output the video feature. This is used as an input to the fusion-net with features extracted from the sensor-based model to make the model classify the operation mode.

3.1   Video-Based Model

The video data can be viewed as image sequence. We first extract the features of each video frame and convert the image data into a sequence of spatial features. We use a model of an autoencoder to automatically extract image features. An autoencoder, a way to learn the representation of the data, consists of an encoder that compresses data and a decoder that reconstructs the data [14, 19, 20]. We use the frames of the task video to train the autoencoder. The objective function of autoencoder is as follows.

$$\mathcal{L}_v^{AE} = \mathcal{L}_r(g_v(f_v(x)), x), \tag{1}$$

where $\mathcal{L}_r : \mathcal{X} \times \mathcal{X}$ is a loss function measuring the reconstruction error from $g_v(f_v(x))$ to x, g_v is a decoder, f_v is an encoder, and x is input data.

The encoder from learned autoencoder is reused in the video-based model to extract the features on a frame-by-frame basis. The sequence of extracted features is used as an input to the time series model and transformed into a feature of the video data. We use a long-short term memory (LSTM) network, one of the recurrent neural networks, as a time series model [21]. In the video-based model, we train it to classify the operation of the equipment by adding a fully-connected layer after the time series model. The objective function of video-based model is shown in Eq. (2). However, when learning fusion network, we try to extract the features of video data by reusing the models except the fully-connected layer in the learned video-based model.

$$\mathcal{L}_v = \mathcal{L}_c(h_v(f_v(x)), y), \tag{2}$$

where $\mathcal{L}_c : \mathcal{Y} \times \mathcal{Y}$ is a loss function measuring the difference between real label y and computed output $h_v(f_v(x))$ and h_v is newly added fully-connected layer.

3.2   Sensor-Based Model

We design a sensor-based model that extracts features by inputting 30 types of sensor values. All sensor values are preprocessed at the same interval by the sliding window technique and the features are extracted by the convolutional neural network. We also inspire the inception module to extract the characteristics of sensor values using various-sized filters [22]. The extracted features are processed by the LSTM model, and the operation mode of the equipment is classified by a fully-connected layer. As in the case of the video-based model, we try to extract the characteristics of the sensor data by reusing the models except the fully-connected layer in the learned sensor-based model when learning the fusion network. Unlike the video-based model, 1D convolutional neural network is used to extract the characteristics of sensor data. The objective function of sensor-based model is as follows.

$$\mathcal{L}_s = \mathcal{L}_c(h_s(f_s(x)), y), \tag{3}$$

where h_s and f_s are a classifier and a feature extractor of sensor-based model, respectively.

3.3   Fusion Network

We propose a fusion network to extract the characteristics of multi-modal data and to classify the operation mode by integrating them. There are some methods to ensemble existing models, such as voting and averaging. However, as can be seen in Figs. 1 and 2, there are limits to conventional methods because the classes that can be classified according to the characteristics of one type of data are different. Therefore, we propose a fusion network that extracts the features of each data, reuses the feature extraction parts of models with different kinds of data, and performs classification by combining them. As depicted in Fig. 5, we concatenate features extracted from the video-based model and the sensor-based model, and use them as inputs to the new fully-connected layer. The newly added layers are trained to classify the operation mode by properly combining the extracted features with the weights of the new fully-connected layer. The objective function of fusion network is described as follows.

$$\mathcal{L}_f = \mathcal{L}_c\big(h_f(f_v(x_v), f_s(x_s)), y\big), \tag{4}$$

where $h_f : \mathcal{H} \times \mathcal{H}$ is a newly added fully-connected layer which gets extracted features of video and sensor data as input, $\mathcal{H}$ is a space of extracted features, x_v is a video data, and x_s is a sensor data.

4 Experiments

4.1 Dataset and Experimental Settings

In order to verify the performance of the proposed model, we collected and used the video and sensor data of the excavator. Some examples of video data and sensor data used are shown in Figs. 1 and 2, respectively. A total of 30 sensor data are composed of 9 sensors associated with the engine, 12 sensors related to pressure, and 9 sensors associated with voltage and current. The frame size of the video data is 1920×1080, and it is reduced to 320×180 for computational convenience.

We performed a total of nine tasks and divided them into four classes. Information about each class is summarized in Table 2. We divided the collected data into 3:1 for training and test. The model distinguishes the operation type by looking at the operation video and sensor values for about 10 s.

Table 2. Details of data used in this paper. The "Time" column is the time of the collected operation mode, and the "Number" column is the number corresponding to the operation mode after the data preprocessing. The "Class" column represents the class that we want to classify.

Operation mode	Time	Number	Class
Leveling up down	25′48″	1519	Leveling
Leveling front-back	33′12″	1963	
Leveling left-right	25′26″	1497	
Digging	27′20″	1604	Excavation
Deep excavation	54′40″	3218	
Excavation	52′05″	3034	
Slope excavation	28′33″	1666	
Rock excavation	1:47′37″	6342	Rock excavation
Drive	58′12″	3463	Drive

4.2 Result Analysis

In this section, we illustrate the performance of the proposed method. We compare the 3D convolutional neural network, which has three-dimensional filters, Conv LSTM, which uses computation as a multiply operation in the typical LSTM, and the long-term recurrent convolutional networks (LRCN) model, which recurrently uses a convolutional neural network [23, 24]. As shown in Fig. 6, the proposed method has better performance than the conventional methods, as well as the models using one kind of data

32   J.-Y. Kim and S.-B. Cho

only. The performance of the fusion network is only 0.18% higher than that of the video-based model, but it has the advantage of complementing the advantages of both models.

We calculate the precision, recall, and F1-score for each class to see which operation the proposed model classifies well as shown in Table 3. The method solves the 4-class classification problem, but when calculating the precision, recall, and F1-score, the classification result is set as one versus rest for each class. In the leveling operation, the sensor-based model was stronger, and the video was better classified to distinguish the type of excavation. We can see the results of the fusion network taking advantage of both video and sensor-based model as a whole. Even in the excavation class, the performance exceeds that of the video and sensor-based models.

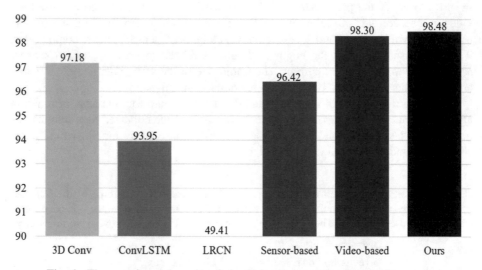

Fig. 6. The experimental results of classifying the operation mode of excavator.

Table 3. Results of classifying the operation with precision, recall and F1-score.

	Leveling	Excavation	Rock exc.	Run	Average
Video-based model					
Precision	0.9745	0.9813	0.9937	0.9838	0.9833
Recall	0.9758	0.9945	0.9994	0.9299	0.9749
F1-score	0.9752	0.9878	**0.9965**	**0.9561**	0.9791
Sensor-based model					
Precision	0.9723	0.9822	0.9226	0.9838	0.9652
Recall	0.9869	0.9540	0.9745	0.9299	0.9613
F1-score	0.9796	0.9679	0.9478	**0.9561**	0.9633
Fusion network					
Precision	0.9766	0.9862	0.9911	0.9828	0.9844
Recall	0.9830	0.9945	0.9994	0.9299	0.9767
F1-score	**0.9798**	**0.9903**	0.9952	**0.9561**	**0.9806**

5   Conclusion

In this paper, we addressed the need to classify the operation mode of excavator. Since there is a limit in classifying the operation mode with only sensor or video data, we constructed a method that performs tasks using both data. After learning the models that classify the excavator operation with each data, only the model that extracts the features was reused in the fusion network. The features extracted from the reused models enter the input of the new fully-connected layer and classify the operation of the excavator. The proposed method has the best performance compared to the conventional models with accuracy of 98.48%. Besides, through the analysis of F1-score, we confirm that the proposed method distinguishes each class complementarily from models using one data.

We will verify whether the reused feature extraction models extract the features of the video or sensor well by extracting the intermediate output. Although the disadvantages of sensor data that cannot distinguish between excavation and rock excavation are supplemented by the use of video data, we need to further collect various data to test whether the sensor data compensate for the disadvantages of the video data that are highly affected by the surrounding environment in the future. Moreover, we limited the number of classes to four, but we will classify all of the detailed classes in the future. Finally, we will apply the proposed model to an embedded board and install it in the real time classification system.

Acknowledgement. This work has been supported by a grant from Doosan infracore, Inc.

References

1. Sikorska, J.Z., Hodkiewicz, M., Ma, L.: Prognostic modeling options for remaining useful life estimation by industry. Mech. Syst. Sig. Process. **25**(5), 1803–1836 (2011)
2. Karpathy, A., Toderici, G., Shetty, S., Leung, T., Sukthankar, R., Fei-Fei, L.: Large-scale video classification with convolutional neural networks. In: Proceedings of IEEE Conference on Computer Vision and Pattern Recognition, pp. 1725–1732 (2014)
3. Wu, Z., Jiang, Y.G., Wang, X., Ye, H., Xue, X.: Multi-stream multi-class fusion of deep networks for video classification. In: Proceedings of ACM on Multimedia Conference, pp. 791–800 (2016)
4. Carreira, J., Zisserman, A.: Quo vadis, action recognition? A new model and the kinetics dataset. In: Proceedings of IEEE Conference on Computer Vision and Pattern Recognition, pp. 4724–4733 (2017)
5. LeCun, Y., Bengio, Y., Hinton, G.: Deep learning. Nature **521**(7553), 436–445 (2015)
6. Sanchez, D., Melin, P., Castillo, O.: Optimization of modular granular neural networks using a firefly algorithm for human recognition. Eng. Appl. Artif. Intell. **64**, 172–186 (2017)
7. Sanchez, D., Melin, P., Castillo, O.: A grey wolf optimizer for modular granular neural networks for human recognition. Comput. Intell. Neurosci. **2017**, 1–26 (2017)
8. Melin, P., Sanchez, D.: Multi-objective optimization for modular granular neural networks applied to pattern recognition. Inf. Sci. **460**, 594–610 (2018)

9. Dao, M., Nguyen, N.H., Nasrabadi, N.M., Tran, T.D.: Collaborative multi-sensor classification via sparsity-based representation. IEEE Trans. Sig. Process. **64**(9), 2400–2415 (2016)
10. Chavez-Garcia, R.O., Aycard, O.: Multiple sensor fusion and classification for moving object detection and tracking. IEEE Trans. Intell. Transp. Syst. **17**(2), 525–534 (2016)
11. Cao, J., Huang, W., Zhao, T., Wang, J., Wang, R.: An enhance excavation equipments classification algorithm based on acoustic spectrum dynamic feature. Multidimension. Syst. Sig. Process. **28**(3), 921–943 (2017)
12. Cao, J., Zhao, T., Wang, J., Wang, R., Chen, Y.: Excavation equipment classification based on improved MFCC features and ELM. Neurocomputing **261**, 231–241 (2017)
13. Choi, S.G., Cho, S.B.: Sensor information fusion by integrated AI to control public emotion in a cyber-physical environment. Sensors **18**(11), 3767–3787 (2018)
14. Kim, J.Y., Cho, S.B.: Electric energy consumption prediction by deep learning with state explainable autoencoder. Energies **12**(4), 739 (2019)
15. Donahue, J., Anne Hendricks, L., Huadarrama, S., Rohrbach, M., Venugopalan, S., Saenko, K., Darrell, T.: Long-term recurrent convolutional networks for visual recognition and description. In: Proceedings of IEEE Conference on Computer Vision and Pattern Recognition, pp. 2625–2634 (2015)
16. Zha, S., Luisier, F., Andrews, W., Srivastava, N., Salakhutdinov, R.: Exploiting image-trained CNN architectures for unconstrained video classification. arXiv preprint arXiv:1503.04144 (2015)
17. Ye, H., Wu, Z., Zhao, R.W., Wang, X., Jiang, Y.G., Xue, X.: Evaluating two-stream CNN for video classification. In: Proceedings of ACM on International Conference on Multimedia Retrieval, pp. 435–442 (2015)
18. Wu, Z., Wang, X., Jiang, Y.G., Ye, H., Xue, X.: Modeling spatial-temporal clues in a hybrid deep learning framework for video classification. In: Proceedings of the ACM International Conference on Multimedia, pp. 461–470 (2015)
19. Han, J., Zhang, D., Wen, S., Guo, L., Liu, T., Li, X.: Two-stage learning to predict human eye fixations via SDAEs. IEEE Trans. Cybern. **46**(2), 487–498 (2016)
20. Kim, J.Y., Cho, S.B., Detecting intrusive malware with a hybrid generative deep learning model. In: International Conference on Intelligent Data Engineering and Automated Learning, pp. 499–507 (2018)
21. Hochreiter, S., Schmidhuber, J.: Long-short term memory. Neural Comput. **9**(8), 1735–1780 (1997)
22. Szegedy, C., Liu, W., Jia, Y., Sermanet, P., Reed, S., Anguelov, D., Erhan, D., Vanhoucke, V., Rabinovich, A.: Going deeper with convolutions. In: Proceedings of IEEE Conference on Computer Vision and Pattern Recognition, pp. 1–9 (2015)
23. Xingjian, S.G.I., Chen, Z., Wang, H., Yeung, D.Y., Wong, W.K., Woo, W.C.: Convolutional LSTM network: a machine learning approach for precipitation nowcasting. In: Advances in Neural Information Processing Systems, pp. 802–810 (2015)
24. Donahue, J., Anne Hendricks, L., Guadarrama, S., Rohrbach, M., Venugopalan, S., Saenko, K., Darrell, T.: Long-term recurrent convolutional networks for visual recognition and description. In: Proceedings of IEEE Conference on Computer Vision and Pattern Recognition, pp. 2625–2634 (2015)

A Study on Trust in Black Box Models and Post-hoc Explanations

Nadia El Bekri[1]([✉]), Jasmin Kling[1], and Marco F. Huber[2,3]

[1] Fraunhofer IOSB, Karlsruhe, Germany
`nadia.elbekri@iosb.fraunhofer.de`
[2] Institute of Industrial Manufacturing and Management IFF,
University of Stuttgart, Stuttgart, Germany
[3] Center for Cyber Cognitive Intelligence (CCI), Fraunhofer IPA, Stuttgart, Germany

Abstract. Machine learning algorithms that construct complex prediction models are increasingly used for decision-making due to their high accuracy, *e.g.*, to decide whether a bank customer should receive a loan or not. Due to the complexity, the models are perceived as black boxes. One approach is to augment the models with post-hoc explainability. In this work, we evaluate three different explanation approaches based on the users' initial trust, the users' trust in the provided explanation, and the established trust in the black box by a within-subject design study.

Keywords: Machine learning · Black box · Explainability · Interpretability · Trust

1 Introduction

Decision-making based on machine learning is used in various applications to assist or even to replace human reasoning. Machines outperform humans when it comes to the amount of data they process in a short time. Furthermore, machines seem to make more consistent decisions since their only source for making decisions is the underlying data [2]. Nevertheless, if the data contains bias, the machines learn this as well. Since the models based on the machine learning algorithms are complex, the reasoning is mostly opaque. Trust has to be established [8] and is considered as one of the motivating aspects of intelligibility [7]. One approach to enhance intelligibility is to explain the decisions made by the complex model without entirely understanding how the model operates. The post-hoc explanations are generated by explainability methods on top of the black box. The human explanation approach is imitated by an explanation system, which can be used to provide explanations for these machine learning algorithms. An under-explored aspect thereby is to examine if the outcomes of explainability approaches really raise the users' trust in black box decisions.

The key contributions of this work is to examine if different explanation approaches really raise the users' trust based on the characteristics competency,

© Springer Nature Switzerland AG 2020
F. Martínez Álvarez et al. (Eds.): SOCO 2019, AISC 950, pp. 35–46, 2020.
https://doi.org/10.1007/978-3-030-20055-8_4

reliability, consistency and understandability. We analyze how different explanation approaches are perceived and moreover examine if explanations are trusted by users and which aspects of the explanation are desirable. A human-grounded evaluation [4] of different explanation approaches to compare explanations based on the measured trust in the explanation and in the black box is performed.

2   Intelligibility and Trust

Since intelligibility is a subjective concept, defining and measuring intelligibility is difficult. Doshi-Velez and Kim [4] characterize three evaluation approaches for intelligibility: application grounded, human grounded and functionality grounded. The application grounded uses the system understanding of a domain expert as the measure. The human grounded evaluation uses simpler experiments performed by an end user, not by a domain expert. The functionality grounded evaluation considers a formal measure.

2.1   Human Subject Studies and Trust Measures

Trust can be defined as "acceptance of an advice or an action, based on sufficient confidence in a positive outcome of that advice or action" [1]. For example, if domain knowledge already exists for a given task, the outcome of a prediction model can be seen as positive when the decision is in accordance to the domain knowledge. Based on this, trust can be measured as the ratio of the number of accepted actions and the number of all actions. It could be the case that faulty actions were accepted. This is why justifiable trust, the ratio of the number of accepted actions that had a positive outcome and the number of all actions, is desired. Poursabzi-Sangdeh *et al.* [9] measure trust by determining the difference between the prediction of the model and the participant's prediction. As a use case, they predict housing prices. The users were given some information about the underlying prediction model and are asked to make a prediction of the housing price on their own. This estimation is compared to the housing prices predicted by the model. Their absolute deviation is a measure of trust whereas smaller values indicate higher trust. Ribeiro *et al.* [10] conducted a user study to measure if the participants will trust the prediction model. At first, the participants are given ten different instances and their black box predictions. Eight out of ten predictions were made correctly by the black box, whereas two instances were misclassified. The participants were asked to answer the following three questions: *"Do you trust this algorithm to work well in the real world? Why? How do you think the algorithm is able to distinguish between the classes?"* Next, the participants were given ten different instances together with their explanations and asked the same questions. These two sets of answers are evaluated against each other to determine whether providing an explanation increases the understanding of the aspects the underlying model uses to make a decision and whether this knowledge influences trust. As a result, it is stated that providing an explanation helps understanding the aspects an model uses for decision making and knowing when to trust it.

2.2 Post-hoc Explanation Approaches

Many different explanation approaches exist, *e.g.*, providing logical statements [16], local models [14], deep explanations [15], rule extraction [6], feature importance [5], and feature tweaking [13]. For the user study, we focus on the post-hoc explanation approaches Local model-agnostic explanations (LIME) [10], CluReFI (extended version of LIME) and treeinterpreter [11]. We focused on those approaches due the fact that they are all based on feature importance and follow a clear concept. LIME and CluReFI both list the most influential features, either for the given instance (LIME) or for the cluster representative (CluReFI). Besides linear classifiers, decision trees are the most widely used classification techniques in real world applications. Therefore, we included treeinterpreter since it's intuitive and easy to understand.

LIME: The approach [10] provides instance explanations that exploit the proximity of the instances to be explained. Sparse linear models are learned for providing local explanations. A local LIME explanation visualizes the prediction probability of the black box and the decision-relevant features together with their importance (see Fig. 3) and the features that provide evidence against it. LIME answers the question *"What information did the model use to make this decision or which not?"*.

Cluster Representatives with LIME (CluReFI): In this paper, LIME was extended with additional information and is named CluReFI. The data was first clustered and the representative of the cluster was explained by LIME. The explanation (see Fig. 1) first assigns an unseen data instance its closest cluster and visualizes this assignment using the range of validity per feature. CluReFI visualizes the feature validity ranges of each cluster for the most important features contributing towards the selected class (in Fig. 1 four features are selected). For example, in Fig. 1 on the top left the validity ranges for the features credit duration is displayed. The actual customer (not the representative) is the orange circle, additionally the group's loan-trustworthy (green) and loan-untrustworthy (blue) are displayed. The selected group for the customer is circled in red. This is done for all important features that contribute towards the selected class. On the bottom left of Fig. 1, the tabular representation of the cluster representative is illustrated. The important features are highlighted to emphasize their importance. The third part of the explanation is the representative's explanation that illustrates a pie chart (see Fig. 1 bottom right) of the most important features contributing towards the class of the representative and their importance. In contrast to LIME, CluReFI illustrates the user only the most important features contributing towards the class for the representative.

Treeinterpreter: The treeinterpreter interprets the predictions made by decision trees (DT) and random forests (RF) [11]. In contrast to LIME treeinterpreter is model-specific only for DTs and RFs. The trained model for the evaluation was a RF. Each instance prediction is decomposed into the prediction

model's bias and the features contributing most to the instance prediction. The
decision path is marked in red (see Fig. 2).

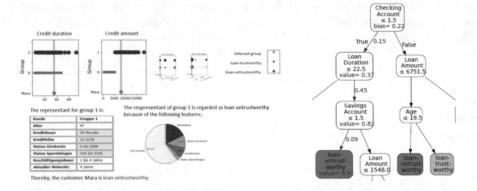

Fig. 1. Explanation for CluReFI

Fig. 2. Explanation for Tree-
Interpreter

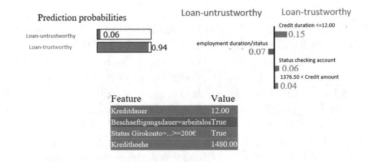

Fig. 3. Explanation for Lime

3   Method

3.1   Participants

28 participants attended the study. The youngest participant was 19 years
old and the oldest 58 years old. No specific background or requirements were
demanded. Half of the participants were younger than 27. In total more men
than women—79% compared to 21%—were surveyed. 43% of the participants
had theoretical experience with machine learning, 14% had no experience at
all, 7% had less than one year of practical experience, 25% had more than one
year but less than 3 years practical experience, 4% had more than three years of
practical experience and 7% specified another experience.

3.2   Materials

The user study was conducted via a graphical user interface [12] and was voluntary. No additional materials were allowed during the evaluation and no restrictions have been made for the users. Every explanation questionnaire was built up the same way (except the visualization of the specific explanation approach) to ensure the consistency.

3.3   Design

The user study is based upon a within-subject design. The setup of the user study was first evaluated in a pre-test before starting the main user study. Self-reports can be a critical source on a questionnaire but as the understanding of explanations is very subjective, we wanted to examine this additionally via self-reports. An interview with the participants in the pre-test phase was conducted, to examine if the measures were actually conform. Based on the German credit data set [3] regarding loan-trustworthiness, the use case is set up as follows: a financial institution is using a black box to classify a new customer as loan-trustworthy or loan-untrustworthy. Overall seven features (age, loan duration, loan amount, amount of checking account, amount of saving account, employment duration and the amount of years living in the present residence) of two exemplary customers are illustrated to the participants. Every customer was explained by the four approaches (three explanation approaches plus baseline explanation). The order of the approaches were randomized to counteract possible order effects and minimize learning across the conditions.

The two illustrated customers were the same for all participants. In the following, the hypotheses about the connection between explainability and trust, as well as the information needs are introduced.

Hypothesis 1: Participants with a bigger initial trust accept black box predictions more likely than participants with a lower initial trust.
Hypothesis 2: Providing an explanation increases the trust in the black box.
Hypothesis 3: Trust in an explanation correlates with the trust in the underlying black box.

Based on these hypotheses, different information needs have to be captured by the study. The variables regarding trust are divided into three categories: initial trust, trust in an explanation approach and trust in the underlying black box. In addition, the understandability of an explanation is measured. The four-point Likert scale was used to capture the participants' opinion on the information need except for the need for *understandability of an explanation*. The gradation of the scale were: *strongly agree, agree, disagree, strongly disagree*.

In the following, the information needs as well as the measurement are introduced and described.

Initial Trust (IT): First the participants' general trust (initial trust) is captured. This initial trust will serve as a base for categorizing participants based on their willingness and ability to trust. Therefore, statements regarding important decisions are listed and the participants are asked to specify their agreement with the given statements.

Understandability of an Explanation (UE): The understandability can be measured by statements about the features that were considered by the explanation and the most important features used. For the understandability of an explanation each question has a certain number of possible answers per explanation. The number of features varies for every explanation approach and displayed customer. When only one choice can be selected, the participant receives one point for the correct answer and zero points for the wrong answer. Whenever multiple choices can be selected, the participant receives one point for each correctly selected choice, but looses one point per incorrectly selected choice. The more points a participant gains, the better he understands the explanation. The amount of possible points depends on the number of features that were used by the explanation approach. The points are scaled and converted to a score between 0 and 1.

Trust in an Explanation (TE): Since self-reported understanding and actual understanding may differ, the explanation understandability metric is used to scale the self-reported understanding. When a participant does not understand the explanation, but reports to do so, his report is weakened based on his actual understanding. Trust in an explanation is queried based on eight statements (overall score between 0 and 24). This score is multiplied by the explanation understandability metric and then normalized to a range from 0 to 1. A final score of 1 indicates a very high trust in the explanation and a score of 0 indicates a very low trust in the explanation.

Trust in the Black Box (TBB): This metric measures the participants' trust in the underlying black box. The participants are queried about their understanding, reliability, consistency and competence of the black box model. The trust in the black box is queried via seven statements and rated on the four point Likert scale, an overall score between 0 and 21 can be reached. This score is scaled down to a range from 0 to 1. A final score of 1 indicates a high trust in the black box and 0 a low trust. The initial trust per participant is measured only once at the beginning of the user study and is grouped into the levels low, medium and high initial trust. The understandability of an explanation, the trust in an explanation, and the trust in the black box are measured once per explanation.

3.4   Procedure

At the beginning, the participants are informed about the topic, length and structure of the user study. Next, the different explanation approaches are introduced in a randomized order and evaluated by each participant. The first explanation only illustrates the prediction outcome (certain class) as well as the prediction probability. This explanation serves as a baseline when evaluating the explanations based on their understandability, trust in the explanation and trust in the black box that is generated by the explanation. Each presentation of the explanation approach is structured the same way. First, the contributions that the explanation approach makes are listed. Second, the explanation approach is described in more detail. Third, the explanation for the first customer is given and the participants' understanding of the explanation is measured. Fourth, the explanation for the second customer is given and the participants' understanding of the explanation is measured. Last, questions are asked about the explanations' explainability and trust, as well as the trust in the black box model that is created by providing the explanations. Thereby, each participant evaluates eight explanations regarding the two exemplary customers.

4   Results

4.1   Trust Variables

Based on the conducted user study and the aforementioned calculation of the different trust metrics, three different categories of trust variables exist: initial trust, trust in an explanation approach and trust in the underlying black box. Since parametric tests require normality, the trust variables were tested for normality to ensure the applicability. Almost one third of the participants (32%) have a low initial trust that is characterized by an initialTrust value less than or equal to 0.2. Half of the participants have an initialTrust value greater than 0.2 but less than or equal to 0.4, belonging to the category medium initial Trust. The remaining 18% have an initial Trust value greater than 0.4 and are characterized by high initial trust.

Influence of Demographic Data: This part uses the collected data to examine if the demographics of the participants have an influence on the trust in black boxes and in explanations. The influence of the different demographic variables *age*, *gender*, and *experience* were tested with Analysis of variance (ANOVA).

Age: H_0 : No statistically significant relationship between the variable *age* and the variables *Trust*.

First, the trust in each explanation was tested. There was no significance for the variables *TEBaseline* with $F(3, 24) = 1.6690$, $p = 0.2000$, *TELime* with $F(3, 24) = 0.4380$, $p = 0.7280$, *TECluReFI* with $F(3, 24) = 0.7800$, $p = 0.5170$, *TETree* with $F(3, 24) = 0.1970$, $p = 0.8970$. Furthermore, the trust in the black box that is created by each explanation was tested. There was no significance for the variables *TBBBaseline* with $F(3, 24) = 2.4710, p = 0.0862$, *TBBLime* with $F(3, 24) = 1.295$, $p = 0.2990$, *TBBCluReFI* with $F(3, 24) = 1.9150$, $p = 0.1540$, *TBBTree* with $F(3, 24) = 0.3120$, $p = 0.8100$.

Gender: $H_0 : \mu_{\text{male}} = \mu_{\text{female}}$.

First, the trust in each explanation was tested. The variables for *TEBaseline* with $F(1, 26) = 0.0780$, $p = 0.7830$, *TELime* with $F(1, 26) = 2.1220$, $p = 0.1570$, *TECluReFI* with $F(1, 26) = 0.1040$, $p = 0.7490$, *TETree* with $F(1, 26) = 0.3130$, $p = 0.5810$ were insignificant. Furthermore, the trust in the black box that is created by each explanation was tested. There was no significance for the variables *TBBLime* with $F(1, 26) = 0.1670$, $p = 0.6870$, *TBBCluReFI* with $F(1, 26) = 0.0020$, $p = 0.9610$, *TBBTree* with $F(1, 26) = 0.2900$, $p = 0.5950$. Only for *TBBBaseline* with $F(1, 26) = 6.4190$, $p = 0.0177 < 0.05$ the results were significant.

Experience: $H_0 : \mu_{\text{no experience}} = \mu_{\text{theoretical experience}} = \mu_{<1 \text{ year practical experience}}$ $= \mu_{1\text{–}3 \text{ years practical experience}} = \mu_{>3 \text{ years practical experience}} = \mu_{\text{other experience}}$.

First, the trust in each explanation was tested. There was no significance for the variables *TEBaseline* with $F(5, 22) = 0.9000$, $p = 0.4990$, *TELime* with $F(5, 22) = 1.9760$, $p = 0.1220$, *TECluReFI* with $F(5, 22) = 0.9100$, $p = 0.4920$, *TETree* with $F(5, 22) = 1.0740$, $p = 0.4020$. Furthermore, the trust in the black box that is created by each explanation was tested. There was no significance for for the variables *TBBBaseline* with $F(5, 22) = 0.9000$, $p = 0.4990$, *TBBLime* with $F(5, 22) = 1.9760$, $p = 0.1220$, *TBBCluReFI* with $F(5, 22) = 0.9100$, $p = 0.4920$. Only for *TBBTree* with $F(5, 22) = 3.5250$, $p = 0.0172 < 0.05$ the results were significant.

Hypothesis 1: The hypothesis was tested with ANOVA. It is assumed that the different categories of the grouped variable *initialTrust* do not have an influence on the trust in black boxes (i) and in an explanation (ii).

$$H_0 : \mu_{\text{low}} = \mu_{\text{medium}} = \mu_{\text{high}}.$$

(i) At a confidence level of $\alpha = 0.01$, H_0 can only be rejected for the variable *TBBLime* with $F(2, 25) = 6.0370$, $p = 0.0073$ (see results in Table 1). This indicates that the factors of initial trust interact very strong with the trust in black boxes generated by the explanation approach LIME. At a confidence level of $\alpha = 0.05$, H_0 can additional be rejected for the variable *TBBTree* with $F(2, 25) = 4.5700$, $p = 0.0204$ (see results in Table 1).

(ii) At a confidence level of $\alpha = 0.05$, H_0 can be rejected for the variable *TELime* with $F(2, 25) = 3.5620$, $p = 0.0435$ and *TETree* with $F(2, 25) = 5.0930$, $p = 0.0140$ (see results in Table 2).

Furthermore, we used the Post-Hoc-Tukey test for *TBBLime*, *TBBTree*, *TELime* and *TETree*. There is a significant difference between the participants with medium initialTrust and high initialTrust.

For the LIME and the treeinterpreter explanation approach, it can be concluded that participants with a medium initial trust are more likely than other

Table 1. Results of Hypothesis 1 (i)

Approach	df	MSE	F	p-value
TBBBaseline				
Between groups	2	0.0023	0.1660	0.8480
Within groups	25	0.0761		
TBBLime				
Between groups	2	0.1447	6.0370	0.0073
Within groups	25	0.0240		
TBBCluReFI				
Between groups	2	0.0543	1.1430	0.3350
Within groups	25	0.0474		
TBBTree				
Between groups	2	0.1220	4.5700	0.0204
Within groups	25	0.0267		

Table 2. Results of Hypothesis 1 (ii)

Approach	df	MSE	F	p-value
TEBaseline				
Between groups	2	0.0181	0.6920	0.5100
Within groups	25	0.0262		
TELime				
Between groups	2	0.1192	3.5620	0.0435
Within groups	25	0.0335		
TECluReFI				
Between groups	2	0.0910	1.8050	0.1850
Within groups	25	0.0504		
TETree				
Between groups	2	0.1860	5.0930	0.0140
Within groups	25	0.0365		

participants with a higher or lower level of initial trust to build trust in the underlying black box. Participants with a low or high level of initial trust are less likely influenced by providing an explanation. The assumption that participants with a higher initial trust accept black box predictions more likely than participants with a lower initial trust could not be confirmed for those variables. There are two explanations that are plausible for this behavior. Either, the initial trust of humans has no influence on their trust in black box predictions, or the collected data the variable *initialTrust* constitutes of is not actually measuring what is regarded as a humans' initial level of trust.

Hypothesis 2: It is assumed that the mean of the variable measuring the baseline for the trust in the black box and the means of the variables measuring trust in the black box which result from providing an explanation are equal.

$$H_0 : \mu_{\text{baseline}} = \mu_{\text{explanation approach}}.$$

Paired sample t-tests are conducted to test this hypothesis. All of the tests were significant at a significance level of $\alpha = 0.01$. Assuming that the mean of the baseline trust and the means of the trust generated by an explanation are equal does not hold. It can be concluded that the explanation increases the trust of the explanation receiver in the black box.

Hypothesis 3: We conducted three Pearson correlation tests to examine.
H_0 : There is no significant relationship between the variable *TE* and the variable *TBB*.

The hypothesis can be rejected for each test: The association between the variable *TELime* and the variable *TBBLime* was significant with $r(26) = 3.7945$, $p = 0.0008 < 0.01$. The pair *TECluReFI* and *TBBCluReFI* was significant with $r(26) = 4.7661$, $p = 0.0000 < 0.01$ and the variables *TETree* and *TBBTree* were significant with $r(26) = 3.6955$, $p = 0.0010 < 0.01$. The correlations for each explanation approach are illustrated in Figs. 4, 5 and 6.

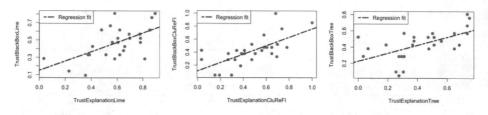

Fig. 4. y = 0.5428x +0.1499, **Fig. 5.** y = 0.6471 +0.1127, **Fig. 6.** y = 0.4942x +0.2255, $R^2 = 0.3564$ $R^2 = 0.4663$ $R^2 = 0.3444$

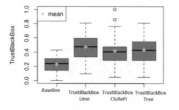
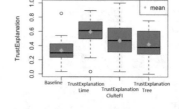

Fig. 7. TBB scores   **Fig. 8.** TE scores

Best Explanation: The best explanation is determined by the variables trust in the black box, trust in the explanation and the provided rankings. To examine the hypothesis ANOVA tests for TBB-Scores (see Fig. 7), TE-Scores (see Fig. 8) and Rankings with Post-Hoc-Tukey tests are performed. For the TBB vector the scores of TBBBaseline, TBBLime, TBBCluReFI and TBBTree were combined. Afterwards the vectors for the groups were created to illustrate the belonging of a group to an approach. That was equally done for the TE-Scores and the rankings. The results were significant with a significance level of $\alpha = 0.01$ for TBB with $F(3, 108) = 10.2400$, $p = 0.0000$, TE with $F(3, 108) = 8.2140$, $p = 0.0000$ and Ranking with $F(3, 108) = 48.5000$, $p = 0.0000$. For the Post-Hoc-Tukey tests pairwise comparisons were conducted. For TBB the pair (CluReFI (C)-Baseline (B)), (LIME (L)-B) and (Tree (T)-B) were significant with $\alpha = 0.01$. For TE the pair (L-B) and (T-L) were significant with $\alpha = 0.01$ and for Ranking the pair (C-B), (L-B), (T-B) and (L-C) were significant with $\alpha = 0.01$. In addition, the participants provided reasons for their ranking. LIME was preferred by 43% of the participants due to its understandability that is based on the familiarity with reading the explanation and a good compromise between level of detail and simplicity. The treeinterpreter was preferred by 39% of the participants due to its clear overview and good visualization. The representation is able to make the decision process, the steps it takes and the limits of the decision visible. Participants preferred CluReFI by 18% due to its informativeness regarding the cluster ranges. Participants felt to grasp the concept of the model and see if there exists a pattern.

5   Conclusion

A common thought is that explanations for black boxes definitely raise the users trust, but little research examines it. Although the prediction model and explanation are independent of each other, trusting the explanation approach is an important aspect of trusting the prediction model. The user study illustrates that creating explainability for the black box is valued by the recipient of the decision. The created explanation leads to a better understanding and thus acceptance of the black box. The user study indicated that for participants understandability and simplicity are essential. It is important to pick an optimal solution between informativeness and simplicity regarding a specific task. Our future work will consider a larger sample size and an improved design of the study design. Furthermore, we will consider participants with a high level of domain knowledge, which is likely to lead to even less trust in the early stages.

References

1. Alexandrov, N.: Explainable AI decisions for human-autonomy interactions. In: 17th AIAA Aviation Technology, Integration, and Operations Conference, p. 3991 (2017)
2. Davenport, T.H., Harris, J.G.: Automated decision making comes of age. MIT Sloan Manag. Rev. **46**(4), 83 (2005)
3. Dheeru, D., Karra Taniskidou, E.: UCI Machine Learning Repository (2017). http://archive.ics.uci.edu/ml
4. Doshi-Velez, F., Kim, B.: Towards a rigorous science of interpretable machine learning. arXiv preprint arXiv:1702.08608 (2017)
5. Goldstein, A., Kapelner, A., Bleich, J., Pitkin, E.: Peeking inside the black box: visualizing statistical learning with plots of individual conditional expectation. J. Comput. Graph. Stat. **24**(1), 44–65 (2015)
6. Lakkaraju, H., Kamar, E., Caruana, R., Leskovec, J.: Interpretable & explorable approximations of black box models. arXiv preprint arXiv:1707.01154 (2017)
7. Lipton, Z.C.: The mythos of model interpretability. arXiv preprint arXiv:1606.03490 (2016)
8. Mohseni, S., Ragan, E.D.: A human-grounded evaluation benchmark for local explanations of machine learning. arXiv preprint arXiv:1801.05075 (2018)
9. Poursabzi-Sangdeh, F., Goldstein, D.G., Hofman, J.M., Vaughan, J.W., Wallach, H.: Manipulating and measuring model interpretability. arXiv preprint arXiv:1802.07810 (2018)
10. Ribeiro, M.T., Singh, S., Guestrin, C.: Why should I trust you?: explaining the predictions of any classifier. In: Proceedings of the 22nd ACM SIGKDD International Conference on Knowledge Discovery and Data Mining, pp. 1135–1144. ACM (2016)
11. Saabas, A.: Treeinterpreter (2015). https://github.com/andosa/treeinterpreter
12. SoSciSurvey: SoSci survey tool (2018). https://www.soscisurvey.de/
13. Tolomei, G., Silvestri, F., Haines, A., Lalmas, M.: Interpretable predictions of tree-based ensembles via actionable feature tweaking. In: Proceedings of the 23rd ACM SIGKDD International Conference on Knowledge Discovery and Data Mining, pp. 465–474. ACM (2017)

14. Turner, R.: A model explanation system. In: 2016 IEEE 26th International Workshop on Machine Learning for Signal Processing (MLSP), pp. 1–6. Vietri sul Mare (2016)
15. Tzeng, F.Y., Ma, K.L.: Opening the black box-data driven visualization of neural networks. In: IEEE Visualization, VIS 2005, pp. 383–390. IEEE (2005)
16. Wang, T., Rudin, C., Doshi-Velez, F., Liu, Y., Klampfl, E., MacNeille, P.: OR's of AND's for interpretable classification, with application to context-aware recommender systems. arXiv preprint arXiv:1504.07614 (2015)

A Study on Hyperparameter Configuration for Human Activity Recognition

Kemilly D. Garcia[1,3]($\boxtimes$), Tiago Carvalho[2], João Mendes-Moreira[2], João M. P. Cardoso[2], and André C. P. L. F. de Carvalho[3]

[1] EWI-DMB, University of Twente, Enschede, The Netherlands
k.dearogarcia@utwente.nl
[2] INESC TEC, Faculty of Engineering, University of Porto, Porto, Portugal
{t.carvalho,jmoreira,jmpc}@fe.up.pt
[3] ICMC, University of São Paulo, São Carlos, SP, Brazil
andre@icmc.usp.br

Abstract. Human Activity Recognition is a machine learning task for the classification of human physical activities. Applications for that task have been extensively researched in recent literature, specially due to the benefits of improving quality of life. Since wearable technologies and smartphones have become more ubiquitous, a large amount of information about a person's life has become available. However, since each person has a unique way of performing physical activities, a Human Activity Recognition system needs to be adapted to the characteristics of a person in order to maintain or improve accuracy. Additionally, when smartphones devices are used to collect data, it is necessary to manage its limited resources, so the system can efficiently work for long periods of time. In this paper, we present a semi-supervised ensemble algorithm and an extensive study of the influence of hyperparameter configuration in classification accuracy. We also investigate how the classification accuracy is affected by the person and the activities performed. Experimental results show that it is possible to maintain classification accuracy by adjusting hyperparameters, like window size and window overlap, depending on the person and activity performed. These results motivate the development of a system able to automatically adapt hyperparameter settings for the activity performed by each person.

Keywords: Human Activity Recognition · Ensemble of classifiers · Semi-supervised learning · Mobile computing

This work was done in the context of the CONTEXTWA project and was partially funded by the ERDF - European Regional Development Fund through the Operational Programme for Competitiveness and Internationalisation - COMPETE 2020 Programme and by National Funds through the Portuguese funding agency, FCT - Fundação para a Ciência e a Tecnologia within project POCI-01-0145-FEDER-016883.

© Springer Nature Switzerland AG 2020
F. Martínez Álvarez et al. (Eds.): SOCO 2019, AISC 950, pp. 47–56, 2020.
https://doi.org/10.1007/978-3-030-20055-8_5

1   Introduction

Advanced mobile devices, such as smartphones, are usually integrated with several sensors capable of any-time sensing and data collection. The different types of motions sensors, such as accelerometers, gyroscopes and magnetometers, allow mobile devices to obtain substantial user-related information by monitoring and tracking movements of their users [3].

Human Activity Recognition (HAR) is a machine learning task focused on the use of sensing technologies to classify human activities and to infer human behavior [1]. Extensive research has been carried out in this area in the last decade [4–7], for applications like health and well-being [2], mobile security [3,9] and elderly care [1].

Most approaches of HAR found in the literature are based on supervised learning algorithms and assume that the data true label is always available. However, this assumption may not be feasible in real online scenarios, when labeled data is rare and the system feedback has to occur at runtime. As an example, in a fall detection system for elderly care, the classification feedback must occur as close as possible to the real moment of the user's fall [1].

Besides, as human beings perform activities differently, dissonant input signals are expected for the same activity [8]. To keep accuracy over time, classification models, used in HAR systems, need to be adapted to the current user. However, due to limitations of most mobile devices, different hardware resources need to be manage, such as battery and execution power, in order to keep the system efficiently working and accurate over time. Thus, there is a trade-off between amount of processed information and the resources available.

This work is based on an ensemble classifier firstly described in [15]. This algorithm has two phases, an offline and an online phase. In the beginning, the offline phase, the ensemble model is trained with labeled data from several users. In the online phase, this ensemble is used as a basic model to classify activities from a specific user, not present in the ensemble training. The ensemble model can be updated online with the user's data, if the classification has a high confidence factor.

The main contributions of this work are the extensive study of two hyperparameters important for HAR classification: the window size and the overlapping between windows (overlap factor). We analyze the impact of these hyperparameters in the model classification accuracy. Additionally, we conducted experiments with an ODROID-XU+E board[1] to evaluate the impact of these hyperparameters regarding energy consumption and execution time in a hardware similar to a smartphone.

This paper is structured as follows. Section 2 presents the related work on HAR and window parameterization. In Sect. 3 we describe the methodology

---

[1] ODROID-XU+E is a board mainly consisting of an Exynos5 Octa SoC, which includes 2 quad cores ARM CPUs and a PowerVR GPU, and a power measurement circuit to measure CPU, GPU and DRAM power consumption. The Exynos5 Octa SoC has been used in a number of families of smartphones.

applied in this study. The results obtained with the experiments are presented and discussed in Sect. 4. Finally, in Sect. 5, we summarize our main conclusions and point out future work directions.

2 Related Work

Dobbins et al. [2] propose an approach that uses personal data to better infer lifestyle choices for its users. Considering only labeled data, they evaluate the predictive performance of 10 supervised HAR classifiers in terms of accuracy and mobile system performance (execution time and energy consumption). Their experimental setup is based on a fixed window size of 512 samples and overlap factor of 0.5, i.e., 256 samples are reused from the previous window and only 256 new samples are used for the current window. They suggest that the sensing data should be processed in the cloud and not in the device. However, personal privacy and Internet connection are not considered. Furthermore, all data used is labeled, which cannot be guaranteed in a real online mobile system. The datasets used in the experiments contain complex activities and different user's data, but the results are not compared in terms of accuracy per user.

Mannini et al. [10] propose an SVM classifier to detect 4 activities from 33 different users. The classifier performance was tested for different window sizes, but not for the overlap between consecutive windows. Also, they do not compare, in terms of execution time, the classification task with different window sizes. The results show large variability among users performing the same activity, due to the problem of different sensor body location.

Window size has also been discussed by other authors. For example, [11–14] compare the predictive performance of classifiers over a set of window sizes. However, most of the studies do not consider the use of overlap factor and the impact of the user on the obtained accuracy.

In [11] it is presented an extensive review of the literature in window size and HAR. The accuracy of several classifiers was analyzed for different window sizes, but not regarding users. Additionally, the experimental setup was not elaborated with a leave-one-user-out, which would be a more realist approach. Instead, they used a cross validation approach, which is more affected by user variability than leave-one-participant-out.

The study conducted in this paper uses the PAMAP2 public dataset [15]. PAMAP2 includes a vast number of sensors and more complex activities than the data used by many of others studies. This dataset allows the study of the impact on HAR accuracy for different window sizes, users and activities.

3 Activity Recognition Overview

The HAR classification task can be split into 4 main steps, as illustrated in Fig. 1. The steps 1, 2 and 3 are the training phase with multiples users. The steps 1, 2 and 4 are used for the online user-specific classification.

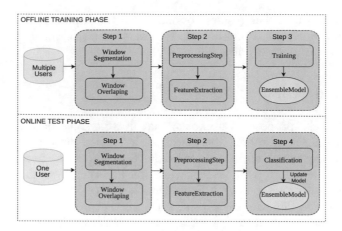

Fig. 1. Overview of the semi-supervised ensemble model for HAR.

Figure 1 shows a batch with raw data extracted from different wearable sensors and/or smartphones. The raw data samples are stored in a sliding window with fixed size. Ideally, a sliding window should contain data from a unique activity. However perfect segmentation is not always feasible, so a between-window overlap factor can be used to include samples from sequential activities. Also the size of the sliding window is reduced by the overlap factor, allowing a reduction of stored data. Thus, the step 1 is the window segmentation of the raw data and the overlapping of sequential windows.

Sensor's data are usually susceptible to noise, especially the accelerometers data [2]. Thus, it is important to process and convert the data into meaningful values. A pre-processing step (step 2) may also include calibration and filtering of the input signals in order to reduce noise. Sequential to that, a Feature Extraction (step 2) is used to calculate a single instance containing features that are then used for building the ensemble model. These features (see, e.g., [16]) include time-domain calculus, specifically mean and standard deviation for each sensor signal and correlation (Pearson correlation) between axes for the 3D sensors.

Each new instance is used to train (step 3) an ensemble model composed by three classifiers: kNN, VFDT and Naive Bayes. The implementation of the ensemble classifier is the combination of Democratic Co-Learning [17] and Tri-Training [18].

After training the ensemble model, in the online phase, sensors data are acquired from a single user. This data is pre-processed and features are extracted from them, similar to the processes (step 1 and 2) described in training phase. Each generated instance is classified by the ensemble (step 4), which classifies the instance and provides a confidence factor for that classification. The instances classified with high confidence, more than 99% value, are used to update the ensemble model.

4   Experimental Results

We conducted several experiments with our approach using the PAMAP2 dataset [15]. The objectives of these experiments are: compare the accuracy of a supervised HAR versus a semi-supervised HAR when using different configurations of the hyperparameters: window size and overlap factor. We also intend to study a HAR system behavior with the different hyperparameters configurations in terms of classification accuracy, energy consumption and execution time.

4.1   The PAMAP2 Dataset

The PAMAP2 [15] is a public dataset for human physical activities[2]. The data was collected from tree devices positioned in different body areas: wrist, chest and ankle. Each device has three sensors embedded: a 3-axis accelerometer, a 3-axis gyroscope and a 3-axis magnetometer.

The PAMAP2 dataset contains 1.926.896 samples of raw sensor data from 9 different users and 18 different activities. The activities executed by the users are divided in basic activities (walking, running, Nordic walking and cycling), posture activities (lying, sitting and standing), everyday activities (ascending and descending stairs), household (ironing and vacuum cleaning) and fitness activities (rope jumping). Also, the users were encouraged to perform optional activities (watching TV, computer work, car driving, folding laundry, house cleaning and playing soccer).

4.2   Experimental Setup

Using the PAMAP2 dataset [15], each ensemble was trained with data from 8 users and tested with one isolated user, not presented in the training process. This approach is called leave-one-user-out.

We conducted four experiments with two ensemble models, each one consisting of three classifiers: kNN, Naive Bayes, and Hoeffding Tree (VFDT), as in [8]. As verified in [2], these three classifiers have good classification performance in HAR problems. Thus, we analyze the accuracy performance of one ensemble model with a semi-supervised approach and another ensemble model with a supervised approach.

The box-plots correspond to the variance in accuracy for different values of window size (from 100 to 1000 with increments of 100), overlap factor (from 0.0 to 0.9 with increments of 0.1) and users (from 1 to 9 with increments of 1 user per experiment).

4.3   HAR Accuracy Results

Figure 2 presents the *accuracy* (axis y) for each value of the overlap factor, *overlapping* (axis x). The semi-supervised model reduces accuracy variance, compared with supervised model, for most of the overlapping and has average accuracy close to 90%. For both models, overlapping has more influence on accuracy

---

[2] http://archive.ics.uci.edu/ml/datasets/pamap2+physical+activity+monitoring.

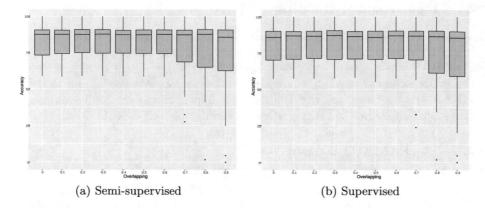

(a) Semi-supervised   (b) Supervised

Fig. 2. Supervised vs semi-supervised ensemble accuracy: overlap factor influence.

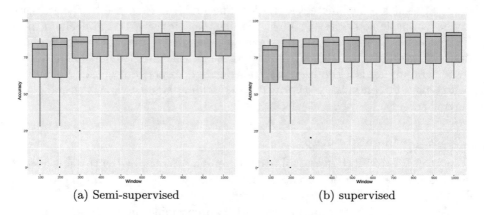

(a) Semi-supervised   (b) supervised

Fig. 3. Supervised vs semi-supervised ensemble accuracy: window size influence.

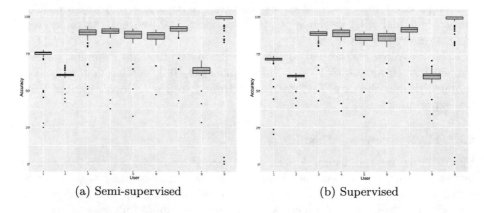

(a) Semi-supervised   (b) Supervised

Fig. 4. Supervised vs semi-supervised ensemble accuracy: user influence.

for values higher than 0.7, but the semi-supervised model is less susceptible to that influence than the supervised model. As shown in Fig. 3, variance of *accuracy* (axis y) and *window size* (axis x), the semi-supervised model reduces accuracy variance for each value of window size. We also notice that windows with small sizes have worse results, especially for sizes of 100 and 200.

We analyze the models accuracy for each user. For that, we analyze the variance of *accuracy* (axis y) when varying the hyperparameters *window size* and *overlapping* for each *user* (axis x). In Fig. 4, for both models, user 5 and 6 have variance higher than users 2 and 1. The semi-supervised model reduces accuracy variance for users 4 and 8. An interesting case to analyze is user 9. For most of the cases, the accuracy is 100%, however user 9 only has instances for the Rope Jumping activity, which means that this user influences the results to higher values. In some cases, the individual accuracy can be lower, as we can see with users 2, 8 and 1, justifying the analyzes by user instead of analyzing all population.

With the results, we can see that window size and overlapping do influence the accuracy of the models. Based on these results and depending on the HAR application, one can decide about the window size and overlap factor level that make possible a certain minimum desired classification accuracy. The exhaustive exploration allows us to also understand the acceptable ranges to explore within a runtime autotuning system, e.g., to keep a minimum accuracy (e.g. 80%). These ranges can be used, at runtime, to search for the best combination of the hyperparameters that provide the best results, e.g., in terms of execution time or energy consumption. The following subsection shows the impact on execution time and energy consumption of different window sizes and overlap factor.

4.4 Execution Time and Energy Consumption

We also analyze the execution time and energy consumption for processing all the data from the PAMAP2 dataset. For that, we conducted experiments in an ODROID-XU+E[3] system running Android. The experiments focus on a single user, user 6, and the execution time and energy required to process 250.096 raw samples.

The first experiment is about the execution time necessary to process all the data of user 6. The execution time was divided into three parts. The first part, *samplingTime*, represents the time required to access all the data from the user and the instantiation of each data window as an instance. The second part, *featureTime* is the feature extraction and the "final instance" instantiation, this part depends on the *window size* and the *overlap* factor used. The last part, *classificationTime* is the total time required to classify all the instances calculated in the feature extraction phase.

In Fig. 5, since the feature extraction depends on the window size, the time to calculate all instances increases as the window size also increases, despite the decreasing number of calculated features. This means that the feature extraction

---

[3] https://www.hardkernel.com/.

phase is sensitive to the number of raw instances to process. Furthermore, as we increase the overlap factor, due to the increased number of instances that are calculated, the execution time also increases. The classification time is rather small and slightly increases as the number of calculated features augments.

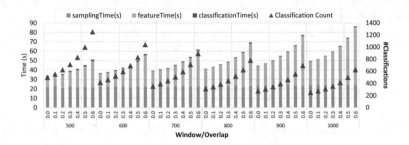

Fig. 5. Total execution time required (left axis) to process the PAMAP2 dataset, per window size and overlap factor, divided in three parts: sampling (data extraction), features extraction and classification. The number of classifications per configuration (right axis) is shown as triangle marks.

The second experiment is presented as a heat map representing the energy, *Joules*, consumed to process raw data from user 6. For the different *window sizes* and *overlap factors*. The colors represent a range of *Joules*, where the red color depicts higher energy consumed and, reversely, green color depicts less energy consumed. The *accuracy* is also shown in the map over each circle depicting the energy color to compare the energy consumed with the classification accuracy.

In Fig. 6, we can see that smaller windows result in less energy consumption than bigger windows. This is due to the increased effort to calculate features for larger window sizes. It is also perceivable that increasing the overlap factor also increases the energy consumed, essentially due to the increased number of feature calculations and classifications to be carried out.

Relating the energy consumption with the accuracy achieved for a given configuration, it is observable that the best accuracy values reside in more "heated" zones, i.e., where energy consumption is higher. Lower window sizes present lower accuracy while higher window sizes provide higher accuracy. For instance, in configurations without overlapping (i.e., with an overlap factor of 0), the accuracy rises from 85% for a window of size 500 to 90% for a window of size 1000.

The overlap shows more fluctuations in terms of accuracy, however with the best factors concentrated between 0.1 and 0.5. This shows that it is not trivial to select a single window size and overlap factor if it is intended to have two possible scenarios, one where accuracy is the most important factor and another one where energy consumption is the top priority but still with a minimum accuracy value in mind.

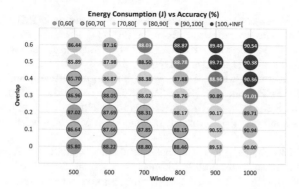

Fig. 6. Energy consumed, in Joules, while processing the PAMAP2 dataset, per window size and overlap factor. Green values represent less energy consumed while red values represent higher energy consumed. The values in each configuration represent the accuracy, in percentage, of that configuration.

5 Conclusion

In this work we presented an analysis of the impact of hyperparameters, as window size and overlap factor, on HAR classification accuracy, execution time and energy consumption. The analysis was focused on a public dataset, which includes raw sensor data from 9 different users and 18 physical activities.

The experimental results confirm the need of adapting the classification model to the current user. Due to the impact of window size and overlap factor, each activity requires a specific configuration of these hyperparameters in order to improve classification accuracy.

Furthermore, the results also motivate the development of a system that is able to adapt the application at runtime when trade-offs between performance accuracy and energy consumption need to be considered. Bearing in mind this, the window size and overlap factor can be used to develop runtime strategies able to adapt these parameters according to the target goals.

As future work, we plan to implement a system able to dynamically adjust at runtime the window size and overlap factor and aware of activities and users. The dynamic adaptation needs to consider an exploration of possible parameter configurations to find the best configurations for each adaptation scenario and thus the experimental results presented in this paper are also part of that exploration phase.

References

1. Krishnan, N.C., Cook, D.J.: Activity recognition on streaming sensor data. Pervasive Mob. Comput. **10**, 138–154 (2014)
2. Dobbins, C., Rawassizadeh, R., Momeni, E.: Detecting physical activity within lifelogs towards preventing obesity and aiding ambient assisted living. Neurocomputing **230**, 110–132 (2017)

3. Miluzzo, E., Varshavsky, A., Balakrishnan, S., Choudhury, R.R.: TapPrints: your finger taps have fingerprints. In: MobiSys (2012)
4. Aggarwal, J.K., Ryoo, M.S.: Human activity analysis: a review. ACM Comput. Surv. **43**(3), 1–43 (2011). Article ID 16
5. Lara, O.D., Labrador, M.A.: A survey on human activity recognition using wearable sensors. IEEE Commun. Surv. Tutor. **15**(3), 1192–1209 (2013)
6. Ramamurthy, S.R., Roy, N.: Recent trends in machine learning for human activity recognition - a survey. Wiley Interdisc. Rev. Data Min. Knowl. Discov. **8**(4), e1254 (2018)
7. Shoaib, M., Bosch, S., Incel, O., Scholten, H., Havinga, P.: A survey of online activity recognition using mobile phones. Sensors **15**, 2059–2085 (2015)
8. Cardoso, H., Mendes-Moreira, J.: Improving human activity classification through online semi-supervised learning. In: Workshop StreamEvolv Co-located with ECML/PKDD 2016, pp. 15–26 (2016)
9. Pisani, P.H., Lorena, A.C.: A systematic review on keystroke dynamics. J. Braz. Comput. Soc. **19**(4), 573–587 (2013)
10. Mannini, A., et al.: Activity recognition using a single accelerometer placed at the wrist or ankle. Med. Sci. Sports Exerc. **45**(11), 2193 (2013)
11. Banos, O., et al.: Window size impact in human activity recognition. Sensors **14**(4), 6474–6499 (2014)
12. Harasimowicz, A., Dziubich, T., Brzeski, A.: Accelerometer-based human activity recognition and the impact of the sample size. In: Proceedings of the 13th International Conference on Artificial Intelligence, Knowledge Engineering and Data Bases, Gdansk, Poland (2014)
13. Baños, O., et al.: Evaluating the effects of signal segmentation on activity recognition. In: IWBBIO (2014)
14. Niazi, A.H., et al.: Statistical analysis of window sizes and sampling rates in human activity recognition. In: HEALTHINF (2017)
15. Reiss, A., Stricker, D.: Creating and benchmarking a new dataset for physical activity monitoring. In: Proceedings of the 5th International Conference on PErvasive Technologies Related to Assistive Environments. ACM (2012)
16. Figo, D., Diniz, P.C., Ferreira, D.R., Cardoso, J.M.P.: Preprocessing techniques for context recognition from accelerometer data. Pers. Ubiquitous Comput. **14**(7), 645–662 (2010)
17. Zhou, Y., Goldman, S.: Democratic co-learning. In: 16th IEEE International Conference on Tools with Artificial Intelligence, pp. 594–602. IEEE Computer Society (2004)
18. Zhou, Z.H., Li, M.: Tri-training: exploiting unlabeled data using three classifiers. IEEE Trans. Knowl. Data Eng. **17**(11), 1529–1541 (2005)

A Fuzzy Approach for Sentences Relevance Assessment in Multi-document Summarization

Eduardo Valladares-Valdés[1], Alfredo Simón-Cuevas[1(✉)] [ID],
José A. Olivas[2] [ID], and Francisco P. Romero[2] [ID]

[1] Universidad Tecnológica de La Habana José Antonio Echeverría, Cujae,
La Habana, Cuba
{evalladaresv, asimon}@ceis.cujae.edu.cu
[2] Universidad de Castilla-La Mancha, Ciudad Real, Spain
{JoseAngel.olivas, FranciscoP.Romero}@uclm.es

Abstract. Text summarization is becoming an indispensable solution for dealing with the exponential growth of textual and unstructured information in digital format. In this paper, an unsupervised method for extractive multi-document summarization is presented. This method combines the use of a semantic graph for representing textual contents and identify the most relevant topics with the processing of several sentences features applying a fuzzy logic perspective. A fuzzy aggregation operator is applied in the sentences relevance assessment process as a contribution to the multi-document summarization process. The method was evaluated with the Spanish and English texts collection of MultiLing 2015. The obtained results were measured through ROUGE metrics and compared with those obtained by other solutions reported from MultiLing2015.

Keywords: Multi-document summarization · Extractive summarization · Semantic graph · Sentence feature · Fuzzy aggregation operator

1 Introduction

The information constitutes one of the most valuable resources for decision making in the development of the current society. However, the exponential growth of textual and unstructured data in digital format, for example on the Internet, have provoked that to distill the most relevant information from the amount of available information constituted a significant challenge to the textual information processing. The development of computational solutions based on the application of natural language processing and text mining techniques emerging as the most promising alternatives to deal with this challenge and the automatic text summarization constitute one of them.

The automatic text summarization has the purpose of condensing the most relevant and essential information contained in one or several text documents into a shorter and more manageable text [5], considering the conciseness criteria from the user [9, 19], and preserving the main content of the information source, as well as its general meaning [5, 19]. The summaries can be automatically obtained through extractive (selecting the most important sentences from the documents) or abstractive methods

F. Martínez Álvarez et al. (Eds.): SOCO 2019, AISC 950, pp. 57–67, 2020.
https://doi.org/10.1007/978-3-030-20055-8_6

(using different words and phrases from those of the original documents) [2]. These problems have been widely addressed considering supervised, unsupervised or semi-supervised processes, but remains a challenging task that requires special attention, fundamentally when the summary should be produced from multi-document [7]. A multi-document summary is a brief representation of the essential contents of a set of related documents.

This paper is focused on the generation of extractive summaries. This process consists basically in the selection of sentences or relevant sentences within an original manuscript and its concatenation in a shorter form [16]. This approach is usually carried out through three fundamental tasks: (1) construction of an intermediate representation of the input texts that expresses the most important aspects contained in the text; (2) relevance evaluation of the sentences according to the representation used (other aspects are also considered); and (3) selection of the summary that includes a certain amount of those sentences [1]. The relevance sentence is strongly based on the statistical and linguistic characteristics of these contents [2]. Extractive methods have been design considering different approaches [16]: graphs based [13, 15, 20, 22], concepts based [20, 22], LSA (Latent Semantic Analysis) based [23], among others.

One of the tools to deal with some crucial issues in extractive summarization like ambiguity, vagueness, incompleteness, approximate reasoning is fuzzy logic. However, a few studies were done in this area [3, 4, 18, 21, 24]. Moreover, these fuzzy-logic approaches are focused on the application of the basic theory of fuzzy sets [3, 4, 18] or fuzzy clustering algorithms [21, 24] to extract topics from documents. Our work presents as an original contribution to the utilization of fuzzy aggregation operators to address some key question in the multi-document summarization process.

In this paper, an unsupervised extractive method that combines the use of a semantic graph with the processing of several sentences features applying a fuzzy logic perspective to improve the multi-document summarization is presented. The proposed method extends the single-document summarization proposal reported in [20], offering a new language-independent approach for the solution of the multi-document summarization problem. In this new approach, a semantic graph is used to represent the conceptualization and the underlying semantic structure of the textual content, through the identification of concepts and semantic relationship between them from WordNet [14]. The generated semantic graphs from each document are automatically merged, and a concept clustering algorithm is applied to identifying the most relevant topics of the documents set. The graph merging process is supported by a disambiguation algorithm, aspect with more significant impact in the multi-document summarization task [6]. The relevance assessment process of the sentences is carried out considering several sentence features, such as: *TF-ISF*, *sentence length*, *sentence-to-sentence similarity*, and *sentence-to-cluster similarity*. In this process, the fuzzy aggregation function reported in [25] is applied to combine the computed results of these features and obtain the sentence relevance score. The proposed method was evaluated with the Spanish and English texts corpus provides from MultiLing 2015 to evaluate the Multilingual Multi-Document Summarization task. The results were measured through the ROUGE metrics (Recall-Oriented Understudy for Gisting Evaluation) [11] and compared with those obtained by the participant systems in MultiLing 2015.

The rest of the paper is organized as follows: Sect. 2 describes the proposed method; Sect. 3 presents the experimental results and the corresponding analysis, and conclusions arrived and future works are given in Sect. 4.

2 Proposed Method

The proposed method constitutes a new approach for automatic extractive multi-document summarization based on the use of a semantic graph to represent the conceptual content and the underlying semantic structure of the documents with the processing of several sentence features, such as: *TF-ISF, sentence length, sentence-to-sentence similarity*, and *sentence-to-cluster similarity*. The semantic processing of the textual content is supported in the use of WordNet [14], which offers the advantage of carried out independent domains texts summarizations process. The relevance assessment process of the sentences in the documents set for constructing the summary is addressed as a fuzzy logic problem. In this sense, the results of the four sentence features are combined applying a fuzzy logic technique to deal with the imprecision, uncertainty, and incompleteness problems that can arise in the processing and relevance assessment of the sentences in the texts summarization solutions [4]; fundamentally, in sentence features based approach. The method has been designed in several phases, as shown in Fig. 1.

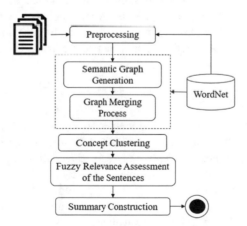

Fig. 1. Overview of the conceived phases in the proposed method

2.1 Preprocessing

In this phase, different basic tasks of natural language processing (NLP) are executed on each document and several sentence features are computed, such as: *TF-ISF (TF-ISF), sentence length (SL)*, and *sentence-to-sentence similarity (SS)*. The resultant score of each feature is assigned as a weight to the sentences in the documents set and will be used in the sentence relevance assessment process proposed. This phase begins extracting the plain text from the texts to be processed, using a Java library that provides support for processing pdf, docx, doc, html, htm, rtf, txt files. Each plain text is segmented into sentences, the tokens from the sentences are extracted, and the POS tagging process is carried out. The terms with very general meaning and not useful when discriminating between relevant and irrelevant sentences for the summary are removed; using a stop words list. These basic NLP tasks are supported using the Freeling syntactic analyzer, which offers the possibility of processing texts in Spanish and English language.

The *TF-ISF* (Term Frequency-Inverse Sentence Frequency) feature measures the relevance of the sentences according to the importance of the words that compose it respect to the document set. TF means to evaluate distribution (frequency) of each word over a document, and ISF means the terms that occur in only a few sentences which are more important than others that occur in many sentences. For each word in the documents set, the *tf-isf* is calculated and a *tf-isf* weight is assigned, which is calculated averaging the obtained tf-isf results of each document in which it appears. For each sentence *S*, the *TF-ISF* score is calculated to by the sum the weight of the words composes it. The *sentence length* (SL) feature useful to have a measure of the quantity information that contributes according to their length, considering that very large or very short sentences are not usually good for summarization. For each sentence *S*, the *SL* feature score is calculated dividing the length of *S* between the length of the longest sentence *L* identified in the documents set. The *sentence-to-sentence similarity* (*SS*) feature measure the similarity among the included sentences in all documents. For each sentence *S*, the similarity between *S* and every other sentence in the documents is computed applying the sentence-to-sentence semantic similarity metric reported in [10] and using WordNet. The obtained similarity values are averaged to obtain the *SS* score for each sentence.

2.2   Semantic Graph Generation

In this phase, a semantic graph of each document is automatically generated, where the conceptualization and the underlying semantic structure of the document are represented through the identified concepts and semantic relationship between them from WordNet. In this graph model, the concepts are represented as vertex and the semantic relationship between them, such as hypernym-hyponym and meronym-holonym (part-whole relation) WordNet relations, are represented as an edge. Initially, the synsets of each concept included in the sentence, as well as, others related synstes with them in a vicinity of radius two are captured from WordNet (considering hypernym-hyponym and meronym-holonym relationship). Next, the relationships between these systems (according to the relations types mentioned above) are also retrieved. Finally, a semantic graph is generated form each document, integrating the synsets and relationship identified. The number of represented concepts in the graph will depend on the degree to which the vocabulary of each document is covered with WordNet. The use of the mentioned relations types and synsets vicinity extraction from WordNet allows to increase the document coverage in this process and produces more connected graphs, respect to other proposals that only consider hypernym relation [20]. On the other hand, a document can include ambiguous concepts, which they are present in more than one synset in WordNet, and these concepts would be present in more than one vertex in the generated semantic graph. However, this ambiguity problem is solved in the next phase to reduce the uncertainty that can generate this problem in natural language text processing.

2.3   Graph Merging Process

The purpose of this process is to obtain an only one semantic graph for representing the included textual content in the documents set, which is a crucial aspect for identifying the more relevant conceptual topics in the next phase. This process is carried out integrating the generated semantic graphs from each document. The graph merging is based on the identification and merging of those concepts that have the same meaning, that is, they are represented through the same synsets in the semantic graphs. Therefore, the ambiguity of the represented concept needs to be previously solved. In this sense, the disambiguation algorithm reported in [8] is applied to each generated semantic graphs. This new approach improves the disambiguation results, by combining the results obtained from the processing of the domain, context and gloss heuristics, for determining the sense of the represented concept in the graph. The combination of these heuristics reduces the weaknesses of applying only one of them to determine the sense of the concepts, as in [20], in which a gloss-based disambiguation algorithm is applied. Next, the processed semantic graphs are refined, through the elimination of the associated vertex to the synsets of the ambiguous terms that received less voting as a result of the disambiguation algorithm, keeping the corresponding ones to the not disambiguated concepts. The inclusion of the disambiguation process in the proposed method constitutes a key piece to improving the results of the multi-document sum-marization, due to not only creates bases to reduce the possible redundancies, but rather also contributes to reduce the effects of other problems that may arise in this solution types, such as: disjointed and unrelated information and the lack of coherence in the resulting summary. Finally, all the generated semantic graph form each document are merged through integrating the vertexes that represent the same synset, obtaining the semantic graph of the documents set.

2.4   Concepts Clustering

In this phase, the represented concepts in the resultant graph of the previous phase are grouped in clusters, which represent the main conceptual topics that are represented in the content of the documents set. The centroid concepts in these clusters are those that provide the most relevant information about each topic. This process is carried out using a connectivity-based clustering algorithm, similar to the applied in [20]. The concepts are clustered according to the connections degree between them, considering that in this type of semantic graph a few nodes are highly connected to each other (hub nodes), while the degree of connection of the remaining vertexes is relatively low.

The clustering algorithm begins by locating the vertexes set in the graph that have higher connections with other vertexes, obtaining a relevance value for each vertex (v_i) through the sum of all the edges that have v_i as origin or destination. The n (determinate by the user) higher relevance vertexes are identified as HUB-Vertex (HUBV) and represent the most connected vertex in the graph. Next, the HUBV are grouped to forming a HVS (HUBV Sets), which are the groups of strongly connected vertexes that will constitute the centroids of the clusters to be constructed. To do this, the clustering algorithm search, for each HUBV and iteratively, most connected HUBV to it, and merges them into a single HVS. Then, for each pair of HVS, the algorithm checks, if their internal connectivity is lower than the connectivity between them, in which case, both HVS are merged. The connectivity between the concepts inside a cluster must be maximized, while the connectivity between concepts of different groups must be minimized. Finally, the assignment process of not classified vertexes as HUBV to the corresponding HVS (with which one has higher connectivity) to obtaining the final concepts clusters is carried out.

2.5   Fuzzy Relevance Assessment of the Sentences

In this phase, the relevance assessment process of each sentence in the documents set is carried out to obtain a sentences ranking. The three-sentence features scores computed in the preprocessing phase, as well as the results of the *sentence-to-cluster similarity* (considered as another sentence feature), are used for measuring the relevance of the sentences according to a fuzzy approach. The *sentence-to-cluster similarity (SC)* feature evaluates the relationship level between the sentences in the documents set and the identified topics, which are represented in the previously constructed concepts cluster (HVS). This process is carried out through the similarity evaluation between the clusters and the graph obtained from the sentence. According to [20], the similarity is calculated using a non-democratic voting mechanism (Eq. 1), where each vertex (v_k) in the sentence graph (S_j) assigns a different number of votes $(w_{k,j})$ to each cluster (C_i), depending on whether v_k belongs or not to the HVS of C_i. Considering that the concepts belongs to the HVS given greater importance than the others, a vote (1.0) is assigns if v_k belongs to that cluster's HVS, a half vote (0.5) if v_k belongs to the cluster but not to its HVS, and no votes (0) otherwise.

$$similarity(C_i, S_j) = \sum_{v_k/v_k \in S_i} w_{k,i} \qquad (1)$$

where $w_{k,i} = 0 \mid v_k \notin C_i$, $w_{k,i} = 1 \mid v_k \in HVS(C_i)$, and $w_{k,i} = 0.5 \mid v_k \notin HVS(C_i)$ and $v_k \in C_i$. The obtained similarity values between the sentence and every cluster are averaged for calculating the *SC* score of each sentence.

The compensatory aggregation function reported in [25] (as described in Eqs. 2–3) and the algebraic t-norm described in Eq. (4), are applied to aggregate the numerical values (s_i) from the four sentence features computed (*TF-ISF, SL, SS,* and *SC*) into a single one relevance score. An overview of the fuzzy relevance assessment proposed is shown in Fig. 2.

$$Z_\gamma(s_1, s_2, \ldots, s_n) = \left(\prod_{i=1}^{n} s_i \right)^{1-\gamma} * \left(1 - \prod_{i=1}^{n} (1 - s_i) \right)^{\gamma} \qquad (2)$$

$$\gamma = \frac{T(s_1, s_2, \ldots, s_n)}{T(s_1, \ldots, s_n) + T(1 - s_1, \ldots, 1 - s_n)} \tag{3}$$

$$T(s_1, s_2, \ldots, s_n) = \prod_{i=1}^{n} s_i \tag{4}$$

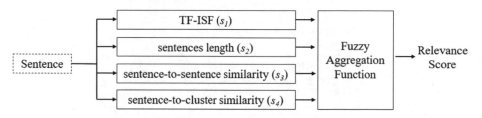

Fig. 2. Overview of the fuzzy relevance assessment

2.6   Summary Construction

Once the relevance of the sentences has been computed in the previous phase, the summary construction process is carried out through selecting the N sentences with a higher relevance score, considering that the N value depends to the desired compression rate (summary size). In this selection process, another two aspects are considered: firstly, the documents order in which the sentence with higher relevance score appear and then the position in which these sentences occur inside the document.

3   Experimental Results

The feasibility of the method is verified with texts corpus in Spanish and English language provides from MultiLing 2015 to evaluate the Multilingual Multi-Document Summarization task (MMS). The evaluation in other languages requires the use of the corresponding WordNet version, nevertheless several WordNet in different languages have been constructed, such as: Italian, German, French, and others (all of them using the same structure). The reliance on WordNet is considered a possible bottleneck, because success can be constrained by the coverage of WordNet [17]. However, in the proposed method the processing of several sentence features were included, for mitigating the effect of this problem in the relevance assessment process of the sentences. MultiLing is a community initiative that promotes the state-of-the-art of automatic summarization, from where data sets are provided and promote research topics in this scientific area. The selected corpus is formed by news articles coming from WikiNews and associated with 15 different topics. The corpus was grouped into two test collections to perform the experiments: ENCol and SPCol, which are characterized in Table 1. The quality of the obtained summaries using the proposed method was measured using precision (P), recall (R) and F-measure (F) metrics, in the context of

ROUGE [11]. Specifically, ROUGE-1 (using unigrams) and ROUGE-2 (using bigrams) were used. Table 2 shows these results and those obtained by the participant systems in the MultiLing 2015 MMS task for English and Spanish languages.

Table 1. Characterization of the test collections.

Characteristics	ENColl	SPColl
Language	English	Spanish
Corpus	15	15
Documents in each corpus	10	10
Sentences	4469	5000
Ave. of sentence in corpus	297.93	333.33
Ave. of document sentences	28.92	33.34

Table 2. Summary Results.

Systems	ROUGE-1 with ENCol			ROUGE-2 with ENCol			ROUGE-1 with SPCol			ROUGE-2 with SPCol		
	P	R	F	P	R	F	P	R	F	P	R	F
SCE-Poly	.21	.20	.20	.13	.12	.12						
BUPT-CIST	.13	.11	.12	.02	.02	.02	.13	.12	.12	.03	.03	.03
BGU-MUSE	.30	.28	.28	**.19**	.17	.18						
NCSR/SCIFY	.15	.14	.14	.05	.04	.05	.23	.20	.21	.07	.06	.06
UJF-Grenoble	.14	.12	.13	.04	.03	.04						
UWB	**.34**	**.33**	**.33**	.19	**.18**	**.18**	**.41**	**.39**	**.40**	**.26**	**.25**	**.25**
ExB	.23	.23	.23	.09	.09	.09				.10	.10	.10
ESIAllSummr	.16	.15	.15	.04	.03	.04	.21	.20	.20	.05	.06	.05
IDAOCCAMS	.23	.23	.23	.07	.07	.07	.25	.24	.25	.09	.08	.08
GiauUngVan	.15	.12	.13	.04	.03	.03						
Proposal	**.39**	**.41**	**.38**	**.13**	**.13**	**.13**	**.47**	**.47**	**.47**	**.20**	**.19**	**.20**

According to the results, the proposed method achieves to increase (although not in a significant way) the values of all metrics in the case of both test collections, when these are calculated using ROUGE-1, respect the other systems. This behavior is not shown in the same way when the ROUGE-2 metric is used. In the case of ENCol, the proposed method can be overcome in all metrics by UWB [23] and BGU-MUSE [12], and in the case of SPCol, is overcome only by the first one. However, in all cases, the differences between the resulting values are not significant. UWB is based on LSA technique for topics modelling in the documents and the length of sentence vectors (the sentence score) is consider as measure for evaluating the sentence relevance within the top topics cluster [23]. Although the obtained results (the best according to ROUGE-2), this strategy suffer from the same drawback as the lexical chains approach because more than one sentence may be required to convey all information pertinent to the topics [17]. The proposed method doesn't suffer of this problem since the topics modelling is carried out in a more granular way, representing and integrating the concept-based semantic structures from the documents and applying a concepts clustering process. BGU-MUSE [12] is a supervised summarizer based on the use of genetic algorithms (trained with collections offered in MultiLing 2015). The main weakness of this method, respect the proposed unsupervised approach, is that requires a training collection usually linked to a specific domain. Therefore, the application of this method in other domains would require a new training process and the collection may not be available. In general, we can consider that the obtained results by the proposed method applied to English and Spanish texts are promising.

4 Conclusions and Future Works

In this work, a new extractive multi-document summarization approach has been presented. Semantic graphs generated from WordNet are used as an intermediate representation of the documents set, reducing the mixture of disjointed and unrelated information in the textual content, as well as the lack of coherence in the resultant summary. The use of a disambiguation algorithm that combines several heuristics to determine the meaning of ambiguous concepts was incorporated, increasing the accuracy of the semantic graph merging process carried out for obtaining a global semantic presentation of the document set. Through this algorithm not only the ambiguity was reduced, but also it contributes to reduce the redundancies in the resulting summary which is a key aspect in multi-document summarization. The concept clustering strategy considered allowed to identify the most relevant conceptual topics in the documents, being other relevant information in multi-document summarization solution. The processing of the different sentence features in the method, combined with the application of the fuzzy aggregation function, contributes to obtain a more robust sentence relevance assessment process. Evaluations of the method were carried out on texts in Spanish and English, using recognized corpus and metrics by the scientific community, and their results were compared with those obtained by other systems. In general, the obtained results were satisfactory, achieves to increase the values of ROUGE-1 metrics in both corpus (respect to the other systems) and similar values of ROUGE-2 metrics to those obtained by the best systems evaluated. These results suggest that the use of the fuzzy aggregation operators constitute another fuzzy technique useful to improve the multi-document summarization process.

In the future works, other more enriched knowledge recourses such as: BabelNet, will be applied to increase the document coverage in the semantic graph generation process. Also, the use of other sentence features and the application of other fuzzy aggregation operators, such as the ordered weighted average (OWA) operator or other compensatory function will be evaluated.

Acknowledgments. This work has been partially supported by FEDER and the State Research Agency (AEI) of the Spanish Ministry of Economy and Competition under grant MERINET: TIN2016-76843-C4-2-R (AEI/FEDER, UE).

References

1. Allahyari, M., Pouriyeh, S., Safaei, S., Trippe, E.D., Gutierrez, J.B., Kochut, K.: Text summarization techniques: a brief survey. Int. J. Adv. Comput. Sci. Appl. **8**(10), 397–405 (2017)
2. Bhatia, N., Jaiswal, A.: Trends in extractive and abstractive techniques in text summarization. Int. J. Comput. Appl. **117**(6), 21–24 (2015)
3. Bhoir, A.S., Gulati, A.: A multi-document hindi text summarization technique using fuzzy logic. Int. J. Adv. Res. Sci. Eng. (IJASE) **4**(1), 468–473 (2015)
4. Chatterjee, N., Yadav, N.: Fuzzy rough set-based sentence similarity measure and its application to text summarization. IETE Tech. Rev. 1–9 (2018)
5. Das, D., Martins, A.F.: A survey on automatic text summarization. Lit. Surv. Lang. Stat. II Course CMU **4**, 192–195 (2007)
6. Ferreira, R., Cabral, L.S., Lins, R.D., Pereira e Silva, G., Freitas, F., Cavalcanti, G.D.C., Lima, R., Simske, S.J., Favaro, L.: Assessing sentence scoring techniques for extractive text summarization. Expert Syst. Appl. **40**, 5755–5764 (2013)
7. Gambhir, M., Gupta, V.: Recent automatic text summarization techniques: a survey. Artif. Intell. Rev. **47**(1), 1–66 (2017)
8. Hojas-Mazo, W., Simón-Cuevas, A., de la Iglesia, M., Romero, F.P., Olivas, J.A.: A concept-based text analysis approach using knowledge graph. In: Communications in Computer and Information Science, vol. 854, 696–708 (2018)
9. Kumar, Y.J., Salim, N.: Automatic multi document summarization approaches. J. Comput. Sci. **8**(1), 133–140 (2012)
10. Li, Y., McLean, D., Bandar, Z.A., O'Shea, J.D., Crockett, K.: Sentence similarity based on semantic nets and corpus statistics. IEEE Trans. Knowl. Data Eng. **8**, 1138–1150 (2006)
11. Lin, C.-Y.: ROUGE: a package for automatic evaluation of summaries. In: Proceedings of the Workshop on Text Summarization Branches Out, Barcelona, España (2004)
12. Marina, L., Vanetik, N., Last, M., Churkin, E.: MUSEEC: a multi-lingual text summarization tool. In: Proceedings of the 54th Annual Meeting of the ACL - System Demonstrations, pp. 73–78 (2016)
13. Mihalcea, R.: Graph-based ranking algorithms for sentence extraction, applied to text summarization. In: Proceedings of the ACL 2004 on Interactive poster and demonstration sessions, ACLdemo 2004 (2004)
14. Miller, G., Fellbaum, C.: WordNet: An Electronic Lexical Database. The MIT Press, Cambridge (1998)
15. Mittal, N., Agarwal, B., Vijay, N., Gupta, A., Upadhyay, N.K.: Semantic enhanced text summarization. Int. J. Comput. Syst. **1**(1), 26–29 (2014)
16. Moratanch, N., Chitrakala, S.: A survey on extractive text summarization. In: IEEE International Conference on Computer, Communication, and Signal Processing (ICCCSP 2017) (2017)
17. Nenkova, A., McKeown, K.: A survey of text summarization techniques. In: Aggarwal, C.C., Zhai, C.X. (eds.) Mining Text Data, pp. 44–76. Springer, Heidelberg (2012)
18. Patil, P.D., Kulkarni, N.J.: Text summarization using fuzzy logic. Int. J. Innov. Res. Adv. Eng. (IJIRAE) **1**(3), 42–45 (2014)
19. Padmapriya, K.D.G., Rajasekaran, V.G.: A view on natural language processing and text summarization. Int. J. Commun. Eng. (2012)
20. Plaza, L., Díaz, A.: Using semantic graphs and word sense disambiguation techniques to improve text summarization. Procesamiento del Lenguaje Natural **47**, 97–105 (2011)

21. Ramyashri, B.N., Ananda, K.R.: A fuzzy relational clustering algorithm for document summarization. Int. J. Adv. Found. Res. Comput. (IJAFRC) **1**(5), 166–172 (2014)
22. Sankarasubramaniam, Y., Ramanathan, K., Ghosh, S.: Text summarization using wikipedia. Inf. Process. Manag. **50**(3), 443–461 (2014)
23. Steinberger, J.: The UWB Summariser at Multiling-2013. In: Proceedings of the MultiLing 2013 Workshop on Multilingual Multi-documents Summarization, ACL, pp. 50–54 (2013)
24. Vishnu, P., Sangeetha, K., Deepa, D.: Extractive text summarization system using fuzzy clustering algorithm for mobile devices. Asian J. Inf. Technol. **15**(5), 933–939 (2016)
25. Zimmermann, H.-J., Zysno, P.: Latent connectives in human decision making. Fuzzy Sets Syst. **4**(1), 37–51 (1980)

Online Estimation of the State of Health of a Rechargeable Battery Through Distal Learning of a Fuzzy Model

Luciano Sánchez[1]([⊠]), José Otero[1], Manuela González[2], David Anseán[2], and Inés Couso[3]

[1] Computer Science Department, Oviedo University, Gijón, Spain
{luciano,jotero}@uniovi.es
[2] Electronic Engineering Department, Oviedo University, Gijón, Spain
mgonzalez@uniovi.es
[3] Statistics Department, Oviedo University, Gijón, Spain
couso@uniovi.es

Abstract. A fuzzy model is presented for detecting changes in the incremental capacity curve of an automotive lithium-ferrophosphate battery through analysis of the data collected while the vehicle is being operated. By means of the proposed model, the state of health of the energy storage system can be estimated on-vehicle. The fuzzy model is derived through distal learning and describes the instantaneous slope of the stored charge with respect to the battery voltage. The scheme has been validated in batteries with different levels of deterioration. It is concluded that the fuzzy model is able to anticipate the most frequent deteriorations of the battery and therefore it can be used to prevent its premature wear.

1   Introduction

The knowledge of the State of Health (SoH) of the rechargeable batteries of an electric vehicle anticipates different types of deterioration, such as the loss of capacity or the increase in impedance, which limits the range and power of the vehicle, and also reduces the risk of destructive phenomena [21]. However, some of the electrochemical variables that affect the health of a battery are not directly observable and, to the best of our knowledge, a procedure for displaying in real time the SoH of the battery in the dashboard of an electric vehicle has not been devised yet. The presence of dendrites, the electrodepositions, the thickness of the film that covers the anode, or the inventory of lithium, to name a few, can only be measured in batteries that have special sensors or through destructive procedures. To circumvent this problem, a fuzzy model is presented in this paper that can infer changes in the hidden health variables of an automotive lithium-ferrophosphate battery through analysis of the set of observable variables that are collected while the vehicle is being operated.

Battery modelling is a mature field or research, and the catalog of battery models is extense. The simplest approches are based on ARIMAX or NARMAX

F. Martínez Álvarez et al. (Eds.): SOCO 2019, AISC 950, pp. 68–77, 2020.
https://doi.org/10.1007/978-3-030-20055-8_7

time series [12] or its Artificial Intelligence-based extensions, such as the neural and fuzzy models of the NARX type (Nonlinear AutoRegressive with eXogenous inputs) [9,11]. Other intelligent technologies include recurrent neural networks [8], rule-based fuzzy systems (for modeling complete systems or isolated non-linearities) [5], fuzzy models in the state space [13], and fuzzy models of the Hammerstein type or Wiener [1]. Hybrid methods combining fuzzy logic and neural networks have been studied too; for instance, neuro-fuzzy systems of the NARX type [3] or recurrent networks [22]. Notwithstanding this, the model that will be proposed in this paper is not oriented to predict the observable variables but it is able to infer the so called "Incremental Capacity" (IC) curve of the battery, from which different silent deteriorations of the battery can be determined. In this respect, the nearest works to ours are described in references [18] and [16]. Also in connection with the derivation of the IC curve there exist precedents of this study in the Wiener type models developed in Reference [17], the LSTM networks (Long Short-Term Memory networks) in [2], and the ESN (Echo State Networks) in [15].

This document consists of the following parts: Sect. 2 introduces IC curves and contains a description of the proposed model. Section 3 describes an empirical validation on batteries with different degree of deterioration. Section 4 concludes this document and suggests lines of future work.

2   Description of the Proposed Model

To the best of our knowledge, there are not previous applications of distal learning for estimating the state of health of a rechargeable battery. The nearest technique followed in the past consists in learning a model from on-vehicle data in an online manner, thus the predictions of the model always mimic the observable variables of the actual battery. Once the model is obtained, those laboratory experiments that would be applied to the battery (in order to measure the SoH) are simulated over this model. The outcome of these simulated laboratory experiments is the estimation of the SoH, valid for the time when the model was last updated.

2.1   IC Curves and Analysis

Among the different diagnosis procedures, the "Incremental Capacity Analysis" or ICA [6] consists in studying of the derivative of the charge stored in the battery with respect to the voltage at its terminals. The voltage at the terminals of a battery presents transitions in certain characteristic voltages, which appear as extrema in the IC curve (see Fig. 1). Both the position and the height of these extrema change when the battery is degraded, and for this reason the deterioration of a cell can be estimated by comparing the incremental capacity curve with the same curve when the battery was new. The differences are interpreted according to the criteria detailed in mentioned references [6].

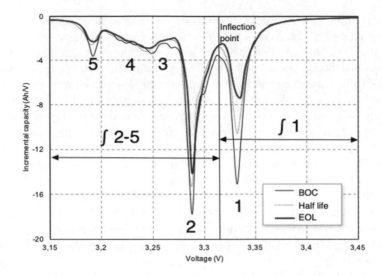

Fig. 1. IC curve and characteristic points for new batteries, half life and end of life

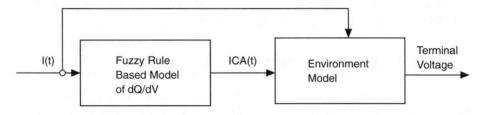

Fig. 2. Block diagram of the distal learning setup for assessing the battery health

IC curves are defined by the position and shape of a small number of characteristic peaks (numbered 1-5 in the Figure), and are suitable for a linguistic fuzzy rule-based description. However, the derivative of the stored charge with respect to the voltage cannot be directly measured at the vehicle, neither conventional supervised rule learning algorithms can be applied for obtaining this IC curve, because there is not a proper set of input-output data pairs. The dependence in time between the terminal voltage of the battery and the discharge current depend on the IC curve, but this dependence is not explicit.

2.2 Proposed Model and Learning Methodology

In this study it is suggested that distal learning [10] can be used for learning the IC curve of a battery. Distal learning is intended for problems in which an unknown dynamical system intervenes between actions and desired outcomes. The estimation of the SoH of a battery falls into this set of problems because the SoH can be inferred from the response of the battery to certain stimuli. Distal learning consists in cascading the desired model with an instrumental model of the environment, so the output of the desired model is the input to the environment model. It is the output of this environment model that is compared to the desired outcomes. The proposed model consists therefore in a fuzzy rule-based model of the ICA curve that is put in series with a model that relates the derivative of the charge and the current with the voltage at the battery terminals, that can be readily measured (see Fig. 2).

The main purpose of this contribution is to define a fuzzy model of the derivative of the stored charge with respect to the voltage. This fuzzy model provides a linguistic description of the position and shape of the characteristic points of the ICA curve, that can be related to the SoH of the battery. The proposed model consists of a fuzzy knowledge base with learning capacity, where the antecedents of the rules match the extrema of the IC curve and the consequents correspond to the heights of the same curve, so the SoH of the battery can be obtained after a simple exploration of the learned rules.

The mathematical expression of the ICA curve is

$$\mathrm{ICA(OCV)} = \frac{dQ}{dOCV} = \frac{dQ(t)/dt}{dOCV(t)/dt} = \frac{I(t)}{dOCV(t)/dt} \tag{1}$$

where OCV stands for "Open Circuit Voltage", which is the voltage at the battery terminals when the current is near zero. It is remarked that the battery voltage V will be lower than OCV when the battery is being discharged, and larger when the battery is being charged. From Eq. 1 it follows that

$$\frac{dOCV(t)}{dt} = \frac{I(t)}{\mathrm{ICA(OCV(t))}} \tag{2}$$

and therefore

$$\mathrm{OCV}(t) = \mathrm{OCV}_0 + \int_0^t \frac{I(\tau)}{\mathrm{ICA(OCV(\tau))}} d\tau. \tag{3}$$

The environment model in Fig. 2 models the difference between the OCV of the battery at time t and the voltage when a current $I(t)$ is circulating through the cell. Observe the environment model is not a combinational model but it has memory, i.e. the value $V(t) - OCV(t)$ do not only depend on $I(t)$ but on the values of $I(\tau)$ from $\tau = 0$ to $\tau = t$.

The difference

$$\mathrm{OVP}(t) = V(t) - \mathrm{OCV}(t) \tag{4}$$

is named "overpotential" in the specialized literature [4], and there are recent works where it is stated that the evolution in time of the overpotential cannot be modelled by a system of ordinary differential equations, but a fractional dynamics element is needed [20]. Because of this, in this work, the environment model consists in an equivalent circuit model that comprises a resistance in parallel with a fractional impedance,

$$\mathrm{OVP}(t) = I_1(t) \cdot R \tag{5}$$

$$C \frac{d^\alpha \mathrm{OVP}(t)}{(dt)^\alpha} = I(t) - I_1(t) \tag{6}$$

thus the overpotential is the solution of the following fractional differential equation:

$$C\frac{d^\alpha \text{OVP}(t)}{(dt)^\alpha} = I(t) - \frac{\text{OVP}(t)}{R}, \tag{7}$$

that depends on three coefficients C, R and α. The simulation of an element with fractional dynamics is computationally expensive, although there are different approaches with good practical results in terms of accuracy and efficiency [23].

2.3   Fuzzy Rule-Based Model

A fuzzy system of the TSK type of order 0 [19], comprising 5 rules (one for each characteristic point) has been used. Rules have the following form:

Rule i: If OCV is $\mu(\text{OCV}, c_i, \sigma_i)$ **then** C is h_i,

where the membership function of the antecedent is the derivative of the softsign function [7], whose form is closer to the peaks of the IC curve than the most usual belongings in this context (triangular or Gaussian). The antecedent of the i-th rule depends on the OCV, has center c_i and width w_i:

$$\mu(\text{OCV}, c_i, \sigma_i) = \left(1 + \left|\frac{\text{OCV} - c_i}{\sigma_i}\right|\right)^{-2}. \tag{8}$$

Lastly, the output of the fuzzy system is:

$$C(\text{OCV}) = \frac{\sum_{i=1}^{5} \mu(\text{OCV}, c_i, \sigma_i) \cdot h_i}{\sum_{i=1}^{5} \mu(\text{OCV}, c_i, \sigma_i)}. \tag{9}$$

There are 18 parameters in the model: the values of c_i, σ_i and h_i for $i = 1, \cdots, 5$, plus the three coefficients R, C and α of the environment model. The fitness function is the mean squared error between the values $V(t)$ produced by the cascade of the fuzzy model and the environment model and the voltage measured at the battery terminals for a given sequence of discharge currents.

3   Empirical Study

The empirical study in this section serves for comparing the IC curve estimates obtained with the proposed fuzzy model and the estimations of that curve in the laboratory. The model parameters are fitted first, by means of a genetic algorithm, to a new battery. After this initial estimation, each model is fitted to an aged battery by means of a greedy algorithm, using the solution found for the new battery as the starting point. Observe that the parameters of the aged batteries are to be estimated for high charge and discharge currents, so that the health estimate will be affected by a certain inaccuracy. In this study, it is assessed whether this inaccuracy is acceptable and if the proposed model can be used reliably in an electric vehicle.

Fig. 3. Image of the measurement equipment used in the experimentation

3.1 Experimental Setup

A battery analysis equipment of the Arbin brand, model BT-2000 and a Memmert environmental chamber were used (see Fig. 3). The ambient temperature is $23\,^{\circ}$C. Cylindrical batteries of the LFP type (A123 Systems brand) of 2.3 Ah capacity, commonly used in electric cars, have been used.

The batteries have been tested at the beginning of its life, at half life (3000 cycles) and at the end of his life (6000 cycles). One of the cells has suffered an abnormal deterioration (electrodeposition). We want to find out if the estimation of the IC curve by the method proposed in this paper is accurate enough to appreciate all the types of deterioration.

The dynamics of the fractional element has been approximated by the Oustaloup method [14], for a range of frequencies between 0.01 and 0.1 Hz. The sampling time is 10 s.

3.2 Numerical Results

Indirect estimates of the IC curves of batteries at the beginning of its life, at 3000 and 6000 cycles are shown in the left part of Figs. 4, 5 and 6, respectively. In these graphs the IC curve has been drawn along with the integral of that curve. The discharge current was high enough for draining the battery in one hour. The mentioned integral has the purpose of measuring the area of the IC curve to the left and to the right of the minimum between peaks 1 and 2, since the evolution of this area is used to diagnose different types of deterioration. In the right part of the same figures, the estimation of the curves obtained was in the laboratory for a slow (25-h) download.

It is remarked that the position of the peaks has been recovered with a good accuracy for the fuzzy model and the high current estimation. The most significant error occurs in the heights. The areas are approximately correct and the

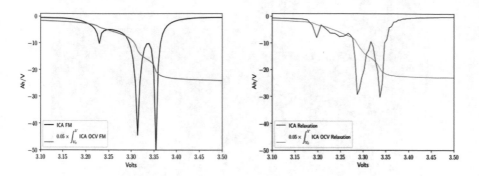

Fig. 4. Left: Indirect estimation (FM, fuzzy model) of the IC curve at the beginning of the battery life. Right: Laboratory estimation (ground truth)

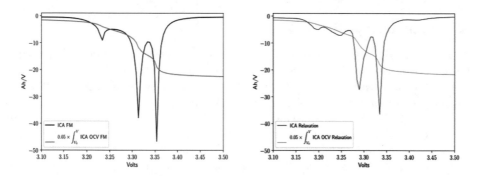

Fig. 5. Left: Indirect estimation (FM, fuzzy model) of the IC curve at the middle of the battery life. Right: Laboratory estimation (ground truth)

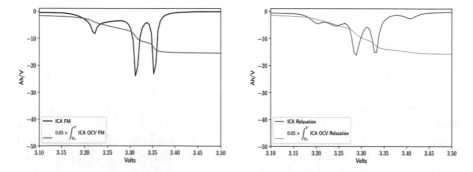

Fig. 6. Left: Indirect estimation (FM, fuzzy model) of the IC curve at the end of the battery life. Right: Laboratory estimation (ground truth)

relative differences between these areas and the heights, when the battery ages, are coherent with the laboratory estimation. Note that the deterioration between cycles 3000 and 6000 is noticeable to the naked eye (height difference between Figs. 5 and 6) but the anomalous peak, due to electrodeposition (deformation in 3.42 V, right part of Fig. 6) was not recovered.

Table 1 includes a numerical comparison between the values measured on the curves of the fuzzy estimation and the curves obtained in the laboratory for the same batteries. It should be noted that the diagnosis made in the laboratory coincides with that provided by the indirect estimation: up to half of the battery life, Peak 1 area is maintained and area 2-5 decreases slightly. Between the middle and the end of life both area 1 and area 2-5 decrease. This is compatible with the deteriorations that are named LLI (Loss of Lithium Inventory) and LAM (NE) (Loss of Active Material - Negative Electrode) in the specialized literature [6].

Table 1. Areas (in Ah) between the different extrema of the incremental capacity curve of the battery. The diagnosis made in the laboratory coincides with the fuzzy estimation: up to half of the battery's life, Peak Area 1 is maintained and area 2-5 decreases slightly. Between the middle and the end of life both area 1 and area 2-5 decrease: there are effects of type "Loss of Lithium Inventory" and "Loss of Active Material in the negative electrode"

	Method	New battery	Half life	End of life
Marker #1	Laboratory	147	151	102
Marker #1	**Fuzzy model**	**158**	**166**	**81**
Markers #2-#5	Laboratory	312	281	211
Markers #2-#5	**Fuzzy model**	**332**	**292**	**230**

4   Concluding Remarks

The estimation of the IC curves of a battery from fast discharges is of great practical interest, since it could be the basis of a health diagnostic system for automotive batteries integrated in the instrument panel of the vehicle. In this study, it has been proposed to employ a fuzzy model and learn its parameters by mean of distal learning. It has been concluded that the system is usable to diagnose certain types of deterioration (loss of inventory of lithium and active material in the electrodes) but that it is not yet possible to identify electrodepositions and that, in general, the heights of the peaks of the curve IC estimated with the method proposed in this communication decrease in a manner consistent with the degradation of the battery, but do not converge to the actual heights of said curve.

In future work we will try to introduce the temperature in the model. Although in the tests at low current this is not a relevant factor (and, on the other hand, these are carried out at a controlled temperature) in the normal use of a vehicle the temperatures have great fluctuations. The OCV curve does not have a large variation with the temperature in LFP cells, and the incremental capacity curves depend on the latter, so it is not expected that these curves depend on the temperature significantly. However, the overpotential varies substantially with the temperature. Therefore, the temperature dependence must be included in the environment model. On the other hand, if it is desired to take advantage of data collected when the battery is completely full or close to emptying, it will also be necessary to take into account the variation of the resistance with the load. For both reasons, in future work the environment model will be implemented with recurrent neural networks.

Acknowledgements. This work has been funded by the Ministry of Science and Innovation of the Government of Spain and by the FEDER funds of the European Community through projects with codes FC-IDI/2018/000226, TEC2016-80700-R (AEI/FEDER, UE) and TIN2017-84804-R.

References

1. Abonyi, J., Babuška, R., Botto, M.A., Szeifert, F., Nagy, L.: Identification and control of nonlinear systems using fuzzy hammerstein models. Ind. Eng. Chem. Res. **39**(11), 4302–4314 (2000)
2. Almansa, E., Anseán, D., Couso, I., Sánchez, L.: Health assessment of automotive batteries through computational intelligence-based soft sensors: an empirical study. In: International Joint Conference SOCO17-CISIS17-ICEUTE17, Proceeding, León, Spain, 6–8 September 2017, pp. 47–56. Springer (2017)
3. Babuška, R., Verbruggen, H.: Neuro-fuzzy methods for nonlinear system identification. Annu. Rev. Control. **27**(1), 73–85 (2003)
4. Bard, A.J., Faulkner, L.R., Leddy, J., Zoski, C.G.: Electrochemical methods: fundamentals and applications, vol. 2. Wiley, New York (1980)
5. Bartczuk, Ł., Przybył, A., Cpałka, K.: A new approach to nonlinear modelling of dynamic systems based on fuzzy rules. Int. J. Appl. Math. Comput. Sci. **26**(3), 603–621 (2016)
6. Dubarry, M., Truchot, C., Liaw, B.Y.: Synthesize battery degradation modes via a diagnostic and prognostic model. J. Power Sources **219**, 204–216 (2012)
7. Glorot, X., Bengio, Y.: Understanding the difficulty of training deep feedforward neural networks. In: Proceedings of the Thirteenth International Conference on Artificial Intelligence and Statistics, pp. 249–256 (2010)
8. Haykin, S.S.: Neural Networks and Learning Machines, vol. 3. Pearson, Upper Saddle River (2009)
9. Jang, J.S.: ANFIS: adaptive-network-based fuzzy inference system. IEEE Trans. Syst. Man Cybern. **23**(3), 665–685 (1993)
10. Jordan, M.I., Rumelhart, D.E.: Forward models: supervised learning with a distal teacher. Cogn. Sci. **16**(3), 307–354 (1992)
11. Lin, T., Horne, B.G., Tino, P., Giles, C.L.: Learning long-term dependencies in NARX recurrent neural networks. IEEE Trans. Neural Netw. **7**(6), 1329–1338 (1996)
12. Ljung, L.: System identification. In: Signal Analysis and Prediction, pp. 163–173. Springer (1998)

13. Massad, E., Ortega, N.R.S., de Barros, L.C., Struchiner, C.J.: Classical dynamical systems with fuzzy rule-based parameters. In: Fuzzy Logic in Action: Applications in Epidemiology and Beyond, pp. 207–223. Springer (2008)
14. Oustaloup, A., Levron, F., Mathieu, B., Nanot, F.M.: Frequency-band complex noninteger differentiator: characterization and synthesis. IEEE Trans. Circuits Syst. I: Fundam. Theory Appl. **47**(1), 25–39 (2000)
15. Sánchez, L., Anseán, D., Otero, J., Couso, I.: Assessing the health of LiFePO4 traction batteries through monotonic echo state networks. Sensors **18**(1), 9 (2018)
16. Sánchez, L., Blanco, C., Antón, J.C., García, V., González, M., Viera, J.C.: A variable effective capacity model for LiFePO4 traction batteries using computational intelligence techniques. IEEE Trans. Ind. Electron. **62**(1), 555–563 (2015)
17. Sánchez, L., Couso, I., Blanco, C.: A class of monotone fuzzy rule-based wiener systems with an application to Li-Ion battery modelling. Eng. Appl. Artif. Intell. **64**, 367–377 (2017)
18. Sánchez, L., Couso, I., Otero, J., Echevarría, Y., Anseán, D.: A model-based virtual sensor for condition monitoring of Li-Ion batteries in cyber-physical vehicle systems. J. Sensors (2017)
19. Takagi, T., Sugeno, M.: Fuzzy identification of systems and its applications to modeling and control. IEEE Trans. Syst. Man Cybern. **1**, 116–132 (1985)
20. Tian, J., Xiong, R., Yu, Q.: Fractional-order model-based incremental capacity analysis for degradation state recognition of lithium-ion batteries. IEEE Trans. Ind. Electron. **66**(2), 1576–1584 (2019)
21. Vetter, J., Novák, P., Wagner, M.R., Veit, C., Möller, K.C., Besenhard, J., Winter, M., Wohlfahrt-Mehrens, M., Vogler, C., Hammouche, A.: Ageing mechanisms in lithium-ion batteries. J. Power Sources **147**(1–2), 269–281 (2005)
22. Zhang, J., Morris, A.J.: Recurrent neuro-fuzzy networks for nonlinear process modeling. IEEE Trans. Neural Netw. **10**(2), 313–326 (1999)
23. Zou, C., Zhang, L., Hu, X., Wang, Z., Wik, T., Pecht, M.: A review of fractional-order techniques applied to lithium-ion batteries, lead-acid batteries, and supercapacitors. J. Power Sources **390**, 286–296 (2018)

A Proposal for the Development
of Lifelong Dialog Systems

David Griol$^{(\boxtimes)}$, Araceli Sanchis, and Jose Manuel Molina

Department of Computer Science, Universidad Carlos III de Madrid, Leganés, Spain
{david.griol,araceli.sanchis,josemanuel.molina}@uc3m.es

Abstract. In this paper we describe a proposal that employs Soft Computing techniques for developing intelligent dialog systems that can improve over time. To do this, our proposal merges statistical dialog management methodologies, intentional and emotional information in order to make dialog managers more efficient and adaptive. The prediction of the user intention and emotion is carried out for each user turn in the dialog by means of specific modules that are conceived as an intermediate phase between natural language understanding and dialog management in the architecture of these systems. We have applied and evaluated our method in the UAH system, for which the evaluation results show that merging both sources of information improves system performance as well as its perceived quality.

Keywords: Conversational interfaces · Spoken interaction ·
Evolving classifiers · Fuzzy-rule based systems ·
Human-machine interaction

1   Introduction and Related Work

Soft Computing (SC) has been proposed as a paradigm that combines fuzzy, neuro, and evolutionary computing for the construction of new generation artificial intelligence and solving nonlinear and mathematically unmodeled systems [11]. Increasingly, these techniques have been playing an important role in advanced knowledge processing to develop human-like communication systems.

Spoken Dialog Systems are computer programs that engage the user in a dialog using natural language that aims to be similar to that between humans [9]. Usually, SDSs carry out five main tasks: Automatic Speech Recognition (ASR), Spoken Language Understanding (SLU), Dialog Management (DM), Natural Language Generation (NLG), and Text-To-Speech Synthesis (TTS).

The spoken dialog industry has currently reached a maturity based on standards that pervade technology to provide high interoperability, which makes it possible to divide the market in a vertical structure of technology vendors,

This work has been partially supported by Spanish projects TEC2017-88048-C2-2-R and TRA2016-78886-C3-1-R.

F. Martínez Álvarez et al. (Eds.): SOCO 2019, AISC 950, pp. 78–88, 2020.
https://doi.org/10.1007/978-3-030-20055-8_8

platform integrators, application developers, and hosting companies [10]. The artificial intelligence of conversational interfaces is also revolutionizing smart business with a number of platforms to develop chatbots[1].

The design practices of conventional commercial chatbots are currently well established in industry. In these practices, voice user interface (VUI) experts [2] handcraft a detailed dialog plan based on their knowledge about the specific task and the business rules. On the other side, spoken dialog research has been moving on a parallel path trying to attain naturalness and freedom of communication.

In the literature we can find different systems and research projects focused on the integration of dialog systems for lifelong or continuous learning (LL, also known as Learning to Learn) [5,8]. As described in [9], current research areas and future trends are focused on assistive, adaptive and proactive system design, dialog management and system-environment interaction. The main guidelines that we propose in this contribution to fulfill the lifelong learning objective are:

- The use of Soft Computing dialog management methodologies to automatically learn adaptive dialog policies that can be easily extended to new domain for solving prediction of system responses and dealing with additional dialog tasks. Section 2 describes our proposal for statistical dialog management based on a classification process that considers the complete history of the dialog to decide the next system response.
- The incorporation of user models in the architecture of the conversational interface as a valuable information that can be considered by the dialog manager to modify and/or adapt the system's responses according to the users' preferences and requirements. Section 3 presents our proposal for user intention recognition, in which each user action is modeled considering the subtask to which the turn contributes, and the semantic information that it provides. Section 4 describes our proposal to recognize users emotional state, which is focused on recognizing negative emotions that might discourage users or even abort an ongoing dialog.
- The combination of our approaches for statistical dialog management, user intention modeling and emotion recognition and for statistical dialog to develop a practical conversational interface. We have applied and evaluated these methodologies in the UAH system, for which the evaluation results show that merging both sources of information improves system performance as well as its perceived quality. Section 5 describes the application of our proposal for this system and Sect. 6 presents the results of its evaluation. Finally, our conclusions and future work are presented in Sect. 7.

---

[1] To mention a few: Aivo (https://aivo.co/), Botsify (https://botsify.com), Chatfuel (https://chatfuel.com/), FlowX0 (https://flowxo.com/), Dialogflow (https://dialogflow.com/), Imperson (http://imperson.com/), ItsAlive (https://itsalive.io/), ManyChat (https://manychat.com/), Pandorabots (https://home.pandorabots.com) (Last access: January 2019).

2   Statistical Dialog Management Methodologies

One of the core aspects of developing dialog systems for lifelong learning is to design flexible dialog management strategies. The dialog strategy defines the system conversational behavior in response to user utterances and contextual information that, for example, can be based on observed events or beliefs.

We have developed a toolkit to develop dialog managers for spoken dialog systems based on evolving Fuzzy-rule-based (FRB) classifiers [7]. Our proposal is focused on slot-filling dialog systems, for which dialog managers use a structure comprised of one slot per piece of information that the system can gather from the user. The system can capture several data at once and the information can be provided in any order (more than one slot can be filled per dialog turn and in any order).

We propose to determine the next system action by means of a classification process. The toolkit uses the *eClass0* (evolving classifier) for the definition of the classification function. These classifiers generate several fuzzy rules per class using an evolving clustering approach to decide when a new rule is created:

$$Rule_i = IF(Feature_1 \ is \ P_1) \ AND \ldots AND(Feature_n \ is \ P_n)THEN \ Class \ = \ T_i$$

where i denotes the number of rule; n is the number of input features (observations corresponding to the different slots defined for the semantic representation of the user's utterances); the vector *Feature* stores the observed features, and the vector P stores the values of the features of one of the prototypes of the corresponding class $c_i \in$ {set of different classes}. Each class is then associated to a specific user action (response).

The following steps are carried out by the developed dialog managers after each user turn:

1. The values of the different slots are provided by the NLU module for the current user turn also taking into account the confidence scores.
2. The state of the dialog is updated with the new values for the slots determined in the previous step.
3. The classifier determines the fuzzy rule to be applied (i.e., next system action).

The *eClass0* model consists of several fuzzy rules per class (the number of rules depends on the heterogeneity of the input data of the same class). During the training process, a set of rules is formed from scratch using an evolving clustering approach to decide when to create new rules. The inference in *eClass0* is produced using the "winner takes all" rule and the membership functions that describe the degree of association with a specific prototype are of Gaussian form.

The *potential* (Cauchy function of the sum of distances between a certain data sample and *all* other data samples in the feature space) is used in the partitioning algorithm. However, in these classifiers, the potential (P) is calculated recursively (which makes the algorithm faster and more efficient). The potential of the kth

data sample (x_k) is calculated by means of Eq. (1). The result of this function represents the *density* of the data that surrounds a certain data sample.

$$P(x_k) = \frac{1}{1 + \frac{\sum_{i=1}^{k-1} distance(x_k, x_i)}{k-1}} \tag{1}$$

where *distance* represents the distance between two samples in the data space.

The potential can be calculated using the euclidean or the cosine distance. In this case, cosine distance (*cosDist*) is used to measure the similarity between two samples; as it is described in Eq. (2).

$$cosDist(x_k, x_p) = 1 - \frac{\sum_{j=1}^{n} x_{kj} x_{pj}}{\sqrt{\sum_{j=1}^{n} x_{kj}^2 \sum_{j=1}^{n} x_{pj}^2}} \tag{2}$$

where x_k and x_p represent the two samples to measure its distance and n represents the number of different attributes in both samples.

The resolution of Eq. (1) requires all the accumulated data sample available to be calculated, which contradicts to the requirement for real-time and on-line application needed in the proposed problem. For this reason, a recursive expression for the cosine distance is used:

$$P_k(z_k) = \frac{1}{2 - \frac{1}{(k-1)\sqrt{\sum_{j=1}^{n}(z_k^j)^2}} B_k} ; k = 2, 3...$$

$$where: \quad B_k = \sum_{j=1}^{n} z_k^j b_k^j \; ; \; b_k^j = b_{(k-1)}^j + \sqrt{\frac{(z_k^j)^2}{\sum_{l=1}^{n}(z_k^l)^2}} \tag{3}$$

$$and \quad b_1^j = \sqrt{\frac{(z_1^j)^2}{\sum_{l=1}^{n}(z_1^l)^2}} \; ; \; j = [1, n+1]; \; P_1(z_1) = 1$$

where z_k represents the kth data sample (x_k) and its corresponding label $(z = [x, Label])$. Using this expression, it is only necessary to calculate $(n+1)$ values where n is the number of different subsequences obtained; this value is represented by b, where $b_k^j, j = [1, n]$ represents the accumulated value for the kth data sample.

3   User Intention Modeling

Research in techniques for user modeling has a long history within the fields of language processing and dialog systems. A thorough literature review on the application of how data mining techniques to user modeling for system personalization can be found in [6,12]. It is possible to classify the different approaches with regard to the level of abstraction at which they model dialog. This can be at either the acoustic level, the word level or the intention-level. In recent years, simulation on the intention-level has been most popular [12].

The methodology that we have developed for modeling the user intention is based on the one proposed in [1]. In this model, each user turn is modeled as a user action defined by a subtask to which the turn contributes, the dialog act of the turn, and its named entities.

For speaker u, DA_i^u denotes the dialog act in the ith turn, and ST_i^u denotes the subtask label to which the ith turn contributes. The interpretation process is modeled in two stages. In the first stage, the dialog act of the clause is determined from the information about the user's turn and the previous dialog context, which is modeled by means of the k previous utterances. This process is shown in Eq. (4).

$$DA_i^u = \operatorname*{argmax}_{d^u \in \mathcal{D}} P(d^u | c_i^u, ST_{i-1}^{i-k}, DA_{i-1}^{i-k}, c_{i-1}^{i-k}) \qquad (4)$$

where c_i^u represents the lexical, syntactic, and semantic information (e.g., words, part of speech tags, predicate-argument structures, and named entities) associated with speaker u's ith turn; ST_{i-1}^{i-k} represents the dialog subtask tags for utterances $i \cdots 1$ to $i \cdots k$; and DA_{i-1}^{i-k} represents the dialog act tags for utterances $i \cdots 1$ to $i \cdots k$.

In a second stage, the dialog subtask is determined from the lexical information, the dialog act computed according to Eq. (4), and the dialog context, as shown in Eq. (5).

$$ST_i^u = \operatorname*{argmax}_{s^u \in \mathcal{S}} P(s^u | DA_i^u, c_i^u, ST_{i-1}^{i-k}, DA_{i-1}^{i-k}, c_{i-1}^{i-k}) \qquad (5)$$

where the different variables have been explained for the previous equations. The conditional distributions shown in Eqs. (4) and (5) can be estimated by means of the general technique of choosing the maximum entropy (MaxEnt) distribution that properly estimates the average of each feature in the training data [1].

4   Emotional State Recognition

With the aim of developing systems capable of maintaining a conversation as natural and rich as a human conversation, emotion is gaining increasing attention from the dialog systems community as it affects the actions that the user chooses to communicate with the system [9]. Emotional information can be useful to improve dialog strategies, predict system errors, adapt dialog management, and provide support to users depending on their emotional state.

In this paper, we have employed a recognition method based on our previous work [4]. This method is focused on recognizing negative emotions that might discourage users from employing the system again or lead them to abort an ongoing dialog. Concretely, we have considered three negative emotions: anger, boredom, and doubtfulness, where the latter refers to a situation in which the user is uncertain about what to do next. The recognizer employs acoustic information to distinguish anger from doubtfulness or boredom, and dialog information to discriminate between doubtfulness and boredom, which are more difficult to discriminate only by using phonetic cues.

In most information providing spoken dialog systems, the application domain is not highly affective, thus a baseline algorithm which always chooses 'neutral' provides a very high accuracy (85% for the UAH system). This rate is difficult to improve by classifying the rest of emotions, which are very subtlety produced. Instead of considering neutral as another emotional class, we calculate the most likely non-neutral category. The dialog manager employs the intention information together with this category to decide whether to treat the user input as emotional or neutral, as will be explained in Sect. 5.

5   The Enhanced UAH Dialog System

Universidad Al Habla (UAH - University on the Line) is a spoken dialog system that provides academic information about the Dept. of Languages and Computer Systems at the University of Granada, Spain. The information that the system provides can be classified in four main groups: subjects, professors, PhD courses and student registration.

A corpus of 100 dialogs was acquired with this system from student telephone calls. The total number of user turns was 422 and the recorded speech has a duration of 150 min. In order to develop an enhanced version of the system that includes our proposals for statical dialog management, intention and emotion recognition, we carried out two types of corpus annotation: intentional and emotional.

On the one hand, we estimated the user intention for each user utterance by using concepts and attribute value pairs. One or more concepts represent the intention of the utterance, and a sequence of attribute-value pairs contains the information about the values provided by the user. We defined four concepts to represent the different queries that the user can perform (*Subject, Lecturers, Doctoral studies,* and *Registration*), three task-independent concepts (*Affirmation, Negation,* and *Not-Understood*), and eight attributes (*Subject-Name, Degree, Group-Name, Subject-Type, Lecturer-Name, Program-Name, Semester,* and *Deadline*).

The labeling of the system turns was similar to that for user turns. To do so, 30 concepts were defined and grouped as task-independent concepts (e.g. *Affirmation* and *Negation*), concepts used to inform the user about the result of a specific query (e.g. *Subject* or *Lecturers*), concepts defined to require the user the attributes that are necessary for a specific query (e.g. *Subject-Name*), and concepts used for the confirmation of concepts and attributes.

On the other hand, we assigned an emotion category (neutral, doubtful, angry, or bored) to each user utterance. Nine annotators tagged the corpus twice and the final emotion for each utterance was assigned by majority voting. A detailed description of the annotation procedure and the intricacies of the calculation of inter-annotator reliability can be found in a previous study [4].

Additionally, we have modified the dialog manager to process the user state information in order to reduce the impact of the user negative states and the user experience on the communication, by adapting the system responses considering user states. The dialog manager tailors the next system answer to the user state by changing the help providing mechanisms, the confirmation strategy and the interaction flexibility. The conciliation strategies adopted are, following the constraints defined in [3], straightforward and well delimited in order not to make the user loose the focus on the task.

If the recognized emotion is doubtful and the user has changed their behavior several times during the dialog, the dialog manager changes to a system-directed initiative and generates a help message describing the available options. This approach is also selected when the user profile indicates that the user is non-expert (or if there is no profile for the current user), and when their first utterances are classified as doubtful.

In the case of anger, if the dialog history shows that there have been many errors during the interaction, the system apologizes and switches to DTMF (Dual-Tone Multi-Frequency) mode. If the user is assumed to be angry but the system is not aware of any error, the system's prompt is rephrased with more agreeable phrases and the user is advised that they can ask for help at any time.

In the case of boredom, if there is information available from other interactions of the same user, the system tries to infer from those dialogs what the most likely objective of the user might be. If the detected objective matches the predicted intention, the system takes the information for granted and uses implicit confirmations. For example, if a student always asks for subjects of the same degree, the system can directly disambiguate a subject if it is in several degrees.

In any other case, the emotion is assumed to be neutral, and the next system prompt is decided only on the basis of the user intention and the user profile (i.e., considering user preferences, previous interactions, and expertise level).

6   Experiments

In order to evaluate our proposal, we have recorded the interactions of 6 recruited users. Four of them recorded 30 dialogs (15 scenarios with the baseline system and 15 with the enhanced system), and two of them recorded 15 dialogs (15 dialogs with the baseline or the enhanced system only). Thus, a total of 150 dialogs were recorded in such a way that there were two dialogs recorded per scenario, three in the case of the five most frequent scenarios of the initial UAH corpus.

As observed in Table 1, on the one hand the success rate for the enhanced system is higher than the baseline. This difference showed a significance of 0.03 in a two-tailed t-test. On the other hand, although the error correction rate is also higher in absolute values in the enhanced system, this improvement is not significant. Both results are explained by the fact that we have not designed a specific strategy to improve the recognition or understanding processes and decrease the error rate. Instead, our proposal for adaptation to the user state overcomes these problems during the dialog once they are produced.

Table 1. Results of the objective evaluation of the systems

Evaluation metrics	Baseline	Enhanced
Dialog success rate	85.0	96.0
Error correction rate	81.0	91.5
Average number of turns per dialog	12.1	8.1
Average number of actions per turn	1.8	1.5
% of different dialogs (intention only)	85.0	83.5
% of different dialogs (intention and emotion)	85.0	88.0
Number of repetitions of the most seen dialog	3.5	6
Number of turns of the most seen dialog	5.5	4.5
Number of turns of the shortest dialog	4.5	4.5
Number of turns of the longest dialog	14.5	12.0

Regarding the number of dialog turns, the enhanced system produced shorter dialogs (with a 0.00 significance value in a two-tailed t-test when compared to the number of turns of the baseline system). As shown in Table 1, this general reduction appears also in the case of the longest, shortest and most seen dialogs for the enhanced system. There is also a slight reduction in the number of actions per turn for the dialogs of the enhanced system (with a 0.00 significance value in the t-test). This might be because users have to explicitly provide and confirm more information using the baseline system, whereas the enhanced system automatically adapted the dialog to the user and the dialog history.

Regarding the percentage of different dialogs obtained, the rate was lower using the enhanced system, due to an increment in the variability of ways in which users can provide the different data required to the enhanced system. This result was significant when the dialogs were considered different only when they differed in the sequence of observed user intentions, and also when even with the same sequence of intentions, two dialogs were considered different if the emotions observed were different. This is consistent with the fact that the number of repetitions of the most observed dialogs is higher for the baseline system.

Regarding dialog style and cooperativeness, Figs. 1 and 2 respectively show the frequency of the most dominant user and system dialog acts in the dialogs collected with the enhanced and baseline systems. On the one hand, Fig. 1 shows that users need to provide less information explicitly using the enhanced system, which explains the higher proportion of queries (significant over 98%). On the other hand, Fig. 2 shows that there is a reduction in the system requests when the enhanced system is used. This explains a higher proportion of system turns to provide information in the enhanced system.

Table 2 shows the average results obtained with respect to the subjective evaluation. As can be observed, both systems correctly understand the different user queries and obtain a similar evaluation regarding the user observed easiness in correcting errors made by the ASR module. However, the enhanced system is judged to be better regarding the user observed easiness in obtaining the data required to fulfill the complete set of objectives defined in the scenario, as well as the suitability of the interaction rate during the dialog.

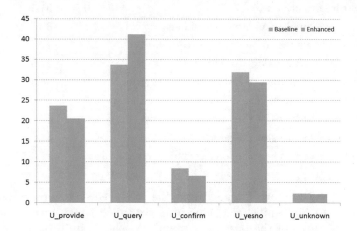

Fig. 1. Histogram of user dialog acts in the enhanced and baseline systems

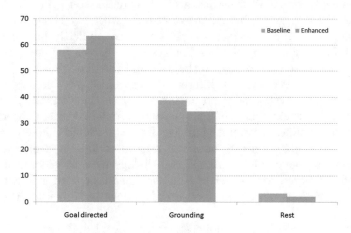

Fig. 2. Histogram of system dialog acts in the enhanced and baseline systems

Table 2. Results of the subjective evaluation of the systems

Questions (1 to 5 scale)	Baseline	Enhanced
How well did the system understand you?	4.6	4.8
How well did you understand the system messages?	3.6	3.9
Was it easy to obtain the requested information?	3.8	4.3
Was the interaction rate adequate?	3.4	4.2
If the system made errors, was it easy for you to correct them?	3.2	3.3

7   Conclusions and Future Work

Soft computing is already a major area of academic research. In this paper, we have contributed a proposal to integrate SC for developing lifelong spoken dialog systems. To do this, a statistical dialog model based on evolving classifiers selects the next system response taking into account the prediction of the user's intention and emotional state, and the history of the dialog up to the current dialog state. Our domain-independent approach is scalable and can help to reduce the dimensionality of complex slot-filling domains with a high number of input parameters. We have evaluated the proposal with the UAH spoken dialog system, implementing the prediction modules between the system's natural language understanding module and dialog manager. Additionally, we have improved the dialog manager to take this information into account in order to compute and adapt the system responses. The results show that this version of the system performs better in terms of duration of the dialogs, number of turns needed for successful dialogs, and number of confirmations and repetitions needed. Additionally, the test users judged the system to be better when it could adapt its behavior to their intentions and emotions.

For future work we plan to apply the proposed technique to other domains in order to see whether it can be used for comparison between several user models and dialog management techniques. We also intend to extend the evaluation of the system considering user satisfaction measures that complement the proposed adaptation.

References

1. Bangalore, S., DiFabbrizio, G., Stent, A.: Learning the structure of task-driven human-human dialogs. IEEE Trans. Audio Speech Lang. Process. **16**(7), 1249–1259 (2008)
2. Barnard, E., Halberstadt, A., Kotelly, C., Phillips, M.: A consistent approach to designing spoken-dialog systems. In: Proceedings of ASRU 1999, pp. 1173–1176 (1999)
3. Burkhardt, F., van Ballegooy, M., Engelbrecht, K., Polzehl, T., Stegmann, J.: Emotion detection in dialog systems - Usecases, strategies and challenges. In: Proceedings of ACII 2009 (2009)
4. Callejas, Z., López-Cózar, R.: Influence of contextual information in emotion annotation for spoken dialogue systems. Speech Commun. **50**(5), 416–433 (2008)
5. Chen, Z., Liu, B.: Lifelong Machine Learning. Morgan & Claypool, San Rafael (2018)
6. Engelbrecht, K.: Estimating Spoken Dialog System Quality with User Models. Springer, Heidelberg (2012)
7. Griol, D., Sanchis, A., Molina, J.: FRB-dialog: a toolkit for automatic learning of fuzzy-rule based (FRB) dialog managers. In: Proceedings of HAIS 2017, pp. 306–317 (2017)
8. Mazumder, S., Ma, N., Liu, B.: Towards a continuous knowledge learning engine for chatbots. Comput. Res. Repos. **2**, 1–11 (2018)

9. McTear, M.F., Callejas, Z., Griol, D.: The Conversational Interface. Talking to Smart Devices. Springer, Heidelberg (2016)
10. Pieraccini, R., Suendermann, D., Dayanidhi, K., Liscombe, J.: Are we there yet? Research in commercial spoken dialog systems. LNCS **5729**, 3–13 (2009)
11. Ravi, K., Ravi, V., Prasad, P.S.R.K.: Fuzzy formal concept analysis based opinion mining for CRM in financial services. Appl. Soft Comput. **60**, 786–807 (2017)
12. Schatzmann, J., Weilhammer, K., Stuttle, M., Young, S.: A survey of statistical user simulation techniques for reinforcement-learning of dialogue management strategies. Knowl. Eng. Rev. **21**(2), 97–126 (2006)

Smart Cities and IOT

Real-Time Big Data Analytics in Smart Cities from LoRa-Based IoT Networks

Antonio M. Fernández, David Gutiérrez-Avilés, Alicia Troncoso,
and Francisco Martínez-Álvarez[✉]

Data Science & Big Data Lab, Pablo de Olavide University, ES-41013 Seville, Spain
fmaralv@upo.es

Abstract. The currently burst of the Internet of Things (IoT) technologies implies the emergence of new lines of investigation regarding
not only to hardware and protocols but also to new methods of produced data analysis satisfying the IoT environment constraints: a real-
time and a big data approach. The Real-time restriction is about the
continuous generation of data provided by the endpoints connected to
an IoT network; due to the connection and scaling capabilities of an IoT
network, the amount of data to process is so high that Big data techniques become essential. In this article, we present a system consisting
of two main modules. In one hand, the infrastructure, a complete LoRa
based network designed, tested and deployment in the Pablo de Olavide
University and, on the other side, the analytics, a big data streaming system that processes the inputs produced by the network to obtain useful,
valid and hidden information.

Keywords: IoT · LoRaWAN · Real-time · Big data · Data streaming

1 Introduction

The current technological reality points to two main lines of research and development. First, the line of industry and services wherein the rise of the advanced
technology to M2M communications or the Internet of Things (from now on,
IoT) [21] is changing our means of production and our service management systems. This fact leads our society to a new industrial revolution, the 4.0 industry
[12]. On the other side, the line of data science and big data [2] emerges as a
consequence of the vast amount of it generating, day by day, in our society.

There is an intimal relation between IoT technology and the data science
and big data. The IoT networks can potentially manage a massive amount of
data depending on the number of endpoints connected to it. Although the management of the data traffic of an IoT network is a critical element, the useful
and efficient treatment of this data is another crucial point to take into account.
Due to the huge amount of data involved in this new framework, issues like data
storage, data buffering appears implying the use of Big data solutions. Furthermore, Data science techniques are needed to analyze the real-time data of the
IoT network and obtain useful, valid and hidden information [14].

© Springer Nature Switzerland AG 2020
F. Martínez Álvarez et al. (Eds.): SOCO 2019, AISC 950, pp. 91–100, 2020.
https://doi.org/10.1007/978-3-030-20055-8_9

In this article, we present the an IoT agent system consisting in an in-production LoRa based IoT network deployed, whose usability has been tested in the Pablo Olavide University (Seville, Spain) and a big data streaming system, based on HDFS and Spark, that analyzes in real time the data provided by the earlier mentioned IoT network.

The rest of the article is structured as follows. A summary of the previous researches related to the paper's topic is presented in Sect. 2. The architecture of the proposed system and the methods used in the development can be found in Sect. 3. The experimental setup carried out, and the yielded results are reported in Sect. 4. Finally, the conclusions and future work are provided in Sect. 5.

2   Related Works

Nowadays, GPRS, Sigfox, Narrowband Internet of Things (NB-IoT), and LoRa are four widely used IoT technologies providing the best coverage for IoT devices. These technologies have been deeply studied in terms of coverage in [13]. The authors simulated the coverage of the previously mentioned IoT networks comparing them in an area of $7800\,km^2$. This study aimed to obtain the technology that provides better coverage for the connected IoT devices concluding that NB-IoT provided a better coverage but having a maximum signal coupling and a signal loss of 164 dB.

By contrast, the authors in [17] demonstrated the low-consumption devices connected to LoRaWAN based IoT networks could transmit the data more efficiently compared with other devices and network servers. The authors carried out field testing with line of sight and no line of sight in an Indonesian University campus.

An architectural study was carried out in [18]. Here, the authors analyzed the LoRa technology and demonstrated the LoRaWAN based network architectures shows a good match with the measurement systems. Furthermore, their experimental results confirmed the capability of a low-cost transceiver to schedule the transmission of frames with a standard uncertainty less than $3\,\mu s$.

Related to the data, we can distinguish between two concepts. On the one hand, we find the discovery or extraction of valid, useful and hidden information from data sets; that discipline is known as data mining, the main step within the knowledge discovering in databases process. The Data Mining covers the data source selection processes, pre-processing, and the application of machine learning algorithms providing us of descriptor, predictor or classifying models of the data. These models extract those mentioned above valid, useful and hidden information that implies a huge number of applications like variable predictions, client segmentation or fraud detection [8], among others.

On the other hand, we find the processing and management of vast volumes of data form a new state-of-the-art discipline called big data [14]. The big data applies to an extensive collection of fields. In [5] a big data environment is developed to electricity market field, managing volumes of 1 TB of data with the aim of detect fraudulent clients. The authors in [6] present a software tool for behavior pattern discovery in vast amounts of biological data and they develop in [7] a methodology to process and evaluate the results of the previous tool that they are equally big.

In this paper we present a new aspect into the big data and the data mining: the analysis of significant flows of data, coming from IoT sensors of a LoRaWAN network, in real time with auto-incremental machine learning algorithms. This field is called big data streaming. There are a few works related with this new approach; however, in [11], the authors propose a comparison between the main frameworks to work in streaming context these are Storm, Flink y Spark Streaming; the parameters of the study are performance and failure tolerance. The authors in [1] conduct an analysis of the performance of the linear regression algorithm with the Spark MLlib library and the Massive Online Analysis platform (MOA). Finally, in [16], a real-time methodology to detect cybercrimes and credit card fraud based on Spark Streaming is presented.

3   System Architecture

The proposed architecture is illustrated in Fig. 1. We can observe that the system is composed of several modules. Each of them is responsible for a task in the system and are connected by input/output links. In a general way, the LoRa based infrastructure is in charge of obtaining and manage the data form several IoT devices. The preprocessing module takes these data as input and decodes and prepares to the following subsystems. The real-time or experimental environment manages the data streams and feed to the big data streaming engine. This engine performs the training of an auto-incremental machine learning algorithm and makes predictions of the measured variables by the sensors.

Next, each module is separately analyzed in the following sections: the LoRa-based infrastructure in Sect. 3.1, the JSON payload buffering and preprocessing in Sect. 3.2 and the real-time environment in Sect. 3.3.

Fig. 1. Proposed IoT architecture.

3.1   LoRa Based Infrastructure

We present our LoRa based infrastructure whose function is to collect the data that will be processed and analyzed by the real-time big data analytics system.

The inputs of the LoRa infrastructure will be de data picked by IoT sensors; specifically, these devices will measure values of two variables, this is pressure and temperature. The outputs of this subsystem will be the raw metrics of pressure and temperature registered in the *LoRaServer* software.

In Fig. 1 we can observe the elements of this module. Firstly, we have three IoT devices with two sensors measuring temperature (in Celsius, C) and pressure (in kilopascals, kPa).

The next element is the LoRa network based on the LoRaWAN standard. LoRaWAN defines an IoT communication protocol and a system architecture for the deployment of a network. The protocols and the architecture of a network are the most influencer elements to determine the life of the batteries, the net capability, the quality of the service and the security [17]. We have chosen the LoRaWAN standard for two main reasons: LoRaWAN is an open standard, it offers excellent performance for our purpose, and there are several open-source software solutions to net managing (*LoRaServer* software).

Then, we can see the group of three LoRa gateways. These devices carry out the management of the access points to the network and, thanks to a piece of software called LoRa Gateway Bridge, transmit the collected data to the controller software. As a part of this transmission, the binary packets from the sensors are transformed into a JSON that can managed by for LoRaServer system. The protocol between LoRa gateways and LoraServer is called Semtech packet forwarder protocol and is included in the LoRa Gateway Bridge.

The last part of the LoRa infrastructure is the LoRaServer system that manages and controls every payload that travels for the IoT net. Three subsystems compose the LoRaServer system: the *LoRa server* module responsible for the management of the gateways and end points. Next, we found the *LoRa App Server* that manage the applications, users, services and devices. Finally, the core of the system is the MQTT broker Eclipse Mosquito that connects the LoRa server with the LoRa Gateway Bridge and the LoRa App Server with any application.

3.2   JSON Payload Buffering and Preprocessing

The inputs of this module will be the LoRa JSON payloads acquired by a MQTT consumer. The outputs will be Kafka producers. Apache Kafka is an open-source software developed and maintained by Apache Software Foundation. The main objective of this software is to provide a unified platform with high performance and low latency for the manipulation of data sources in real-time environments. It can be interpreted like a message queue, developed as a register of distributed transactions using the publisher-subscription pattern. Apache Kafka is a large scale scalable, partitioned and replicated platform [3, 10, 15].

The general function of this module is to receive the JSON payloads from the LoRa network, then performing a JSON parsing to extract the *data* field. This information is encrypted by the Base64 method; then, the next step will be decrypting the data for, afterward, carry out device-driven decoding. After these steps, we will have the measures of pressure and temperature in a legible form (in kPa and C respectively).

Finally, the module checks the environment of the system. For this particular proposal, a real-time environment is used. Each W-dependent instance is sent to its corresponding module via Kafka producer.

3.3   Real-Time Environment

This module carries out the set up of the environment needed for training and testing the auto-incremental machine learning algorithm. In this module a real-time environment is settled; therefore, every time that a new instance with a W learning window is building, it will arrive at it using a Kafka consumer channel.

The elements of this module can be observed in Fig. 2. There, we can see how the Kafka consumer channel forwards the instances to a software layer called Kappa architecture [14]. It is a software architecture pattern whose purpose is to process streams of data in real-time and to store the results as mentioned earlier.

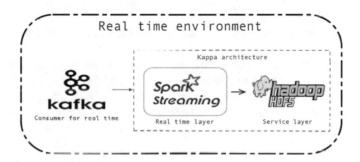

Fig. 2. Real-time big data analytics environment.

The Kappa architecture has two different elements. On the one hand, the real-time layer, implemented by Apache Spark Streaming is in charge of processing the instances coming from the Kafka channel and adapts the data for machine learning algorithm training and testing. On the other hand, we found the service layer, implemented by Apache Hadoop Distributed File System (HDFS) file system [19] which is in charge of storing the results of machine learning processing.

At this point, a brief explanation of Spark Streaming and Apache Hadoop HDFS is given:

- *Apache Spark Streaming*: It is an API of Apache Spark whose purpose is enabling scalable, high-throughput, fault-tolerant stream processing of live data streams. The streams of data can come from multiple sources (Apache Kafka, in our case). The processing results can be stored, likewise, in multiples systems like databases, dashboards, and, (in our case) HDFS.

– *Apache Hadoop HDFS*: It is one of the modules of Hadoop. HDFS is a distributed file system designed to run on commodity hardware. Furthermore, it is highly fault-tolerant and is designed to be deployed on low-cost hardware and provides high throughput access to application data and is suitable for applications that have large datasets.

3.4   Big Data Streaming Engine

This section describes the engine used in order to forecast big data streaming.

Data are received continuously, and therefore the prediction of a given time horizon, h, must be made in real time. The Apache Spark Streaming machine learning library (MLlib) is proposed for this purpose. In particular, the linear regression algorithm [9] is used to forecast future values for the target variables (temperature and pressure in our case).

MLlib's linear regression algorithm does not support multi-step forecasting and, for this reason, the algorithm must be adapted as previously done in [4, 20]. Therefore, the problem that must be solved is:

$$[x(t+1), x(t+2), \ldots, x(t+h)] = f(x(t), x(t-1), \ldots, x(t-w-1)) \quad (1)$$

where w represents the window of past values considered for predicting the h future values.

For each data stream received from any sensor at the instant t, a forecasting model M_t is generated at the training stage. This model is incrementally updated along with new coming data, thus generating new models at different time stamps M_{t+1}, $M_{t+2},...,M_{t+n}$. When a prediction is made, the last generated model is used.

4   Results

This section describes the experimentation that has been carried out for the extraction of information with the input data and application of a big data streaming algorithm.

4.1   Dataset Construction and Linear Regression Parametrization

The data used are those obtained by the sensors of the LoRaWAN network. The measurement of the data has been performed every ten seconds for a period of one month, having a raw data set of approximately seventy thousand records.

The dataset contained erroneous records, so a pre-processing of the data was applied, checking that the data had a correct measurement and eliminating the erroneous measurements. At the end of the cleaning pre-processing, five thousand seven hundred and twenty-six records are obtained.

The linear regression algorithm with gradient descent [9], used for experimental study in streaming, requires the optimization of two main parameters as indicated in the previous section:

- α. It is the size of the step that moves the gradient in the downward direction.
- σ. The number of iterations necessary for the method to converge.

To obtain the optimum value of these two parameters, an exhaustive search algorithm has been developed. Finally, the optimum parameters obtained are shown in Table 1, when the mean relative error (MRE) is minimized.

Table 1. Optimum parameters obtained for the linear regression algorithm.

Data	σ	α (stepSize)	MRE
Pressure	10	3.33E-11	1.27E-05
Temperature	15	3.61E-05	4.18E-05

4.2 Experimental Setup

After analyzing the input data, building the datasets, and obtaining the optimal execution parameters for the linear regression model, the following experimental design is proposed to test the effectiveness of the Apache Spark streaming linear regression model in a real streaming situation.

The aim of this study is to test the MRE variation when the self-incremental learning model is fed with new learning data through the streaming channel to which it is connected.

Thus, for each variable measured $V = \{P, T\}$ by the pressure and temperature sensor and for each time window $w_i = \{w_3, w_6, w_{12}, w_{24}, w_{90}, w_{180}\}$ the following process is performed:

1. Activate the streaming system.
2. Parametrize the incremental Linear regression algorithm with the optimal parameters $\langle \alpha, \sigma \rangle$ for V and w_i.
3. Inject training dataset into the training channel with a time lapse of 5 s, generating a M_i model for each dataset.
4. For each M_i model a prediction of the complete test set is made, measuring the associated MRE.

Once this process is executed, the MRE is obtained for each model generated $M_1, M_2, M_3, ..., M_{20}$. At the same time, the observed values will be compared with the predicted values, highlighting the models of the streaming channel that reaches lower MRE in the prediction.

4.3 Analysis

In this section different experiments are carried out in order to perform an exhaustive analysis of the incremental learning of the different online models that are generated to estimate the predictions in a streaming environment. First,

98   A. M. Fernández et al.

the section on errors evaluates the quality of the prediction, in terms of the
average relative error, while adding LP to the training package used to obtain
the prediction model. Finally, the predictions section presents the results of
the prediction of the set of tests obtained with the best model for each time
window. To obtain the online models, the training data set has been divided
into 30 batches. Each of these subsets will be injected one at a time at 5 s
intervals. In this way, 30 prediction models will be obtained in real time in an
incremental way to predict the test set. These models are obtained using the
optimal configuration of the parameters α and σ, necessary for linear regression,
which is shown in Table 1.

Table 2. MRE for pressure sensor data stream.

	w_3	w_6	w_{12}	w_{24}	w_{90}	w_{180}
M_1	5.08E-03	9.25E-03	5.76E-03	5.74E-03	5.80E-03	**4.48E-03**
M_2	4.95E-03	4.54E-03	4.55E-03	4.49E-03	4.44E-03	4.52E-03
M_3	4.89E-03	4.66E-03	4.65E-03	4.31E-03	4.70E-03	4.52E-03
M_4	4.99E-03	4.59E-03	4.40E-03	4.56E-03	4.44E-03	6.29E-03
M_5	4.89E-03	**4.51E-03**	4.63E-03	**4.31E-03**	4.65E-03	4.92E-03
M_6	4.96E-03	4.52E-03	4.38E-03	4.60E-03	**4.42E-03**	5.13E-03
M_7	4.98E-03	4.52E-03	4.58E-03	4.32E-03	4.49E-03	4.73E-03
M_8	4.89E-03	4.53E-03	4.40E-03	4.48E-03	4.81E-03	5.31E-03
M_9	4.96E-03	4.54E-03	4.36E-03	4.61E-03	4.42E-03	5.40E-03
M_{10}	4.90E-03	4.69E-03	4.56E-03	4.32E-03	4.53E-03	4.72E-03
M_{11}	**4.88E-03**	4.68E-03	4.65E-03	4.52E-03	5.41E-03	5.13E-03
M_{12}	4.98E-03	4.67E-03	4.39E-03	4.33E-03	4.59E-03	5.21E-03
M_{13}	4.90E-03	4.64E-03	**4.36E-03**	4.31E-03	5.17E-03	4.73E-03
M_{14}	4.89E-03	4.64E-03	4.60E-03	4.55E-03	4.42E-03	5.25E-03
M_{15}	4.99E-03	4.54E-03	4.70E-03	4.34E-03	4.46E-03	4.76E-03
M_{16}	4.89E-03	4.54E-03	4.37E-03	4.31E-03	4.50E-03	5.44E-03
M_{17}	4.96E-03	4.55E-03	4.59E-03	4.49E-03	5.23E-03	4.71E-03
M_{18}	5.00E-03	4.62E-03	4.37E-03	4.32E-03	4.59E-03	5.88E-03
M_{19}	4.90E-03	4.74E-03	4.56E-03	4.49E-03	4.98E-03	4.75E-03
M_{20}	4.89E-03	4.53E-03	4.63E-03	4.32E-03	4.50E-03	5.26E-03

The errors made in the prediction of the test set when using the different
models that are generated in an incremental way are now discussed. Table 2
shows the MRE made when predicting the test set for the pressure sensor, using
each of the models obtained online for different lengths of historical data. In
bold type, the models that have obtained the minimum error for each window
are highlighted. Similar results are reported for both pressure and temperature
but, due to space limitations, they are not shown here.

5   Conclusions

In this article we propose a complete system to collect information from the environment through the use of a IoT architecture using LoRa technology and the LoRaWAN standard. The system has been successfully deployed at Pablo de Olavide University (Seville, Spain). The implementation of an architecture for data analysis, called Kappa architecture, for real-time analysis and development of experimentation is also described, and it is based on the underlying pache Spark Streaming and HDFS technologies. Experiments carried out using the physical IoT network and real sensors are reported in order to evaluate the incremental models and the quality of the predictions made.

Acknowledgments. We would like to thank the Spanish Ministry of Economy and Competitiveness for the support under project TIN2017-88209-C2-1-R. Additionally, we want to express our gratitude to Enrique Parrilla, Lantia IoT's CEO, since all the equipment has been provided by him. The T-Systems Iberia company is also acknowledged since all experiments have been carried out on its Open Telekom Cloud Platform based on the OpenStack open source.

References

1. Akgün, B., Öğüdücü, Ş.G.: Streaming linear regression on Spark MLlib and MOA. In: Proceedings of the IEEE/ACM International Conference on Advances in Social Networks Analysis and Mining, pp. 1244–1247 (2015)
2. Chen, M., Mao, S., Liu, Y.: Big data: a survey. Mob. Netw. Appl. **19**(2), 171–209 (2014)
3. D'Silva, G.M., Khan, A., Gaurav, Bari, S.: Real-time processing of IoT events with historic data using Apache Kafka and Apache Spark with dashing framework. In: Proceedings of the IEEE International Conference on Recent Trends in Electronics, Information Communication Technology, pp. 1804–1809 (2017)
4. Galicia, A., Talavera-Llames, R., Troncoso, A., Koprinska, I., Martínez-Álvarez, F.: Multi-step forecasting for big data time series based on ensemble learning. Knowl. Based-Syst. **163**, 830–841 (2019)
5. Gutiérrez-Avilés, D., Fábregas, J.A., Tejedor, J., Martínez-Álvarez, F., Troncoso, A., Arcos, A., Riquelme, J.C.: SmartFD: a real big data application for electrical fraud detection. Lect. Notes Artif. Intell. **10870**, 120–130 (2018)
6. Gutiérrez-Avilés, D., Rubio-Escudero, C., Martínez-Álvarez, F., Riquelme, J.: Trigen: a genetic algorithm to mine triclusters in temporal gene expression data. Neurocomputing **132**, 42–53 (2014)
7. Gutiérrez-Avilés, D., Giráldez, R., Gil-Cumbreras, F.J., Rubio-Escudero, C.: TRIQ: a new method to evaluate triclusters. BioData Min. **11**, id15 (2018)
8. Han, J., Pei, J., Kamber, M.: Data Mining: Concepts and Techniques. Elsevier, Amsterdam (2011)
9. Hu, T., Wu, Q., Zhou, D.X.: Convergence of gradient descent for minimum error entropy principle in linear regression. IEEE Trans. Signal Process. **64**(24), 6571–6579 (2016)
10. Ichinose, A., Takefusa, A., Nakada, H., Oguchi, M.: A study of a video analysis framework using Kafka and Spark Streaming. In: Proceedings of the IEEE International Conference on Big Data, pp. 2396–2401 (2017)

11. Karakaya, Z., Yazici, A., Alayyoub, M.: A comparison of stream processing frameworks. In: Proceedings of the International Conference on Computer and Applications, pp. 1–12 (2017)
12. Lasi, H., Fettke, P., Kemper, H.G., Feld, T., Hoffmann, M.: Industry 4.0. Bus. Inf. Syst. Eng. **6**(4), 239–242 (2014)
13. Lauridsen, M., Nguyen, H., Vejlgaard, B., Kovacs, I.Z., Mogensen, P., Sorensen, M.: Coverage Comparison of GPRS, NB-IoT, LoRa, and SigFox in a 7800 km^2 Area. In: Proceedings of the IEEE Vehicular Technology Conference, pp. 1–5 (2017)
14. Marz, N., Warren, J.: Big Data: Principles and Best Practices of Scalable Real-time Data Systems. Manning Publications Co., Shelter Island (2015)
15. Noac'h, P.L., Costan, A., Bougé, L.: A performance evaluation of Apache Kafka in support of big data streaming applications. In: Proceedings of the IEEE International Conference on Big Data, pp. 4803–4806 (2017)
16. Pallaprolu, S.C., Sankineni, R., Thevar, M., Karabatis, G., Wang, J.: Zero-day attack identification in streaming data using semantics and Spark. In: Proceedings of the IEEE International Congress on Big Data, pp. 121–128 (2017)
17. Rahman, A., Suryanegara, M.: The development of IoT LoRa: a performance evaluation on LoS and Non-LoS environment at 915 MHz ISM frequency. In: Proceedings of the International Conference on Signals and Systems, pp. 163–167 (2017)
18. Rizzi, M., Ferrari, P., Flammini, A., Sisinni, E.: Evaluation of the IoT LoRaWAN solution for distributed measurement applications. IEEE Trans. Instrum. Meas. **66**(12), 3340–3349 (2017)
19. Shvachko, K., Kuang, H., Radia, S., Chansler, R.: The Hadoop distributed file system. In: Proceedings of the IEEE Symposium on Mass Storage Systems and Technologies, pp. 1–10 (2010)
20. Torres, J.F., Galicia, A., Troncoso, A., Martínez-Álvarez, F.: A scalable approach based on deep learning for big data time series forecasting. Integr. Comput.-Aided Eng. **25**(4), 335–348 (2018)
21. Wortmann, F., Flüchter, K.: Internet of things. Bus. Inf. Syst. Eng. **57**(3), 221–224 (2015)

Deep Learning in Modeling Energy Cost of Buildings in the Public Sector

Marijana Zekić-Sušac[1]([⊠]), Marinela Knežević[1],
and Rudolf Scitovski[2]

[1] Faculty of Economics, University of Josip Juraj Strossmayer in Osijek,
Trg Ljudevita Gaja 7, 31000 Osijek, Croatia
{marijana,marinela.knezevic}@efos.hr
[2] Department of Mathematics, University of Josip Juraj Strossmayer in Osijek,
Trg Ljudevita Gaja 6, 31000 Osijek, Croatia
scitowsk@mathos.hr

Abstract. The cost of energy consumed in educational, health, public administration, military, and other types of public buildings constitutes a substantial proportion of the total expenditure of the public sector. Due to a large number of attributes that influence the energy cost of a building, most of the models developed in the literature use only a subset of predictors, often neglect occupational data, and do not exploit enough the potential of deep learning methods. In this paper a real data from Croatian public sector is used including constructional, energetic, geographical, occupational and other attributes. Algorithms for data preprocessing and for deep learning modelling procedure are suggested. The number of hidden units in the deep neural network is optimized by a cross-validation procedure, while the sigmoid activation function was tested with Adam optimization algorithm. The feature selection was conducted using the recursive feature elimination method with a regression random forest kernel. The aims were to identify the subset of relevant predictors of energy cost in public buildings that could assist decision makers in determining the priority of reconstruction measures as well as to test the potential of deep learning in predicting the yearly energy cost. The results have shown that the deep learning network with three hidden layers was the most successful in predicting energy cost using the wrapper-based method of feature extraction. The selection of features confirms the importance of occupational data, as well as heating, cooling, electricity lightning, and constructional attributes for estimating the total energy cost. Those predictors can be used in decision making on allocating resources in public buildings reconstructions. The model implementation could improve public sector energy efficiency, save costs and contribute to the concepts of smart buildings and smart cities.

Keywords: Energy cost · Deep learning · Neural networks · Public sector

© Springer Nature Switzerland AG 2020
F. Martínez Álvarez et al. (Eds.): SOCO 2019, AISC 950, pp. 101–110, 2020.
https://doi.org/10.1007/978-3-030-20055-8_10

1 Introduction

Creating efficient models that will be able to extract features and predict the cost of energy consumption of a building is vital in energy management. However, the previous research lacks an extensive usage and integration of machine learning methods on real energy data. EU directives (2012/27/EU and 2010/31/EU) state that 40% of all energy consumption belongs to the building sector which is itself expanding. Public buildings are usually larger in dimension than residential buildings, thus potentially larger energy consumers, and efficient models for managing energy in those buildings could make a substantial reduction in energy consumption of a state. Croatia is among the highest ten energy intensity countries in EU [10].

Although the methodology of deep learning is introduced two decades ago [1], it was not exploited enough in the area of energy management and especially in the area of cost savings [16]. Internet of things (IoT) and predictive models were suggested as components of an intelligent system for energy management in the context of managing one building or a smart city [8], but not on the public sector level. Deep learning showed its potential in time seria prediction of electrical energy consumption [15]. However, the predictive ability of deep learning in modeling energy cost was not investigated in public sector. In this paper, a real dataset from the Croatian energy management information system – EMIS was collected to create prediction models by deep neural networks. The aims of the paper are: (1) to test the potential of deep learning by measuring its accuracy in predicting the total energy cost, (2) to identify important predictors of energy cost of public buildings. Separate algorithms for data preprocessing and deep learning modeling are suggested to fulfill the above goals. Deep learning modeling was conducted using Keras library for deep neural network in R software tool, with TensorFlow as a backend. The paper contributes to the methodology by providing the algorithms for a systematic approach to data preprocessing and deep learning modeling of prediction problems, and brings a new knowledge to the theory of key performance indicators of buildings energy consumption. Those indicators can be incorporated in a smart city system architecture as suggested by [8].

2 Literature Review

Calvillo et al. [2] emphasized that energy management is one of the most demanding issues within smart city concept where a high quality of life should be provided to its inhabitants through optimal resource management. Analysis of existing prediction models of energy consumption of buildings provided by Pérez-Lombard et al. [11] and Zekić-Sušac [19] shows that authors use three groups of methods in this domain: (1) statistical methods such as multiple linear regression, (2) machine learning methods, such as artificial neural networks, decision trees [9], clustering [19], as well as (3) simulation modeling. Torres et al. [15] suggested an algorithm for dealing with multi-horizon time-series forecasting by deep learning within Big Data environment on Spanish electricity data. Most of the papers in energy modeling use data from residential buildings, or from the building sector in general, while public buildings were rarely investigated [19]. The attributes of buildings relevant for modeling the cost of energy consumption usually includes meteorological data, building attributes available in energy certificates, or data collected by authors' surveys. Besides those "static" data, some previous authors emphasize the importance of including occupancy and usage data [7, 18]. It is interesting that Liang et al. [6] have found that occupancy data did not significantly improve the accuracy of the baseline model. In our experiments, an extensive set of input attributes is used which is larger in dimension than the data used in known previous research, and we were challenged to test the efficiency of deep learning architectures in dealing with such data as well as the selection of important features.

3 Data and Sampling

The research is conducted on a real dataset from the Croatian energy management information system – EMIS which collects and manages data about energy consumption of buildings in the Croatian public sector. Out of more than 17 000 records in the database, only 1584 buildings had available all attributes of their geospatial, construction, heating, cooling, meteorological, occupational, and energetic characteristics, as well their energy consumption. The model aimed to predict average annual cost of energy with taxes based on the 62 building attributes as inputs (12 categorical (factor) and 50 numeric) computed in the period from 2015 to 2017. The first stage in the preprocessing procedure was data cleansing, which included several steps: (1) replacing missing values, (2) removing outliers using the DBSCAN algorithm and Sym-DIRECT method according to Viswanath and Babu [17] modified by Scitovski et al. [12], (3) removing records with attribute values that were out of possible range (which are likely to be human errors in data entry), as determined by Krstić and Teni [4]. The second stage consisted of pre-modeling filtered-based variable reduction, which was conducted based on $\chi 2$ test of independence for factor variables, correlations for continuous (numeric) variables. This procedure extracted 32 numeric and 7 factor variables, and after transforming factor variables into binary categories, there were 49 inputs left after the interdependence filter. In order to furtherly reduce the size of input vector, a wrapping method was performed, which selects predictors on the basis of their influence to the output variable during the modeling phase. For that purpose a recursive feature elimination method (RFE) with a non-linear regression random forest kernel was used. This method is chosen due to its ability to deal with numeric output. It has extracted 12 important predictors, which were used in DNN modeling, and compared to the DNN model with all available inputs. The steps of data preprocessing are presented in the suggested Algorithm 1. The Output1 of the Algorithm 1 was used to create DNN Model 1- with filter-based selected features, while the Output2 of the preprocessing algorithm was used to create DNN Model 2 – with wrapper-based selected features. The outlier removal using the DBSCAN algorithm was performed in Mathematica software, while all other computations were performed in the R software package using Keras library for deep learning.

Algorithm 1. Data preprocessing

Input: initial dataset matrix $X \in R^{m \cdot n}$

1. Determine cases x_{ij} $(i, j \in N, \ i \leq m, j \leq n)$ with missing values. If variable j is numerical, replace missing values x_{ij} with the mean of remaining values in vector X_j, otherwise replace missing values x_{ij} with the mode of vector X_j
2. Determine outliers in vectors X_j. Remove outliers using the DBSCAN algorithm (according to [12], [17]).
3. Determine and remove cases x_{ij} with values out of possible range (human errors in data entry) [4].
4. Filter-based variable reduction on input space – based on $\chi 2$ test of inter-dependence for factor variables, and inter-correlations for numeric attributes

Output1: preprocessed dataset matrix $X' \in R^{k_1 \cdot l_1}, k_1, l_1 \in N, \ k_1 \leq m, \ l_1 \leq n$

5. Wrapper-based variable reduction – based on recursive feature elimination (RFE) method with a regression random forest kernel

Output2: preprocessed dataset matrix $X'' \in R^{k_1 \cdot l_2}, k_1, l_2 \in N, \ l_2 \leq l_1$

After the above steps, the final data sample consisted of 547 observations (buildings). The sample was divided into a sub-sample for modeling, i.e., train data (80%), and a sub-sample for validation, i.e., test data (20%). For the purpose of optimizing DNN topology (i.e., number of hidden layers and number of hidden units in each layer), the train sub-sample in DNN modeling was further divided into the DNN train (80% of train) and DNN test data (20% of train). Table 1 presents the list of 12 attributes extracted as significant by RFE method by their ranking (from the most important to the least important).

Table 1. Rank of selected features and their descriptive statistics

Rank	Attribute description	Attribute descriptive statistics
1.	Number of employees	Min.: 1.0, Mean: 39.1, Max.: 439.0, St.dev.: 57.1
2.	Heated surface of the building [m^2]	Min.: 24.5, Mean: 2295.1, Max.: 21122.7, St.dev.: 2771.6
3.	Total heating power [kW]	Min.: 0.0, Mean: 72.3, Max.: 3149.6, St.dev.: 228.2
4.	Total installed cooling power of cooling bodies [kW]	Min.: 0.0, Mean: 12.13, Max.: 840.0, St.dev.: 44.4
5.	Total installed power of other heating bodies [kW]	Min.: 0.0, Mean: 73.3, Max.: 1694.5, St.dev.: 155.4
6.	Total installed power of other consumers	Min.: 0.0, Mean: 9.822, Max.: 570.0, St.dev.: 44.11
7.	Total power of lamps with incandescent lamps [kW]	Min.: 0.0, Mean: 1.3, Max.: 59.4, St.dev.: 4.4

(continued)

Table 1. (*continued*)

Rank	Attribute description	Attribute descriptive statistics
8.	Total number of lamps with incandescent lamps	Min.: 0.0, Mean: 11.26, Max.: 647, St.dev.: 43.2
9.	Number of working days per year	Min.: 0.0, Mean: 246.1, Max.: 365, St.dev.: 75.6
10.	Total number of lamps with fluorescent tubes	Min.: 0.0, Mean: 46.14, Max.: 1557.00, St.dev.: 141.83
11.	Number of floors	Min.: 1.0, Mean: 1.9, Max.: 8.0, St.dev.: 1.2
12.	Factor of building shape	Min.: 0.1, Mean: 0.8, Max.: 1.7, St.dev.: 0.3
Output	Annual energy cost of a building including tax – output variable	Min.: 6.6, Mean: 66792.4, Max.: 794250.2, St.dev.: 110938.0

The data used to model DNN were scaled to distance between the maximum and minimum value of each attribute before training, and unscaled before computing the test errors.

4 Methodology

LeCun, Bengio, and Hinton [5] define deep learning (DL) as computational models composed of multiple processing layers that are able to learn complex data representations and are suitable for solving high-dimensional problems. The most frequent types of problems solved by deep learning are computer vision through prediction, classification, and pattern recognition, image and video recognition, speech recognition, and other complex problems. Deep learning is part of machine learning, where methods that learn from previous examples, such as artificial neural networks (ANNs), tree-partitioning methods (CART decision trees, random forest, etc.), support vector machine and other methods aim to approximate function between input and output, through multiple layers. In this paper we use deep learning ANNs, or DNNs. A number of hidden layers in DNNs allow multiple non-linear transformations and in some types of networks involve the selection of features through layers. The most frequently used ANN types in DL are multi-layer perceptrons (MLP), recurrent neural networks (RNN), and recently, convolutional neural networks (CNN). The basic structure of a DNN is shown in Fig. 1. A feed-forward multi-layer perceptron (MLP) DNN is used, originally suggested by Werbos in 1974, improved by Rumelhart et al. and later modified for deep learning [1, 5]. The computation of a DNN MLP with two fully connected hidden layers can be expressed by:

$$y_c = f_2\left(\sum_{j=1}^{m} f_1\left(\sum_{i=1}^{n} w_i^{(1)} x_i\right) w_j^{(2)}\right) \tag{1}$$

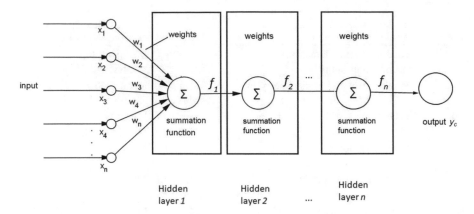

Fig. 1. Basic structure of a deep neural network

where y_c is the computed output, x_i are the elements of the input vector, $w_i^{(1)}$ are the elements of the weight vector of the first hidden layer, $w_j^{(2)}$ are the elements of the weight vector of the second hidden layer, f_1 is an activation function used in the first hidden layer, while f_2 is used in the second hidden layer.

In each hidden layer in our experiments, the number of hidden units is optimized by a cross-validation procedure, such that 20 randomly generated number of hidden units in each layer were trained and tested. Instead of classical gradient descent algorithm or BFGS used in "shallow" ANNs, the Adam optimization algorithm introduced by Kingma and Ba [3] was used in our experiments with 0.001 learning rate. It is proven that this method outperforms other optimization methods for a variety of models and datasets, and that it is convenient for large-scale high-dimensional machine learning problems, such as the one in this research [3]. In order to prevent overfitting, a stochastic dropout procedure was added into both hidden layers, with rate = 0.1, meaning that 10% of randomly selected weights be dropped out of further processing as suggested by Srivastava [13]. The maximum number of training epochs was 200.

The accuracy of both DNN models was evaluated by using the normalized root mean square error (NRMSE) and the symmetric mean average percentage error (SMAPE) according to [14]:

$$SMAPE = 100 \frac{1}{n} \sum_{i=1}^{n} \frac{|y_t - y_c|}{|y_t| + |y_c|} \tag{2}$$

where y_t is the target (real) output, while y_c is the computed output and n is the number of observations in the sample. All DNN architectures were trained on the same training subsample, and finally tested on the same hold-out test sample in order to enable the comparison of results. The procedure of training and validating DNN architectures with two and three hidden layers is described in Algorithm 2.

Algorithm 2. Modeling DNN architecture

Input: (1) Output1 of the Algorithm1, and (2) Output2 of the Algorithm2

1. Scale data. Divide dataset into train and test.
2. Generate arrays hid1, hid2, hid3 of 20 random integer numbers from the interval 1 to 100.
3. for $i=1$ to 20 do
   - (a) Set random seed to a constant value.
   - (b) Create sequential DNN model with (1) two and (2) three hidden layers:
     - (i) hidden layer 1 with *hid1[i]* hidden units, sigmoid function, adam optimization, and dropout noise for weight regularization (rate=0.1)
     - (ii) hidden layer 2 with *hid2[i]* hidden units, sigmoid function, adam optimization, and dropout noise for weight regularization (rate=0.1)
     - (iii) hidden layer 3 with *hid3[i]* hidden units, sigmoid function, adam optimization, and dropout noise for weight regularization (rate=0.1)
     - (iv) set output layer with 1 hidden unit (numeric output)
   - (c) Train sequential DNN model using 200 epochs, stopping rule, MSE error as a loss function on the validation set (validation split=0.2)
   - (d) Evaluate sequential DNN model on the test set, compute predictions
   - (e) Compute and save NRMSE and SMAPE on the test set for the architecture of hidden units *hid1[i], hid2[i] , hid3[i]*
4. end for
5. Select the architecture with min SMAPE.

Output: (1) Model 1 and (2) Model 2 with the best architecture of *i* hidden units in the hidden layer1, *j* hidden units in the hidden layer 2.

Following the Algorithm 2, 40 architectures were tested for Model 1 and 40 for Model 2 (20 architectures with 2 hidden layer, and 20 with 3 hidden layers for each model) and their results are compared.

5   Results and Discussion

The two datasets created by Algorithm 1 were used for DNN modelling according to the procedure described in Algorithm 2. Due to the lack of space, only the best selected architectures for each model with two and three hidden layers are presented in Table 2. It can be seen from Table 2 that the most accurate model was Model 2 with three hidden layers (12-50-55-61-1) with wrapper-based feature selection which produced the SMAPE of 32.2099%. In both models, the additional hidden layer improved the accuracy, therefore revealing the potential of deep learning for this problem. The results also reveal that the wrapper-based variable reduction has generated more successful subset of predictors than the filter-based method.

Table 2. Results of deep learning modeling on the validation sample

Model	DNN topology	MSE on the validation sample (scaled)	NRMSE on the test sample (original data)	SMAPE on the test sample (%)
Model 1 – filter-based feature selection				
2 hidden layers	49-27-13-1	0.0061	63.90	37.2106
3 hidden layers	49-66-82-74-1	0.0067	65.00	36.9722
Model 2 – wrapper-based feature selection				
2 hidden layers	12-28-20-1	0.0061	66.80	33.3539
3 hidden layers	12-50-55-61-1	0.0065	67.80	**32.2099**

The predictions of the DNN models and the real output (ave.yearly energy cost) obtained on the test sample are shown in Fig. 2.

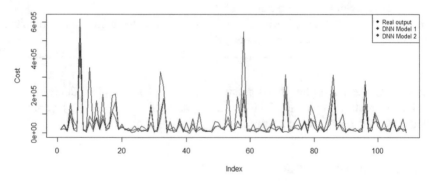

Fig. 2. Comparison of real output and DNN predictions of Model 1 and 2 on the test sample

The wrapper-based RFE method has selected 12 predictors (see Table 2). The five most relevant predictors were: *Number of employees, Heated surface of the building [m²], Total heating power [kW], Total installed cooling power of cooling bodies [kW],* and *Total installed power of other heating bodies [kW]*. This finding is consistent with suggestions of previous authors to include occupational variables into the model [18] while it is not consistent with the finding of Liang et al. [6] whose results showed that occupancy did not significantly improve the accuracy of the baseline model. Our research revealed that the *Number of employees* has the highest impact on the energy cost, while *Number of working days per year* as another occupational attribute was also found important. Three heating variables were extracted among the first five, indicating that the heating energy plays substantial role in the energy cost. The attribute related to cooling (*Total installed cooling power of cooling bodies [kW]*) is also placed among the first five predictors, which is also consistent with previous research in which heating and cooling loads were extracted as the most important [6]. Other features extracted by RFE belong to the group of attributes related to electricity and lightning (see Table 1), while the last group represents constructional variables (*Number of floors* and *Factor of building shape*).

6 Conclusion

The paper deals with predicting the cost of energy consumed in public buildings by deep learning methodology. A real dataset of public buildings was used from the Croatian public sector database. The input vector dimension was reduced using the filter-based and wrapper-based methods. The two algorithms were suggested: (1) for data preprocessing including replacement of missing values, dealing with outliers, and variable reduction, and (2) for deep learning modeling. A large number of DNN architectures were trained and tested in order to find the most successful model using two and three hidden layers and optimizing the number of hidden units in a cross-validation procedure. The results have shown that the wrapper-based feature extraction has produced higher accuracy than the models based on pure filtering of input variables. The model reveals a potential of deep learning to deal with prediction type of problems with a numeric output variable that has a large interval of possible values, which is not an easy task to handle. The results confirm the relevance of occupational data for predicting energy cost, which was extracted as more important than heating, cooling, lightning and constructional attributes. In future research, the dataset should be expanded to a larger number of buildings in order to create models on Big Data platforms with more advanced algorithms that will be less time-consuming. Such models could be implemented in public sector intelligence system to improve energy efficiency, save costs and manage future consumption within smart buildings and smart cities.

Acknowledgments. This work was supported by Croatian Science Foundation through research grant IP-2016-06-8350 and research grant IP-2016-06-6545.

References

1. Bengio, Y.: Learning deep architectures for AI. Found. Trends Mach. Learn. **2**, 1–127 (2009). https://doi.org/10.1561/2200000006
2. Calvillo, C.F., Sánchez-Miralles, A., Villar, J.: Energy management and planning in smart cities. Renew. Sustain. Energy Rev. **55**, 273–287 (2016). https://doi.org/10.1016/j.rser.2015.10.133

3. Kingma, D.P., Ba, M.: Adam: a method for stochastic optimization. In: 3rd International Conference on Learning Representations (2014). https://arxiv.org/abs/1412.6980
4. Krstić, H., Teni, M.: Algorithm for constructional characteristics data cleansing of large-scale public buildings database. In: High Performance and Optimum Design of Structures and Materials III, WIT Transactions on The Built Environment, vol. 175, pp. 213–224 (2018). https://doi.org/10.2495/HPSM180221
5. LeCun, Y., Bengio, Y., Hinton, G.: Deep learning. Nature **521**, 436–444 (2015). https://doi.org/10.1038/nature14539
6. Liang, X., Hong, T.Z., Shen, G.Q.: Improving the accuracy of energy baseline models for commercial buildings with occupancy data. Appl. Energy **179**, 247–260 (2016). https://doi.org/10.1016/j.apenergy.2016.06.141
7. Mangold, M., Osterbring, M., Wallbaum, H.: Handling data uncertainties when using Swedish energy performance certificate data to describe energy usage in the building stock. Energy Build. **102**, 328–336 (2015). https://doi.org/10.1016/j.enbuild.2015.05.045
8. Marinakis, V., Doukas, H.: An advanced IoT-based system for intelligent energy management in buildings. Sensors **18**(2/610), 1–16 (2018). https://doi.org/10.3390/s18020610
9. Naji, S., Shamshirband, S., Basser, H., Alengaram, U.J., Jumaat, M.Z., Amirmojahedi, M.: Soft computing methodologies for estimation of energy consumption in buildings with different envelope parameters. Energ. Effi. **9**(2), 435–453 (2016). https://doi.org/10.1007/s12053-015-9373-z
10. Odyssée-Mure: Key indicators (2016). http://www.indicators.odyssee-mure.eu/online-indicators.html. Accessed 03 Jan 2019
11. Pérez-Lombard, L., Ortiz, J., Pout, C.: A review on buildings energy consumption information. Energy Build. **40**(3), 394–398 (2008). https://doi.org/10.1016/j.enbuild.2007.03.007
12. Scitovski, R., Zekić-Sušac, M., Has, A.: Searching for an optimal partition of incomplete data with application in modeling energy efficiency of public buildings. Croatian Oper. Res. Rev. **9**(2), 255–268 (2018). https://doi.org/10.17535/crorr.2018.0020
13. Srivastava, N.: Dropout: a simple way to prevent neural networks from overfitting. J. Mach. Learn. Res. **15**, 1929–1958 (2014)
14. Tofallis, C.: A better measure of relative prediction accuracy for model selection and model estimation. J. Oper. Res. Soc. **66**, 1352–1362 (2015). https://doi.org/10.1057/jors.2014.103
15. Torres, J., Fernandez, A., Troncoso, A., Martínez-Álvarez, F.: Deep Learning-Based Approach for Time Series Forecasting with Application to Electricity Load. Lecture Notes in Computer Science, vol. 10338, pp. 203–212 (2017). https://doi.org/10.1007/978-3-319-59773-7_21
16. Touzani, S., Ravache, B., Crowe, E., Granderson, J.: Statistical change detection of building energy consumption: applications to savings estimation. Energy Build. **185**, 123–136 (2019). https://doi.org/10.1016/j.enbuild.2018.12.020
17. Viswanath, P., Babu, V.S.: Rough-DBSCAN: a fast hybrid density based clustering method for large data sets. Pattern Recognit. Lett. **30**, 1477–1488 (2009). https://doi.org/10.1016/j.patrec.2009.08.008
18. Wang, Z.X., Ding, Y.: An occupant-based energy consumption prediction model for office equipment. Energy Build. **109**, 12–22 (2015). https://doi.org/10.1016/j.enbuild.2015.10.002
19. Zekić-Sušac, M., Scitovski, R., Has, A.: Cluster analysis and artificial neural networks in predicting energy efficiency of public buildings as a cost-saving approach. Croatian Rev. Econ. Bus. Soc. Stat. **4**(2), 57–66 (2018). https://doi.org/10.2478/crebss-2018-0013

Framework for the Detection of Physiological Parameters with Musical Stimuli Based on IoT

Mario Alcántara-Garrote, Ana B. Gil-González[✉],
Ana de Luis Reboredo, María N. Moreno, and Belén Pérez-Lancho

Department of Computing and Automation,
University of Salamanca–Sciences Faculty, Plaza de la Merced s/n,
37008 Salamanca, Spain
{malcantag, abg, adeluis, mmg, lancho}@usal.es

Abstract. Music is a universal language present in all cultures throughout history, and its influence on human psychology is amply demonstrated. Music provides a direct link to emotions. Numerous investigations study the relationship between musical characteristics and physiological parameters produced or even induced. This paper presents the development of a framework that includes support for sensorization infrastructure based on IO technologies. The tool allows to monitor in real time the physiological response of the user to musical stimuli and to store such data. In this way, external stimuli, as well as fluctuations in the physiological response, can be treated and analysed and, ultimately, the emotional state of the user can be determined.

Keywords: Emotions recognition · Physiology · Biosignal acquisition ·
Musical stimulation · MQTT protocol · Sensorization · Wearables

1 Introduction

The technological advances of recent years have meant a complete revolution in the way we view technology and the way we relate to it. The rise of mobile technologies and the unstoppable growth of the internet have led the human being to be constantly connected, receiving and sending information in a massive and global network. This constant interaction opens an immense range of possibilities for companies, which seek to provide more personalized and higher quality services, but also for users, who have access to new services that will allow them to improve their quality of life.

Each day more and more devices are designed to monitor different physiological parameters of the user. These data provide very relevant indications about the health status of the person, their needs or responses to certain stimuli. Information of this type has potentially hundreds of uses, but many of them are being researched or have not yet been discovered.

In the next years, advances in biometrics and medicine will allow new lines of research and provide more utilities for all these data. However, in order to facilitate this development and expansion of utilities based on biometric data, tools are needed that allow the reception, analysis and processing of the data received through the sensors.

© Springer Nature Switzerland AG 2020
F. Martínez Álvarez et al. (Eds.): SOCO 2019, AISC 950, pp. 111–120, 2020.
https://doi.org/10.1007/978-3-030-20055-8_11

On the other hand, music has been and will always be an important part in the development and life of human beings. It arises from the need to communicate and has, since its inception, a function of social cohesion that could facilitate demographic and territorial expansion. It is a code, a universal language present in all cultures of the history of mankind, and its influence on the psychology of the human being is amply demonstrated. Historically, music had therapeutic application. Some papyri found in Egypt explain how music was used as a remedy in temples and sanatoriums: in fact, the hieroglyphic sign for music is the same as for well-being and joy [16]. In its evolution as an artistic discipline, this strong link between music and emotions is maintained.

This work presents the development of a framework based on IoT technologies, which allows to monitor in real time the physiological response of the user to musical stimuli and store it to obtain information analysing external stimuli and how they cause fluctuations in the physiological response, and at the end, the emotional state from the user.

The rest of the paper is organized as follows: Sect. 2 includes a short survey of theories and works on the link between emotions and the physiological parameters of the human being before external stimuli and the possibility of managing said biosignals with solutions based on in IoT technology. The proposed tool is described in Sect. 3. The case study carried out with this Framework is explained along point 4. Finally, the conclusions and the lines of work future are given in Sect. 5.

2 Physiological Parameters, Emotions and External Stimuli

Thanks to a multitude of research and experiments with current media, we know that music has a remarkable effect on people's emotions.

At this point it would be necessary to define what an emotion is. After hundreds of years looking for a way to define it, the human being has not yet come up with a unanimous response. Currently, our concept of emotion, based on a large number of theories that have emerged over the years, is as follows: A subjective reaction that occurs as a result of physiological and psychological changes that influence thinking and behaviour.

In summary, we know that emotions are linked to the physiological changes experienced by human beings, although we do not know with certainty the way in which both concepts are related but it is also amply demonstrated that music has an impact on emotional states. We could assume then that there must be a direct relationship between the music and the physiological response.

2.1 From the Theories of Emotion to Affective Computing

The first studies of Darwin about the expression of emotions in humans and animals have motivated a more exhaustive study of them. Biopsychology arises from these investigations and its objective is to explain how emotions occur and to try to unravel their meaning [12].

Advances in anatomy, neuroscience and chemistry have led to a deeper understanding of the brain areas involved in emotional expression and to a better understanding of the chemical and physiological reactions derived from emotional changes.

These discoveries have given rise to a multitude of theories encompassed within the scope of biopsychology. The two most relevant for being initial theories, are the theory of James-Lange [4] and the Cannon-Bard [6]. The postulate advocated by James and Lange is that in response to experiences and stimuli, the autonomic nervous system creates physiological responses (muscular tension, tearing, cardiorespiratory accelera-tion …) from which emotions are created. The James-Lange theory established that it is the cognitive component of an emotion, which would originate the physiological changes associated with emotion. Although initially both theories emerged as an opposite approaches, it was shown that both positions were correct. Therefore, in any way, the attention to physiological responses can bring us closer to knowing the emotion in the human being.

Many more are the theories on which to work around the measurement of emotions. Especially important is Russell's dimensional model [14, 15] where the arousal and valence levels represent different aspects but related to the emotional response. According to the relationship between excitement levels and mood, listening to music has a direct effect on the levels of excitation in the state of mind [5].

Affective Computing [9] consists in the study and development of intelligent systems and devices capable of providing us with the ability to recognize, interpret process and stimulate human emotions. It is an interdisciplinary field that embraces multiple branches of science, biology, psychology and computing. Since the term Affective Computing was coined at the end of the 90 s, originated by Rosalind Picard [10, 11], numerous works have appeared in the literature focused on the automatic extraction of different forms of emotions and in different areas, such as [13, 19]. To date, systems have been developed that recognize emotions with high success rates, through different human channels, such as facial expressions [8], physiological signals [2], voice [1], etc.

2.2 Physiological Parameters and Emotional States

Humans exhibit an electrical profile, which can be measured through biosignals [3, 7, 18]; e.g., electromyography or muscle tension (EMG), electrodermal activity (EDA), skin temperature, etc. When various psychological and/or physiological processes produce an alteration of these signals, we could say that these can reveal our emotions.

Normally, when we talk about the use of a sensor to measure emotions, it is not strictly correct, since what is really measured is the physiological response to a stimulus or set of stimuli. This response is the reflection of psico-physico-chemical changes, which are magnitudes that the sensors can register. Afterwards, these data are stored, and from the study of them we can extrapolated conclusions and/or determine patterns. Specifically biosignals have the advantage that they are free from social masking and have the potential of being measured by non-invasive sensors, making them suited for a wide range of applications.

2.3   IoT and Biosensors

The physiological parameters described in previous sections are collected by the so-called biosensors. These biosensors are more or less complex devices that measure these physiological magnitudes. Through these devices the physical magnitudes are measured and electrical signals are generated which are translated into values that allow these magnitudes to be represented.

Many are the commercial solutions that are appearing for measurement and monitoring through biosensors that offer their own tools, such as BitBrain[1], Movisens[2] or Affectiva[3] among many others. However, these sensors and their corresponding applications are into closed commercial kits, where in most cases there are no options for user management, stimulus parallelization, etc.

The sensory framework of our work focuses on monitoring the environment of the activity and the person to extract all those data that allow the recognition of emotions or moods. The difficulties presented by the implementation of these measures have been mitigated by the new technologies equipped with wireless connection, as well as by their cost reduction. The possibility in the sensorization improves with the Internet of the things allowing the communication between sensor and computer practically from any place.

The translation of the values read by sensors is specific to each sensor and also for each person. Therefore, fixed values cannot be established for the physiological response of people to stimuli. It should be possible to establish normal, maximum and minimum values for each sensor. The connection interface of a sensor with a person varies depending on each sensor and the type of magnitude that is measured.

This motivates us to create our own sensor, using the Arduino technology and the MQTT (*Message Queue Telemetry Transport*) transmission protocol [17], which allows the development of wearables. Thus, it is possible to have a sensor independent of a specific application, with the ability to transmit data in real time. The main sensors used in this work are an EDA sensor and a skin temperature sensor. The EDA sensor uses two electrodes that connect to two fingers of a hand to measure the electrical current that passes through the skin. Data were collected through two ring-shaped aluminium surfaces that are placed on a pair of fingers of the hand and measure the potential difference between the two as shown in Fig. 1. The temperature of the skin is measured on the user's wrist. The collected signals are processed on an Arduino board (ESP8266) attached to the device, which has an integrated battery. The plate and the battery are housed in a box made with a 3D printer, in which all the elements fits so that it can be placed on the wrist.

The board has a Wi-Fi connection and uses the MQTT protocol to transmit the processed value through the network to a server. Following this protocol, our application retrieves the data and store them into the system.

---

[1] https://www.bitbrain.com/.

[2] https://www.movisens.com/.

[3] https://www.affectiva.com/.

Fig. 1. Positioning of the developed sensor

3 Smoodsically. An Overview of the Framework Proposed

A web application is proposed for the storage and real-time visualization of biometric sensor data, as shown by the architecture of Fig. 2. This framework, which we call Smoodsically, allows working with open-code sensors which allows this application can be accessed by any type of sensor in any type of system. This allows independence between the sensors and applications detected and explained in previous points.

The communication of the Server with the Interface is based on HTTP, JSON and XML standards. The server receives the data from the sensors, stores it in the database and manages the requests of the interface. This is followed by an architecture based entirely on the Model - View - Controller pattern. With sensors the operations are asynchronous and a simultaneous communication between components is used, so that it allows a modular structure adapted to future extensions. The senders are integrated through the use of the MQTT protocol. Whole application is based on open source tools.

The user interface is a set of views in which the user interacts and sends information to the server. The design of the interface has been developed mainly in the construction and transition phases, and has focused on achieving a simple and clear handling.

The tool has different menus that integrate its functionalities, for the monitoring of biosignals, storage and management of songs in our case, to develop the studies, as shown in Fig. 3. The functionalities are neither many nor very complex. However, the large number of technologies needed for communication with the server makes the implementation of the interface complex.

Another functionality of great importance is that any monitored experiment can be exported in CSV format, as shown in Fig. 4, allowing its treatment from any other specialized data software.

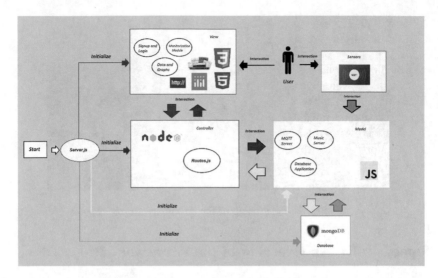

Fig. 2. Smoodsically framework architecture

Fig. 3. Monitoring interface for biometric data stored (L) EDA (R) temperature

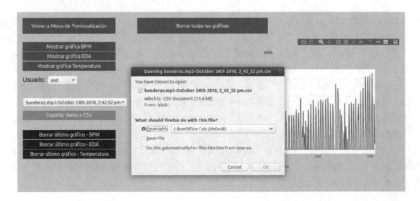

Fig. 4. Export of data to documents CSV

4   Case Study

Using the Framework presented, we conducted the following light case study as proof of concept. In addition, this section served as a test section for the system. For the experiment, there were five subjects of varying ages, who were invited to sit comfortably while the sensor was placed on their wrist. Each of the subjects wore audio helmets to listen to music and were asked to register in the system and become familiar with the interface. After this step, users were asked to begin listening to the song "Can't stop, won't stop," one of the songs stored in the system's library. We used the same song for all users, in order to draw conclusions about similarities and significant changes based on different users.

During the time that the song lasts, all the data sent by the participants through their sensors are collected and once finished, they are saved for later revision, aligning this data with those of the song itself in time.

4.1   EDA and Temperature Results

As detailed previously in the theoretical concepts section, the EDA measures the sweating and electrical potential of the skin and it varies significantly with external stimuli. Higher values provide indicative of high activation of the nervous system, caused by emotional changes. The lowest values and the lowest variations are associated with relaxation states.

The results of the experiment for two analysed subjects are shown in Fig. 5. In general terms, we can observe an increase in the EDA values from half of the follow-up in both graphs. It is relevant that the uptrend observed in both graphs coincides with the section of higher intensity of the song used in the case study. From the second 100 begins an ascending motif that is activated in the chorus of the song, at which time the graphs reach their moment of greatest average value.

As we see, the values of subject 1, called Marta are greater than those of subject 2, called Julia, and this may be due to many factors. For example, one of those factors can be age, Julia is older than Marta, and the monitoring was done in different months, although at similar times. Perhaps with a greater number of samples one could see what this difference is due to and if it is relevant for the measurements.

In summary, we have obtained very relevant conclusions by looking at the trend of the graph in a general way. If we analyse the values in detail, that is, if we record the graphs with a smaller sampling interval or we applied statistical functions to the data by adjusting the sensor, we could obtain many more conclusions in next future.

As indicated in previous sections, the EDA gives very relevant information about the health and the state of the individual. Smoodsically, the Framework presented, allows registering accurately the EDA variations that occur while compare between different measurements of the subjects. In this case, we have monitored the variations produced by music, but we could use the framework with all kinds of external stimuli or situations to which we put through the subjects.

Regarding the temperature measurements, we are going to see all the overlapping graphs, but because the variations are very small and they do not appreciate well if we see the whole monitoring, we are going to focus only on the part where we saw an increase in the values of EDA before. If we look at the graphs of subjects 3 and 5 (Julia and Clemente), at Fig. 6, we can see that there is an upward trend from the second 100. This effect coincides with the trend of his EDA graphs. It does not directly indicate that there is any type of relationship, but it is a very relevant fact and it is possible that they are interrelated in some way.

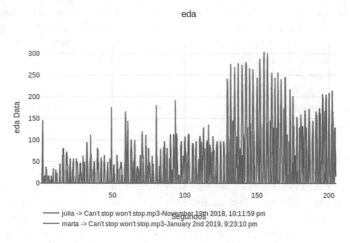

Fig. 5. Graphics EDA from Julia y Marta measures

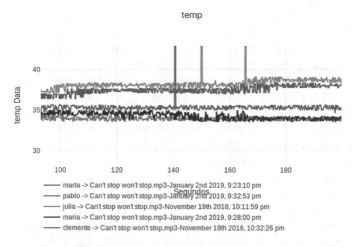

Fig. 6. Temperature graphs of all subjects

5   Conclusion and Future Work

The main objective of the project has been to develop a computer tool is an open source web application capable of working with any sensor using a standardized transmission protocol, based on compatible MQTT, in order to obtain and manage biosignals data in real time. In addition, the tool stores data permanently for consultation and treatment. This platform simplifies the task of connecting any sensor with the application itself. It allows sensor monitoring, storage and graphical data visualization.

For the case of the study presented two sensors and an Arduino board are integrated on a blacelet than simplifies the connection tasks. Using a standardized transmission protocol, based on compatible MQTT, collected data are sent to a server. The device provides a simple and easy to configure interface for experimentation on the emotional effects of music.

The system has been tested on several users listening to a song and results indicate that the subjects exhibited substantial variation in emotions depending on specific parts of the song. This case study has allowed validating theoretical studies on music, emotions and physiological parameters. It is concluded that the ability to measure emotions in real time with musical stimulus provides a useful foundation for designing and managing another experiences.

As future lines of work, once the platform is developed the next work will consist of the experimentation phase. Several studies will be planned to collect data from more users and types of musical stimuli, adapted to different sensors and with more types of external stimuli. On the other hand, the possibility of reproducing music based on physiological responses and contour aspects is a line in which we are also working. The most relevant is the current work in which the tool may be able to identify real emotional states based on the physiological parameters extracted. Many lines of work allow the potential of the developed tool shown.

Acknowledgments. This work has been supported by project "IOTEC: Development of Technological Capacities around the Industrial Application of Internet of Things (IoT)". 0123_IOTEC_3_E. Project financed with FEDER funds, Interreg Spain- Portugal (PocTep).

References

1. Agarwal, G., Maheshkar, V., Maheshkar, S., Gupta, S.: Recognition of emotions of speech and mood of music: a review. In: International Conference on Wireless Intelligent and Distributed Environment for Communication, pp. 181–197. Springer, Cham (2018)
2. Ali, M., Mosa, A.H., Al Machot, F., Kyamakya, K.: Emotion recognition involving physiological and speech signals: a comprehensive review. In: Recent Advances in Nonlinear Dynamics and Synchronization, pp. 287–302. Springer, Cham (2018)
3. Ayata, D.D., Yaslan, Y., Kamaşak, M.: Emotion recognition via galvanic skin response: comparison of machine learning algorithms and feature extraction methods. Istanb. Univ.-J. Electr. Electron. Eng. **17**(1), 3147–3156 (2017)
4. Cannon, W.B.: The James-Lange theory of emotions: a critical examination and an alternative theory. Am. J. Psychol. **39**(1/4), 106–124 (1927)
5. Chanel, G., Ansari-Asl, K., Pun, T.: Valence-arousal evaluation using physiological signals in an emotion recall paradigm. In: IEEE International Conference on Systems, Man and Cybernetics, ISIC, pp. 2662–2667. IEEE (2007)
6. Dror, O.E.: The cannon-bard thalamic theory of emotions: a brief genealogy and reappraisal. Emot. Rev. **6**(1), 13–20 (2014)
7. Eleftheriou, G., Fatouros, P., Tsirmpas, C.: U.S. Patent Application No. 15/648,730 (2018)

8. Goyal, S.J., Upadhyay, A.K., Jadon, R.S., Goyal, R.: Real-life facial expression recognition systems: a review. In: Smart Computing and Informatics, pp. 311–331. Springer, Singapore (2018)
9. Lisetti, C.L.: Affective computing (1998)
10. Picard, R.W.: Affective computing (1995)
11. Picard, R.W.: Affective computing: challenges. Int. J. Hum.-Comput. Stud. **59**(1–2), 55–64 (2003)
12. Pinel, J.P.: Biopsychology. Pearson Education, London (2009)
13. Rincón, J.A., Costa, A., Novais, P., Julián, V., Carrascosa, C.: Using non-invasive wearables for detecting emotions with intelligent agents. In: International Joint Conference SOCO 2016-CISIS 2016-ICEUTE 2016, pp. 73–84. Springer, Cham (2016)
14. Russell, J.A.: A circumplex model of affect. J. Pers. Soc. Psychol. **39**, 1161–1178 (1980)
15. Yik, M., Russell, J.A., Steiger, J.H.: A 12-point circumplex structure of core affect. Emotion **11**(4), 705 (2011)
16. Sachs, C.: The Rise of Music in the Ancient World, East and West. Courier Corporation, Chelmsford (2008)
17. Singh, M., Rajan, M.A., Shivraj, V.L., Balamuralidhar, P.: Secure MQTT for internet of things (IoT). In: 2015 Fifth International Conference on Communication Systems and Network Technologies (CSNT), pp. 746–751. IEEE (2015)
18. Van den Broek, E.L., Lisý, V., Janssen, J.H., Westerink, J.H.D.M., Schut, M.H., Tuinenbreijer, K.: Affective man-machine interface: unveiling human emotions through biosignals. In: Fred A., Filipe J., Gamboa H. (eds.) Biomedical Engineering Systems and Technologies. BIOSTEC 2009. Communications in Computer and Information Science, vol. 52. Springer, Heidelberg (2010)
19. Vicente, J.G., Gil, A.B., de Luis Reboredo, A., Sánchez-Moreno, D., Moreno-García, M.N.: Moodsically. Personal music management tool with automatic classification of emotions. In: International Symposium on Distributed Computing and Artificial Intelligence, pp. 112–119. Springer, Cham (2018)

Edge Computing Architectures in Industry 4.0: A General Survey and Comparison

Inés Sittón-Candanedo$^{(\boxtimes)}$, Ricardo S. Alonso, Sara Rodríguez-González,
José Alberto García Coria, and Fernando De La Prieta

IoT Digital Innovation HUB, University of Salamanca, Salamanca, Spain
{isittonc,ralorin,srg,jalberto,fer}@usal.es

Abstract. Edge Computing represents the computing and networking tasks that IoT (Internet of Things) devices perform at the Edge of the network in communication with the remote Cloud. In this sense, recent researches try to demonstrate that Edge Computing architectures represent optimal solutions in order to minimize latency, improve privacy and reduce bandwidth and related costs in IoT-based scenarios, such as Smart Cities, Smart Energy, Smart Farming or Industry 4.0. This work is a review of the main existing Edge Computing reference architectures aimed at Industry 4.0 proposed by the Edge Computing Consortium, the FAR-Edge Project and the Industrial Internet Consortium for Industry 4.0. This paper includes a comparison among these reference architectures, as well as their most important features in order to build a new Edge Computing Reference Architecture as future work.

Keywords: Edge Computing · Internet of Things ·
Cloud Computing · Reference architectures · Industry 4.0

1 Introduction

The application of Internet of Things (IoT) devices, services and applications in complex scenarios such as Smart Cities [1], Smart Energy [2], Smart Farming or Industry 4.0 [3] has generated new challenges for solutions based on Cloud Computing. Before the emergence of Edge Computing paradigm, Cloud Computing had a centralized structure where computing and storage were only deployed in a centralized or distributed server on the remote Cloud [4]. Nonetheless, Cloud Computing approaches present some drawbacks when dealing with IoT applications, as these usually require reduced latency and response in real-time. Furthermore, some estimations predict that by 2019 data produced by human and machine entities will reach 500 zettabytes, and that by 2020 there will be 50 billion IoT devices connected to the Internet [5]. The Edge Computing paradigm emerged as a solution to these problems, being presented as a set of technologies that help to optimize computing and storage processes performed until then in

© Springer Nature Switzerland AG 2020
F. Martínez Álvarez et al. (Eds.): SOCO 2019, AISC 950, pp. 121–131, 2020.
https://doi.org/10.1007/978-3-030-20055-8_12

the Cloud. This is achieved by pre-processing the data collected by devices prior to being sent to the remote set of centralized or distributed servers deployed in the Cloud [6].

Likewise in the Edge Computing paradigm an important part of the computing and even storage tasks no longer take place in the Cloud, but in the *"Edge"*, putting these computing processes closer to the places where IoT devices work. This allows releasing an important part of the computational load from the Cloud servers, avoiding the network traffic overload and costs, and reducing the response time required for new IoT based applications [7]. This distributed computing approach has demonstrated their benefits in a wide range of real scenarios, including Smart homes, Healthcare, Smart Cities or Smart grids [8]. In this sense, recent researches propose the use of the Edge Computing paradigm as a complement to improve the performance of only Cloud Computing based applications [9]. Nowadays, there are several independent organizations and entities focused on proposing reference architectures as framework for implementing Edge Computing based solutions [10–12].

Next section describes the narrow relation between Internet of Things technologies and the Edge Computing paradigm. Section 3 presents the main three Edge Computing reference architectures aimed at Industry 4.0 and based on the ISO/IEC/IEEE 2010:2011 standard [13]. After that, Sect. 4 compares and evaluates the different Reference Architectures presented, and remarks the key features of each of them with the aim to design a new and improved Reference Architecture in the future. Finally, the conclusions and future work are depicted in the Sect. 5.

2   Internet of Things and Edge Computing

In Edge Computing paradigm, a relevant portion of the managed data is processed at the Edge of the network [9]. This computing paradigm has been driven by the continuously growing proliferation of Internet of Things (IoT) devices. The Internet of Things (IoT) can be seen as the interaction and communication carried out between the devices that generate and exchange data with the *things* or objects in the real world [14]. Among the main characteristics of the IoT we have: heterogeneity of devices and networks [15], a large volume of events, and that data is generated by objects or *things* in the real world [16]. IoT devices allow the digitization of the physical world by the measurement and transmission of data that can be processed and analyzed by machines, transforming them into information used to build intelligent systems in a wide range of scenarios such as healthcare, transportation, or smart energy [17], among many others.

Nonetheless, the requirements for the development of applications in *only-Cloud* scenarios managing a large amount of IoT devices are becoming increasingly difficult to meet. The large volume of data generated by IoT devices to be processed by Data Analytics or Machine Learning techniques is a task that requires an extremely high network bandwidth [18] and associated costs by the Cloud service providers. Edge Computing paradigm allow to find solutions that comply with its requirements and can make use to the fullest of the potential and capabilities of IoT. In this sense, in Edge Computing the Big Data generated by the different IoT objects can be filtered and pre-processed at the *Edge* of the network instead of being transmitted to the Cloud to be processed there [19]. This enables faster services and reduced response times compared to only Cloud Computing based scenarios [20].

Even though there are several formal definitions for Edge Computing, all of them usually include the following three terms: *open platform, enabling technologies and computer resources*. For Shi *et al.* Edge Computing refers to the enabling technologies that allow performing computation at the network edge so that computing happens closer to data sources [6]. Satyanarayanan considers that Edge Computing is a new paradigm in which substantial computing and storage resources, also referred to as *cloudlets*, micro data centres or even *fog* nodes, are placed at the Internet's edge, close to sensing devices [21]. The Edge Computing Consortium defines Edge Computing as a distributed open platform at the network edge, close to the things or data sources, and integrating the capabilities of networks, storage, and applications [11]. By delivering edge intelligence services, Edge Computing meets the key requirements of industry digitization for agile connectivity, real-time services, data optimization, application intelligence, security and privacy protection [11]. For the Industrial Internet Consortium (IIC) Edge Computing, or simply the *Edge*, brings processing close to the data source, and it does not need to be sent to a remote Cloud or other remote centralized or distributed systems to be further processed. By eliminating the distance and time necessary to transmit data from sources to the Cloud, the speed and performance of transportation of data are improved [12].

Therefore, Edge Computing based architectures are capable of shifting a portion of the computing capacity that is performed in the Cloud to the nodes located at the Edge of the network [22]. By shifting processing capacity to nodes, Edge architectures offer the following advantages [23]:

- Data streams coming from different data sources are processed by nodes to filter information of no value. This allows to save bandwidth and storage resources.
- Proximity and low latency thanks to information being processes close to its source of origin.
- Decentralized storage and processing enhance scalability.
- The nodes of Edge architectures provide each node of the network with isolation and privacy.

3   Edge Computing Reference Architectures

A *Reference Architecture* is a document or set of documents that propose recommended structures, products and services to form a solution. A reference architecture incorporates industry-accepted best practices, usually suggesting the delivery method or specific optimal technologies. Reference architectures help project managers, software developers, enterprise designers, and IT managers to collaborate and communicate effectively around an implementation project [24]. For this work we have chosen those Edge Computing reference architectures that have been designed basing on the ISO/IEC/IEEE 42010:2011 standard [13].

3.1 FAR-Edge RA

The FAR-Edge RA has been developed as part of the H2020 FAR-Edge project as a challenge to the adoption of decentralized automation architectures in industrial scenarios [10]. This reference architecture is a conceptual framework for the design and implementation of the FAR-Edge project platform based on Edge Computing and Distributed Ledger Technologies (DLT), close related to blockchain [25]. The two principal concepts of FAR-Edge are scopes and tiers [22]. On the one hand, *scopes* refer to elements in a plant or its ecosystems such as machinery, field devices, workstation, SCADA (Supervisory Control And Data Acquisition), MES (Manufacturing Execution System) and ERP (Enterprise Resource Planning) systems. On the other hand, *tiers* provide information about system components and the relationships between them.

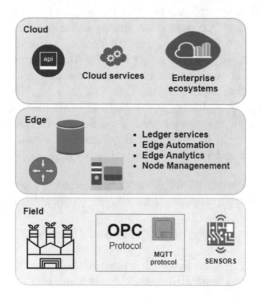

Fig. 1. FAR-Edge RA.

Figure 1 depicts the principal components of this architecture, including its layered structure:

– *Field layer:* it is the lowest layer of the architecture and it is formed by different types of devices, ranging from a smart machine to a sensor or an actuator and representing the *Edge node.*

- *Edge layer:* the software designed to run on several Edges nodes is located here. This tier contains the Edge gateways (EGs), computing devices that operate as gateways between the field tier and the digital world and aimed at performing data analytics in real time. A Ledger by means of *smart contracts* based on blockchain technologies is included to provide security support to any layer of the architecture [25].
- Cloud layer: consists of Cloud Servers that host the software responsible for planning, monitoring and managing the resources, dealing with the logical execution of the functional components of the architecture.

The FAR-Edge reference architecture is notable for having implemented Edge Computing while guaranteeing information security through blockchain. By including automation and analysis through smart contracts, this reference architecture can be used as a reference for cases of use of automation, analysis and digital simulation in industrial environments [22].

3.2 Edge Computing RA 2.0

The Edge Computing Consortium (ECC) was created in 2016 by several organizations such as Huawei, Shenyang Institute of Automation (SIA) of Chinese Academy of Science, Intel, ARM, iSoftStone and the China Academy of Information and Communications Technology (CAICT). Moreover, the Chinese Academy of Telecommunications Research (CATR) together with the Ministry of Industry and Information Technology (MIIT) formed in 2016 the Industrial Internet Alliance (IIA) aimed at the development of industrial Internet in China [11]. After that, the Edge Computing Reference Architecture (EC-RA) 2.0 was proposed by the joint work between the ECC and the AII, basing on international standards such as ISO/IEC/IEEE 42010:2011.

Vertically the Edge Computing Reference Architecture 2.0 uses the following services: *management, data life-cycle and security,* focused on intelligent services throughout the life cycle. Horizontally, the Edge Computing Reference Architecture (EC-RA) 2.0 follows a layers model with open interfaces, as can be seen in Fig. 2 [11]:

- *Smart Services:* this layer is based on a model-driven service framework. Intelligent coordination between service development and deployment is achieved through the *Development service framework* and the *Deployment and operation service framework.* These frameworks enable coherent software development interfaces and automatic implementation and operations.
- *Service Fabric (SF):* defines the tasks, technological processes, path plans, and control parameters of the processing and assembly phases, implementing the fast deployment of service policies and fast processing of multiple types of products.
- *Connectivity and Computing Fabric (CCF):* the Operation, Information and Communications Technology (OICT) infrastructure is responsible for deploying operations and coordinating between the computational resources services and the needs of the organization.

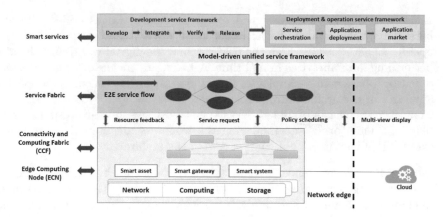

Fig. 2. Edge Computing RA 2.0.

– *Edge Computing Node (ECN):* in this layer the intelligent Edge Computing Nodes (ECNs) provide with real time processing and response capacities, being compatible with diverse heterogeneous connections. The security is integrated into the hardware and the software.

The architecture was developed based on the international standards defined by ISO/IEC/IEEE 42010:2011, which strengthen its use as a reference for the development of industrial solutions based on the Edge Computing paradigm.

3.3   Industrial Internet Consortium RA

As the Edge Computing Consortium, the Industrial Internet Consortium has also developed its reference architecture using the ISO/IEC/IEEE 42010:2011 standard. This standard is the principal guide to identify conventions, principles and best practices for consistent architectures and frameworks based on the Internet of Things. The ISO/IEC/IEEE 42010:2011 standard facilitates the evaluation, communication, documentation and systematic or effective resolution in a reference architecture [12]. Figure 3 presents the Industrial Internet Consortium Reference Architecture (IIC-RA) and its three principal layers [12]:

– *Edge:* this layer collects data from the Edge nodes through a proximity network. The main architectural features of this layer include breadth of distribution, location, scope of governance and nature of the proximity network. Each of these characteristics will change according to each use case.
– *Platform:* this layer is responsible for processing and sending control commands from the third layer (*Enterprise*) to the Edge layer. Its main function is to group the processes and analyze the data flows from the Edge layer and the upper layers. It manages the active devices in the Edge layer for data consultation and analysis through domain services.

– *Enterprise:* finally, this layer hosts specific applications such as decision support systems [26], end user interfaces, or operation management, among others. This layer generates control commands to be sent to the Platform and Edge layers and also receives data flows from them.

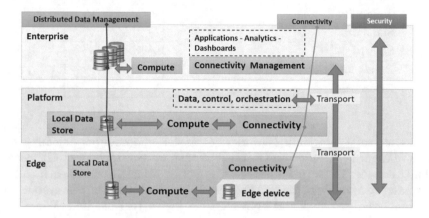

Fig. 3. Industrial Internet Consortium Reference Architecture.

4 Evaluation of the Edge Reference Architectures

The three reference architectures described in this work are based on a three-layer model for Edge Computing, which does not replace Cloud services, but complement them by means of their integration as the last level of the architectures. Wherever the volume of gathered IoT data is high, Edge nodes in these models are the first stage for processing, controlling and reducing the amount of data transferred to those services running in the Cloud. Therefore, they allow reducing storage requirements, decreasing latency and providing real-time responses to end users and applications.

Table 1 summarizes the main characteristics of the three Edge Computing reference architectures compared in this work.

One of the main objectives of this comparison is to evaluate and guess the main strengths of each analyzed architecture in order to set the basis for the future design of a new Edge Computing Reference Architecture based on some components coming from them. Therefore, focusing on the key strengths of each reference architecture, we have:

– FAR-Edge: Although the three architectures include a large number of advantages for applications and environments in which large volumes of data are collected and processed, from the architectures described, only FAR-Edge RA

Table 1. Summary of the Edge Computing reference architectures' main features.

Characteristics	FAR-Edge	ECC	IIC
Connectivity and communication	✓	✓	✓
Device management	✓	✓	✓
Data collection, analysis and performance	✓	✓	✓
Scalability	✓	✓	✓
Standards		✓	✓
Security	✓	✓	✓
Data encryption			
Blockchain	✓		

takes into consideration the security of transferred data proposing an intermediate Ledger layer based on the concept of smart contracts over a blockchain. Thus, services of Ledger layer are transaction-oriented, where each call that requests to modify or make changes in a shared state of the system must be evaluated and approved by peer nodes before its execution. These ledger services consist of codes that are executed simultaneously on all nodes that are active at the time a service is requested [22]

- Edge Computing RA 2.0: For the Edge Computing Consortium it is important the use of International Standards such as ISO/IEC/IEEE 42020:2011 in order to present Edge solutions and frameworks to industries. Therefore, the Edge Computing RA 2.0 is based on the IEC 61499 standard, which proposes the design of architectures through *Functional Blocks* (FBs) defined as the basic structural unit of the models. Each block is characterized by its inputs, outputs and internal functions, as is the case with classic block diagrams. Functional Blocks are not executed until they receive an input signal, so they remain in a *rest* state in order to allow the portability and the reuse of Functional Blocks, behaving as a unit completely independent of the rest.
- IIC: In fact FAR-Edge RA is in turn based on this architecture. An important element of this reference architecture is the Enterprise layer, in which specific processes and applications are hosted by the private or public Cloud, including Decision Support Systems, operation management, Big Data Analytics, model training, predictions, as well as business analytics.

5   Conclusions and Future Work

Industry 4.0 is experiencing a continuous growth mainly motivated by the emergence of innovative technologies such as smart sensors, devices and intelligent objects, tablets, telephones, computers and others. This exponential growth is fostering an unexpected explosion in the volume and heterogeneity of data produced, which has motivated enterprises to adopt innovative technological solutions based on more robust architectures that allow a faster access and improved data visualization putting together information from different systems regardless they are interconnected or not. In this sense, three main Edge Computing Reference Architectures aimed at Industry 4.0 and based on the ISO/IEC/IEEE 2010:2011 standard proposed by some of the largest and most influential technological task forces were presented and analyzed. These three main reference architectures focus on using Edge Computing as a solution to reinforce the capabilities of their Cloud-based technology implementations when dealing with Big Data coming from massive IoT scenarios.

The authors have identified a gap in the development of Edge Computing reference architectures applied to other non-Industry 4.0 scenarios. This motivates as future work the design of a new Edge Computing Reference Architecture taking into consideration the main strengths of the three reference architectures analyzed. This new reference architecture will have as main objectives to achieve a real-time analysis of data at the level of local devices and edge nodes and not in the cloud; to lower operating and management costs by means of the reduction of traffic and data transfer between the Edge and the Cloud; to improve application performance, as applications that tolerate latency can reach lower levels of latency at the limits of the network, compared to the Cloud; and to improve security through blockchain technologies that are incorporated into the architecture from the bottom IoT layers to the top layers at the Cloud.

Acknowledgements. This work was developed as part of the project *"Virtual-Ledgers: Tecnologías DLT/Blockchain y Cripto-IOT sobre organizaciones virtuales de agentes ligeros y su aplicación en la eficiencia en el transporte de última milla"*, ID SA267P18, project co-financed by Junta de Castilla y León, Consejería de Educación (Ministry of Education of the Government of Castile and León, Spain), and FEDER funds. Inés Sittón has been supported by IFARHU – SENACYT scholarship program (Government of Panama).

References

1. Chamoso, P., Prieta, F.D.L.: Swarm-based smart city platform: a traffic application. ADCAIJ: Adv. Distrib. Comput. Artif. Intell. J. **4**(2), 89–98 (2015)
2. García, O., Chamoso, P., Prieto, J., Rodríguez, S., de la Prieta, F.: A serious game to reduce consumption in smart buildings. In: Highlights of Practical Applications of Cyber-physical Multi-agent Systems. Communications in Computer and Information Science, pp. 481–493. Springer (2017)
3. Sittón-Candanedo, I., Rodríguez, S.: Pattern extraction for the design of predictive models in industry 4.0., pp. 258–261 (2018)
4. De La Prieta, F., Corchado, J.M.: Cloud Computing and Multiagent Systems, a Promising Relationship. Springer, Cham (2016)
5. Cisco: Cisco Global Cloud Index: Forecast and Methodology, 2016–2021 (2018). https://www.cisco.com/c/en/us/solutions/collateral/service-provider/global-cloud-index-gci%20/white-paper-c11-738085.html#wp9000816. Accesed 20 Nov 2018
6. Shi, W., Cao, J., Zhang, Q., Li, Y., Xu, L.: Edge computing: vision and challenges. IEEE Internet Things J. **3**(5), 637–646 (2016)

7. Garcia, P., Montresor, A., Epema, D., Datta, A., Higashino, T., Iamnitchi, A., Barcellos, M., Felber, P., Riviere, E.: Edge-centric computing: vision and challenges. ACM SIGCOMM Comput. Commun. Rev. **45**(5), 37–42 (2015)
8. Chamoso, P., González-Briones, A., Rodríguez, S., Corchado, J.M.: Tendencies of technologies and platforms in smart cities: a state-of-the-art review. Wirel. Commun. Mob. Comput. **2018** (2018). https://doi.org/10.1155/2018/3086854
9. Taleb, T., Samdanis, K., Mada, B., Flinck, H., Dutta, S., Sabella, D.: On multi-access edge computing: a survey of the emerging 5G network edge cloud architecture and orchestration. IEEE Commun. Surv. Tutor. **19**(3), 1657–1681 (2017)
10. FAR-EDGE Project: FAR-EDGE Project H2020 (2017). http://far-edge.eu/#/. Accessed 20 Nov 2018
11. Edge Computing Consortium, Alliance of Industrial Internet: Edge Computing Reference Architecture 2.0. Technical report, Edge Computing Consortium (2017). http://en.ecconsortium.net/Uploads/file/20180328/1522232376480704.pdf. Accessed 20 Nov 2018
12. Tseng, M., Canaran, T.E., Canaran, L.: Introduction to edge computing in IIoT. Technical report, Industrial Internet Consortium (2018). https://www.iiconsortium.org/pdf/Introduction_to_Edge_Computing_in_IIoT_2018--06--18.pdf. Accessed 20 Nov 2018
13. ISO/IEC/IEEE 42010: Systems and software engineering - engineering. Technical report, ISO/IEC/IEEE 42010 (2011)
14. Ganz, F., Puschmann, D., Barnaghi, P., Carrez, F.: A practical evaluation of information processing and abstraction techniques for the internet of things. IEEE Internet Things J. **2**(4), 340–354 (2015)
15. Alonso, R.S., Tapia, D.I., Bajo, J., García, Ó., de Paz, J.F., Corchado, J.M.: Implementing a hardware-embedded reactive agents platform based on a service-oriented architecture over heterogeneous wireless sensor networks. Ad Hoc Netw. **11**(1), 151–166 (2013)
16. Razzaque, M., Milojevic-Jevric, M., Palade, A., Clarke, S.: Middleware for internet of things: a survey. IEEE Internet Things J. **3**(1), 70–95 (2016)
17. García, Ó., Alonso, R.S., Prieto, J., Corchado, J.M.: Energy efficiency in public buildings through context-aware social computing. Sensors **17**(4), 826 (2017)
18. Jing, Q., Vasilakos, A.V., Wan, J., Lu, J., Qiu, D.: Security of the Internet of Things: perspectives and challenges. Wirel. Netw. **20**(8), 2481–2501 (2014)
19. Brogi, A., Forti, S.: QoS-aware deployment of IoT applications through the fog. IEEE Internet Things J. **4**(5), 1–8 (2017)
20. Lin, J., Yu, W., Zhang, N., Yang, X., Zhang, H., Zhao, W.: A survey on internet of things: architecture, enabling technologies, security and privacy, and applications. IEEE Internet Things J. **4**(5), 1125–1142 (2017)
21. Satyanarayanan, M.: The emergence of edge computing. Computer **50**(1), 30–39 (2017)
22. Isaja, M., Soldatos, J., Gezer, V.: Combining edge computing and blockchains for flexibility and performance in industrial automation. In: International Conference on Mobile Ubiquitous Computing, Systems, Services and Technologies (UBICOMM) (c), pp. 159–164 (2017)
23. Shi, W., Schahram, D.: The promise of edge computing. Computer **49**(0018), 78–81 (2016)
24. Moghaddam, M., Cadavid, M.N., Kenley, C.R., Deshmukh, A.V.: Reference architectures for smart manufacturing: a critical review. J. Manuf. Syst. **49**, 215–225 (2018)

25. Khan, M.A., Salah, K.: IoT security: review, blockchain solutions, and open challenges. Futur. Gener. Comput. Syst. **82**, 395–411 (2018)
26. Faia, R., Pinto, T., Vale, Z.: Dynamic fuzzy clustering method for decision support in electricity markets negotiation. ADCAIJ: Adv. Distrib. Comput. Artif. Intell. J. **5**(1), 23–35 (2016)

Predictive Maintenance from Event Logs Using Wavelet-Based Features: An Industrial Application

Stéphane Bonnevay[1], Jairo Cugliari[1(✉)], and Victoria Granger[2]

[1] ERIC EA3083, Université de Lyon, 5 av. Pierre Mendès France, 69676 Bron, France
Jairo.Cugliari@univ-lyon2.fr
[2] ENEDIS, 124 boulevard Marius Vivier Merle, Lyon, France

Abstract. In industrial context, event logging is a widely accepted concept supported by most applications, services, network devices, and other IT systems. Event logs usually provide important information about security incidents, system faults or performance issues. In this way, the analysis of data from event logs is essential to extract key information in order to highlight features and patterns to understand and identify reasons of failures or faults. Our objective is to help anticipate equipment failures to allow for advance scheduling of corrective maintenance. We propose a supervised approach to predict faults from an event log dataset using wavelets features as input of a random forest which is an ensemble learning method.

1 Introduction

Smart electric devices automatically monitor information about energy consumption or production, they are defined by the ability to connect to a network and to operate remotely. They report meaningful and appropriate information to relevant parties (consumers, energy distribution system operators or energy providers) and their systems. Modern electric smart devices produce enormous amount of data. The first one is the inherently primary data associated to the devices' main activity and implemented features. Its exploration and use involve privacy issues which have been largely debated and are beyond the scope of this work. In addition to this, the second category of transmitted information is about events, a relatively new category of data, the value of which has yet not been assessed. Event is basically a notification that originates from a electrical device and contains the information regarding the object, action or process to which the event is related. Events are issued while monitoring different aspects of the system and give an overview about equipment communications, devices' secondary non-core functionalities, network intrusions or activity on the grid.

We believe that event logs could be processed and analysed to unveil useful information, in addition to devices' primary data. More precisely, we assume that these data can be useful to inform about the device's operative state and eventually to predict device failure. However, event logs concern a wide range of

F. Martínez Álvarez et al. (Eds.): SOCO 2019, AISC 950, pp. 132–141, 2020.
https://doi.org/10.1007/978-3-030-20055-8_13

uses and the difficulty comes from the volume and variety of logs received. The main challenge is to analyse this data and extract useful knowledge from the unremitting flow of notifications. The issue therefore is to identify appropriate events containing helpful information. Furthermore, it is essential to detect a shift or an alteration in the patterns of these specific events which could alert users about a fault occurrence.

In literature, patterns from event logs are defined in various ways, for example as partial orders of a process [1], or considered as Petri nets [2]. Also as repeated sequences that capture process models from event logs in order to improve their detection [3]. Let us mention some works dealing with predictive maintenance: a general classification-based failure prediction method which has been tested on real ATM run-time event logs data [4], or event logs data extracted from medical equipment used to treat a multi-instance learning task [5]. Also, a Cox proportional hazard model has been used to provide a prediction of system failures based on the time-to-failure data extracted from the event sequences [6].

In this work, we consider the event distribution over time as a function of time. Our first objective is to extract characteristic features from the time series, which will then be presented to a learning algorithm. In order to make this step as automatic as possible, we decided to perform the Discrete Wavelet Transform (DWT) which is an appropriate tool for noise filtering, data reduction, and singularity detection, and thus it a good choice for time series and signal processing. The decomposition coefficients obtained from the DWT are then used as input of a supervised learning algorithm. To predict the fault occurrence and to measure variable importance in order to select the best set of features.

2   From Event to Time Functions

Our study is based on events monitored on electrical devices installed on ENEDIS network, the French Distribution System Operator. An example of our logs is displayed in Table 1. Besides the recorded attributes (first three columns), we define an additional feature, a 13 group code representing an hierarchical level of event codes (using domain experts knowledge). None of these notifications have any level of criticality or priority.

Table 1. Event logs data.

timestamp	deviceId	eventCode	groupCode
2014-01-24 17:49:44.537	001	A3	A
2014-01-24 15:09:35.970	001	A23	A
2014-01-25 03:55:56.872	002	A3	A
2014-01-27 00:14:42.463	002	B8	B
2014-01-27 08:10:25.470	002	A23	A
. . .	. . .	. . .	. . .

Records were collected between over a year and events were aggregated on a daily basis. In total about 1.25 millions of events were recorded on 2623 devices presenting similar settings and technical specifications. A fault occurrence is considered when the device fails to provide its main function.

We focus on the number of logs effectively observed over a reasonable period of length δ using a given time resolution (e.g. hours, days) for each type of event. We then consider

$$(N(t_1), \ldots, N(t_\delta)), \tag{1}$$

where $N(t_j)$ is the number of events at time t_j. To fix ideas, say that δ may span over two weeks and using a daily resolution the vector of counts would have length equal to 14. Figure 1 plots four cases of these trajectories.

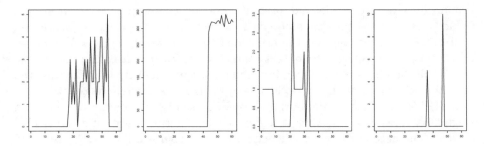

Fig. 1. Examples of trajectories from event logs data. Tracking is done daily over 64 days. Cases (a) and (b) are from working devices; and cases (c) and (d) are from faulty ones.

This vector constitutes the building block of our approach since we create instances of this vector for both normal and abnormal regimes (*cf.* Sect. 3). Actually, each instance is the tracking of a device along a period of length δ. Then the couple device × time should not be view as tracking over contemporaneous instants but as snapshots of the life time of the devices.

If we now consider that K different type of events exists, then we have

$$(N_k(t_1), \ldots, N_k(t_\delta)), k = 1, ..., K,$$

that is each device × time is a set of K counting vectors. From the mathematical point of view, we may look at vectors (1) as time series trajectories. And since there exists K of them, we have a multivariate time series where there may be some dependence structure between components of the vectors as well as time dependence within components. One way to cope with time dependence is to see each trajectory as a discrete sampling, eventually with some noise, of the time function $z_k(t), t \in [0, \delta], k = 1, \ldots, K$. Notice again that time should be consider as relative to period δ and not as an absolute quantity.

Let us illustrate the construction by choosing events from only one code to assemble a sample of trajectories containing both faulty and working devices.

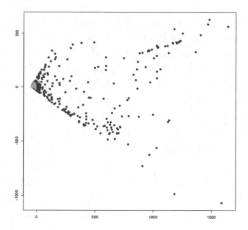

Fig. 2. Multidimensional scaling of trajectories from one event code. Each point represents a trajectory from a working device (in gray) or a faulty device (in red).

These are represented with a multidimensional scaling in Fig. 2 using the euclidean distance on standardized versions of the trajectories. Notice that since no information about the class is used this technique is essentially unsupervised. However we add a colour reference (grey: working devices, red: faulty devices) to the scatter plot in order to visualize eventual differences. Even if the sample is very unbalanced, a clear distinction between both classes is appreciated. Distances between working devices are relatively small with reference to distances between faulty devices. Other conclusion we can draw is that a (eventually non linear) reduction of the dimension may suffice to extract the useful information on the signals. Taking into account the time dependent structure of the functions is necessary to obtain an appropriate construction that yields on a dimension reduction.

3   Methods

We describe here the methods we use to construct our solution: the wavelets transform and the random forest classifier.

3.1   Wavelets Transform

Wavelets are a domain transformation technique that allows one to represent time domain signals into a bivariate location-scale domain [7]. While location in the new domain is connected to the original time domain, scales can be associated to Fourier frequencies and both with good localization properties. That is, the transform gives local information on locations connected to only a time span (not the global time) and scales connected to only some frequencies (and not all of them).

Moreover we use the Discrete Wavelet Transform (DWT) which provides an orthonormal basis of the space, allowing us to encode all the available information on a signal without any loss of information [8].

Consider the signal $z(t)$ which is an univariate function defined on the time domain $\mathcal{T}$, for example $\mathcal{T} = [0,1]$. The DWT will provide two terms: a global approximation of the signal $\mathcal{S}(t)$ and the ensemble of details $\mathcal{D}(t)$ well localized both in time and frequency. If $z \in L_2([0,1])$, then the DWT provides us with a basis of the functional space. The basis is created by simple transformations of a scaling function $\phi(t)$ and a wavelet mother $\psi(t)$ which are associated to the orthogonal multi resolution analysis of $L_2([0,1])$. Indeed, we consider the family $\{\phi_{j,k}(x) = 2^{-j}\phi(2^{-j}x - k)\}_{j,k}$ which is obtained by dilatations of a factor 2^j and by integer translations on the new scale. Similar operations are done to get the family $\{\psi_{j,k}(x)\}_{j,k}$. Then, a finite energy signal z can be expressed as

$$z(t) = \underbrace{\sum_{k=0}^{2^{j_0}-1} c_{j_0,k}\phi_{j_0,k}(t)}_{S_{j_0}(t)} + \underbrace{\sum_{j\geq 0}\sum_{k=0}^{2^j-1} d_{j,k}\psi_{j,k}(t)}_{\mathcal{D}(t)}, \qquad (2)$$

where $c_{j,k} = <z,\phi_{j,k}>, d_{j,k} = <z,\phi_{j,k}>$ are the scale coefficients and wavelet coefficients respectively. The scale j_0 separates the two terms. The first one, gives a smooth approximation at resolution 2^{j_0}. The second one, keeps all the details of the curves on a hierarchical structure depending on scales and locations. The approximation coefficients $c_{j_0,k}$ retains the information of the local (at location k) mean level of the curve, while the detail coefficients $d_{j,k}$ code the information of discontinuities and other singularities.

With finite data $\{z(t_i), i = 1,\ldots,N\}$, the signal $z(t)$ can only be approximated by a truncation at some maximum scale level $J = \log_2(N)$, that is we approximate (2) by

$$z_J(t) = c_0\phi_{0,0}(t) + \sum_{j=0}^{J-1}\sum_{k=0}^{2^j-1} d_{j,k}\psi_{j,k}(t). \qquad (3)$$

Notice that we have also fixed the approximation part at the coarsest resolution $j_0 = 0$ which means that only one scaling coefficient is used to approximate this term. For convenience we choose the number of sampling points per curve, N, to be a power of 2. The maximum number of scales J is then an integer. With this, we are in conditions to use the highly efficient Mallat's pyramidal algorithm [7] to obtain both the scaling and wavelet coefficients. If the sampling grid $\{i/N, i = 1,\ldots,N\}$ is not regular or N is not a power of 2, then one can choose a finer regular grid and use any interpolation scheme to meet our choices.

Haar wavelet leads to a easy and clear intuition on the wavelets coefficients. The only scaling coefficients we retain is proportional to mean level of the whole signal. The approximation term is then a constant function $S_0(t) = c_{0,0}\psi(t)$ proportional to the mean function of the signal. The detail coefficients are the difference on the constant approximations between two juxtaposed halves. We interpret them as the change observed at some resolution (related to the scale j) and at some time (related to the location k).

In what follows we will need to reduce the number of coefficients we use in order to keep the calculations into a reasonable time. With this, we are further truncating the approximation on (3) into smaller values of J. Since finer approximations may capture only the signal's noise, the changes on these scales would reflect random fluctuations not necessarily connected to the structure of the signal. For this, one should only retain coarsest scales. and the detail coefficients $d_{0,0}, d_{0,1}$.

3.2   Random Forest

Very popular in statistical machine learning, random forests (RF) are an ensemble method [9]. It builds up on specific versions of CART (Classification And Regression Trees) [10], which is an algorithm that constructs binary tree-based predictors. With respect to individual predictors, the aggregate one aims to augment robustness, variance reduction and improve prediction performance.

For this, RF add two layers of randomness. First, each tree-based predictors is trained only on a different bootstrap sample from the data. Second, only a strictly subset of variables are randomly chosen as candidates at each split of the trees' construction. Note that the trees are constructed up to its maximal size and they are not pruned. While using a stopping criterion and pruning are usual in CART, these versions of tree-predictors sacrifices generalization power by a better in-sample fit – at least on each bootstrapped sample – and introduces bias by considering only partial information from available variables. With this, individual trees tend to be less dependent between them which is useful under an aggregation scheme. RF is then the resulting predictor obtained by some aggregation rule of the individual prediction of the so described trees. Usual choices of the aggregation rule are majority vote for classification and mean average for regression.

Variable Importance Measure. Different approaches can be used to determine the importance of a feature for the construction of the forest. In this work, a variable is considered more important if it participates more to the decrease of some impurity notion (e.g. the Gini index). Then, we can track over the individual trees where each variable participates on each node split and record the decrease on the Gini coefficient. Then a plot like the one in Fig. 4 where the variables are represented in lines sorted decreasingly on the mean Gini reduction. Most important variables on the construction of the classifier are on top of the plot.

Observations Proximity. Two observations are closer if they are classified within the same terminal node by more and more individuals trees. Then, the proximity is normalized to be between 0 and 1. If we call $p_{ii'}$ the proximity between observation i and i', then we obtain a dissimilarity measure $1 - p_{ii'}$.

While the change is trivial, it allows us now to perform a multidimensional scaling on the proximity matrix associated to the proximity measurements. This yields on a natural representation of the individuals that analogously to discriminant analysis represents in a low dimensional space how the classifier 'sees' the individuals.

4   Experiments

For each gap before fault, we create a data set of positive and negative failure occurrences. From faulty devices, δ time points (albeit the gap period) before fault occurrence were considered to compute K event vectors. Among working devices, a period of length δ is drawn randomly per device to compute K event vectors per device. Notice that each device at some point of the time is described as a number of 39 features, that is 3 wavelets coefficients per group of events' code, with a total of 13 event codes. In all our experiments we use the open source R software. DWT is performed using `wavethresh` and `randomForest` is used to learn random forests using the default options.

4.1   Predictive Performance

We apply a random forest classifier for each of 16 datasets composed of 39 wavelets coefficients. Two week event profile (for each device) is characterized by 3 coefficients for each of the 13 groups of events. We compute both false negative rate (FNR) representing the percentage of faulty devices classified as working devices and false positive rate (FPR) as the percentage of working devices predicted as prone to failure. We also compute the global model error, summarizing the percentage of observations which are classified wrong and resuming model global accuracy.

The performance scores of random forest models are displayed in Fig. 3, results are presented in relation to the predictive gap before failure occurrence. Global model accuracy ranges from 79%, when the predictive gap equals to 15 days, up to 89% when detecting fault the day of occurrence. This performance displays evidence of meaningful information in the event logs.

Overall, the predicting error rate is higher for faulty devices and it is easier to decide on a working status on the basis of resumed event profile of a device, independently of the prediction gap. The error rate is lower when classifying working devices, as observations of negative fault occurrence dominate the learning error. The result is consistent with the fact that random forests tend to maximize the model global accuracy, keeping a low error rate on larger classes (working devices) while allowing the smaller classes have a larger error rate.

The smaller the temporal gap is, the more precise it is to predict both fault occurrence or devices' normal regime by event data, for example the FPR for working regimes being equal to 3.74% for 1 day-ahead prediction and 7.15% for a 10-day-ahead horizon, see Fig. 3.

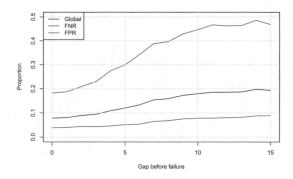

Fig. 3. Predictions performance at various gaps, in days, to failure. Red curve shows classification results of failed devices, blue curve shows classification results of working devices and black curve global error.

4.2 Variable Importance

Figure 4 shows importance ranking of attributes in classification for a 0 days predictive gap, variable importance displays similar results for all of the 16 models for different values of the predictive gap (results not shown). 3 groups of events appear relevant when predicting fault occurrence: A, B and J. First and third wavelet components of B group appear to be the red flag for an abnormal regime leading to a failure. A different level of these events for a device and an alteration of the number of received events can be seen as an alarm for failure occurrence. Overall, we observe the same pattern for all of three groups of events: the faulty devices' average level of events is generally higher than working devices' event frequencies. Moreover, there is a substantial gap between the event regime 7 days before failure occurrence and the week before that. This is particularly interesting, as events are related to low level communication on the grid. We suppose that failure affects the ability of devices to interact with other devices on the network.

More helpful, Fig. 4 shows that a considerable amount of information received and processed by the system are not relevant for revealing devices' operative status. Independently of their number or frequency, events of group C, G, K, D and F, seem to carry very little information about a possible failure of an equipment. This is to be expected as these events monitor different software activity of various devices of the grid. In a predictive maintenance framework, the monitoring and processing of these categories presents no interest, events have no correlation with the fault occurrence.

4.3 Observations Proximity

As in Fig. 2, we use a multidimensional scaling to represent observation proximity in Fig. 4. Recall that now the distances on plot are the ones implicitly learned by the classifier so it is effectively using the information on the labels (coded in

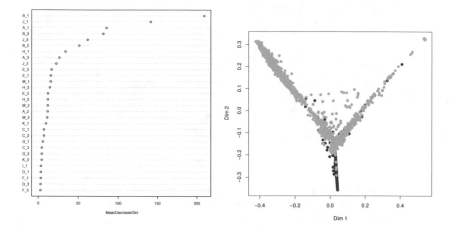

Fig. 4. Random forest variable importance output for a 0 day temporal gap (left) and MDS from RF (right): each point represents a trajectory from a working device (in gray) or a faulty device (in red).

colours on the plot). Two important differences are to be highlighted. First, the classes are now better clearly separated even if with some overlap. The class of faulty devices (smaller, in red) forms now a compact group that aligned along a straight pattern. Second, the class of normal operation, that is without fault, presents a two arm structure. This means that while connected by some elements that are close to both arms in the middle of the plan, the structure suggest that this class is actually formed by two subclasses which are homogeneous for each of them. From a technical point of view, this result also indicates that working devices present two distinctive event profiles, which shift to a single highly abnormal regime when failure emerges. This outcome is of a particular interest, as experts do not have any *a priori* knowledge about this singularity.

5   Discussion and Conclusion

From the modeling point of view, the use of wavelets and random forests gave several benefits. First, the proposed approach is general in the sense that it is not specific for predictive maintenance. Actually, it may be used on different kind of anomaly detection from event logs such as intrusion detection, outage occurrence, *etc.*, as long as one disposes with a way to construct the two class learning data set. Second, wavelets allows an important dimension reduction while keeping discriminatory power. With this, up-scaling the procedure is feasible since the processing needed to pass from functions to wavelet coefficients may be done independently (and so using parallel or distributed computing schemes) for each device. Last, random forest gives interest insights through feature selection and observations proximity. The former benefits from localized coefficients that gives nice interpretation properties to the DWT. The latter can be used together with graphical displays to unveil hidden patterns in the data.

To sum up, classification results show that abnormal dynamics in specific events, can be considered, to a certain extent, forerunner of a future fault. For a long term preventive strategy, there is an obvious need to cross the profiles of identified group of events with other sources of information to increase model accuracy. Geographical situation the grid, power demand, voltage quality, or environmental factors could affect gradually devices leading potentially to a failure. Primary data and the monitoring of information resulting of the implemented features could allow to enhance the predictive capacity of events. Information on other components of the grid could offer complementary perspectives on the network activity leading to devices usury.

Acknowledgements. We would like to acknowledge ENEDIS for this collaboration and we especially thank Pierre Achaichia, Paul Mersy and Thomas Pilaud from ENEDIS for rich discussions.

References

1. Diamantini, C., Genga, L., Potena, D.: Behavioral process mining for unstructured processes. J. Intell. Inf. Syst. **47**(1), 5–32 (2016)
2. Mannhardt, F., de Leoni, M., Reijers, H.A., van der Aalst, W.M.P., Toussaint, P.J.: From low-level events to activities - a pattern-based approach. In: Business Process Management. Springer, Cham (2016)
3. Bose, R.P.J.C., Aalst, W.M.: Abstractions in process mining: a taxonomy of patterns. In: Proceedings of the 7th International Conference on Business Process Management, BPM 2009, pp. 159–175. Springer (2009)
4. Wang, J., Li, C., Han, S., Sarkar, S., Zhou, X.: Predictive maintenance based on event-log analysis: a case study. IBM J. Res. Dev. **61**(1), 11121–11132 (2017)
5. Sipos, R., Fradkin, D., Moerchen, F., Wang, Z.: Log-based predictive maintenance. In: Proceedings of the 20th ACM SIGKDD International Conference on Knowledge Discovery and Data Mining, KDD 2014, pp. 1867–1876. ACM (2014)
6. Li, Z., Zhou, S., Choubey, S., Sievenpiper, C.: Failure event prediction using the cox proportional hazard model driven by frequent failure signatures. IIE Trans. **39**(3), 303–315 (2007)
7. Stephane, M.: A Wavelet Tour of Signal Processing: The Sparse Way. Academic Press, Cambridge (2008)
8. Nason, G.: Wavelet Methods in Statistics with R. Springer, Heidelberg (2010)
9. Breiman, L.: Random forests. Mach. Learn. **45**(1), 5–32 (2001)
10. Breiman, L., Friedman, J., Stone, C.J., Olshen, R.A.: Classification and Regression Trees. CRC Press, Boca Raton (1984)

Building Robust Prediction Models for Defective Sensor Data Using Artificial Neural Networks

Cláudio Rebelo de Sá[1]([✉]), Arvind Kumar Shekar[2], Hugo Ferreira[3], and Carlos Soares[4]

[1] Twente University, Enschede, Netherlands
c.f.pinhorebelodesa@utwente.nl
[2] Robert Bosch GmbH, Stuttgart, Germany
[3] INESC TEC, Porto, Portugal
[4] Faculdade de Engenharia, Universidade do Porto, Porto, Portugal

Abstract. Sensors are susceptible to failure when exposed to extreme conditions over long periods of time. Besides they can be affected by noise or electrical interference. Models (Machine Learning or others) obtained from these faulty and noisy sensors may be less reliable. In this paper, we propose a data augmentation approach for making neural networks more robust to missing and faulty sensor data. This approach is shown to be effective in a real life industrial application that uses data of various sensors to predict the wear of an automotive fuel-system component. Empirical results show that the proposed approach leads to more robust neural network in this particular application than existing methods.

1 Introduction

Predicting the wear of components is pivotal in various domains such as the automotive, health and aerospace industries [2,16,17]. Robust and accurate predictions have a great potential for preventing unanticipated equipment failures and increasing productivity. With the recent widespread adoption of the Internet-of-Things (IoT), many sensor signals are now readily accessible for predicting the wear of components.

In the automotive industry, one often encounters datasets with several hundreds of sensor measurements from vehicles [17]. These can be used for predicting the state of health of components. Several off-the-shelf algorithms, such as Random forests and Support Vector Machines (SVM), were previously tested on fuel system data from different vehicles. It was observed that all algorithms selected a similar subset of attributes as the most relevant ones.

A problem arises in the case when one, or more, of these selected (relevant) attributes are invalid due to malfunctioning sensors. When malfunctioning, sensors produce incorrect measurements e.g. a constant value (e.g. stuck-at-zero condition [7]) or none at all. Additionally, sensors are electrical devices that are prone to be affected by noise. Therefore, they can generate distorted measurements amidst valid values. For example, the magnetic field generated by the

F. Martínez Álvarez et al. (Eds.): SOCO 2019, AISC 950, pp. 142–153, 2020.
https://doi.org/10.1007/978-3-030-20055-8_14

ignition system of a vehicle can affect other sensors [6]. If such a malfunctioning sensor represents a relevant attribute for the target prediction, it can lead to less reliable predictions.

Industries spend millions of dollars to remove this type of noise from the signals [14]. Additionally, the manual data cleansing process is laborious, time consuming and error-prone [8,22]. It is therefore essential to train models that are robust to faulty measurements. The challenge is to train a prediction model that is both robust to missing and noisy measurements. Solving this problem is an important challenge for predicting the state of health of vehicle components.

Preprocessing techniques for handling noisy and missing input attributes have been of great interest in the data mining community [22]. However, in real world applications, we do not know the type of noise that can interfere with the measurements of the sensors. Applying imputation techniques to extrapolate these values as in the case of a missing value problem is not always acceptable [22].

To solve this problem, we propose *NoiseDrop*, a data augmentation technique for Artificial Neural Networks (ANNs) models. NoiseDrop enhances the robustness of the predictions in spite of missing and/or faulty attributes. It basically replaces real measurements in the data with random values from a Gaussian distribution. Therefore the models must learn not to trust every single measurement alone and use most of the information available.

We compare NoiseDrop with similar concepts, such as Dropout [19] and additive zero-centered Gaussian noise [9]. Dropout is a well-known regularization technique used in the training of ANNs. By testing different levels of dropout in the input layer, we simulate the missing sensors by randomly setting different measurements to zero per instance. Additive zero-centered Gaussian noise, on the other hand, reportedly enhance the generalization of ANNs by forcing more hidden units to be used [18]. These two techniques either add small levels of noise to the existing real measurements or simply remove them from the input. NoiseDrop, on the other hand, completely replaces these measurements by random values. In other words, the essential difference is that the sensor values are totally distorted instead of being slightly distorted or removed.

The approach was tested on labelled automotive data representing the state of health of Bosch fuel systems. The results show that NoiseDrop makes models more robust than Dropout and added Gaussian noise both for missing and faulty sensor measurements.

2   Proposed Approach

Artificial Neural Networks (ANNs) have been widely used in machine learning tasks such as classification [4,21] and regression [15]. In ANNs, information is processed using a set of highly interconnected nodes (also referred to as neurons). Based on the flow of information between the nodes, there are two main types of ANNs, Feed-forward Neural Networks (FNN) and Recurrent Neural Networks (RNN) [5]. In FNNs, the flow of information through the hidden layers is acyclic. On the other hand, with RNN, the flow of information in the hidden layers can be bi-directional or cyclic. FNNs have been used in many different domains such as the prediction of medical outcomes [20], environmental problems [10], stock market index predictions [12] and the wear out of machines [1].

In this work, we propose a new data augmentation for ANNs, *NoiseDrop*, for addressing the problems of having missing and noisy attributes. We also compare this method with an alternate data augmentation strategy referred to as input drop. Data augmentation is a concept introduced from the literature of image classification [3]. It involves transforming the original data (e.g., rotation, zoom, rescaling and cropping) to avoid over-fitting [13]. This data is then used to train the machine learning models.

Let us represent a subset of defective sensors defined as $M \subset \mathcal{F}$, where $\mathcal{F} = \{a_1, \cdots, a_d\}$ is a d-dimensional attribute space. This means that each attribute $a \in M$ is stuck at zero (continuously generates null values) or measures values that do not represent actual measurements.

2.1 NoiseDrop

NoiseDrop consists in replacing a certain percentage of randomly selected attributes in the dataset with noise. More formally, for each *instance* of the attribute space $\mathcal{F}$, a random attribute subset of size $\alpha \in \mathbb{Z}$ (where, $0 \le \alpha < |\mathcal{F}|$) is selected and replaced with random values from a particular distribution. Despite the many choices of distributions one can have, in this work we use a Gaussian distribution (c.f. Algorithm 1), with mean zero and standard deviation of one, i.e., $\mathcal{N}(0,1)$, because it adequately models the effect of many random processes that occur in nature [11]. That is, NoiseDrop deliberately introduce noise to the original training data so we can train models to be robust to faulty and missing measurements.

Algorithm 1. NoiseDrop augmentation

Input: $\mathcal{F}, \alpha \in \{0, 20, 40, \cdots, 140\}$
 1: $\mathcal{I} = \{1, ..., 149\}$ ▷ Set of attribute indices
 2: **for each** *Instance i* **do**
 3:     *Select random subset of attribute indices* $\mathcal{I}' \subset \mathcal{I}$, *where,* $|\mathcal{I}'| = \alpha$
 4:     *Replace attributes* $a_j \in \mathcal{F} \mid \forall j \in \mathcal{I}'$ *with values from* $\mathcal{N}(0,1)$ on instance i
 5: **end for**

This approach is illustrated in Table 2. Highlighted in bold are the values replaced by NoiseDrop in the original dataset in Table 1.

Table 1. Original data.

id	a1	a2	a3	a4	Target
1	0.1	1.5	−0.1	−0.1	A
2	0.2	2.5	−0.2	−0.2	A
3	0.3	3.5	−0.3	−0.3	B

Table 2. NoiseDrop augmentation.

id	a1	a2	a3	a4	Target
1	0.1	**1.2**	−0.1	**−0.4**	A
2	**0.8**	**−1.0**	−0.2	−0.2	A
3	**0.0**	3.5	**0.7**	−0.3	B

2.2   Input Drop

Dropout is a regularization technique commonly used in ANNs [19]. The concept consists in randomly removing a certain percentage of neurons from a layer at each iteration. Technically, this prevents the units from co-adapting too much and consequently avoids over-fitting while training the network.

In practice, dropping a neuron is setting its activation function to zero. Therefore, when dropout is used between the input layer and the first hidden layer, it is hiding a certain percentage of features during training. This emulates the situation of missing sensors and consists of randomly setting a few different measurements to zero per instance. Hence, we expect the network to be more robust to missing sensors when using dropout in the input layer. For simplicity, we refer to *Input drop* for the use of dropout in the input layer. More formally, given an *instance* of the attribute space $\mathcal{F}$, a random attribute subset of size $\alpha \in \mathbb{Z}$ (where, $0 \leq \alpha < |\mathcal{F}|$) is replaced with zero.

3   Experimental Setup

We address the challenges defined in Sect. 1 with data augmentation approaches. With this experimental setup, we hope to understand which approach makes the models more robust to both missing and noisy sensors. At the same time, we would also like to understand what is a reasonable percentage of noise to be used.

All experiments used a FNN architecture with: an input layer of 149 neurons, three hidden layers of 128, 256 and 128 neurons, and an output layer of 7 neurons. A fixed dropout rate of 50% was also used in the hidden layers for regularization purposes. The selected architecture, the regularization method and its hyperparameters were manually selected to generate accurate and robust models for the untainted dataset.

3.1   Data

In this work, we use a real automotive dataset. It consists of a high-dimensional attribute space $\mathcal{F} = \{a_1, \cdots, a_{149}\}$ of 149 attributes and 4 million instances, which we define as $\mathcal{D} = \{\mathcal{F}, Y\}$. The attributes are obtained from various sensor sources present in vehicles. It also includes signals that are calculated by the vehicles' hardware using the sensor measurements. The target attribute Y, has nominal values which represent the health state of the automotive fuel system. Table 3 shows the distribution of the different classes in the dataset. As the data for each health state was obtained from different vehicles, we treat every instance as a snapshot of the fuel system.

The dataset was split into two parts for training and testing purposes based on the chronology of the data collection. The training data, $D0$ corresponds to data collected in the more distant time and the testing data $DTest0$ to data collected more recently. Both train and test datasets were standardized by subtracting the mean and dividing by the standard deviation (separately).

Table 3. Distribution of the classes in the dataset

Class	Health state	Class distribution
Class 1	0%	10.0%
Class 2	20%	4.8%
Class 3	40%	6.3%
Class 4	60%	17.4%
Class 5	80%	50.8%
Class 6	100%	10.6%

Using the Input drop augmentation we transformed the original dataset, $D0$, into 7 different datasets: $D2, D4, \ldots, D14$ with varying Input drop levels, $20, 40, \ldots, 140$ respectively. We also transformed $DTest0$ into $DTest2$, $DTest4$, $\ldots, DTest14$ subjected to the same Input drop process.

With the NoiseDrop augmentation we generated another 7 training datasets from $D0$. We denote the different variants of training data $V0, V2, V4, \ldots, V14$ where $\alpha \in \{0, 20, 40 \cdots, 140\}$, respectively. For example, $V2$ represents a dataset where 20 randomly selected attributes of the training data are replaced by random numbers from a Gaussian distribution per instance. We note that $V0$ is the same as $D0$. The corresponding transformation is also applied to the test data and is denoted as $VTest0, VTest2, VTest4, \ldots, VTest14$.

3.2   Results

We trained multiple models with different levels of Input drop and NoiseDrop. These models are tested and compared and their predictions measured in terms of accuracy. We evaluated the performance of the models in terms of classification accuracy and F1 measure.

Input Drop Models. We study the effect of training different models with different levels of Input drop augmentation. We trained 8 ANNs on the datasets $D0, D2, D4, \ldots, D14$, therefore obtaining 8 different models $Model\ D0$, $Model\ D2$, $Model\ D4$, $\ldots, Model\ D14$, respectively. This implicitly means that each model is trained with a different number of missing sensors per instance. For example, $Model\ D2$ denotes an ANN model with 20 attributes set to zero per instance.

Each of these models $Model\ D0, Model\ D2, Model\ D4, \cdots, D14$ was evaluated on test datasets, $DTest0, DTest2, \ldots, DTest14$. The results are illustrated in Fig. 1.

We can observe that, the network trained with the original data, i.e., $Model\ D0$, is accurate when tested on datasets with low or no Input drop, i.e., $DTest0$ and $DTest2$. However, its accuracy declines steeply when the number of missing data increases. It decreases until it reaches an accuracy of only 0.5 for $DTest14$.

As for the remaining models, $Model\ D2, ..., Model\ D14$, we observe that these are comparatively more robust to missing inputs. Most of them even have higher accuracy that $Model\ D0$, at $DTest0$, when data shows no missing values. Moreover, they also maintain a high accuracy on test datasets that have more missing values than the datasets used for their own training. We also observe that the curves of $Model\ D6$ and $Model\ D8$ are highest in comparison to the other models. They are therefore much more robust than $Model\ D0$ in the presence of missing data.

In terms of the F1 measure, $Model\ D6$ scored 0.77 in $DTest0$, 0.76 in $DTest2$ and 0.75 in $DTest4$. $Model\ D0$, on the other hand, scored 0.71, 0.70 and 0.67 respectively, which confirms our previous analysis.

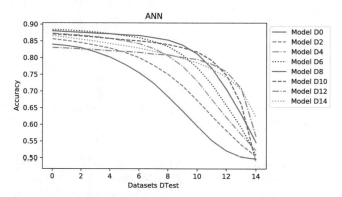

Fig. 1. Input drop with missing sensor data: Accuracy (y axis) of the different models. The accuracy was calculated for each test data (x axis) with different levels of Input drop ($DTest0, ..., DTest14$).

Let us consider $Model\ D8$ being used in a real world scenario to predict the health of the fuel system. Despite the failure of 100 sensors ($DTest10$), the predictions still have an approximate accuracy of 0.85. This is significantly better than the baseline ($Model\ D0$) accuracy of 0.60 at the same level of sensor degradation and comparable to the baseline's performance when evaluated with no sensor failures. Hence, the *Input drop* helps us to tackle the problem of failed sensors in real world prediction systems.

Now we test the performance of $Model\ D0, Model\ D2, ..., D14$ in datasets with faulty measurements $VTest0, VTest2, ..., VTest14$. The behavior of the accuracy of the models is represented in Fig. 2. In comparison to the previous experiment (c.f. Fig. 1), we can observe that most models are not as robust to noisy data as they are for missing values. This is not surprising because the training was performed without noise. Nevertheless, by comparing the behavior of $Model\ D0$ with $Model\ D6$ and $Model\ D8$ we observe that training models with Input drop is still making the models more robust to noisy measurements.

NoiseDrop Models. In this section, we test the proposed data augmentation, NoiseDrop, and analyze the influence of different noise levels in the training.

To make the network robust to noisy sensors, we trained our models with the augmented dataset variants $V0, V2, \ldots, V14$. The corresponding networks trained using these datasets are denoted as $Model\ V0, Model\ V2, \ldots, Model\ V14$.

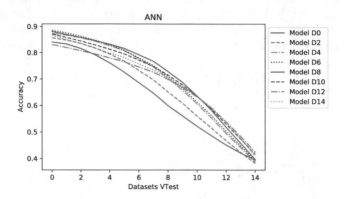

Fig. 2. Input drop with corrupted sensor data: Accuracy (y axis) of different models trained with input drop data. The accuracy was measured for each test data (x axis) with different levels of noise $(VTest0, \ldots, VTest14)$.

We test these models with datasets $VTest0, VTest2, \ldots, VTest14$ that underwent a similar transformation (c.f. Algorithm 1). The results can be seen in Fig. 3. $Model\ V6$ seems to have the best performance. It is able to predict with an accuracy of 0.88 even when around 25% of the entire set of inputs (40 sensor measurements) are noisy. This is significantly better than the baseline $Model\ V0$ that has an accuracy of 0.74 at this same level of sensor failure. As with the previous case of input drop, most models trained with NoiseDrop have better performance than the the baseline $(Model\ V0)$ when no sensors are noisy. As expected, in the most extreme case $(VTest14)$, no model can adequately predict the correct health state of the automotive fuel system.

In terms of the F1 measure, $Model\ V6$ scored 0.77 in $VTest0$, 0.76 in $VTest2$ and 0.74 in $VTest4$. On the other hand, $Model\ D6$ (Sect. 3.2) scored 0.77, 0.74 and 0.69 respectively. These results indicate that the best way to deal with noisy sensors is by training the ANNs with a NoiseDrop augmentation

In Fig. 6 we see how the same models behave, but this time tested in datasets $DTest0, \ldots, DTest14$. Interestingly, we observe that $Model\ V6$ and $Model\ V8$, besides performing quite well in the presence of noise are also quite robust to missing data. In terms of the F1 measure, $Model\ V6$ scored 0.77 with dataset $DTest0$, 0.78 with $DTest2$ and 0.78 with $DTest4$. This means that the Noise-Drop $Model\ V6$ performed better than the Dropout $Model\ D6$ which scored 0.77, 0.76 and 0.75, respectively.

The confusion matrix of $Model\ V6$ with dataset $DTest4$ can be seen in Fig. 5, side by side with the confusion matrix of $Model\ V0$ tested without missing data $DTest0$, Fig. 4 (Each row was normalized by the number of instances per class and rounded to two decimal places). A closer look into this table tells us that most misclassification errors happen with adjacent classes. For example, a reasonable number of *Class* 3 are predicted as *Class* 4. Comparing the two matrices, we observe that *Model V*6 in the presence of missing data still performs better than *Model V*0 with all information.

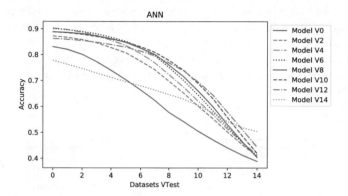

Fig. 3. NoiseDrop with corrupted sensor data: Accuracy (y axis) of different models trained with input noise data. The accuracy was measured for each test data (x axis) transformed with the same input noise approach, with different levels of noise (*VTest*0, . . . , *VTest*14).

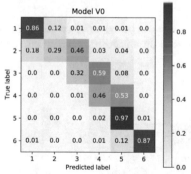
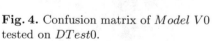
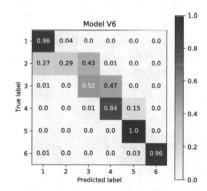

Fig. 4. Confusion matrix of *Model V*0 tested on *DTest*0.

Fig. 5. Confusion matrix of *Model V*6 tested on *DTest*4.

By comparing Fig. 2 to Fig. 3 we observe that all these models trained with noise are significantly more robust than the baseline models *Model D*0, *Model D*2, . . . , *Model D*14. In particular, models trained with NoiseDrop have the best overall performance and are therefore a good option when dealing with both missing and faulty sensor measurements.

Finally, we trained models using a third data augmentation approach where zero-centered Gaussian noise was randomly added to a certain number of real sensors' measurements. The models trained with this approach are denoted as *Model W*0, *Model W*2, ..., *Model W*14. On input drop datasets, we observed a general improvement of 3% in accuracy when using *Model W*4, *Model W*6, and *Model W*8, in comparison to the model trained without additive noise (*Model W*0). On NoiseDrop datasets, *VTest*2, ..., *VTest*14, there was also an improvement of roughly 2%, using the same models. For the sake of space, we do not provide a detailed overview of the results. However, none of these models outperformed the models trained with NoiseDrop and Input drop. In Fig. 7, we see how *Model W*8 behaves with test data *VTest*2. This table can be compared with the table in Fig. 8, from *Model V*6, which reported the best results in the discussion presented above.

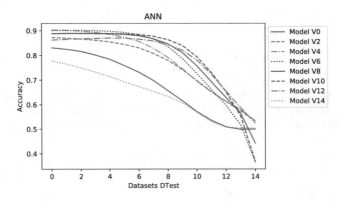

Fig. 6. NoiseDrop with missing sensor data: Accuracy (y axis) of networks trained using various levels of noise in training data and tested on datasets with varying levels of input dropout.

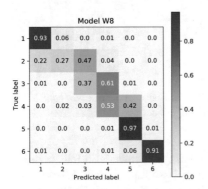
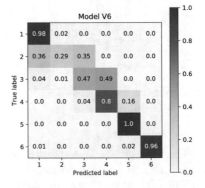

Fig. 7. Confusion matrix of *Model W*8 tested on *VTest*2.

Fig. 8. Confusion matrix of *Model V*6 tested on *VTest*2.

In terms of the F1 measure, *Model W*8 scored 0.71 with dataset *DTest*0, 0.70 with *DTest*2 and 0.68 with *DTest*4. In NoiseDrop datasets, *Model W*8 scored 0.71 with dataset *VTest*0, 0.68 with *VTest*2 and 0.62 with *VTest*4.

4 Conclusions

In this work we introduced NoiseDrop, a data augmentation technique that inserts random noise into the training data, which can be seen as a way of simulating noisy sensors. Our goal was to determine if the proposed data augmentation approach could successfully generate neural network models that are robust to both missing sensor data and noisy sensors. Based on the results of this case study, we believe that NoiseDrop really helped ANN to generalize better, therefore making them robust to missing and noisy sensors. We observed that the NoiseDrop augmentation technique was more effective than Dropout and additive zero-centered Gaussian noise in helping the models to be more robust for both missing and faulty measurements in our problem. This seems to indicate that, at least in our case study, NoiseDrop helps the models be more robust to both missing and faulty measurements. However, even though Noise-Drop solved this particular problem, more experiments are needed to ascertain the effectiveness of the approach for other problems.

Based on our observations, the best values for Dropout are between 60 to 80 attributes (i.e., between 40% and 50% of the attributes). This results are consistent with the recommendations in [19]. As for the right level of NoiseDrop, results indicate that model *V*6 (i.e., around 40% of attributes) is ideal, both in terms of noisy and missing sensor data.

As future work, NoiseDrop needs to be tested on other application domains (where sensor data are partially missing or faulty) and in other neural network architectures (including other regularization techniques). Future work could also study how different noise distributions lead to the similar results.

Acknowledgments. We gratefully acknowledge the support of NVIDIA Corporation with the donation of the Titan X Pascal GPU used for this research.

References

1. Ali, J.B., Chebel-Morello, B., Saidi, L., Malinowski, S., Fnaiech, F.: Accurate bearing remaining useful life prediction based on weibull distribution and artificial neural network. Mech. Syst. Signal Process. **56**(57), 150–172 (2015)
2. Allred, D., Harvey, J.M., Berardo, M., Clark, G.M.: Prognostic and predictive factors in breast cancer by immunohistochemical analysis. Mod. Pathol.: Off. J. US Can. Acad. Pathol. Inc. **11**(2), 155–168 (1998)
3. Arandjelović, R., Zisserman, A.: Three things everyone should know to improve object retrieval. In: 2012 IEEE Conference on Computer Vision and Pattern Recognition (CVPR), pp. 2911–2918. IEEE (2012)
4. Baxt, W.G.: Use of an artificial neural network for data analysis in clinical decision-making: the diagnosis of acute coronary occlusion. Neural Comput. **2**(4), 480–489 (1990)

5. Cho, K., van Merrienboer, B., Gülçehre, Ç., Bahdanau, D., Bougares, F., Schwenk, H., Bengio, Y.: Learning phrase representations using RNN encoder-decoder for statistical machine translation. In: Proceedings of the 2014 Conference on Empirical Methods in Natural Language Processing, EMNLP 2014, 25–29 October 2014, Doha, Qatar, A meeting of SIGDAT, a Special Interest Group of the ACL, pp. 1724–1734 (2014)
6. Dziubiński, M., Drozd, A., Adamiec, M., Siemionek, E.: Electromagnetic interference in electrical systems of motor vehicles. In: IOP Conference Series: Materials Science and Engineering, vol. 148, p. 012036. IOP Publishing (2016)
7. Elleithy, K., Sobh, T.: Innovations and Advances in Computer, Information, Systems Sciences, and Engineering, vol. 152. Springer, Heidelberg (2012)
8. Jäger, G., Zug, S., Brade, T., Dietrich, A., Steup, C., Moewes, C., Cretu, A.M.: Assessing neural networks for sensor fault detection. In: 2014 IEEE International Conference on Computational Intelligence and Virtual Environments for Measurement Systems and Applications (CIVEMSA), pp. 70–75, May 2014
9. Klambauer, G., Unterthiner, T., Mayr, A., Hochreiter, S.: Self-normalizing neural networks. In: Advances in Neural Information Processing Systems 30: Annual Conference on Neural Information Processing Systems 2017, 4-9 December 2017, Long Beach, CA, USA, pp. 972–981 (2017)
10. Maier, H.R., Dandy, G.C.: Neural networks for the prediction and forecasting of water resources variables: a review of modelling issues and applications. Environ. Model. Softw. 15(1), 101–124 (2000)
11. Maybeck, P.: Stochastic Models, Estimation, and Control. Mathematics in Science and Engineering. Elsevier Science, Amsterdam (1982)
12. Moghaddam, A.H., Moghaddam, M.H., Esfandyari, M.: Stock market index prediction using artificial neural network. J. Econ. Financ. Adm. Sci. 21(41), 89–93 (2016)
13. Perez, L., Wang, J.: The effectiveness of data augmentation in image classification using deep learning. ArXiv e-prints, December 2017
14. Redman, T.C., Blanton, A.: Data Quality for the Information Age. Artech House, Inc., London (1997)
15. Refenes, A.N., Zapranis, A., Francis, G.: Stock performance modeling using neural networks: a comparative study with regression models. Neural Netw. 7(2), 375–388 (1994)
16. Reuss, P., Stram, R., Althoff, K., Henkel, W., Henning, F.: Knowledge engineering for decision support on diagnosis and maintenance in the aircraft domain. In: Synergies Between Knowledge Engineering and Software Engineering, pp. 173–196. Springer (2018)
17. Shekar, A.K., Bocklisch, T., Sánchez, P.I., Straehle, C.N., Müller, E.: Including multi-feature interactions and redundancy for feature ranking in mixed datasets. In: Machine Learning and Knowledge Discovery in Databases - European Conference, ECML PKDD 2017, Skopje, Macedonia, 18–22 September 2017, Proceedings, Part I, pp. 239–255 (2017)
18. Sietsma, J., Dow, R.J.: Creating artificial neural networks that generalize. Neural Netw. 4(1), 67–79 (1991)
19. Srivastava, N., Hinton, G.E., Krizhevsky, A., Sutskever, I., Salakhutdinov, R.: Dropout: a simple way to prevent neural networks from overfitting. J. Mach. Learn. Res. 15(1), 1929–1958 (2014)
20. Tu, J.V.: Advantages and disadvantages of using artificial neural networks versus logistic regression for predicting medical outcomes. J. Clin. Epidemiol. 49(11), 1225–1231 (1996)

21. Widrow, B., Rumelhart, D.E., Lehr, M.A.: Neural networks: applications in industry, business and science. Commun. ACM **37**(3), 93–105 (1994)
22. Zhu, X., Wu, X.: Class noise vs. attribute noise: a quantitative study. Artif. Intell. Rev. **22**(3), 177–210 (2004)

Temporal Data Analysis

Ensemble Deep Learning for Forecasting ^{222}Rn Radiation Level at Canfranc Underground Laboratory

Miguel Cárdenas-Montes$^{(\boxtimes)}$ ⓘ and Iván Méndez-Jiménez

Centro de Investigaciones Energéticas Medioambientales y Tecnológicas,
Madrid, Spain
{miguel.cardenas,ivan.mendez}@ciemat.es

Abstract. Ensemble Deep Learning Architectures have demonstrated to improve the performance in comparison with the individual architectures composing the ensemble. In the current work, an ensemble of variants of Convolutional and Recurrent Neural Networks architectures are applied to the prediction of the ^{222}Rn level at the Canfranc Underground Laboratory (Spain). To predict the low-level periods allows appropriately scheduling the maintenance operations in the experiments hosted in the laboratory. As a consequence of the application of Ensemble Deep Learning, an improvement of the forecasting capacity is stated. Furthermore, the learned lessons from this work can be extrapolated to other underground laboratories around the world.

Keywords: Ensemble · Time series analysis · Deep learning ·
Forecasting · Convolutional Neural Networks ·
Recurrent Neural Networks ·
Seasonal and Trend Decomposition Using Loess

1 Introduction

Scientific laboratories generate large data volumes with relevant information about the environment in which the experiments are hosted. These data embody information that should be timely processed and provided to the experiments managers.

The modelling and forecasting of ^{222}Rn in underground laboratories are very relevant tasks. ^{222}Rn is a radionuclide produced by the ^{238}U and ^{232}Th decay chains. Being gas at room temperature, it can be emanated by the rocks and concrete of the underground laboratory, diffusing in the experimental hall. This contamination in the air is a potential source of background, both directly and through the long life radioactive daughters produced in the decay chain, which can stick to the experimental surfaces. The ^{222}Rn contamination in air can be reduced by orders of magnitude only in limited closed areas, flushing pure N_2 or "Rn-free" air produced by dedicated structures. In the deep underground

© Springer Nature Switzerland AG 2020
F. Martínez Álvarez et al. (Eds.): SOCO 2019, AISC 950, pp. 157–167, 2020.
https://doi.org/10.1007/978-3-030-20055-8_15

laboratories the average activity depends on the local conditions and must be constantly monitored. It typically ranges from tens to hundreds of Bq/m^3, with periodic and non-periodic variations. Seasonal dependence has been observed in some cases [1,15]. A detailed understanding of the ^{222}Rn periodicity can be fundamental for a precise comprehension of the background of rare-event search experiments. This is particularly true in case of the dark matter direct searches, whose distinctive feature is the annual modulation of the signal foreseen by the hypothesis of a weakly interactive massive particle (WIMP) halo model. At the same time, the prediction of the evolution of the ^{222}Rn concentration in the laboratory is relevant in order to correctly organize the operations foreseeing the exposure of the detector materials to the air, minimizing, in such a way, the deposition of the radionuclide on the surfaces.

The Canfranc Underground Laboratory (LSC) is composed of diverse halls for hosting scientific experiments with requirements of very low-background. The two main halls, Hall A and Hall B—which are contiguous, have instruments for measuring the level of ^{222}Rn, particularly there is an Alphaguard P30 in each hall recording the radioactivity level every 10 min, with an accumulated record from July 2013 to June 2017 (Fig. 1).

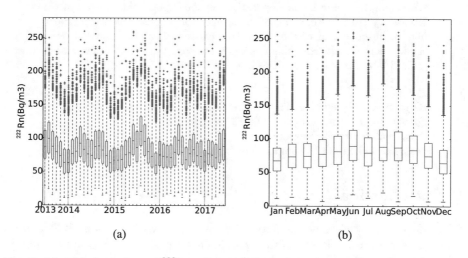

(a)   (b)

Fig. 1. Monthly box-plots of ^{222}Rn level at Hall A of the LSC, by year (Fig. 1(a)) and gathering the months independently of the year (Fig. 1(b)). Data taking corresponds to the period from July 2013 to June 2017. Hereafter, the monthly medians are the values used for creating the time series, and therefore, for further analyses.

With regard to the measurements, few missing values are in the data set, as well as gaps of several days in some years. The larger gaps appear in July 2014 with 913 missing values, in June 2015 with 1053, and in January 2016 with 585. In the worst case, the gap spans over a week (7.3 days). However, the missing values are not representative in comparison with the total number of observations (more than 200,000), nor the number of observations per month ($\approx$4,000).

^{222}Rn time series is very noisy. Only the monthly median exhibits a certain modulation (Fig. 1). Therefore, the monthly medians of the ^{222}Rn level has been selected as the monthly representative value. This critically penalizes the data volume accessible, reducing it to 48 values.

In the past, Convolutional Neural Networks (CNN) has been applied for time series classification [23,24] and for time series prediction [5,11,21], including ensembles [17]; as well as Recurrent Neural Networks (RNN) [2,8,20]. In [9] an review of these deep learning architectures is presented.

In this work, an ensemble of deep learning architectures (EDL) variants of Convolutional and Recurrent Neural Networks architectures are used to improve the forecasting the ^{222}Rn monthly level at LSC. The ensemble is composed of RNN, Bidirectional Recurrent Neural Networks (BRNN), CNN, and a variant of CNN, termed CNN+STL, in which the original observations used as input are replaced by the components generated by Seasonal and Trend decomposition using Loess (STL): trend, seasonal and remainder components [16].

Concerning the previous efforts in the analysis of the ^{222}Rn level at LSC, in [15] the initial efforts for preprocessing and modelling this time series using classical and deep-learning-based approaches are shown. In this paper, the times series is modelled using CNN and Recurrent Neural Networks, being the main focus on the forecasting capacity for scheduling maintenance operations of the experiment hosted at LSC, and the characterization of the annual modulation of observations.

The rest of the paper is organized as follows: a brief description the deep learning algorithms used for building the ensemble are presented in Sect. 2. The models comparison and the results obtained are presented and analysed in Sect. 3. Finally, Sect. 4 contains the conclusions of this work.

2 Methods and Materials

2.1 Convolutional Neural Networks

Convolutional Neural Networks (CNN) [12,13] are specialized Neural Networks with special emphasis in image processing [9], although nowadays they are also employed in time series analysis [5,22]. The CNN consists of a sequence of convolutional layers, the output of which is connected only to local regions in the input. These layers alternate convolutional, non-linear and pooling-based layers which allow extracting the relevant features of the class of objects, independently of their placement in the data example. The CNN allows the model to learn filters that are able to recognize specific patterns in the time series, and therefore it can capture a richer information from the series than other methods. It also embodies three features which provide advantages over the Multilayer Perceptrons (MLP): sparse interactions, parameter sharing and equivariance to translation [9].

Although Convolutional Neural Networks are frequently associated to image or audio classification—2D grid examples—or video sequence—3D grid examples, it can also be applied to time series analysis—1D grid examples. When processing time series, instead of a set of images, the series has to be divided in overlapping contiguous time windows. These windows constitute the examples, where the CNN aims at finding patterns. At the same time, the application to time series modelling requires the application of 1D convolutional operators, which weights are optimized during the training process.

One of the most identifiable feature of CNN is parameter sharing. Parameter sharing allows extending and applying the model to examples of different form. Conversely, if sequence-based specialization is used, e.g. Multilayer Perceptrons (MLP), then separated parameters are generated for each value of the time index. This leads to the impossibility to generalize to sequence lengths not seen during the training process, nor share statistical strength across different sequence lengths and across different positions in time.

In this work, Keras [3] has been used for implementing the CNN and RNN architectures. The CNN employed are composed of two convolutional layers of 32 and 64 filters with `relu` as activation function, `MaxPooling1D` with size 2, and an output layer composed of a dense layer with a single neuron with linear activation function, and trained with 10 epochs. In all the algorithms checked, the loop-back parameter is configured to 12 values of the time series, the Mean Squared Error has been selected as the `loss function`, and the weights are optimized by using `Adam optimizer`.

2.2 Seasonal and Trend Decomposition Using Loess

The intuition behind the time series decomposition is that the time series is the composition of three more elementary series, $Y_t = T_t + S_t + R_t$. On the one hand, a trend (T_t), which is responsible of long-term increase or decrease of data. It does not have to be linear. On the other hand, a seasonal pattern is the second component (S_t). It is influenced by seasonal factors, such as: the month, the day of the week, or the quarter of the year. It has mean null in the seasonal period. Finally, the third component is the remainder or random component (R_t).

Diverse techniques for time series decomposition have been proposed. STL, *Seasonal and Trend decomposition using Loess* [4], was proposed taking into account the limitations of previous classical decomposition methods, for example X-12-ARIMA. In contrast with X-12-ARIMA, STL can handle any type of seasonality, not only monthly or quarterly; and the seasonal component can change over time, being the amount of the allowed change controlled by a parameter of the algorithm. Besides, the smoothness of the trend component can be also controlled by the algorithm.

In CNN+STL strategy, the ^{222}Rn time series is predicted with CNN using as input the three series resulting from the STL decomposition, in contrast with the use as input of the original observations [16]. This approach implies to handle three CNN for independent predictions of each of the three components time series from STL, with final merging for obtaining the prediction.

2.3   Recurrent Neural Networks

Recurrent Neural Networks (RNN) are a set of neural networks which the main purpose is to process sequential data $x^1, \ldots, x^\tau$ [9,14,18]. Whereas CNN aims at processing grid of data, such as images, the RNN are specialized networks for processing a sequence of data. Some example of the application of RNN to time series analysis can be found in [2,8,20]. Similarly to CNN, the RNN can process sequences of different length without sequence-based specialization, such as MLP.

With regards to parameter sharing, when CNN are used for analysing time series, strong similarities appear with RNN. However, in comparison with RNN, shared parameters in CNN are considered as shallow. In the convolution operation, a network shares parameters across the time among a reduced number of neighbouring members of the input. The concept of parameter sharing appears in the application of the same convolution kernel at each time step. In RNN, each member of the output is a function of the previous members of the output, and it is produced by using the same rule which has been applied to the previous outputs (Eq. 1).

$$h = \sigma(W_{hh}\, h_{i-1} + W_{hx}\, x_i + b_h)$$
$$\hat{y}_i = W_{yh}\, h_i \tag{1}$$

where x_i is the input vector, y_i is the output vector, h_i is the hidden state, W_{hx} is input-to-hidden weights, W_{hh} is hidden-to-hidden weights, W_{yh} is hidden-to-output weights, and $\hat{y}$ is the predicted values. In the current work, a single hidden layer is used, with 24 LSTM elements, and trained with 100 epoch. Hyperbolic tangent is used as activation function. The output layer is composed of a single dense layer of a neuron with linear activation function.

Long Short-Term Memory (LSTM) introduces self-loops in the RNN schema, allowing these self-loops be conditioned on the context [10]. This architecture has the same inputs and outputs as an ordinary RNN, although it has more parameters and a system of gating units that controls the flow of information.

Bidirectional Recurrent Neural Networks (BRNN) combines two RNN, a first one that moves forward in the temporal sequence, and a second one that moves back [19]. This permits to compute the prediction based on past and future observations. RNN and BRNN have been also included in the Ensemble.

2.4   Statistics

In order to ascertain if the proposed forecasting methods applied to the test set improve the prediction, two different types of tests can be applied: parametric and non-parametric. The difference between both relies on the assumption that data is normally distributed for parametric tests, whereas non explicit conditions are assumed in non-parametric tests. For this reason, the latter is recommended when the statistical model of data is unknown [6,7]. Statistical inference is used in this work to infer which model produces better results, and if the differences are significant or not.

The Kruskal-Wallis test is a non-parametric test used to compare three or more groups of sample data. For this test, the null hypothesis assumes that the samples are from identical populations.

3   Experimental Results and Models Comparison

The collected data are divided into two sets. The training set—including the three firsts years, from July 2013 to June 2016, and the testing data set, which includes the last twelve months, from July 2016 to June 2017.

In Fig. 2 and Table 1, the Mean Squared Error (MSE) and the Mean Absolute Error (MAE) after 25 independent runs for the proposed architectures are shown. As can be appreciated, the EDL produces, for both metrics, the best prediction for the test set, and therefore, the lowest error.

The application of the Kruskal-Wallis test to the MSE and the MAE indicates that the differences between the medians are significant for a confidence level of 95% (p-value under 0.05), which means that the differences are unlikely to have occurred by chance with a probability of 95%.

Table 1. Mean squeared error (MSE) and mean absolute error (MAE) for the deep architectures evaluated for 25 independent runs.

	MSE	MAE
RNN	65 ± 6	6.3 ± 0.2
CNN	60 ± 5	6.2 ± 0.4
CNN+STL	54 ± 5	5.8 ± 1.2
BRNN	54 ± 5	6.1 ± 0.3
EDL	45 ± 4	5.3 ± 0.4

In Fig. 3(a), the real and predicted values of the test set are shown. As can be observed the methods based on deep architectures and EDL reproduce appreciably well the test set. If comparisons with other time series forecasting methods, not involving deep architectures (Fig. 3(b)), they are critically outperformed as it is appreciated [15].

A key point is the month of August 2016. This month behaves differently that in the previous years. In general, this month has high levels of ^{222}Rn, except for the year 2016, where the value is much lower. Deep learning architectures, including EDL, are able to capture enough information from the previous values, and predict a closer value to the real observation of this month.

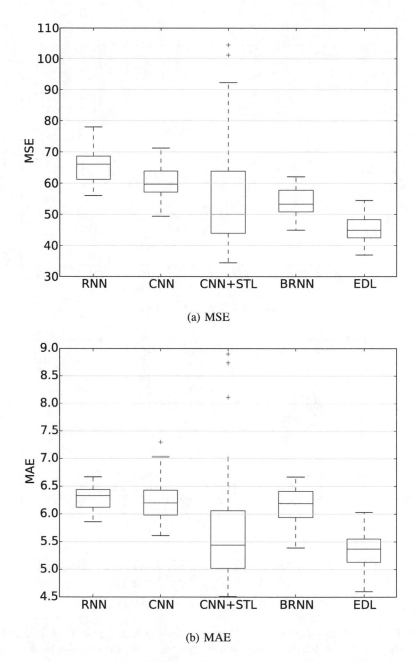

(a) MSE

(b) MAE

Fig. 2. Mean squeared error (MSE) and mean absolute error (MAE) for the deep architectures evaluated for 25 independent runs.

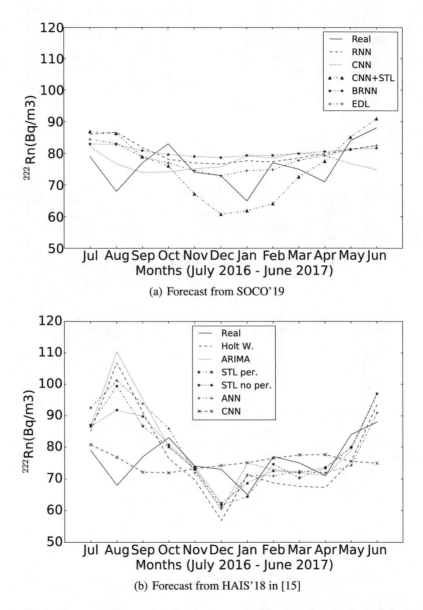

(a) Forecast from SOCO'19

(b) Forecast from HAIS'18 in [15]

Fig. 3. Real values and forecasting for the test set—the fourth year, from July 2016 to June 2017—for the methods used in this study (Fig. 3(a)), and for the methods used in [15] (Fig. 3(b)).

4   Conclusions

In this paper, an Ensemble Deep Learning approach is proposed to improve the prediction capacity of the ^{222}Rn level at Canfranc Underground Laboratory. The Ensemble is composed of CNN and RNN variants. They include Recurrent Neural Networks, Bidirectional Recurrent Neural Networks, Convolutional Neural Networks and STL Convlutional Neural Networks, which uses as input the series generated by the STL decomposition instead of the original observations.

The results and the statistical analysis state that the proposed Ensemble Deep Learning significantly improves the accuracy of the prediction in comparison with classical techniques used in previous works, such as ARIMA, Holt-Winters Exponential Smoothing and Seasonal and Trend Decomposition using Loess –based, but also in comparison with deep learning-based, such as Convolutional Neural Networks, Recurrent-, and Bidirectional Recurrent Neural Networks.

Finally, the analysis of the performance of alternative and more elaborated deep architectures over five years data, and their collective performance when integrating ensembles, are proposed as Future Work.

Acknowledgment. The research leading to these results has received funding by the Spanish Ministry of Economy and Competitiveness (MINECO) for funding support through the grant FPA2016-80994-C2-1-R, and "Unidad de Excelencia María de Maeztu": CIEMAT - FÍSICA DE PARTÍCULAS through the grant MDM-2015-0509.

IMJ is co-funded in a 91.89 percent by the European Social Fund within the Youth Employment Operating Program, for the programming period 2014–2020, as well as Youth Employment Initiative (IEJ). IMJ is also co-funded through the Grants for the Promotion of Youth Employment and Implantation of Youth Guarantee in Research and Development and Innovation (I+D+i) from the MINECO.

The authors would like to thank Roberto Santorelli, Pablo García Abia and Vicente Pesudo for useful comments regarding the Physics related aspects of this work, and the Underground Laboratory of Canfranc by providing valuable feedback.

References

1. Bettini, A.: New underground laboratories: Europe, Asia and the Americas. Phys. Dark Universe **4**(Suppl. C), 36–40 (2014). https://doi.org/10.1016/j.dark.2014.05.006. dARK TAUP2013
2. Chniti, G., Bakir, H., Zaher, H.: E-commerce time series forecasting using LSTM neural network and support vector regression. In: Proceedings of the International Conference on Big Data and Internet of Thing, BDIOT 2017, pp. 80–84. ACM, New York (2017). https://doi.org/10.1145/3175684.3175695
3. Chollet, F., et al.: Keras (2015). https://github.com/fchollet/keras
4. Cleveland, R.B., Cleveland, W.S., McRae, J., Terpenning, I.: STL: a seasonal-trend decomposition procedure based on loess. J. Off. Stat. **3**, 3–73 (1990)
5. Gamboa, J.C.B.: Deep learning for time-series analysis. CoRR abs/1701.01887 (2017). http://arxiv.org/abs/1701.01887

6. García, S., Fernández, A., Luengo, J., Herrera, F.: A study of statistical techniques and performance measures for genetics-based machine learning: accuracy and interpretability. Soft Comput. **13**(10), 959–977 (2009)
7. García, S., Molina, D., Lozano, M., Herrera, F.: A study on the use of non-parametric tests for analyzing the evolutionary algorithms' behaviour: a case study on the CEC 2005 special session on real parameter optimization. J. Heuristics **15**(6), 617–644 (2009)
8. Garcia-Pedrero, A., Gomez-Gil, P.: Time series forecasting using recurrent neural networks and wavelet reconstructed signals. In: 2010 20th International Conference on Electronics Communications and Computers (CONIELECOMP), pp. 169–173, February 2010. https://doi.org/10.1109/CONIELECOMP.2010.5440775
9. Goodfellow, I., Bengio, Y., Courville, A.: Deep Learning. MIT Press, Cambridge (2016)
10. Hochreiter, S., Schmidhuber, J.: Long short-term memory. Neural Comput. **9**(8), 1735–1780 (1997). https://doi.org/10.1162/neco.1997.9.8.1735
11. Lago, J., Ridder, F.D., Schutter, B.D.: Forecasting spot electricity prices: deep learning approaches and empirical comparison of traditional algorithms. Appl. Energy **221**, 386–405 (2018). https://doi.org/10.1016/j.apenergy.2018.02.069, http://www.sciencedirect.com/science/article/pii/S030626191830196X
12. LeCun, Y.: Generalization and network design strategies. University of Toronto, Technical report (1989)
13. Lecun, Y., Bottou, L., Bengio, Y., Haffner, P.: Gradient-based learning applied to document recognition. Proc. IEEE **86**(11), 2278–2324 (1998). https://doi.org/10.1109/5.726791
14. Lipton, Z.C.: A critical review of recurrent neural networks for sequence learning. CoRR abs/1506.00019 (2015). http://arxiv.org/abs/1506.00019
15. Méndez-Jiménez, I., Cárdenas-Montes, M.: Modelling and forecasting of the ^{222}Rn radiation level time series at the Canfranc Underground Laboratory. In: Proceedings of Hybrid Artificial Intelligent Systems - 13th International Conference, HAIS 2018, Oviedo, Spain, 20–22 June 2018. Lecture Notes in Computer Science, vol. 10870, pp. 158–170. Springer (2018)
16. Méndez-Jiménez, I., Cárdenas-Montes, M.: Time series decomposition for improving the forecasting performance of convolutional neural networks. In: Proceedings of Advances in Artificial Intelligence - 18th Conference of the Spanish Association for Artificial Intelligence, CAEPIA 2018, Granada, Spain, 23–26 October 2018. Lecture Notes in Computer Science, vol. 11160, pp. 87–97. Springer (2018). https://doi.org/10.1007/978-3-030-00374-6_9
17. Qiu, X., Zhang, L., Ren, Y., Suganthan, P.N., Amaratunga, G.A.J.: Ensemble deep learning for regression and time series forecasting. In: 2014 IEEE Symposium on Computational Intelligence in Ensemble Learning, CIEL 2014, Orlando, FL, USA, 9–12 December 2014, pp. 21–26 (2014). https://doi.org/10.1109/CIEL.2014.7015739
18. Rumelhart, D.E., Hinton, G.E., Williams, R.J.: Learning representations by back-propagating errors. Nature **323**(6088), 533–536 (1986). https://doi.org/10.1038/323533a0
19. Schuster, M., Paliwal, K.: Bidirectional recurrent neural networks. Trans. Sig. Proc. **45**(11), 2673–2681 (1997). https://doi.org/10.1109/78.650093
20. Walid, A.: Recurrent neural network for forecasting time series with long memory pattern. J. Phys.: Conf. Ser. **824**(1), 012038 (2017). http://stacks.iop.org/1742-6596/824/i=1/a=012038

21. Wang, H.Z., Li, G.Q., Wang, G.B., Peng, J.C., Jiang, H., Liu, Y.T.: Deep learning based ensemble approach for probabilistic wind power forecasting. Appl. Energy **188**, 56–70 (2017). https://doi.org/10.1016/j.apenergy.2016.11.111. http://www.sciencedirect.com/science/article/pii/S0306261916317421
22. Wang, Z., Yan, W., Oates, T.: Time series classification from scratch with deep neural networks: a strong baseline. CoRR abs/1611.06455 (2016). http://arxiv.org/abs/1611.06455
23. Zheng, Y., Liu, Q., Chen, E., Ge, Y., Zhao, J.L.: Time series classification using multi-channels deep convolutional neural networks. In: Li, F., Li, G., Hwang, S.W., Yao, B., Zhang, Z. (eds.) Web-Age Information Management, pp. 298–310. Springer, Cham (2014)
24. Zheng, Y., Liu, Q., Chen, E., Ge, Y., Zhao, J.L.: Exploiting multi-channels deep convolutional neural networks for multivariate time series classification. Front. Comput. Sci. **10**(1), 96–112 (2016). https://doi.org/10.1007/s11704-015-4478-2

Search of Extreme Episodes in Urban Ozone Maps

Miguel Cárdenas-Montes[(✉)] [iD]

Centro de Investigaciones Energéticas Medioambientales y Tecnológicas,
Madrid, Spain
miguel.cardenas@ciemat.es

Abstract. Nowadays numerous urban areas have deployed a network of sensors for monitoring multiple variables of air quality. The measurements of these sensors can be treated individually—as time series—or collectively. Collectively, a variable monitored by a network of sensors can be transformed into a map embodying the same information, but converting numerical information into visual one. Once the numerical information has been transformed into maps, they can be used as images for the usual purposes of machine learning algorithms, and specially for clustering and outlier detection. Air quality is one of the main concerns in urban areas. In this work, firstly the numerical information of 12 monitoring station measuring the concentration of Ozone in Madrid (Spain) is transformed into daily maps. For this purpose a methodology for converting numerical information from a geographically distributed network of sensors into grey-scaled maps is proposed. Later, these maps are investigated for searching outliers—extreme episodes—with *Density-based spatial clustering of applications with noise*. Also the sensitivity of the search of extreme episodes to the methodology for transforming numerical information into maps is investigated.

Keywords: DBSCAN · Outlier detection · Air quality · Madrid

1 Introduction

Air quality is a mayor concern in urban areas with a critical impact on public health [3,4,8]. European cities, including Madrid (Spain), have undertaken efforts to reduce the concentration of pollutants, such as: nitrogen oxides, suspended particulates and volatile organic compounds, which are precursors of ozone [1].

Air quality can be analysed and predicted with time series analyses [9,10]. Alternatively, numerical information of a network of sensors can be transformed in maps, and by handling these maps as images to analyse them.

In this work, the mean daily values of O_3 concentration of 12 monitoring stations from the Air Quality Monitoring Network of Madrid are used [2]. The station identifications are: *Plaza del Carmen, Barrio del Pilar, Escuelas Aguirre,*

© Springer Nature Switzerland AG 2020
F. Martínez Álvarez et al. (Eds.): SOCO 2019, AISC 950, pp. 168–178, 2020.
https://doi.org/10.1007/978-3-030-20055-8_16

Arturo Soria, Farolillo, Casa de Campo, Barajas Pueblo, Parque del Retiro, Ensanche de Vallecas, Plaza Fernández Ladreda, El Pardo, and *Parque Juan Carlos I*. In Fig. 1, the map of the monitoring stations used in this work and their relative position is shown. These monitoring stations include the three station categories: suburban (term for stations in parks in urban areas), traffic (term for stations affected by traffic and close to a principal street or road), and background (term for urban background station affected by both traffic and background pollution)[1].

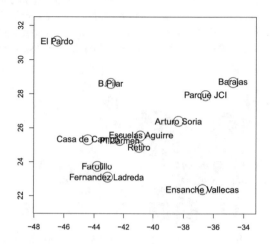

Fig. 1. Map with the location of the monitoring stations: x-coordinate is the longitude in minutes plus 3° West, and y-coordinate is the latitude in minutes plus 40° North.

Whereas other pollutants, such as NO_2 and CO have strong a relationship with human activities. traffic pollution and heating system, and therefore restrictions to traffic with private cars are applied when the alert threshold is exceeded; the highest concentrations of O_3 are produced in summer, without a clear relationship with human activities. In these episodes, sport activities in urban parks should be limited, with the associated and appropriate alert advertising system to citizens.

The values of O_3 concentrations used in this work expand from 1-1-2010 to 31-05-2018, thus a total of 3195 maps is generated. Maps are generated in png format, with a size of 400 × 400 pixels with a resolution of 150 ppi. The mean daily values of the 12 monitoring stations are transformed from a numerical format to a visual one. For this purposed, the 12 values of a single day are converted into a single map. For each monitoring station the numerical value is transformed into a grey-scaled circle with grey intensity and radius proportional to the value

---

[1] Following the Directive 2008/50/ Royal Decree 102/2011, the minimum number of sampling points for O_3 for the population of Madrid are 5 stations being at least 3 suburban. Air Quality Monitoring Network in Madrid is composed of 14 monitoring stations, with 3 suburban.

of the O_3 concentration. Besides, when overlapping occurs among circles, the most intense—the highest value of the O_3 concentration—will overlap the other circles with lower values. Through this mechanism the numerical information is transformed into visual one with three main factors: grey-intensity, circle size and overlapping position. Maps are generated without additional elements, such as axis, labels, title, etc. As a consequence of this size, grey-scale and position governed by the value of the station, the maps can vary their global appearance. This appearance will mark the similariaty or dissimilarity among them.

Some missing values might appear in the stations. These missing values are not imputed by any mechanism. Therefore, some circles could disappear in some maps. In Fig. 2, examples of maps of high- (low-quality air) and low-concentration (high-quality air) values of monitoring stations are shown.

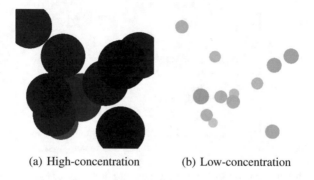

(a) High-concentration   (b) Low-concentration

Fig. 2. Examples of maps of high- (low-quality air) and low-concentration (high-quality air) values of O_3 of 12 monitoring stations in Madrid.

In order to reduce the dimensionality of the problem and to avoid the sparsity in the search space, the loaded maps are reduced to a size of 50×50 pixels before feeding the clustering algorithm. This allows maintaining an affordable number of features in the problem.

After the maps generation, they are used to search extreme episodes, the seek of outliers. These outliers might correspond to the extreme maps, both low-quality days—very high values in the concentrations of O_3 for multiple stations, and high-quality days—low concentrations of O_3. In this work, the search of extreme maps is undertaken through the use of DBSCAN algorithm (Density-based spatial clustering of applications with noise) [5,7].

The rest of the paper is organised as follows: the methodology is described in Sect. 2. The Results and the Analysis are shown in Sect. 3. Finally, Sect. 4 contains the conclusions of this work.

2   Methods and Materials

2.1   DBSCAN

Differently to other algorithms based on distance, such as K-Means; or on probabilistic models of the examples, such as Gaussian Mixture Models (GMM); DBSCAN is a clustering algorithm based on density [5]. All of them are able to generate clusters of similar examples, and distinct of the examples of other clusters. However not all of them are able to label examples as anomalous or outliers. For example, K-Means has not the capacity to point examples as outlier.

In general, density-based methods follow an incremental approach growing the clusters as far as the density of the cluster exceeds a certain threshold. Rawly, for each example in a cluster it has a certain number of neighbours in a stated distance. This allows discovering clusters with arbitrary shapes and, due to the breach of the previous conditions on the distance and on the number of examples in the neighbouring, labelling examples as outliers.

Conversely to K-Means and to GMM, in DBSCAN the number of clusters is not a predefined parameter. The algorithm has the capacity to find the most appropriate number of clusters for the dataset. The parameters of DBSCAN are the number of points in the neighbourhood of a point to be considered as a core point, and a minimum distance between two points to be in the same neighbourhood.

The algorithm does not particularise the distance to be used. Two distance metrics or norms have been evaluated in this work, L^1-norm and L^2-norm (see Sect. 2.2). Independently of the norm used, the algorithm normalizes the distances between the pair of examples, in such a way that the largest distance for the pairs of examples is rescaled to the unit.

In this work, an implementation of DBSCAN based on Scikit-Learn [11] has been used.

2.2   Distance Metrics

In supervised and non-supervised learning, the concept of distance is central to evaluate the similarity or dissimilarity among the examples. Dissimilarity metrics are based on the abstraction of the notion of distance between two points [6]. The inverse of a distance as well as the subtraction to the unit of a normalized distance is a similarity metric. As was previously mentioned, DBSCAN does not impose a particular distance metric or norm in its implementation.

As defined in [12] a dissimilarity metrics is:

Definition 1. *A dissimilarity on a set S is a function $d : S^2 \to \mathbb{R}_{\geq 0}$ satisfying:*
(DIS_1), $d(x, x) = 0 \ \forall x \in S$; and
(DIS_2), $d(x, y) = d(y, x) \ \forall x, y \in S$.

The properties of a such dissimilarity metrics are:

1. $d(x, y) = 0$ implies $d(x, z) = d(y, z)$, $\forall x, y, z \ in \ S$; **evenness**

2. $d(x, y) = 0$ implies $x = y$, $\forall x, y\ in\ S$; **definiteness**
3. $d(x, y) \leq d(x, z) + d(z, y)$, $\forall x, y, z$; **triangular inequality**

Based on the previous definition, Euclidean metric or L^2-norm (Eq. 1), and Manhattan metric or L^1-norm (Eq. 2) can be defined. L^1-norm and L^2-norm are used in conjunction with DBSCAN to find the outliers in the map set created from the O_3 monitoring network in Madrid.

$$L^2 - norm \rightarrow |\boldsymbol{x}|_2 = \left(\sum_{i=1}^{N}(x_i)^2\right)^{1/2} \tag{1}$$

$$L^1 - norm \rightarrow |\boldsymbol{x}|_1 = \sum_{i=1}^{N}|x_i| \tag{2}$$

3   Experimental Results

3.1   Outliers Detection with DBSCAN and L^2 Norm

Low-quality air episodes in urban areas can be detected by using outlier detection algorithms. DBSCAN is a clustering algorithm with capacity to generate clusters based on the similarity of the inputs and to point certain inputs as outliers [5]. Inputs in the same cluster should exhibit a high similarity among them, and a high dissimilarity with inputs in other clusters. Besides, inputs labelled as outliers should exhibit a high dissimilarity with all the other inputs.

In this work diverse tests have been performed with DBSCAN to evaluate how many maps and which maps are pointed as outliers. The tests show a certain dependence with the configuration of DBSCAN, and the choice of the metric of similarity or distance. However, as later shown, the maps labelled as outlier in the most restrictive configurations (the lowest number of outliers) are also in the the outlier set of the less restrictive configurations.

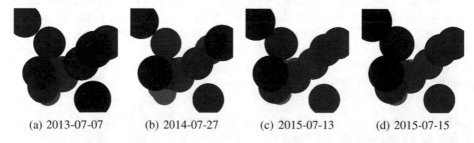

(a) 2013-07-07   (b) 2014-07-27   (c) 2015-07-13   (d) 2015-07-15

Fig. 3. The outlier maps when using DBSCAN with a configuration of a minimum of 400 points (maps) in the neighbourhood to be considered as a core point, and a minimum distance between two points to be in the same neighbourhood of 12. This configuration generates a single cluster with 4 outliers from the total of 3195 maps, 0.13%.

In Fig. 3, the four maps labelled as outlier by DBSCAN implementation with a configuration of a minimum of 400 points (maps) in the neighbourhood to be considered as a core point, and a minimum distance between two points to be in the same neighbourhood of 12 units in normalized distance are shown. As can be appreciated, the daily maps labelled as outlier correspond to extreme episodes of low-quality air, at the same time that high-quality air maps are not selected as outlier. The difference with an example of high-quality map is notable (see Fig. 2(b)). These four extreme maps are only the 0.13% of the total maps, thus they should be considered as the extreme cases of the period analysed.

Obviously the number of maps labelled as outlier strongly relies on the configuration of DBSCAN. Extreme cases, and specially low-quality air episodes, are very relevant when considering air quality, and how to find more or less outlier maps should be investigated. For this reason, the configuration of DBSCAN is tuned for finding few extreme outliers: minimum number of objects for a cluster is 400, and a minimum distance between two points to be in the same neighbourhood of 12 units in normalized distance. Visually, the maps labelled as outlier for this particular configuration correspond for very high-level of O_3 episodes (Fig. 3).

When modifying the configuration of DBSCAN using clusters of at least 400 points and a distance for core points of 10, instead for 12, the number of maps labelled as outlier increases from 4 to 15 (Fig. 4). All the maps labelled as outlier in the previous configuration (Fig. 3) are included in the new one (Fig. 4).

Other configurations lead to the absence of outliers. For example, when the minimum number of points in the cluster is 100, and the distance for core point is 12 or 15, no outliers are pointed; whereas when the distance for core point is 10, only one outlier is pointed: 2015-07-15 (Fig. 3(d)). If the number of core points goes down until 5, then 28 maps are labelled as outlier, a 0.9% of the total maps.

Robustness Against the Image Size. One of the open questions is the robustness of the proposed procedure for converting numerical information in visual one. For this reason the number of maps labelled as outlier is tested if the the configuration used in DBSCAN is a minimum of 400 points (maps) in the neighbourhood to be considered as a core point, a minimum distance between two points to be in the same neighbourhood of 12, and the maps are generated with alternative sizes for the images: 100×100 pixels, and also 20×20 pixels. In all the cases, the same four maps that are pointed for the configuration 50×50 pixels, are labelled as outlier for the alternative image size configurations (Fig. 3).

Conversely to the previous case, when using the configuration of minimum distance of 10, instead of 12, the number of outliers varies with the configuration of the image size. In comparison with the image size of 50×50 pixels, the use of a configuration with a shorter image, 20×20 pixels, produces the same number of outliers but eliminating the map corresponding to 2014-05-17 (Fig. 4(e)), and adding to the outlier set the map corresponding to 2014-07-26 (Fig. 5). In relation to the outlier set of 50×50 pixels configuration (Fig. 4), when using an image

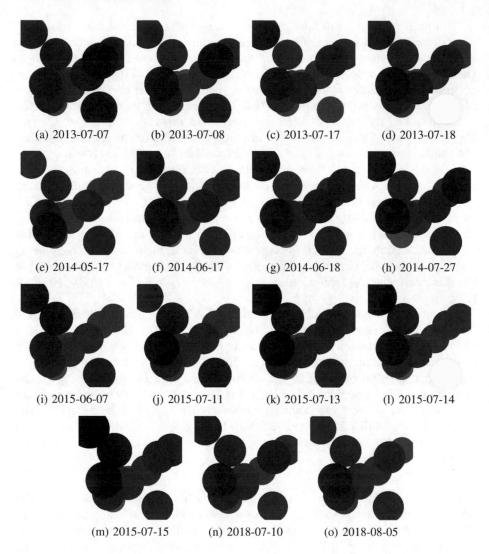

(a) 2013-07-07   (b) 2013-07-08   (c) 2013-07-17   (d) 2013-07-18

(e) 2014-05-17   (f) 2014-06-17   (g) 2014-06-18   (h) 2014-07-27

(i) 2015-06-07   (j) 2015-07-11   (k) 2015-07-13   (l) 2015-07-14

(m) 2015-07-15   (n) 2018-07-10   (o) 2018-08-05

Fig. 4. The outlier maps when using DBSCAN with a configuration of a minimum of 400 points (maps) in the neighbourhood to be considered as a core point, and a minimum distance between two points to be in the same neighbourhood of 10. This configuration generates a single cluster with 15 outliers from the total of 3195 maps, 0.47%.

size configuration of 100 × 100 pixels, 16 maps are pointed as outlier. The map corresponding to 2017-05-18 (Fig. 6) is incorporated to the outlier set of the configuration 50 × 50 (Fig. 4).

Finally, the configuration of a minimum number of points in the cluster of 100 and the distance for core point of 10 leads to a single outlier for image size

Fig. 5. Additional outlier map pointed when using a configuration of image size 20×20 pixels, instead of 50×50 pixels, 10 core points and a minimum of 400 points (maps) in the neighbourhood to be considered as a core point.

Fig. 6. Additional outlier map pointed when using a configuration of image size 100×100 pixels, instead of 50×50 pixels, 10 core points and a minimum of 400 points (maps) in the neighbourhood to be considered as a core point.

configurations of 100×100 pixels and 50×50 pixels: 2015-07-15 (Fig. 3(d)); whereas no outliers are pointed for 20×20 pixels image-size configuration.

As can be appreciated, the proposed approach offers a high robustness in relation to the size of the images. The number of and the maps pointed as outlier are relatively independent of the image-size configuration. This allows the generation of similar outliers set while handling input images of different sizes, even very reduced ones (20×20 pixels) avoiding large memory consumption and saving processing time.

3.2   Outliers Detection with DBSCAN and L^1-Norm

As previously mentioned, DBSCAN does not force to use a particular distance or similarity metric. In the previous tests, L^2-norm (Eq. 1) has been used to calculate the distances between the points. In this section, L^2-norm is replaced by L^1-norm (Eq. 2). L^1-norm is also known as Manhattan or city block distance.

In comparison with L^2-norm, the implementation of L^1-norm in DBSCAN, with a configuration of a minimum of 400 points (maps) in the neighbourhood to be considered as a core point and a minimum distance between two points to be in the same neighbourhood of 12, leads to a larger number of outliers. This configuration produces 12 outliers in L^1-norm (Figs. 4(a), (b), (c), (d), (e), (f), (g), (h), (i), (j), (k), (l), (m) and (n)), and only 4 in L^2-norm (Fig. 3).

Finally, for a configuration of a minimum of 400 points (maps) in the neighbourhood to be considered as a core point and a minimum distance between two points to be in the same neighbourhood of 10, up to 21 maps are labelled as outlier in L^1-norm. For the sake of the brevity, they are skipped.

In Fig. 7, the outliers pointed by DBSCAN with L^1-norm when the minimum number of points in the cluster is 100 and the distance for core point is 12 are shown. For the same configuration, L^2-norm does not point any outlier.

(a) 2013-07-07   (b) 2015-07-15

Fig. 7. The outlier maps when using DBSCAN with L^1-norm and a configuration of a minimum of 100 points (maps) in the neighbourhood to be considered as a core point, and a minimum distance between two points to be in the same neighbourhood of 12. This configuration generates 2 outliers from the total of 3195 maps, 0.06%.

Besides, when the minimum number of points in the cluster is 100 and the distance for core point is 10 points, the L^1-norm points four maps as outlier. They are the same that those obtained with L^2-norm and a configuration of a minimum of 400 points (maps) in the neighbourhood to be considered as a core point, and a minimum distance between two points to be in the same neighbourhood of 12 (Fig. 3).

As observed from this study for the same configuration of DBSCAN, L^1-norm tends to generate more outliers that L^2-norm.

4   Conclusions

In this work, the first steps to transform numerical information into visual, and then to use it with supervised and non-supervised algorithms have been presented. The case study corresponds to the Air-Quality Monitoring Network of Madrid, and concretely to the O_3 concentration for 12 monitoring stations. The relevance of the problem stems from the importance for human health in urban areas of the concentration of this and other pollutants, such as CO or NO_2. As part of the contribution of this work, a methodology for transforming the numerical values of a network of sensors into visual maps has been proposed.

Once the maps has been created, they are used in the usual activities of supervised and non-supervised algorithms. This includes the search of extreme examples, in this case extreme maps, with DBSCAN. These extreme maps can identify episodes of very low-quality air.

Alternative methodologies for transforming numerical information of a network of sensors into coloured maps opens the possibility of incorporating more information at the same time that using Deep Learning architectures over these maps. For example, following the proposed methodology with grey-scaled maps, the concentration of CO, NO_2 and O_3 can be used to form coloured maps, and then to search extreme episodes in these maps. Therefore, the evaluation of these methodologies are proposed as Future Work.

Acknowledgment. The research leading to these results has received funding by the Spanish Ministry of Economy and Competitiveness (MINECO) for funding support through the grant FPA2016-80994-C2-1-R, and "Unidad de Excelencia María de Maeztu": CIEMAT - FÍSICA DE PARTÍCULAS through the grant MDM-2015-0509.

References

1. Madrid air quality plan 2011–2015 (2012)
2. Open data Madrid, August 2018. https://datos.madrid.es/portal/site/egob
3. Alberdi Odriozola, J.C., Díaz Jiménez, J., Montero Rubio, J.C., Mirón Pérez, I.J., Pajares Ortíz, M.S., Ribera Rodrigues, P.: Air pollution and mortality in Madrid, Spain: a time-series analysis. Int. Arch. Occup. Environ. Health **71**(8), 543–549 (1998). https://doi.org/10.1007/s004200050321
4. Díaz, J., García, R., Ribera, P., Alberdi, J.C., Hernández, E., Pajares, M.S., Otero, A.: Modeling of air pollution and its relationship with mortality and morbidity in Madrid, Spain. Int. Arch. Occup. Environ. Health **72**(6), 366–376 (1999). https://doi.org/10.1007/s004200050388
5. Ester, M., Kriegel, H.P., Sander, J., Xu, X.: A density-based algorithm for discovering clusters a density-based algorithm for discovering clusters in large spatial databases with noise. In: Proceedings of the Second International Conference on Knowledge Discovery and Data Mining, KDD 1996, pp. 226–231. AAAI Press (1996)
6. Fréchet, M.: Sur quelques points du calcul fonctionnel. Rendiconti del Circolo Matematico di Palermo **22**, 1–47 (1906)
7. Han, J., Kamber, M., Pei, J.: Data Mining Concepts and Techniques, 3rd edn. Morgan Kaufmann Publishers, Waltham (2012)
8. Linares, C., Díaz, J., Tobías, A., Miguel, J.M.D., Otero, A.: Impact of urban air pollutants and noise levels over daily hospital admissions in children in Madrid: a time series analysis. Int. Arch. Occup. Environ. Health **79**(2), 143–152 (2006). https://doi.org/10.1007/s00420-005-0032-0
9. Méndez-Jiménez, I., Cárdenas-Montes, M.: Modelling and forecasting of the ^{222}Rn radiation level time series at the Canfranc underground laboratory. In: Hybrid Artificial Intelligent Systems - 13th International Conference, HAIS 2018, Oviedo, Spain, 20–22 June 2018. Lecture Notes in Computer Science, vol. 10870, pp. 158–170. Springer (2018). https://doi.org/10.1007/978-3-319-92639-1_14
10. Méndez-Jiménez, I., Cárdenas-Montes, M.: Time series decomposition for improving the forecasting performance of convolutional neural networks. In: Advances in Artificial Intelligence - 18th Conference of the Spanish Association for Artificial Intelligence, CAEPIA 2018, Granada, Spain, 23–26 October 2018. Lecture Notes in Computer Science, vol. 11160, pp. 87–97. Springer (2018). https://doi.org/10.1007/978-3-030-00374-6_9

11. Pedregosa, F., Varoquaux, G., Gramfort, A., Michel, V., Thirion, B., Grisel, O., Blondel, M., Prettenhofer, P., Weiss, R., Dubourg, V., Vanderplas, J., Passos, A., Cournapeau, D., Brucher, M., Perrot, M., Duchesnay, E.: Scikit-learn: machine learning in Python. J. Mach. Learn. Res. **12**, 2825–2830 (2011)
12. Simovici, D.A., Djeraba, C.: Mathematical Tools for Data Mining - Set Theory, Partial Orders, Combinatorics. Advanced Information and Knowledge Processing. Springer (2008). https://doi.org/10.1007/978-1-84800-201-2

A Novel Heuristic Approach for the Simultaneous Selection of the Optimal Clustering Method and Its Internal Parameters for Time Series Data

Adriana Navajas-Guerrero[1]([✉]), Diana Manjarres[1], Eva Portillo[2], and Itziar Landa-Torres[3]

[1] Tecnalia Research and Innovation,
Parque Tecnológico de Bizkaia, 48160 Derio, Spain
{adriana.navajas,diana.manjarres}@tecnalia.com
[2] Department of Automatic Control and System Engineering,
School of Engineering, University of the Basque Country, UPV/EHU,
48013 Bilbao, Spain
eva.portillo@ehu.eus
[3] Petronor Innovación S.L., 48550 Muskiz, Spain
itziar.landa@repsol.com

Abstract. Clustering methods have become popular in the last years due to the need of analyzing the high amount of collected data from different fields of knowledge. Nevertheless, the main drawback of clustering is the selection of the optimal method along with its internal parameters in an unsupervised environment. In the present paper, a novel heuristic approach based on the Harmony Search algorithm aided with a local search procedure is presented for simultaneously optimizing the best clustering algorithm (K-means, DBSCAN and Hierarchical clustering) and its optimal internal parameters based on the Silhouette index. Extensive simulation results show that the presented approach outperforms the standard clustering configurations and also other works in the literature in different Time Series and synthetic databases.

Keywords: Harmony Search · Clustering ·
Internal parameters configuration · Optimization ·
Time series clustering

1 Introduction

During last decades the amount of data collected from different fields (i.e. Manufacturing and Service Industry, Computer Science, Biology and Medicine, among others) has experimented an outstanding growth due to the deployment of new data warehouse technologies, along with the installation of advanced equipment and sensors. One of the most relevant challenges is to be capable of properly analyzing and interpretating such amount of data in order to infer valuable

© Springer Nature Switzerland AG 2020
F. Martínez Álvarez et al. (Eds.): SOCO 2019, AISC 950, pp. 179–189, 2020.
https://doi.org/10.1007/978-3-030-20055-8_17

information. In this context, the analysis of Time Series (TS), which represent a sequence of measurements of a collected variable over time, have attracted notable interest by the research community.

In this regard, clustering techniques have been widely employed to efficiently manage TS data and make groups (clusters) according to different similarity measures [1]. As widely known, these techniques are commonly applied to unsupervised learning approaches, i.e problems in which there is no knowledge about the real groups distribution. Within the large amount of TS clustering techniques proposed in the literature, authors in [2] identify four main clustering categories, such as: partitioning (i.e. k-Means [3] and K-medoids (PAM) [4]), hierarchical [5], density-based (i.e. DBSCAN [6]) and grid-based approaches (i.e. STING [7] and Wavecluster [8]).

Nevertheless, the problem of determining the optimum internal parameters of the clustering methods constitutes an important drawback for implementing them in real applications. Regarding K-means, it is required to know the number of clusters (K) in advance. Likewise, DBSCAN needs two parameters to be fixed (*min_samples* and *epsilon*). In order to face this problem, many works have upsurged in the literature to optimize the internal parameters for clustering methods. Some contributions are based on statistical methods and another approaches utilize optimization algorithms. For the former, authors in [9] present a method for obtaining the parameters (*min_samples*, *epsilon*) for DBSCAN by iteratively increasing the value of *epsilon* until the members of the clusters remain in the same cluster. Once it happens, the value of *min_samples* is automatically fixed to the number of k-neighbors of each cluster. Similarly, the work [10] proposes another methodology to compute *epsilon* and *min_samples* parameters. First the distance matrix for all the elements of the dataset is calculated and then the matrix is sorted in ascending order. The value of *epsilon* is calculated with the geometrical mean of each column. Then, *min_samples* parameter is obtained as the mean of the total number of neighbors for each point of the cluster.

For the latter, different optimization algorithms are employed in order to automatically obtain the optimum internal parameters, such as GA (Genetic Algorithms) [11], PSO (Particle Swarm Optimization) [12], HS (Harmony Search) [13], DE (Differential Evolution) [14], among others. Specifically, PSO is frequently used for determining the optimal number of clusters for K-means algorithm [15] by means of optimizing the Davies Bouldin Index (DB) [16], the ADDC (average distance of documents to the cluster centroid) [17], an squared error function [18], a specific fitness function [19], the Silhoutte Index [20], the Adjusted rand Index [21] or the Dunn Index [22], among others. In the same line of research, a recently proposed PSO algorithm for DBSCAN parameters' optimization can be found in [23]. It employs different metric functions depending on the type of problem as classification (supervised learning environment) or clustering (unsupervised approach) requiring different optimization process, and also includes penalty functions to minimize the amount of noise and control the number of obtained clusters. Likewise, several works employ DE in order to obtain the best internal parameter values for DBSCAN [24] and K-means [25]. Similarly, HS algorithm is also used for calculating the potential centroids in [26] for a K-means by optimizing the ADDC metric and its solution outperforms a GA, a GM (Grey Model) and a PSO approach.

At this point it is interesting to remark that the optimal selection of the clustering method and its internal parameters is nowadays a computationally expensive process and by our sake of knowledge there is no proposed technique yet that simultaneously copes up with both issues (i.e. the selection of the clustering method and the optimization of its internal parameters). This paper proposes a novel solution based on the Harmony Search algorithm for Optimal Clustering Configuration (named HSOCC hereafter) for simultaneously optimizing these two relevant aspects based on the widely known Silhouette index. Thus, the proposed solution is capable of determining the best clustering technique among the selected K-means, DBSCAN and Hierarchical clustering approaches, along with their best internal configuration parameters. In order to enhance the outcome of the HS algorithm, a novel local search procedure is also proposed. Simulation results evince the practical applicability of the proposed approach in different TS databases [27] for unsupervised learning problems and in synthetic datasets [28] in which Silhouette performance results are compared towards the implementation of all the presented clustering methods with their commonly known optimal internal parameters and towards the results obtained in [29]. The rest of this paper is organized as follows: Sect. 2 describes the HSOCC algorithm, whereas Sect. 3 presents the experimental results. Finally, Sect. 4 concludes the paper.

2   Proposed Harmony Search Algorithm for Optimal Clustering Configuration (HSOCC)

Harmony Search (HS) is a meta-heuristic algorithm which is based on the harmony improvisation of musicians when searching for the best combination of music pitches in order to obtain the most harmonious melody. The HS is a population-based algorithm that improves the fitness of the solution vector in an iterative fashion by applying several improvisation operators to a set of solutions, stored in the Harmony Memory (HM).

2.1   Encoding Solution

Before delving into the description of the steps of the proposed HSOCC, the encoding solution is presented. The HSOCC comprises five different notes which meanings are related to the employed clustering method. The identification of the clustering method is made by the last note of the harmony, i.e. 0 refers to DBSCAN, 1 to K-means and 2 to HAC. The remaining notes consider the internal parameters for each specific clustering method and are depicted below.

- DBSCAN: [*distance_metric, min_samples, epsilon*, 0, 0]

- K-means: $[0, K, 0, 0, 1]$
- HAC: $[distance\_metric, method, criterion, t, 2]$

The first note, named *distance_metric*, symbolizes the metric used by the clustering method: 0 for the Euclidean distance and 1 for the Dynamic Time Warping (DTW) metric [30]. Regarding DBSCAN, *min_samples* refers to the minimum number of neighbors a point should have so as to be included into a cluster and *epsilon* specifies how close points should be to each other to be considered a part of a cluster. In case of the K-means algorithm, the second note represents the internal parameter K, i.e. the number of clusters. Finally, for HAC clustering method, *method* denotes the linkage method used to calculate the distance between two clusters. Four different linkage methods are considered for both Euclidean and DTW metrics: 0 denotes the *Single* method (minimum distance strategy), 1 the *Complete* (maximum distance strategy), 2 the *Average* or UPGMA (unweighted average distance strategy) and 3 the *Weighted* or WPGMA (weighted average distance strategy) [31]. The third note for HAC clustering method represents the *criterion* that the algorithm uses to form the clusters: *distance* and *maxclust* are selected with 0 and 1 values, respectively. The former creates flat clusters so that the dendrogram is cut at the distance Y_{max}/t, where t defines the fraction in which Y_{max} is divided, and Y_{max} is the maximum distance between the ground and the last cluster. The latter, *maxclust*, forms flat clusters at a distance so that the number of clusters obtained corresponds to the value of the parameter t. Parameter t is defined in the fourth note and, as mentioned, takes different values depending on the *criterion* selected in the previous note.

2.2   Steps of the HSOCC Algorithm

The flow diagram of the proposed HSOCC algorithm is schematically shown in Fig. 1 and can be summarized in five steps: (i) initialization of the HM; (ii) improvisation of a new harmony; (iii) local search procedure; (iv) selection of the best harmonies and update the HM memory, and; (v) repeat until the maximum number of iterations is satisfied.

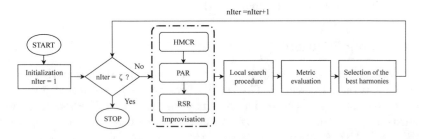

Fig. 1. Flow diagram of the proposed HSOCC algorithm.

Initialization. This step is only performed at the first iteration. Here the harmonies of the HM are created randomly among the feasible values of each note. As above mentioned, the last note of the harmony denotes the type of clustering method (DBSCAN, K-means or HAC) and the remaining notes have a different meaning based on this first note value (see Sect. 2.1).

The value for the *min_samples* parameter is randomly chosen from a range between $-30\%Ln(N) <$ min_samples $< 30\%Ln(N)$, being N the total number of samples in the dataset [32]. Similarly, the *epsilon* parameter is randomly selected from a range between the 30% of its optimal value calculated with the Elbow method [32]. In case the harmony corresponds to a K-means algorithm (i.e. last note to 1), the second note, which symbolizes the parameter K, is randomly selected from a range between 0 and N. Finally, for HAC solutions (i.e. last note to 2), the second note corresponds to the employed linkage method and its value is randomly chosen from 0 to 3. Regarding the third note, that refers to the criterion, two possible values can be selected (0 for *distance* and 1 for *maxclust*). When *distance* criterion is selected, the value for the t parameter (fourth note) is randomly selected from the range $\in [1.5, 1.75, \ldots, 3]$. In case *maxclust* criterion is selected the value for t is randomly chosen from a range between 0 and N.

Improvisation Procedure. In the improvisation procedure, a new Harmony Memory is generated by means of the application of three different probabilistic parameters:

- The Harmony Memory Considering Rate (HMCR) $\in [0,1]$ sets the probability that the new value for a note is drawn uniformly from the values of this same note in the same clustering methods.
- The Pitch Adjusting Rate (PAR) $\in [0,1]$ refers to the probability that the new value for a note is randomly taken from its neighborhood values. Table 1 depicts the criteria followed for the PAR implementation. It can be observed that the new value for the note when implementing PAR procedure depends on the clustering method and also on the internal parameter to which it refers.
- The Random Selection Rate (RSR) $\in [0,1]$ denotes the probability to take a random value for the new note from its doable values.

Local Search Procedure. Once the improvisation operators have been applied, the proposed HSOCC proceeds by performing a local search procedure every τ_{LS} iterations. This procedure aims at improving the fitness of the best candidate solutions of the HM. It is applied to half of the solutions from the HM and the implementation differs in case the note follows a continuous alphabet (range $\in \pm 20\%$) or a discrete one (new value ± 0.5 or 0.25):

1. For *distance_metric* the local search procedure chooses the metric (Euclidean or DTW) that obtains best fitness value.

Table 1. PAR implementation.

Clustering method	Note	Parameter	PAR
DBSCAN	2	min_samples	new value $\in \pm 20\%$
DBSCAN	3	epsilon	new value ± 0.5
K-means	2	K	new value $\in \pm 20\%$
HAC	2	method	new value $\in \{0, 1, 2, 3\}$
HAC	4	t (*distance* criterion)	new value ± 0.25
HAC	4	t (*maxclust* criterion)	new value $\in \pm 20\%$

2. For notes that the new value for PAR is between a range $\in \pm 20\%$ (see Table 1 *min_samples*, *K* and *t* for the *maxclust* criterion) the local search procedure randomly chooses 5 values among this range and keeps the value that achieves the best fitness.
3. For notes that the new value for PAR is generating by adding or substracting a quantity (i.e ± 0.5 or 0.25, see Table 1 *epsilon* and *t* for the *distance* criterion), the local search procedure tests the harmony by adding or substracting this quantity and keeps the value that achieves the best fitness.
4. For the note related to the method of the HAC algorithm, the value that renders the best performance among $\{0, 1, 2, 3\}$ is selected.

By means of this local search procedure the best solutions of the HM are substantially improved. A comparative analysis with regard to not applying this local search procedure is realized below in Sect. 3.

Metric Evaluation. The new generated candidate solutions are then evaluated and the Harmony Memory is updated based on the global Silhouette metric presented in Eq. 1 [33]. This metric evaluates the inter-cluster and intra-cluster distance which indicates the consistency of the formed clusters. In order to achieve a good cohesion between the members of a cluster, the intra-cluster distance should be low. By contrast, for a better differentiation between clusters, the inter-cluster distance should take a high value.

$$S_G = \frac{1}{C} \sum_{j=1}^{C} S_C(j), \quad where \quad S_C = \frac{1}{M} \sum_{i=1}^{M} s(i) \quad and \quad s(i) = \frac{b(i) - a(i)}{max\{a(i), b(i)\}}$$

(1)

where $a(i)$ is the average distance from a point i of a cluster and all the resting points in the same cluster, and $b(i)$ the average distance between i and all the points in the nearest cluster. S_G denotes the global Silhouette that is the average value of the Silhouette per cluster S_C, being C the number of clusters and M the number of points in cluster j. This global Silhouette metric S_G yields values between $[-1, 1]$, being 1 the best clustering distribution according intra and inter-cluster distances.

Sorting and Selection of the Best Harmonies. In this step the Harmony Memory is updated based on the S_G metric. To this end, only those harmonies improving the fitness with respect to those from the previous iteration are included in the next HM. That being so, the Harmony Memory is sorted in ascending order of the fitness values of its compounding melodies and the best ones are kept in the HM. This procedure is repeated until a fixed number of iterations τ is achieved.

3   Simulation Results

In order to assess the effectiveness of the proposed HSOCC algorithm, different computer simulations over synthetic scenarios from the UCR Time Series Classification Archive [27] (*Mixed Shapes Regular Train* (MS), *Medical Images* (MI), *50 Words* (50W), *Arrow Head* (AR), *ECGFive* (ECG), *Ham* (HAM) and *Sony AIBORobot* (SAR)) and from the UCI Irvine Machine Learning Repository [28] (*Iris* (IR), *Wine* (WN), *Glass* (GL) and *Haberman* (HB)) are presented below.

The experimental results are obtained with a number of iterations where the convergence is achieved, in this case 30 iterations, and a Harmony Memory of 50. Regarding the parameters setup, a preliminary simulation campaign has been conducted to choose the most effective configurations: PAR is set fixed to 0.3 whereas the values for both HMCR and RSR operators follow the Eq. 2 in which $OP_{\{HMCR\}_{init}}$ is set to 0.3, $OP_{\{HMCR\}_{end}}$ to 0.9, $OP_{\{RSR\}_{init}}$ to 0.05 and $OP_{\{RSR\}_{end}}$ to 0.3. Therefore, the values for HMCR and RSR gradually increase throughout the iterative process [34].

$$OP_{\{HMCR,RSR\}_i} \mathrel{+}= nIter \frac{(OP_{\{HMCR,RSR\}_{end}} - (OP_{\{HMCR,RSR\}_{init}})}{\tau} \quad (2)$$

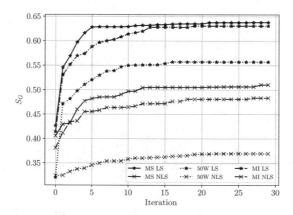

Fig. 2. Evolution of the averaged global Silhouette (S_G) along the iterative process.

Figure 2 depicts the evolution of the global Silhouette S_G averaged over 20 Monte Carlo along the iterations for the MS, 50W and MI datasets and with (LS) and without (NLS) the local search procedure. As it can be observed, the application of LS in all the selected datasets substantially improves the performance of S_G, especially during the first iterations. More specifically, Table 2 presents the (max/mean/std) statistics achieved averaged over the 20 Monte Carlo for the selected datasets with and without the local search process. The improvement obtained by the application of the local search is presented. Note that for the *50 Words* database the improvement slightly exceeds 50%.

The results of the HSOCC are then compared with those obtained by utilising the standard clustering configuration for DBSCAN, K-means and HAC (see Table 3). For K-means, the optimal K value is calculated by means of the widely known Elbow method [35]. For DBSCAN parameters' configuration, *epsilon* is set with the Elbow method and *min_samples* with the neperian logarithm of the database length. Finally, due to the lack of literature about fixing the optimal parameters' values for HAC, all the possible combinations are analyzed and the best values are presented.

Summarizing, in Table 3 it can be observed that HSOCC renders better S_G results than the standard clustering configuration with an improvement between 25% and 83.5%.

Table 2. Statistics of the averaged global Silhouette (S_G) with local (LS) and no local search (NLS) implementation for the selected datasets.

	S_G NLS			S_G LS			Improvement
Dataset	max	mean	std	max	mean	std	
MS	0.701	0.510	0.087	0.701	0.637	0.065	24.96%
MI	0.567	0.483	0.058	0.630	0.630	0	30.45%
50W	0.369	0.369	0	0.557	0.557	0	50.79%

Table 3. Comparative performance of the best results achieved by means of a standard clustering configuration and by the proposed HSOCC algorithm.

	S_G				Improvement
Dataset	DBSCAN	K-means	HAC	HSOCC	
MS	0.180	0.260	0.561	**0.702**	83.5 %
MI	0.278	0.228	0.429	**0.631**	47.02 %
50W	0.303	0.148	0.223	**0.557**	25.12 %

Furthermore, the proposed approach HSOCC is compared against the results achieved in [29]. In Tables 4 and 5 the global Silhouette index (S_G) for eight different databases is shown. As can be observed, our HSOCC algorithm outperforms the counterpart algorithms in six of the eight experiments.

Table 4. Comparative S_G performance of our proposed HSOCC algorithm and the results achieved in [29] for UCR databases.

Dataset	HSOCC	PAM	METACOC	Clues	PAMK	METACOC-K
SAR	**0.284**	0.078	0.105	0.062	0.158	0.166
ECG	**0.465**	0.403	0.403	0.209	0.403	0.404
AR	0.430	0.219	0.222	0.115	0.470	**0.746**
HAM	**0.306**	0.110	0.103	0.066	0.110	0.258

Table 5. Comparative S_G performance of our proposed HSOCC algorithm and the results achieved in [29] for UCI datasets.

Dataset	HSOCC	K-means	ACOC	PAM	METACOC	EMBIC	Clues	PAMK	METACOC-K
IR	**0.722**	0.562	0.562	0.564	0.562	0.707	0.557	0.709	0.711
WN	0.700	0.629	0.636	0.637	0.636	0.553	0.578	0.700	**0.728**
GL	**0.805**	0.537	0.317	0.281	0.250	0.033	0.129	0.675	0.697
HB	**0.660**	0.470	0.471	0.472	0.474	0.193	0.173	0.472	0.609

4 Conclusions and Future Work

This paper has presented a novel heuristic method based on the HS algorithm which is further aided by a LS procedure for the simultaneous optimization of the clustering method and its internal parameters. The proposed approach has been applied to eleven different datasets (i.e. TS and synthetic databases) and clearly outperforms the clustering methods with the standard parameters' configuration and related work of the literature. Current ongoing research is being conducted towards extending this work for optimizing the configuration parameters on a supervised learning environment using external indexes applied to TS databases.

Acknowledgments. This research has been supported by a TECNALIA Research and Innovation PhD Scholarship, ELKARTEK program (SENDANEU KK-2018/00032) and the HAZITEK program (DATALYSE ZL-2018/00765) of the Basque Government.

References

1. Lampert, T., Lafabregue, B., Serrette, N., Forestier, G., Crémilleux, B., Vrain, C., Gançarski, P., et al.: Constrained distance based clustering for time-series: a comparative and experimental study. Data Min. Knowl. Disc. **32**(6), 1663–1707 (2018)
2. Aghabozorgi, S., Shirkhorshidi, A.S., Wah, T.Y.: Time-series clustering-a decade review. Inf. Syst. **53**, 16–38 (2015)
3. MacQueen, J., et al.: Some methods for classification and analysis of multivariate observations. In: Proceedings of the Fifth Berkeley Symposium on Mathematical Statistics and Probability, Oakland, CA, USA, vol. 1, pp. 281–297 (1967)
4. Kaufman, L., Rousseeuw, P.J.: Finding Groups in Data: An Introduction to Cluster Analysis, vol. 344. Wiley, Hoboken (2009)

5. Karypis, G., Han, E.H., Kumar, V.: Chameleon: hierarchical clustering using dynamic modeling. Computer **32**(8), 68–75 (1999)
6. Ester, M., Kriegel, H.P., Sander, J., Xu, X., et al.: A density-based algorithm for discovering clusters in large spatial databases with noise. In: KDD, vol. 96, pp. 226–231 (1996)
7. Wang, W., Yang, J., Muntz, R., et al.: STING: a statistical information grid approach to spatial data mining. In: VLDB, vol. 97, pp. 186–195 (1997)
8. Sheikholeslami, G., Chatterjee, S., Zhang, A.: WaveCluster: a multi-resolution clustering approach for very large spatial databases. In: VLDB, vol. 98, pp. 428–439 (1998)
9. Thang, T.M., Kim, J.: The anomaly detection by using DBSCAN clustering with multiple parameters. In: 2011 International Conference on Information Science and Applications (ICISA), pp. 1–5. IEEE (2011)
10. Zhou, H., Wang, P., Li, H.: Research on adaptive parameters determination in DBSCAN algorithm. J. Inf. Comput. Sci. **9**(7), 1967–1973 (2012)
11. Davis, L.: Handbook of Genetic Algorithms (1991)
12. Shi, Y., et al.: Particle swarm optimization: developments, applications and resources. In: Proceedings of the 2001 Congress on Evolutionary Computation, vol. 1, pp. 81–86. IEEE (2001)
13. Geem, Z.W., Kim, J.H., Loganathan, G.V.: A new heuristic optimization algorithm: harmony search. Simulation **76**(2), 60–68 (2001)
14. Storn, R., Price, K.: Differential evolution-a simple and efficient heuristic for global optimization over continuous spaces. J. Global Optim. **11**(4), 341–359 (1997)
15. Sun, J., Chen, W., Fang, W., Wun, X., Xu, W.: Gene expression data analysis with the clustering method based on an improved quantum-behaved particle swarm optimization. Eng. Appl. Artif. Intell. **25**(2), 376–391 (2012)
16. Kao, Y., Lee, S.Y.: Combining k-means and particle swarm optimization for dynamic data clustering problems. In: 2009 IEEE International Conference on Intelligent Computing and Intelligent Systems, ICIS 2009, vol. 1, pp. 757–761. IEEE (2009)
17. Cui, X., Potok, T.E., Palathingal, P.: Document clustering using particle swarm optimization. In: Proceedings 2005 IEEE on Swarm Intelligence Symposium, SIS 2005, pp. 185–191. IEEE (2005)
18. Xiao, L., Shao, Z., Liu, G.: K-means algorithm based on particle swarm optimization algorithm for anomaly intrusion detection. In: 2006 The Sixth World Congress on Intelligent Control and Automation, WCICA 2006, vol. 2, pp. 5854–5858. IEEE (2006)
19. Ahmadyfard, A., Modares, H.: Combining PSO and k-means to enhance data clustering. In: 2008 International Symposium on Telecommunications, IST 2008, pp. 688–691. IEEE (2008)
20. Rousseeuw, P.J.: Silhouettes: a graphical aid to the interpretation and validation of cluster analysis. J. Comput. Appl. Math. **20**, 53–65 (1987)
21. Steinley, D.: Properties of the Hubert-Arable adjusted Rand index. Psychol. Methods **9**(3), 386 (2004)
22. Alswaitti, M., Albughdadi, M., Isa, N.A.M.: Density-based particle swarm optimization algorithm for data clustering. Expert Syst. Appl. **91**, 170–186 (2018)
23. Guan, C., Yuen, K.K.F., Coenen, F.: Particle swarm optimized density-based clustering and classification: supervised and unsupervised learning approaches. Swarm Evol. Comput. **44**, 876–896 (2018)
24. Karami, A., Johansson, R.: Choosing DBSCAN parameters automatically using differential evolution. Int. J. Comput. Appl. **91**(7), 1–11 (2014)

25. Cai, Z., Gong, W., Ling, C.X., Zhang, H.: A clustering-based differential evolution for global optimization. Appl. Soft Comput. **11**(1), 1363–1379 (2011)
26. Mahdavi, M., Abolhassani, H.: Harmony k-means algorithm for document clustering. Data Min. Knowl. Disc. **18**(3), 370–391 (2009)
27. Dau, H.A., Keogh, E., Kamgar, K., Yeh, C.C.M., Zhu, Y., Gharghabi, S., Ratanamahatana, C.A., Chen, Y., Hu, B., Begum, N., Bagnall, A., Mueen, A., Batista, G.: The UCR time series classification archive, October 2018. https:// www.cs.ucr.edu/~eamonn/time_series_data_2018/
28. Dua, D., Graff, C.: UCI machine learning repository (2017). http://archive.ics.uci. edu/ml
29. Menéndez, H.D., Otero, F.E., Camacho, D.: Medoid-based clustering using ant colony optimization. Swarm Intell. **10**(2), 123–145 (2016)
30. Liao, T.W.: Clustering of time series data–a survey. Pattern Recogn. **38**(11), 1857– 1874 (2005)
31. Müllner, D.: Modern hierarchical, agglomerative clustering algorithms. arXiv preprint arXiv:1109.2378 (2011)
32. Birant, D., Kut, A.: ST-DBSCAN: an algorithm for clustering spatial-temporal data. Data Knowl. Eng. **60**(1), 208–221 (2007)
33. Almeida, H., Guedes, D., Meira, W., Zaki, M.J.: Is there a best quality metric for graph clusters? In: Joint European Conference on Machine Learning and Knowledge Discovery in Databases, pp. 44–59. Springer (2011)
34. Kumar, V., Chhabra, J.K., Kumar, D.: Effect of harmony search parameters' variation in clustering. Procedia Technol. **6**, 265–274 (2012)
35. Kodinariya, T.M., Makwana, P.R.: Review on determining number of cluster in k-means clustering. Int. J. **1**(6), 90–95 (2013)

A Hybrid Approach for Short-Term NO$_2$ Forecasting: Case Study of Bay of Algeciras (Spain)

Steffanie Van Roode[1,2]($\boxtimes$), Juan Jesus Ruiz-Aguilar[1,3],
Javier González-Enrique[1,2], and Ignacio J. Turias[1,2]

[1] Intelligent Modelling of Systems Research Group,
Polytechnic School of Engineering, University of Cádiz, 11202 Algeciras, Spain
steffanie.vanroode@gm.uca.es
[2] Department of Computer Science Engineering,
Polytechnic School of Engineering, University of Cádiz, 11202 Algeciras, Spain
[3] Department of Civil and Industrial Engineering,
Polytechnic School of Engineering, University of Cádiz, 11202 Algeciras, Spain

Abstract. A hybrid model is proposed in this research in order to forecast concentration values of NO$_2$ with one-hour prediction horizon in the air quality monitoring network of the Bay of Algeciras area (Spain). Air pollution is an important environmental problem these days and it requires control. However, it is not an easy task. The main problem is that air pollution data series are non-lineal and non-stationary. Thus, techniques based on regression and simple models are not able to entirely capture the phenomenon behaviour. A LASSO-ANN hybrid model is proposed. The first step has been to predict the linear part of the time-series performing a least absolute shrinkage and selection operator (LASSO) model. Later, an artificial neural network (ANN) model has been performed to predict the residual sequence, the unexplained part of the LASSO model. The chaotic residual behaviour has been smoothed using an autoregressive moving window and applying the window median. The last step has been to aggregate the predicted NO$_2$ value and its predicted residual. The model has been validated and tested using cross-validation based on R correlation coefficient, MSE, MAE and d index of agreement, and also Friedman test and LSD test. In addition, the proposed approach has been compared to a simple ANN model. The results reveal that hybrid model presents a better performance than a multiple linear regression and also a simple ANN model. The main purpose is to develop a forecasting model capable of capturing the non-linear information of the variable and increase the accuracy of the outputs.

Keywords: Air pollution forecast · Artificial neural networks ·
Hybrid models · LASSO

1 Introduction

Air pollution has become one of the main environmental problems. The main cause is the growing human activities. These activities are linked to global warming and the greenhouse effect encouraging air pollution and vice versa. Its impact on human health

© Springer Nature Switzerland AG 2020
F. Martínez Álvarez et al. (Eds.): SOCO 2019, AISC 950, pp. 190–198, 2020.
https://doi.org/10.1007/978-3-030-20055-8_18

and natural environmental is widely known. Global organizations have established safety and prevention measures in terms of air quality. These guidelines set the production, reduction, control, measurements and information dissemination.

Thus, monitoring, estimation and prediction air pollution are very important tasks. Nevertheless, prediction is not an easy task. Air pollution data series are non-lineal and non-stationary. In general terms, models are sensitive to the given data and often underestimate pollution peaks, as techniques based on regression models that tend to minimize the error [1].

A hybrid model has been proposed to reduce the aforementioned problem. Many authors decide to develop hybrid [2, 3] and ensemble models [4] to predict air pollution. There is no unique solution and it is necessary to test the performance of different approaches in a certain location. The different experts are able to yield more information on the analysed variable than just one model. Other authors prefer to use additional exogenous information to the information of the NO$_2$ time series, such as meteorological variables [5, 6]. However, exogenous information is not always available. Here, we assume that only NO$_2$ time series are at our disposal and we aim to formulate a data-driven approach to produce one step ahead prediction.

On the one hand, the idea of this work is to predict the linear part of the variable with a linear regression model and, on the other hand, is to predict the residual (the unexplained part of the variable) with a non-linear model. The linear approach has been performed via a least absolute shrinkage and selection operator (LASSO) model. Multiple linear regression techniques are widely used to estimate and predict air pollution in the most cases for benchmark purposes. The non-linear approach has been performed using an artificial neural network (ANN). ANNs are very commonly used in this field [7, 8]. Also, different authors use hybrid models to predict chaotic time series in which the residual error is considered [9].

This paper is organized as follows: Sect. 2 describes area and data. Section 3 discusses the applied methodology. Section 4 describes the experimental procedure performed. Section 5 discusses the results obtained and, finally, Sect. 6 concludes the paper.

2 Area and Data Description

The study area is located in the Bay of Algeciras in the South of Spain. It constitutes a complex scenario in which different factors are involved. The area is formed of several cities and a total population of about 300,000 inhabitants. It is one of the most industrialized areas in the South and, besides, the most important trade port of the Mediterranean Sea is located in this area. As a result, the different towns that spread out over the Bay cause heavy traffic flows. The A-7 highway, which links the entire area, supports a daily intensity of 70,000 vehicles/day (DGT 2015).

Air pollution data have been provided by the Andalusian Government (research project TIN2014–58516–C2–2–R supported by MICINN Ministerio de Economía y Competitividad – Spain). The monitoring network of the Bay of Algeciras is composed for 14 air pollution monitoring stations distributed throughout the region. Data have been collected with one-hour sampling resolution and cover the period from 2010 to 2015. The data of the period from 2010 to 2014 have been used as training and validation subset and the data of the year 2015 have been used as testing subset in order to measure the generalization capability of the models.

3   Methods

This work proposes a two stage LASSO-ANN hybrid model in order to forecast concentration values of NO_2 with one-hour prediction horizon increasing the accuracy of the outputs. First, a linear approximation is performed with a regression LASSO model. Later, the residual time series obtained as the difference between the real value and the predicted value are modelled using an ANN. Finally, the models outputs are aggregated to compose a single output. In addition, the proposed approach has been compared to a simple ANN model used to predict NO_2 concentrations.

3.1   LASSO

A linear stage has been performed via a multiple regression procedure. The regression has been used over the training data set considering the data collected the previous eight hours. LASSO has been selected to be used in this work [10]. It is a singular multiple regression method that uses a regularization technique in order to avoid overfitting problems. LASSO reduces the degrees of freedom imposing a penalty term equals to the sum of the absolute values. The method solves an optimization problem, as shown in Eq. (1).

$$\min_\beta \left[\frac{1}{2N} \sum_{i=1}^{N} (y_i - \beta x_i)^2 + \lambda \alpha \sum_{j=1}^{k} |\beta_i| \right] \qquad (1)$$

where y_i is the dependent variable, $x_i = (x_{i1}, x_{i2}, \ldots, x_{ip})^T$ is the independent variable vector for $i = 1, 2, \ldots, N$ and $\beta = (\beta_1, \beta_2, \ldots, \beta_k)^T$ is the model coefficient vector for $j = 1, 2, \ldots, k$. Besides, the method performs a variable selection controlled by the hyperparameter λ. Higher the λ value, more coefficients are reduced to zero and removed from the model.

3.2   ANNs

A non-linear approach has been performed via a shallow artificial neural network (ANN). A shallow neural network is used to describe an ANN that usually have only one hidden layer. ANNs are widely used to perform any non-linear relationship between variables [11]. These are able to detect patterns, trends, and learn from data and their interactions with the environment.

The most used shallow ANN models are the Feedforward Multilayer Perceptron using backpropagation learning [12], as shown in Eq. (2).

$$Z_k = \sum_{j=1}^{0} w'_{kj} f\left(\sum_{i=1}^{n} w_{ji} x_i - \theta_j\right) - \theta'_k \qquad (2)$$

where x_i are the network inputs, y_i are the hidden layer outputs and z_k are the final layer outputs which are compared to the objective outputs. In addition, w_{ij} represent the hidden layer weights, θ_j their corresponding threshold, w'_{kj} the output layer weights and θ'_k their corresponding threshold.

4 Experimental Procedure

The aim of this work is to predict the NO$_2$ concentrations with one-hour prediction horizon in the different air pollution monitoring stations in the study area. A LASSO-ANN hybrid approach was used as prediction method. Also, a simple ANN approach was tested. The main points of the experimental procedure are described as follows:

1. A shallow ANN has been used to predict the NO$_2$ values in the monitoring stations. A backpropagation feedforward multilayer neural network (BPNN) has been used. For each station, the data collected during the previous 8 h have been used as model inputs. The net configuration was settled down to one only hidden layer from 1 to 20 units. Friedman and LSD tests have been performed to determine the optimal number of hidden units.
2. LASSO has been used to predict NO$_2$ values in the monitoring stations. This is a first linear approximation of the output value. For each station, the data recorded during the previous 8 h have been used as model inputs. The λ that presents the minimum *MSE* has been selected.
3. Errors or residuals between the real value and the LASSO predicted value are calculated. These errors have a non-linear and non-stationary behaviour. This chaotic behaviour has been smoothed using an autoregressive moving window and calculating its window median.
4. A shallow ANN has been used to predict the residuals (b). This is a second non-linear approximation of the output value. For each station, the errors calculated during the previous 8 h have been used as model inputs. The procedure has been carried out in a similar way as in the step 1.
5. The database has been divided into two subsets: a training-validation subset (data from 2010 to 2014) and a testing subset (data of year 2015). A 2-fold cross validation has been performed with the training-validation subset in order to determine the parameters of each model and then the results have been computed using the unseen test set in order to comparing the generalization results of the different models.
6. Finally, the outputs of both models are composed to a single output. NO$_2$ concentrations have been predicted for the whole year 2015 and the quality indexes R, *MSE*, *MAE* and d have been calculated in order to assess the generalization of the hybrid approach.

5 Results and Discussion

The results of the experimental procedure are presented in this section. As commented above, NO_2 value concentrations have been predicted using a shallow ANN. The records of the 8 previous hours have been used as input. The optimal number of hidden units of the model of each monitoring station have been determined using Friedman and LSD tests and following the criterion of Occam's razor. The performance index results are shown in Table (1). Then, the proposed two-stage approach has been carried out to predict a single output of the NO_2 concentration for each monitoring station.

Table 1. LASSO-ANN hybrid approach vs. LASSO and ANN models performance indexes for each monitoring station (test set = year 2015)

Station		ANN Global	Stage-1: LASSO	Stage-2: ANN	Proposed Hybrid LASSO + ANN
1	R	0.854	0.851	0.579	**0.873**
	MSE	151.048	154.062	39.474	**137.212**
	MAE	8.216	8.350	3.841	**7.043**
	D	0.916	0.916	0.702	**0.933**
2	R	0.869	0.868	0.593	**0.892**
	MSE	49.519	49.880	13.623	**42.085**
	MAE	4.483	4.582	2.207	**3.634**
	D	0.926	0.925	0.716	**0.943**
3	R	0.863	0.863	0.558	**0.886**
	MSE	26.306	26.337	6.804	**22.162**
	MAE	3.080	3.292	1.387	**2.459**
	d	0.925	0.921	0.724	**0.940**
4	R	0.856	0.854	0.615	**0.875**
	MSE	68.701	69.895	14.521	**62.710**
	MAE	4.619	4.771	2.082	**3.817**
	d	0.919	0.917	0.727	**0.934**
5	R	0.855	0.855	0.574	**0.878**
	MSE	54.490	54.684	12.901	**47.974**
	MAE	4.360	4.371	1.980	**3.748**
	d	0.916	0.917	0.708	**0.935**
6	R	0.849	0.846	0.582	**0.870**
	MSE	63.576	64.599	15.442	**57.459**
	MAE	4.352	4.393	1.999	**3.682**
	d	0.913	0.912	0.718	**0.931**
7	R	0.863	0.861	0.600	**0.883**
	MSE	87.745	88.476	19.588	**77.848**
	MAE	5.887	5.951	2.640	**4.982**
	d	0.922	0.922	0.728	**0.939**

(*continued*)

Table 1. (*continued*)

Station		ANN Global	Stage-1: LASSO	Stage-2: ANN	Proposed Hybrid LASSO + ANN
8	R	0.850	0.848	0.628	**0.874**
	MSE	95.035	96.370	21.333	**83.062**
	MAE	6.254	6.426	2.754	**5.280**
	d	0.915	0.913	0.755	**0.934**
9	R	0.884	0.884	0.636	**0.906**
	MSE	39.813	40.053	9.860	**33.357**
	MAE	3.506	3.540	1.621	**2.894**
	d	0.935	0.934	0.749	**0.951**
10	R	0.856	0.855	0.625	**0.878**
	MSE	63.139	63.691	14.772	**55.936**
	MAE	5.053	5.143	2.278	**4.240**
	d	0.919	0.917	0.743	**0.936**
11	R	0.856	0.855	0.611	**0.881**
	MSE	52.991	53.242	13.073	**45.184**
	MAE	4.440	4.501	2.205	**3.595**
	d	0.917	0.916	0.717	**0.937**
12	R	0.843	0.843	0.613	**0.874**
	MSE	107.535	107.729	27.309	**90.858**
	MAE	6.900	6.916	3.134	**5.700**
	d	0.909	0.909	0.735	**0.933**
13	R	0.871	0.870	0.634	**0.893**
	MSE	122.849	123.742	29.969	**104.931**
	MAE	7.131	7.251	3.274	**5.989**
	d	0.925	0.925	0.744	**0.944**
14	R	0.844	0.843	0.628	**0.872**
	MSE	45.341	45.587	10.356	**38.858**
	MAE	3.913	3.930	1.757	**3.227**
	d	0.909	0.908	0.748	**0.931**

At the first stage, LASSO based on lambda with minimum *MSE* have been used to calculate the NO$_2$ concentration of the next hour in each monitoring station. The values of the 8 previous hours (lagged) have been selected as inputs. Different sizes of the prediction window were tested but the performances were worst. Table (1) shows the performance indexes obtained in the first stage. The results show a good performance of the model. Table (2) shows the parameters B of the model. The parameters B represent the model weights where the first coefficient corresponds to the point in time $t - 7$ and the last corresponds to the instant t. In general terms, the variables that most influence are the measured value for the instant t, followed by the value for the instant $t - 1$ and the instant $t - 7$. The remaining coefficients hardly influence the prediction.

At the second stage, ANN based on backpropagation has been used to predict the residual of the LASSO prediction of the next hour in each monitoring station. Residuals have been smoothed in order to soften their chaotic behaviour. The errors of the 8 previous predictions have been used as input. Table (1) shows the performance indexes obtained in the second stage. The optimal number of hidden units have been selected as commented above.

Finally, the LASSO predicted NO_2 value and the predicted residual using an ANN model are added to compose a final output. Table (1) shows the final performance indexes and compares to the simple model results. For each station, the best model is marked in bold in Table (1). The LASSO-ANN hybrid model outperforms the simple LASSO model and the simple ANN model for all parameters in each station increasing the R value by four-five hundredths and decreasing the MSE by 20–25%. Simple LASSO and simple ANN present similar results. Figure (1) shows the measured values versus predicted values with LASSO model and LASSO-ANN hybrid model. The results of station 9 (best results) have been represented for any week of the year 2015. Models are able to fit adequately the real values adapting to the different peaks and valleys that the time series represents. As commented above, residuals have been smoothed before predicting since it is impossible to predict such chaotic variable. Autoregressive moving windows of different sizes have been used and the subset median has been applied. Larger the size of the windows, greater the smoothing of the subset. Finally, a three-hour autoregressive window has been used where values from $t - 1$ to $t + 1$ are considered. Figure (2) shows the real LASSO residual versus the smoothed LASSO residual and the predicted smoothed LASSO residual. The results of station 9 (best results) have been represented for the same previous period of time. Likewise, the ANN model fits properly the residuals.

Table 2. LASSO model coefficients (B) (test data = year 2015)

Station	t − 7	t − 6	t − 5	t − 4	t − 3	t − 2	t − 1	t
1	**0,089**	−0,012	−0,001	−0,012	−0,005	0,037	**−0,106**	**0,905**
2	**0,066**	0,001	0	−0,017	0,017	0,028	**−0,075**	**0,875**
3	0,020	−0,011	0,004	−0,002	0,013	0,037	**−0,118**	**0,934**
4	**0,081**	−0,017	0,001	−0,002	−0,010	0,009	**−0,149**	**0,966**
5	0,042	−0,003	−0,007	−0,006	0,012	0,007	**−0,172**	**0,994**
6	0,041	−0,009	−0,007	0,001	−0,013	0,026	**−0,130**	**0,944**
7	0,046	−0,011	−0,002	0,006	−0,012	0,028	**−0,097**	**0,922**
8	0,045	−0,001	0	0	0,005	0	**−0,058**	**0,877**
9	0,017	0,004	−0,013	0,003	−0,003	0,029	**−0,156**	**0,983**
10	**0,062**	0	0,002	−0,007	−0,006	0,032	**−0,130**	**0,925**
11	0,041	−0,013	0,015	−0,003	−0,010	0,049	**−0,088**	**0,874**
12	0,037	−0,016	0	0,004	0,013	0,034	**−0,077**	**0,860**
13	**0,055**	0,005	0,010	−0,001	−0,011	0,030	**−0,071**	**0,862**
14	**0,058**	−0,009	0	−0,002	−0,012	0,024	**−0,147**	**0,927**

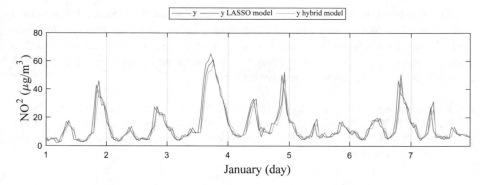

Fig. 1. Measured values (y) vs. predicted values (LASSO and LASSO-ANN model) at station 9 for a period of a week (January 2015).

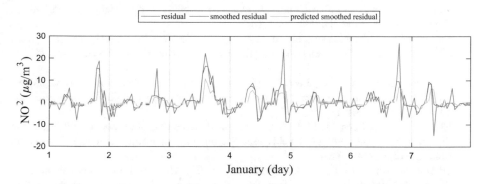

Fig. 2. Real LASSO residual and smoothed LASSO residual vs. predicted smoothed LASSO residual at station 9 for a week (January 2015). Residuals have been predicted with an ANN.

6 Conclusions

The following concluding remarks can be made from the results discussed above:

- LASSO model presents good prediction performances indexes predicting considering only the NO$_2$ values for the previous 8 h. The variables that most influence the prediction are the measured value for the instant t, followed by the value for the instant $t - 1$ and the instant $t - 7$.
- Hybrid approach outperforms simple models in all stations. The hybrid model increases the R value and d value by four-five hundredths and, decreases approximately the *MSE* and *MAE* by 20–25%.
- It is quite difficult to produce a reliable prediction of the LASSO residual because of its chaotic behaviour. Therefore, a smoothing procedure has been applied. The joint two-stage model is able to produce better prediction results combining both models.

The one-hour or two-hours predictions are very useful for citizens. People could make immediate decisions with a high level of security using for example the forecasting made by a website or an app. It would be very useful for daily activities such as running, walking, etc. On the other hand, we believe the air pollution prediction is such a critical task that any effort is really worth it.

Acknowledgements. This work is part of the coordinated research projects TIN2014-58516-C2-1-R and TIN2014-58516-C2-2-R supported by MICINN (Ministerio de Economía y Competitividad - Spain). Monitoring data have been kindly provided by the Environmental Agency of the Andalusian Government.

References

1. Gong, B., Ordieres-Meré, J.: Prediction of daily maximum ozone threshold exceedances by preprocessing and ensemble artificial intelligence techniques: case study of Hong Kong. Environ. Model Softw. **84**, 290–303 (2016)
2. Jiang, P., Li, C., Li, R., Yang, H.: An innovative hybrid air pollution early-warning system based on pollutants forecasting and Extenics evaluation. Knowl.-Based Syst. **164**, 174–192 (2018)
3. Cabaneros, S.M.S., Calautit, J.K.S., Hughes, B.R.: Hybrid artificial neural network models for effective prediction and mitigation of urban roadside NO2 pollution. Energy Procedia **142**, 3524–3530 (2017)
4. Cheng, S., Li, L., Chen, D., Li, J.: A neural network based ensemble approach for improving the accuracy of meteorological fields used for regional air quality modeling. J. Environ. Manag. **112**, 404–414 (2012)
5. González-Enrique, J., Turias, I.J., Ruiz-Aguilar, J.J., Moscoso-López, J.A., Franco, L.: Spatial and meteorological relevance in NO2 estimations. A case study in the Bay of Algeciras (Spain). Stoch. Environ. Res. Risk Assess. **33**, 801–815 (2019)
6. Zhang, Z., et al.: Evolution of surface O3 and PM2.5 concentrations and their relationships with meteorological conditions over the last decade in Beijing. Atmos. Environ. **108**, 67–75 (2015)
7. Turias, I.J., González, F.J., Martin, M.L., Galindo, P.L.: Prediction models of CO, SPM and SO2 concentrations in the Campo de Gibraltar Region, Spain: a multiple comparison strategy. Environ. Monit. Assess. **143**(1–3), 131–146 (2008)
8. Cisneros, M.A.P., Morán, L.J.M., Arreola, A.G.: Artificial neural networks applied in the forecast of pollutants into the Rio Santiago, based on the sample of a pollutant, by data fusion. In: 2016 IEEE 11th Conference on Industrial Electronics and Applications (ICIEA), pp. 1135–1138 (2016)
9. Ardalani-Farsa, M., Zolfaghari, S.: Chaotic time series prediction with residual analysis method using hybrid Elman-NARX neural networks. Neurocomputing **73**(13–15), 2540–2553 (2010)
10. Tibshirani, R.: Regression shrinkage and selection via the lasso: a retrospective. J. R. Stat. Soc. Ser. B Stat. Methodol. **73**(3), 273–282 (2011)
11. Hornik, K., Stinchcombe, M., White, H.: Multilayer feedforward networks are universal approximators. Neural Netw. **2**(5), 359–366 (1989)
12. Rumelhart, D., Hinton, G., Williams, R.: Learning internal representations by error propagation. In: Parallel Distributed Processing, pp. 318–362. MIT Press, Cambridge (1986)

Context-Aware Data Mining vs Classical Data Mining: Case Study on Predicting Soil Moisture

Anca Avram[(✉)], Oliviu Matei, Camelia-M. Pintea, Petrica C. Pop,
and Carmen Ana Anton

Technical University of Cluj-Napoca, North University Center at Baia Mare,
Baia Mare, Romania
{anca.avram,dr.camelia.pintea}@ieee.org, oliviu.matei@holisun.com,
{petrica.pop,carmen.anton}@cunbm.utcluj.ro

Abstract. The purpose of this article is to investigate whether or not including context data (context-awareness) in a classical data mining process would enhance the overall results. For that, the efficiency of the predictions was analyzed and compared: in a classical data mining process versus a context-aware data mining process. The two processes were applied on existing data collected from more weather stations to predict the future soil moisture. The classic data mining process considers historical data on soil moisture and temperature in a time interval, while the context-aware process also adds collected context information on air temperature for that location. The obtained results show advantages of CADM over classical DM.

Keywords: Context-aware data mining · Classical data mining ·
Soil moisture predictions · Machine learning

1 Introduction

Collecting and analyzing data is a process that will provide useful insights, no matter the domain. A way of finding the pattern in data and predicting the trends or customer preferences is provided by data mining. As Linoff and Berry specify in [8], "merely finding patterns is not enough. You must respond to the patterns and act on them, ultimately turning data into information, information into action, and action into value. This is the virtuous cycle of data mining in a nutshell."

According to FAO (Food and Agriculture Organization), agriculture is accountable on average, for usage of 70% of global freshwater withdrawals. Also, as specified in [3], in the last three decades, food production has increased by more than 100% and is estimated to grow another 60% until 2050. So, knowing in advance the most probable value for the soil moisture is an important information that can be later on used by farmers to plan the harvesting, planting or irrigation for the crops in an effective, water-saving manner.

© Springer Nature Switzerland AG 2020
F. Martínez Álvarez et al. (Eds.): SOCO 2019, AISC 950, pp. 199–208, 2020.
https://doi.org/10.1007/978-3-030-20055-8_19

The article wants to thoroughly compare the results of predicting the soil moisture for some specific location, in a classical data mining process versus a context-aware data mining process, using the same set of data and the same machine learning algorithms. Even though there have been some studies in the area [10,11], the novelty of this case-study is that it effectively goes through all the data mining and context-aware data mining stages, models a process for each approach and concludes on the best process and best machine learning algorithms based on the obtained results.

The introduction Sect. 1.1 presents the current status of the CADM related research, followed by the a theoretical overview of the data mining (DM), context-aware data mining (CADM) terms and stages in Sect. 1.2. Section 2 presents the setup of the data used in the experiment and the preliminary decisions taken prior implementing the processes. It is continued with the practical presentation of the DM and CADM processes and the summary of the results in Sect. 3. The conclusions in Sect. 4 present a short overview of the obtained results and what would be the possible future work in applying context-aware data mining.

1.1 Related Work

Lee and Chang in [7] define a context-aware system as a system "that actively and autonomously adapts its operation and provides the most appropriate functionalities to users, using the advantage of people's contextual information". As [6] specifies, the current status of the research in knowledge context is oriented towards capturing and using the context data for obtaining knowledge that can then be acted upon. Context that is valuable in one application, might not be of any use in another application.

Scholtze et al. [13] come with a new proposal of applying context awareness for creating context-sensitive decision support services in an eco-process engineering system setting. Matei et al. [9] made a proposal for an architecture for data mining in the context of Internet of things.

Scholze et al. in [12] want to demonstrate how context sensitivity can be used to realize a holistic solution for (self-)optimization of discrete flexible manufacturing systems. The solution is thought around a context extractor and an optimizer, that both are continuously learning and improving. Scholze proposes a similar approach also in [14]: having a Context Extractor module for dynamic context extraction, processing and storage; an Adapter module and a self-learning module.

In [15] is specified the fact that context sensitivity enhances the possibility to observe the parameters in which a system is operating, observations that afterwards will allow the system to dynamically adapt to varying conditions. Vajirkar et al. [16] come with a proposal regarding a context aware framework, that they use in a medical application, installed on wireless devices.

Matei et al. [11] presented a successful data mining system that predicted the moisture of the soil in real time, by using a data mining system, based on data gathered from ten weather stations. As a continuity in this area, Matei et al. [10] tried improving the accuracy of the predictions by using context data.

The current research adds to the existing research in [10]. While the previous research focused on exploring the impact of the context on the data mining processes and proving the fact that the existence and quality of the context is a factor that influences the data mining results, the goal of the current research is to go step by step through the entire data mining processes on a contiguous set of data, to conclude on the advantages of context-aware processes and on the best machine learning algorithms that could be used.

1.2   Context-Aware DM vs Classical DM Concepts

Currently data mining is widely recognized as the process of discovering relevant patterns in large sets of data, patterns that can be later used. In order to have these patterns discovered, intelligent methods are applied.

Data mining can be applied to any kind of data, as long as it delivers meaningful results to a target application [5]. Fayyad et al. in [4] present a bunch of real-world applications for data mining: in marketing (to predict customer behavior based on previous behavior), fraud detection, investments, in manufacturing, telecommunications, data cleaning.

Context Awareness is a concept that is relatively new and is more and more exploited in the knowledge management research. Dey has provided in [2] a definition of the context that is widely accepted: "Context is any information that can be used to characterize the situation of an entity. An entity is a person, place, or object that is considered relevant to the interaction between a user and an application, including the user and applications themselves."

Classical Data Mining. The basic stages of a classic data mining process are: (1) reading the data; (2) preprocessing the data; (3) applying a machine learning algorithm. Machine learning provides the technical basis of data mining [17].

Context Aware Data Mining. Context aware data mining (CADM) respects the same steps as classical data mining, just that it integrates some context data in the process (Fig. 1). One very important step in integrating the context is the context modeling. Defining the context so that context awareness can be applied is not an easy task to do.

Context information can be stored in various ways starting from key-value storage, to more complex data models (object-oriented model, graphical model, ontology model). The current trend in industry and also in academy is oriented mostly towards modeling the context using Ontology [6,12,14,16].

Lee et al. [7] have observed the fact that although there might be more types of context aware systems, generally, a context aware system has the following steps to process context awareness: acquisition of context information; storing acquired context information; controlling context abstraction level; utilizing the context information for services or applications.

Fig. 1. The CADM process [10]

2   Experimental Setup

The purpose of the experiments performed is to investigate whether or not including the context in a classical data mining process of predicting the soil moisture for a location, in a specific time frame, would improve the overall results. For that, two processes were modeled: a classical data mining process and a context-aware data mining process. The classic data mining process considers historical data on soil moisture and temperature in a time interval, while the context-aware process also adds collected information on air temperature for that location. In this specific case, the context-related data is the air temperature, which is expected to have some influence on the variations of the soil moisture. Its specific influence, however, is shown and explained in Table 4. For our experiments, no other context data was available.

2.1   Description (Reading) of the Existing Data

The data that served as a baseline for this study comes from more weather stations located in the Transylvanian plain. The data obtained from these weather stations was collected using the Smart Temp HOBO temperature sensors from the locations. Data was collected between 2008 and 2015. One stream of data (dtf files) provided the following information: timestamp; soil temperatures at three depths (10 cm, 30 cm and 50 cm): T10, T30, T50; moisture of the soil; precipitations (water content). Another data stream (hobo files) comes with information about the air temperature taken every 10 min for a location: timestamp; air temperature; other events related to the sensor.

2.2   Preprocessing the Data

The quality of the data is given by attributes like: completeness, consistency, accuracy, timeliness, believability and interpretability [5]. In order to reach a reliable data source in terms of quality, preprocessing techniques were applied.

Cleaning Up Noisy Data. Since data was saved from time to time, some periods overlapped, and the overlapping periods were removed from the analysis. This consisted in removing more than 30% of the data imported from the raw data files. Another step in cleaning noise data was removing invalid data (like −888 Fahrenheit degrees for example), indicating more probably an issue with the sensor and, in case it was left, a source of noise, that would affect the predictions.

Data Reduction and Storage. In order to avoid going back and forth between the previously created database and Rapid Miner,[1] data was imported in the local repository.

– One large repository with entire data after duplicates removal, in the format presented in Table 1.

Table 1. Large repository format

Location	Moisture	Date	T_10	T_30	T_50
Cojocna	0.3112	Thu Mar 05 15:00:00 EET 2009	36.554	35.298	34.027
Cojocna	0.3126	Thu Mar 05 15:10:00 EET 2009	36.554	35.298	34.124
Cojocna	0.3133	Thu Mar 05 15:20:00 EET 2009	36.554	35.346	34.174
Cojocna	0.3148	Thu Mar 05 15:30:00 EET 2009	36.554	35.346	34.223

– One compact repository containing the average (A) and standard deviation (SD) of moisture and temperature for each day as in Table 2.

Table 2. Compact repository format

Date	A M	SD M	A T_10	SD T_10	A T_30	SD T_30	A T_50	SD T_50
Mar 05 2009	0.3197	0.0024	36.6194	0.0513	35.5743	0.144	35.4418	0.7106
Mar 06 2009	0.3188	0.0005	36.8852	0.1204	36.4449	0.4084	38.0245	1.8358
Mar 07 2009	0.3243	0.0054	37.4792	0.1818	38.0693	0.2659	39.7873	1.0411
Mar 08 2009	0.3241	0.0006	38.1608	0.1671	38.6494	0.2686	38.0091	1.4811

Both repositories contained data from all the locations. The number of entries in the large repository, containing entries added every 10 min was 3.285.342 entries, while the compact version had 23.431 entries. The difference in size is given by the fact that instead of having 144 entries per day, the compact repository stores only one entry with the daily average.

---

[1] https://rapidminer.com/.

2.3   Preliminary Decisions Before Implementing the Data Mining Processes

Data Source. As a preliminary step, before proceeding with the machine learning processes, an analysis was performed in order to decide what repository is more suitable in the process of predicting the soil moisture for the next day. The results shown that the prediction accuracy was less than 20% and also the execution time was increased in case of the large repository, proving too much noise data, hence the decision was to use the compact repository as starting point for the machine learning process.

Locations and Time Intervals to Be Studied. The interval that was monitored was between 2008 and 2016, but the amount of data per location was not at all constant, hence analysis was performed on the distribution of data. Since the purpose of the research was to find whether data mining can be improved by adding context to the process, there were chosen three locations that proved to have both information on the sole temperatures. First a classical data mining process was applied for the locations and results were extracted. Second, the context (information on the air temperature) was added in the process and the new obtained results were observed. The locations chosen as objects of study are presented in Table 3.

Table 3. Locations analyzed

Location	Interval considered in the analysis
Dipsa	03/01/2008–31/12/2010
Cojocna	03/01/2008–31/12/2010
Braniste	01/01/2011–31/12/2011

Chosen Machine Learning Algorithms for Comparing Methods. The algorithms that were tested for the locations chosen, with both classical DM and CADM are: k-NN (k-nearest neighbor), NN (Neural Net), LPR (Local Polynomial Regression) and SVM (Support Vector Machine).

Tools and Techniques. From this point on, the Rapid Miner tool was used to design and "mine" the data to obtain predictions of the soil moisture for the following day. To predict the soil moisture, the time windowing technique was applied. By using this technique, more values in a time frame are considered to predict the next values. Windowing allows taking time series data and transform it into a cross-sectional format. As configuration for windowing, for most of the experiments the following combination was used: a window size of 7, a step size of 1 and 2 as value for the horizon. For location Cojocna, promising results were obtained with 10 for window size, 1 for step size and 4 for horizon.

3 Experiment Implementation and Results

3.1 Classical DM vs CADM Process Implementation

As a way of working, for each of the machine algorithms chosen to be analyzed, two processes were modeled in Rapid Miner, in order to obtain the prediction for the next day moisture: one classical process, having as input only the soil details known for the specified interval (moisture, temperatures per day); one context-aware process, considering also the piece of context known for that day and location related to the average air temperature. The general format for the classical data mining process applied on the data is as presented in Fig. 2.

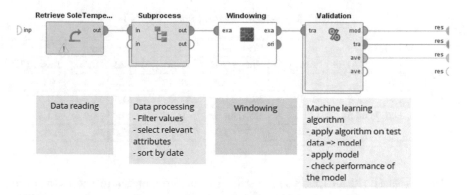

Fig. 2. Classical data mining process for predicting soil moisture

The general format for the context aware data mining process applied on the data is as presented in Fig. 3. The context is the air temperature stored for the considered location. In order to apply the machine learning algorithm, soil moisture data is combined and then the same machine learning algorithm is applied.

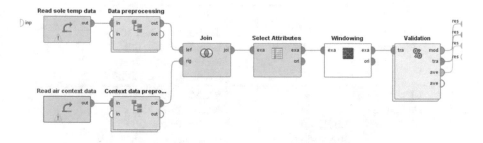

Fig. 3. Context-aware data mining process for predicting soil moisture

3.2   DM vs CADM Results

Analysis of the results was performed after applying the selected algorithms (with optimized parameters) for the specified locations and in the chosen time intervals. Table 4 presents the trend accuracy prediction for each location, for the classic data mining (DM) versus context-aware data mining (CADM) process. The improvements for each location are shown in columns `Impr`. Prediction trend accuracy measures the average of times a regression prediction was able to correctly predict the trend of the regression.

The best predictions for a specific location and a specific algorithm are emphasized.

Table 4. Prediction trend accuracy (%)

Algorithm	Dipsa			Braniste			Cojocna		
	DM	CADM	Impr. (%)	DM	CADM	Impr. (%)	DM	CADM	Impr. (%)
k-NN	70.4	80	**13.6**	72.6	87.2	**20.1**	68.3	65.2	−0.04
NN	44	57.8	**31.3**	35.1	41.2	**17.3**	41.7	53.7	**28.7**
LPR	45.9	46.8	**0.2**	47	46.4	−0.01	49.6	51.6	**0.04**
SVM	48.3	46.5	−0.04	58.7	52.9	−0.1	48.6	51	**0.5**

In 75% of the situations the contribution of the context in the data mining process brings an improvement of the trend accuracy (between 0.05 and 31%), while on the other situations the CADM results are comparable with the results obtained by applying classic DM on the data. Three out of fours algorithms show obvious improvements in the case of CADM relative to classical DM. The only exception is SVM, for which there are no global improvements because it constructs a hyperplane or set of hyperplanes in the solution space for separating the classes of solutions [1]. However, between those classes, there is no clear gap

Table 5. Parameters used when running the algorithms

k-NN	**NN**
k: 2	Training cycles: 30
Measure: Euclidean distance	Learning rate: 0.3
	Momentum: 0.2
	Error epsilon: 1.00E–04
LPR	**SVM**
Degree: 2	Kernel type: dot
Ridge factor: 1.00E–09	Kernel cache: 200
Measure: Euclidean distance	Converge epsilon: 0.001
Neighbourhood type: Fixed number	Max iterations: 100000
k: 5	
Smoothing kernel: Triweight	

(hyperplane). On the other hand, k-NN performs worse on CADM for Cojocna because the available data in this case is much less than for Dipsa and Braniste.

The predictions were obtained with the configurations of the algorithms presented in Table 5 and has been set based on a tuning process for obtaining the best mix of parameter values.

4   Conclusions

The research reported in this article is the starting point in developing an entire framework around context aware data mining for soil moisture. The framework could be organized around the algorithms that have proven to be the source for the best results.

The current article is a study case oriented towards measuring the efficiency of a context-aware process versus a classical data mining process for predicting the soil moisture in some specific locations. In order to apply the machine learning algorithms, data is gathered, cleaned, reduced and transformed. The same machine learning algorithms are applied in both processes and performance is observed after choosing the optimal parameter combination for the run. Some conclusions that could be drawn from the experiment:

- Local Polynomial Regression and Support Vector Machine algorithms, applied in the context-aware data mining process did not provide a significant improvement.
- The best results, in terms of percentage (15 to 31%) were obtained when the Neural Net algorithm on the data.
- Promising results were obtained also with the k-Nearest-Neighbor algorithm.
- Overall, the context-aware process leads to better predictions.

There might seem that there is no big difference between the two concepts. Still, the thing that would differentiate CADM from classical data mining techniques is the fact that the context is something that can be exploited on its own (for example for predicting temperatures in our experiment). Also the classical process could very much survive on its own, giving concluding results, but combining these two gives a better overview of the entire process and, as the article tries to emphasize, for most situations, better results.

As future development, more research on the quality and quantity of the context, respectively its correlations with the results need to be identified, measured along with their impact on influencing the prediction results.

References

1. Chang, C.-C., Hsu, C.-W., Lin, C.-J.: The analysis of decomposition methods for support vector machines. IEEE Trans. Neural Networks **11**(4), 1003–1008 (2000)
2. Dey, A.K.: Understanding and using context. Pers. Ubiquit. Comput. **5**(1), 4–7 (2001)

3. FAO: Water for sustainable food and agriculture. A report produced for the G20 presidency of Germany (2017)
4. Fayyad, U., Piatetsky-Shapiro, G., Smyth, P.: From data mining to knowledge discovery in databases. AI Mag. **17**(3), 37 (1996)
5. Han, J., Kamber, M., Pei, J.: Data mining: concepts and techniques. In: The Morgan Kaufmann Series in Data Management Systems, pp. 230–240 (2006)
6. Kotte, O., Elorriaga, A., Stokic, D., Scholze, S.: Context sensitive solution for collaborative decision making on quality assurance in software development processes. In: Intelligent Decision Technologies: Proceedings of the 5th KES International Conference on Intelligent Decision Technologies (KES-IDT 2013), vol. 255, p. 130. IOS Press (2013)
7. Lee, S., Chang, J., Lee, S.-G.: Survey and trend analysis of context-aware systems. Inf. An Int. Interdiscip. J. **14**(2), 527–548 (2011)
8. Lino, G.S., Berry, M.J.: Data Mining Techniques: For Marketing, Sales, and Customer Relationship Management. Wiley, New York (2011)
9. Matei, O., Anton, C., Bozga, A., Pop, P.: Multi-layered architecture for soil moisture prediction in agriculture 4.0. In: Proceedings of International Conference on Computers and Industrial Engineering, CIE, no. 2, p. 15 (2017)
10. Matei, O., Rusu, T., Bozga, A., Pop-Sitar, P., Anton, C.: Context-aware data mining: embedding external data sources in a machine learning process. In: International Conference on Hybrid Artificial Intelligence Systems, pp. 415–426. Springer (2017)
11. Matei, O., Rusu, T., Petrovan, A., Mihuţ, G.: A data mining system for real time soil moisture prediction. Procedia Eng. **181**, 837–844 (2017)
12. Scholze, S., Barata, J., Stokic, D.: Holistic context-sensitivity for run-time optimization of flexible manufacturing systems. Sensors **17**(3), 455 (2017)
13. Scholze, S., Kotte, O., Stokic, D., Grama, C.: Context-sensitive decision support for improved sustainability of product lifecycle. In: Intelligent Decision Technologies: Proceedings of the 5th KES International Conference on Intelligent Decision Technologies (KES-IDT 2013), vol. 255, p. 140. Courier Corporation (2013)
14. Scholze, S., Stokic, D., Kotte, O., Barata, J., Di Orio, G., Candido, G.: Reliable self-learning production systems based on context aware services. In: IEEE International Conference on Systems, Man, and Cybernetics (SMC), pp. 4872–4877. IEEE (2013)
15. Stokic, D., Scholze, S., Kotte, O.: Generic self-learning context sensitive solution for adaptive manufacturing and decision making systems. In: Proceedings of the ICONS14—The Ninth International Conference on Systems, Nice, pp. 23–27 (2014)
16. Vajirkar, P., Singh, S., Lee, Y.: Context-aware data mining framework for wireless medical application. In: International Conference on Database and Expert Systems Applications, pp. 381–391. Springer (2003)
17. Witten, I.H., Frank, E., Hall, M.A.: Data mining: practical machine learning tools and techniques. In: Morgan Kaufmann Series in Data Management Systems. Morgan Kaufmann, vol. 104, p. 113 (2005)

DTW as Alignment Function in the Context of Time Series Balancing

Enrique de la Cal[1,3]($\boxtimes$), José Ramón Villar[1,3], and Javier Sedano[2,3]

[1] EIMEM, University of Oviedo, Independencia 13, 33004 Oviedo, Spain
delacal@uniovi.es
[2] Instituto Tecnol'ogico de Castilla y León,
Polg. Ind. Villalonquejar, L'opez Bravo 70, 09001 Burgos, Spain
[3] Department of Civil Engineering, University of Burgos, Burgos, Spain

Abstract. Frequently, big data problems deal with Time Series (TS) datasets. This identification of positive events (falls, epilepsy crisis, stroke crisis, etc.) in this kind of problems needs a suitable number of the positive class TS. Consequently, the machine learning algorithms need to cope with the TS data balancing problem, which has not been studied in great depth in the literature. In one of our previous works we presented a TS extension (TS_SMOTE) of the well-known SMOTE algorithm for balancing datasets. The specification of this new algorithm consider two TS distance functions: (i) one used (as KNN distance measure) to choose the parents TS for each new synthetic TS, and (ii) a second distance function used as alignment function of the parents to obtain the new synthetic TS. As there are plenty of alignment functions in the literature here we have chosen the well-known DTW distance and compared against the EUCLIDEAN distance. Thus, current work presents an analysis of different parameters involved in our TS_SMOTE algorithm in order to prove the validity of DTW as good alignment measurement in the context of TS balancing. Some of the parameters analysed are: (a) criteria to choose the parents for new TS in the KNN algorithm, (b) the k parameter of KNN and finally (c) how the level of un alignment between the TS in the dataset affects the TS_SMOTE results. We can conclude that DTW has a very good behaviour obtaining new balanced datasets when the original TS are aligned but not quite good when the TS are unaligned. Also, the EUCLIDEAN distance get very poor results in all the experiments.

Keywords: Dataset balancing algorithms · SMOTE · Time Series · Time Series distances · Time Series alignment

1 Introduction

Technologies involved in BigData produce massive amounts of data that also have to be gathered. In the majority of the cases, these big data are in the form of Time Series (TS). Such cases include the management of sensory systems located on wearable devices, as in the problems of human activity recognition and

© Springer Nature Switzerland AG 2020
F. Martínez Álvarez et al. (Eds.): SOCO 2019, AISC 950, pp. 209–218, 2020.
https://doi.org/10.1007/978-3-030-20055-8_20

abnormal movement detection [13,14]. Furthermore, TS datasets have become multivariate datasets, which makes the data analysis even more complex.

In this context, when leaning models for the detection of some complex events, the problem of unbalanced data arises: there are many more TS segments belonging to normal class than those belonging to the abnormal class to be detected. For instance, in the problem of epilepsy seizure detection [13], the occurrence of a seizure might be once in a month or even less.

Most previous work on the dataset balancing problem is focused on classical datasets, where a sample includes an atomic value for each of the features. These balancing techniques rely on oversampling the minority class (mC) or undersampling the majority classes (MC); however, as long as oversampling does not produce information losses, it is preferred over under-sampling.

Some valid alternatives have also been published, coping with imbalanced problems specific algorithms [9], or proposing ensembles for the mC together with a kind of undersampling of the MC [7]. Examples of oversampling techniques include well-known algorithms such as SMOTE (Synthetic Minority Oversampling Technique [1,5]) and ADASYN (ADAptive SYNthetic Sampling, [8]) However, the problem of balancing TS datasets has received scant attention from the scientific community up to now. The published approaches focused on univariate TS problems [10], where the known data sequence labels are clearly biased to the MC. Therefore, the solutions rely on drawing new synthetic atomic values based on any of the above mentioned algorithms. Besides, in a multivariate TS dataset problem, each sample in the TS dataset includes a TS for each feature. Moreover, the sample is assigned a class, but also a TS is attached as the labelling TS for that sample. From now on, we consider all the TS features from a sample with the same length and sampling frequency; however, the variability in these factors needs further study.

In our previous work [4], a balancing algorithm for multivariate TS dataset called TS_SMOTE was presented.

This study includes a complete analysis of different parameters involved in our TS_SMOTE algorithm in order to prove the validity of DTW as good alignment measurement in the context of TS balancing. And the work is structured as follows. Next section outlines the proposal of elements to analyze in TS_SMOTE including the description of this algorithm. Experimentation and the discussion on the obtained results are coped in Sect. 3. Finally, the main conclusions are drawn in Sect. 4.

2   The Proposal

2.1   The TS_SMOTE Algorithm

The original SMOTE algorithm is an oversampling method [5], where each sample from the minority class is randomly combined with each of its nearest neighbors to balance the dataset. This method assumes a two-class problem: mC and MC stand for minority and majority class respectively. However, it can be easily extended to a multi-class problem [11].

In [4] an extension for multivariate TS of the original SMOTE (TS_SMOTE) was presented (see Algorithm 1).

In this proposal we still consider the two class problem {mC, MC}. So, Then, the mC needs to be SMOTEd provided that we are dealing with a TS dataset with different data sources (multivariate). The first difference is that the new samples are generated without mixing the data sources which is introduced as a different generation per source in line 1.

Then, two TSs are chosen from the population of TSs. The first one $\overline{ts}_i^s$ is selected randomly from the dataset. And a second one $\overline{ts}_k^s$ that must be in the nearest neighbor considering the original SMOTE algorithm and using the Euclidean distance function as neighborhood distance. Thirdly, the generation of the new TS is performed, which represents the main novelty of this proposal. To do so, a random merging value, α is also chosen. Merging these two TS can be done in the Euclidean space, as in SMOTE, this α is used for the weighted sum of the two TS or can be done considering the shape of the TS using a suitable distance functions like DTW. This issue will be tackled in the section.

Now on, the generic distance functions used as neighborhood and merging distances will be called **KnnD** and **MergD** respectively.

Algorithm 1. The extension of the SMOTE algorithm for dealing with TS datasets. Three parameters are needed: the minority class to smote (mC), the number of samples to generate(max_iter), and the TS dataset($\{\overline{TS}_s\}$).

TS_SMOTE(mC, max_iter, $\{\overline{TS}_s\}$)

```
 1: for s=1:S do
 2:     e ← 0
 3:     knn_distances ← knn(ts̄ˢ, KnnD)
 4:     sorted_knn_distances ← sort(knn_distances)
 5:     while e < max_iter do
 6:         i ← random[1 ... |mCₛ|] points to ts̄ᵢˢ
 7:         n_k ← |MCₛ| / |mCₛ| − 1
 8:         j ← 0
 9:         while e < max_iter and j < n_k do
10:             k ← random[1 ... |sorted_knn_distancesₖᴷ(tsᵢˢ)|] points to ts̄ₖˢ
11:             α = random[−1.0..1.0]
12:             new_tsˢ ← []
13:             {new_tsˢ, C̄ₛ} ← MergeD(tsᵢˢ, tsₖˢ)
14:             {TS̄ₛ} ← {TS̄ₛ} ∪ new_ts
15:             {C̄ₛ} ← {C̄ₛ} ∪ C̄ₙₑw
16:             {cₛ} ← {cₛ} ∪ mC
17:             j ← j + 1
18:         end while
19:         e ← e + 1
20:     end while
21: end for
22: function TimePoint(tᵢ, tₖ, ts̄ᵢ,ⱼˢ, ts̄ₖ,ⱼˢ, α)
23:     gpx ← min(tᵢ, tₖ) + (tᵢ + tₖ)/2
24:     d ← |ts̄ᵢ,ⱼ,ₜᵢˢ − ts̄ₖ,ⱼ,ₜₖˢ|
25:     m ← min(ts̄ᵢ,ⱼ,ₜᵢˢ, ts̄ₖ,ⱼ,ₜₖˢ) + (ts̄ᵢ,ⱼ,ₜᵢˢ + ts̄ₖ,ⱼ,ₜₖˢ)/2
26:     gpy = m + α × d return (gpx, gpy)
27: end function
```

2.2   The Distance Functions and Alignment Issues in Balancing TS Problems

Thus, at least two main concerns have to be solved in order to allow the original SMOTE algorithm to cope with TS datasets. The first one is related to the method for choosing the parents TS samples to mate, while the second focuses on the generation of the new TS sample.

The KNN Issues. When choosing the two TS samples that will be used for generating the new TS offspring, TS grouping according to some measurement should be considered. The original SMOTE algorithm randomly selects the closest parents in nearest neighborhood (using KNN algorithm) for mating among those belonging to the mC. In our proposal we have decided using only one alternative as Knn distance (**KnnD**): the EUCLIDEAN Distance (EU).

Concerning the selection of the parents, our proposal will follow the same strategy as the original SMOTE but with two variants: (i) the original random criteria (RANDOM), and (ii) the farthest parents (FAR). The reason to check the strategy (ii) is that we want to check the behavior of TS_SMOTE when there a high deviation in the distance of the TS.

The Merging Issues. The generation of a new TS when oversampling is not a simple task: as long as multivariate TS are considered, the new TS sample will need a TS for each one of the available features. For each feature to be generated, a combination of the parents' same feature should be performed. Furthermore, the combination must be coherent for all the features considered as a single sample. Finally, the TS class has to be generated as well, which is much of a compromise.

We've found out that merging the TS in the Euclidean space might lead to unsuitable outcomes. Therefore, we propose the use of shape factors in the merging of TS. The underlying idea is to find corresponding points in the two TS, merging in the Euclidean space in the related intervals.

For each input feature we apply as **MergD** by default the distance DTW [2,3] to obtain the sequence of matching pairs, each pair is the corresponding timestamps from each of the two TS to merge. For each pair a new value is estimated as shown in function *TimePoint*. Once a TS is obtained for each feature, the class TS is also computed and the TS dataset and the class data set are updated.

Figure 1 graphically explains how to compute each of the time points to add to each TS. Besides, the EUCLIDEAN distance will be used as **MergD** in order to compare its performance against the DTW distance. So, considering the $KnnD = EU$ and $MergD = EU or DTW$ we will define two TS_SMOTE versions: TS_SMOTE EU-EU and EU-DTW.

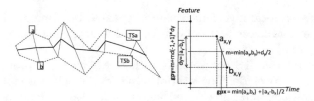

Fig. 1. Random time point proposal: DTW distance example (Left), Time Point calculation proposal (Right).

2.3 Building an Unaligned Dataset

Frequently datasets come from public/private repositories or have been pre-processed, so it's normal they are aligned. Thus, this is not a desirable issue since that doesn't allow to treat the data as if they were gathered from a real source. The dataset on epilepsy seizures that is described in Sect. 3.1 comes from a public repository of our servers, and we gathered these data with very strict protocol of timing (they are realistic epilepsy seizures, not real), so they were aligned in their origin. In this section we propose a very simple method to misalign a TS dataset in order to maximize the average distance. Let's consider that the TS we are going to tackle belong to the mC class, so each TS will have a part of the timestamp labels representing the minority class (Fall, Epilepsy crisis, Stroke, ...) and the remaining timestamp labels of the TS will belong to the majority class (MC). So, the proposal is to arrange all the TS we want to misalign in a timeline of length of the longest TS, shifting each TS in order to get the same distance between each pair of contiguous TS (see Fig. 2).

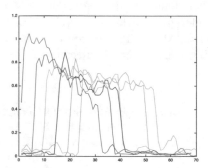

Fig. 2. Misaligned TS after shifting each one in order to get the same distance between each pair of contiguous TS.

3   Experiments and Results

3.1   Materials and Methods

For this experimentation, a real world TS dataset obtained from the simulation of epileptic seizures is used; this dataset is publicly available at [12,13]. In this study, we focus on focal myoclonic seizure, repeated bursting movements of one limb, the upper and lower limbs of one body side or a combination of limb and facial movements.

The above referred TS dataset was gathered following a previously defined and very strict protocol, defining a set of activities, namely, the simulation of the epileptic convulsions and three activities: running, sawing and walking, either gesturing while walking slowly or normal walking at different paces. A wearable triaxial accelerometer sensor (3DACM) included in a bracelet placed on the affected wrist measured the participant movements.

The bracelets have wireless data sampling capabilities at a rate of 16 Hz, having the 3DACM a range of $2 \times g$. Up to 6 healthy participants, all of whom remained anonymous, successfully completed this experiment, each running 10 trials of each activity. The ages of the participants ranged from 22 to 47, with four participants of around 40 years old. One participant out of six was female, and the eldest was left-handed. An identification number was given to each Time Series (TS), including information fields on participant ID, the number of trials, the activity, etc.

The acceleration has been filtered and processed, becoming into a three variable TS dataset: the Signal Magnitude Area (SMA), the Amount of Movement (AoM) and the Time between Peaks (TbP). The complete pre-processing have been described in [13].

This TS dataset, consisting on TS samples of three TS each, SMA, AoM and TbP, $\{\overline{TS}_s\}$, with the label for each activity $\{c_s\}$ for each TS and with each timestamp label $\{\overline{C}_s\}$, has been used in this experimentation.

To select the number of TS samples to be added to the dataset, the following criteria was used. In an imbalanced dataset, there exists approximately $R = 3$ times more examples belonging to the MC (NO_EPILEPSY) class than to the mC (EPILEPSY) class for the s data source (see Table 1). So, balancing the number of samples for both classes means injecting $(R1) \times |mC_s|$ new TS samples belonging to the mC.

Table 1. Number of TSs for each subject from the UNIOVI-Epilepsy dataset. Dataset column refers to the dataset: EPILEPSY and NO EPILEPSY (ADL) refer to the TSs from the UNIOVI-Epilepsy dataset, and TS_SMOTE refer to the new synthetic TSs computed from the EPILEPSY original dataset after running the TS_SMOTE algorithm, Size is the number of TSs in the corresponding dataset and R stands for Imbalance Ratio.

Dataset	Subject											
	1		2		3		4		5		6	
	Size	R	Size	R	Size	R	Size	R	Size	R	Size	R
EPILEPSY	15	2.47	10	4.7	10	2.47	10	3.1	13	2.46	10	3.0
NO EPILEPSY	37		47		31		31		32		30	
TS_SMOTE	22	–	37	–	21	–	21	–	11	–	20	–

Finally, the goal is to determine if new balanced datasets obtained with TS_SMOTE with all the parameters combinations for both datasets (ALIG and UNALIG) keep the statistical distribution of the original dataset ALIG or UNALIG. The way to compare the distribution of the ORIGINAL and the SMOTEd datasets will based in the correlation between the features and the class for each TS in each dataset. The correlation shows whether the relationships between the features and the class remain the same after the new synthetic TS samples have been introduced. In order to assess the similarity between the distribution of the original TS dataset and the balanced TS dataset, we made use of the Pearson Correlation ($\rho_{X,Y}$) and the Mutual Information ($MI(X,Y)$).

3.2 Numerical Results

As it was stated in Sect. 2.2 we have defined two versions of TS_SMOTE: EU-EU and EU-DTW. And the parameters considered in order to evaluate TS_SMOTE are:

- KnnSelec: a criteria to choose the parents for new TS in the KNN algorithm (RANDOM or FARHEST).
- K: the k parameter of KNN algorithm (5, 7 and 9).

In addition, let's remember that in Sect. 2.3 it was defined a method to misalign datasets so as the original dataset (ALIG) considered here is aligned, we have obtained a second dataset with the same TS but unaligned (UNALIG).

Therefore, both versions of TS_SMOTE have been run on both datasets (ALIG and UNALIG) considering all the combinations of the above parameters.

The correlation measurements, ρ and MI, have been calculated between each feature and the class for each TS, for both the ORIG and the SMOTEd. For the sake of space, only the boxplots obtained for Ts of the participant number 1 concerning ρ is depicted in Fig. 3 with Knn criteria RANDOM, K − 5 and taking as original imbalanced dataset ALIG. If we compare each pair of boxplots (features SMA, TbP and AoM from left to right for each subpicture), ORIGINAL-* and SMOTE-*, for the four subpictures we can state visually that SMOTed results are very similar to the ORIGINAL ones. May be, the So, we can say results show that both versions TS_SMOTE EU-EU and EU-DTW generate a TS dataset rather similar to the original one in terms of the correlation between each feature and the class.

Thus, we have computed the Wilcoxon signed-rank test (see Table 2) to check if the Pearson and Mutual Information correlation values from the ORIG and from the SMOTE datasets belong to the same distribution; clearly, there is no evidence against the null hypothesis of the two series belonging to the same distribution of the cases studied for EU-DTW version. Only three out of the 36 Wilcoxon tests ORIG-EU-DTW using Pearson as well as Mutual Information reject the null hypothesis. However, in the case of ORIG-EU-EU 18 out of 36 Wilcoxon tests reject the null hypothesis. Therefore, the behavior of the TS_SMOTE EU-DTW seems to be valid to calculate new samples in order to balance the TS dataset and keeps the distribution of the original dataset better than TS_SMOTE EU-EU for this combination of parameters.

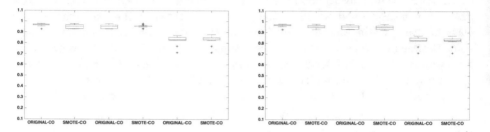

Fig. 3. Boxplot of the Pearson Correlation for participant 1: Each pair ORIGINAL-SMOTE corresponds to the following features from left to right: SMA, TbP and AoM, with Knn criteria RANDOM, k = 5 and taking as original imbalanced dataset ALIG. (left) $\rho_{X,Y}$ EU-EU, (right) EU-DTW $\rho_{X,Y}$

A complete analysis of both versions of TS_SMOTE for all the combinations of parameters is included in Table 3. For the sake of shortness this table contains the number of accepted Wilcoxon null hypothesis for each set of parameters.

Table 2. Wilcoxon signed-rank test P-values at a significance level of 0.05. The null hypothesis ($N0$) is that the data obtained for the $\rho(feature, class)$ and $MI(feature, class)$ calculated for the ORIG dataset and for the SMT dataset (computed with the two variants, EU-EU and EU-DTW) belong to the same distribution. The parameter k in KNN strategy to select parents for SMOTE Algorithm is 5, and the knn criteria is RANDOM.

Part.	EU-EU (ρ)			EU-EU(MI)			EU-DTW(ρ)			EU-DTW(MI)		
	SMA	AoM	TbP	SMA	AoM	TbP	SMA	AoM	TbP	SMA	AoM	TbP
1	0.0037	0.7519	0.2664	0.2017	0.0128	0.3750	0.0051	0.8004	0.8477	0.4883	0.3254	0.9275
2	0.0004	0.0183	0.8241	0.0410	0.0242	0.7913	0.2893	0.5498	0.0523	0.9414	0.6823	0.1503
3	0.1749	0.0014	0.0033	0.4008	0.0907	0.0368	0.4209	0.6597	0.1670	0.9414	0.6823	0.1503
4	0.0272	0.0033	0.0459	0.0620	0.0142	0.0077	0.7498	0.3867	0.0112	0.2298	0.9394	0.1071
5	0.0028	0.0396	0.1525	0.9599	0.5469	0.8018	0.0141	0.1761	0.1394	0.2184	0.2539	0.5391
6	0.0127	0.9126	0.7419	0.0845	0.4596	0.0343	0.1841	0.6061	0.5423	0.4499	0.7304	0.1739
$N0$ = true	7/18			11/18			15/18			18/18		

Finally, the numerical results comparing the statistical distribution of the obtained balanced datasets with the original dataset (ALIG) show that EU-DTW outperform clearly EU-EU in all the cases, but the results for the UNALIG dataset are not so good as we expected since DTW may be is not the suitable alignment function.

Table 3. Resume of Wilcoxon signed-rank test P-values at a significance level of 0.05 for the comparison of the distribution of the ORIGINAL datasets and the two TS_SMOTE variants: EU-EU and EU-DTW. This table includes the sum of accepted null hypothesis ($N0$) obtained with each combination of parameters: KnnSelect (KNN parents selection criteria in TS_SMOTE): RANDOM or FARHEST, Correlation measure: ρ or MI, Aligning of the TS: ALIG or UNALIG. The block TOTAL represents the percentage of accepted null hypothesis obtained with ρ or MI respect the to the total (36 per each combination of parameters).

| | | RANDOM | | | | FARHEST | | | | TOTAL | | | |
| | | ALIG | | UNALIG | | ALIG | | UNALIG | | RANDOM | | FARHEST | |
	k	ρ	MI	ρ	MI	ρ	MI	ρ	MI	ALIG	UNALIG	ALIG	UNALIG
EU-EU	5	7	11	5	13	5	7	1	3	50.0	50.0	33.3	11.1
EU-DTW	5	15	18	7	13	14	16	5	9	91.7	55.6	83.3	38.9
EU-EU	7	11	13	3	11	6	9	0	1	66.7	38.9	41.7	2.8
EU-DTW	7	15	18	5	12	10	15	3	9	91.7	47.2	69.4	33.3
EU-EU	9	11	13	3	6	7	7	2	0	66.7	25.0	38.9	5.6
EU-DTW	9	15	18	6	9	11	16	2	4	91.7	41.7	75.0	16.7

4   Conclusions

This research presented an analysis of the DTW function as alignment measurement in the context of TS balancing using our TS_SMOTE algorithm. As the original dataset (ALIG) used in the experiment section only includes aligned TS, a simple method was defined in order to obtain an unaligned TS dataset (UNALIG) as if it would be gathered from a real source.

As, TS_SMOTE algorithm tackles two distance functions: (i) the KNN distance (KnnD) function to choose the parents for each new synthetic TS, and (ii) the merging function(MergD) to combine the parents to obtain the new synthetic TS, DTW and EUCLIDEAN distances have been analysed as alternatives for (ii) and for the sake of space EUCLIDEAN (EU) distance has been the one only function analyzed as KNN distance function (i). Hence TS_SMOTE has two versions considering two combinations of the EUCLIDEAN and DTW distances for both kinds of distance functions (KnnD-MergD): TS_SMOTE EU-EU and TS_SMOTE EU-DTW.

Besides the parameters analysed for the KNN algorithm were: (a) criteria to choose the parents for new TS in the KNN algorithm (RANDOM or FARHEST) and (b) the k parameter of KNN algorithm (5, 7 and 9).

Therefore, both versions of TS_SMOTE have been run on both datasets (ALIG and UNALIG) considering all the combinations of the above parameters. And the numerical results comparing the statistical distribution of the obtained balanced datasets with the original dataset (ALIG) show that EU-DTW outperform clearly EU-EU in all the cases, but the results for the UNALIG dataset are not so good as we expected since DTW may be is not the suitable alignment function.

We can conclude that alignment is a critical factor in TS balancing, so it's necessary to consider specific alignment functions independent of shifting like

OSB, CDTW and DTW-D [6]. Also, an extension of the experiments to other multivariate problems is necessary.

Acknowledgment. This research has been funded by the Spanish Ministry of Economy, Industry and Competitiveness (MINECO), under grant TIN2017-84804-R.

References

1. Batista, G., Prati, R., Monard, M.: A study of the behavior of several methods for balancing machine learning training data. SIGKDD Explor. **6**, 20–29 (2004)
2. Berndt, D.J., Clifford, J.: Using dynamic time warping to find patterns in time series. In: Proceedings of the 3rd International Conference on Knowledge Discovery and Data Mining, AAAIWS 1994, pp. 359–370. AAAI Press (1994)
3. de la Cal, E., Villar, J., Vergara, P., Sedano, J., Herrero, A.: A smote extension for balancing multivariate epilepsy-related time series datasets. In: Proceedings of 12th International Conference on Soft Computing Models in Industrial and Environmental Applications (SOCO 2017), pp. 439–448 (2017)
4. de la Cal, E.A., Villar, J.R., Vergara, P.M., Herrero, A., Sedano, J.: Design issues in time series dataset balancing algorithms. Neural Comput. Appl. (2019). www.scopus.com
5. Chawla, N.V., Bowyer, K.W., Hall, L.O., Kegelmeyer, W.P.: SMOTE: synthetic minority over-sampling technique. J. Artif. Intell. Res. **16**, 321–357 (2002)
6. Folgado, D., Barandas, M., Matias, R., Martins, R., Carvalho, M., Gamboa, H.: Time alignment measurement for time series. Pattern Recogn. **81**, 268–279 (2018)
7. Galar, M., Fernández, A., Barrenechea, E., Herrera, F.: EUSBoost: enhancing ensembles for highly imbalanced data-sets by evolutionary undersampling. Pattern Recogn. **46**(12), 3460–3471 (2013)
8. He, H., Bai, Y., Garcia, E., Li, S., et al.: ADASYN: adaptive synthetic sampling approach for imbalanced learning. In: 2008 IEEE International Joint Conference on Neural Networks (IEEE World Congress on Computational Intelligence), JCNN 2008, pp. 1322–1328. IEEE (2008)
9. López, V., Fernández, A., del Jesus, M., Herrera, F.: A hierarchical genetic fuzzy system based on genetic programming for addressing classification with highly imbalanced and borderline data-sets. Knowl.-Based Syst. **38**, 85–104 (2013)
10. Moses, D., et al.: A survey of data mining algorithms used in cardiovascular disease diagnosis from multi-lead ECG data. Kuwait J. Sci. **42**(2) (2015)
11. Sáez, J.A., Krawczyk, B., Woźniak, M.: Analyzing the oversampling of different classes and types of examples in multi-class imbalanced datasets. Pattern Recogn. **57**, 164–178 (2016)
12. Villar, J.R.: Researcher's web page (2017). http://www.di.uniovi.es/~villar
13. Villar, J.R., Vergara, P., Menéndez, M., de la Cal, E., González, V.M., Sedano, J.: Generalized models for the classification of abnormal movements in daily life and its applicability to epilepsy convulsion recognition. Int. J. Neural Syst. **26**(6), 1650037 (2016)
14. Villar, J.R., González, S., Sedano, J., Chira, C., Trejo-Gabriel-Galán, J.M.: Improving human activity recognition and its application in early stroke diagnosis. Int. J. Neural Syst. **25**(4), 1450036–1450055 (2015)

Feature Clustering to Improve Fall Detection: A Preliminary Study

Mirko Fáñez[1], José Ramón Villar[2], Enrique de la Cal[2(✉)],
Víctor M. González[3], and Javier Sedano[1]

[1] Instituto Tecnológico de Castilla y León,
Pol. Ind. Villalonquejar, 09001 Burgos, Spain
mirko@mirkoo.es, javier.sedano@itcl.es
[2] Computer Science Department, EIMEM,
University of Oviedo, Oviedo, Spain
{villarjose,delacal}@uniovi.es
[3] Control and Automatica Department, EPI,
University of Oviedo, Gijón, Spain
vmsuarez@uniovi.es

Abstract. In this study, the fall detection method is carried out as stated on [1,11]; a simple finite state machine is used to process acceleration data in sliding windows and whenever a fall-like event is found, features are extracted from this data. Using some clustering and classification algorithms described here, the event is classified as FALL or NOT_FALL. This research evaluates the performance of different proposed clustering and classification methods. It makes use of a new dataset, with data gathered by a wearable device placed on the wrist and used by several members of the research team and an emergency rescue training manikin under different fall scenarios to simulate the falls. A 10-fold cross-validation is also made to evaluate these methods on unseen data.

Keywords: Fall detection · Clustering · Classification · Wearable devices

1 Introduction

Fall Detection (FD) is a major challenge with many applications to healthcare, work safety, etc. The best rated commercial products only reach an 80% of success [14].

Different solutions have been widely studied to perform the fall event detection. Some of them are context-aware systems (like video systems [19]) and others make use of wearable devices (WD) [9], which allows care institutions to optimize their services at lower cost and be available to a wider part of the population. They are crucial because the high percentage of non-completely self-sufficient people (e.g. older people), and their desire to live autonomously in their

© Springer Nature Switzerland AG 2020
F. Martínez Álvarez et al. (Eds.): SOCO 2019, AISC 950, pp. 219–228, 2020.
https://doi.org/10.1007/978-3-030-20055-8_21

own house. Wearable-based solutions may combine different sensors, but triaxial accelerometers (3DACC) is by far the most chosen option [3,7,8,18,20].

Although much of the proposed solutions report good performances, they are just machine learning applied to the focused problem. For instance, a feature extraction stage and Support Vector Machines have been applied directly in [18, 20], using some transformations and thresholds with very simple rules [3,8,10]. A comparison of different classification algorithms has been presented in [7], and several threshold-based fall detection algorithms using 3DACC data were presented in [3,5,6].

There are also studies concerned with the dynamics in a fall event [2,4]. Additionally, Abbate et al. proposed the use of these dynamics as the basis of the FD algorithm [1], with moderate computational constraints but a high number of thresholds to tune.

The common characteristic in all these solutions is that the WDs were placed on the waist or in the chest. In [17], the sensors were placed on the wrist, using Abbate et al. solution plus an over-sampling data balancing stage (using the Synthetic Minority Over-sampling Technique - SMOTE) and a feed-forward Neural Network.

In this study the event detection method proposed in [1,11] is used to detect those high movement activities. Thus, the features described in the mentioned works are extracted. Then, a clustering stage is performed to keep the most interesting patterns; different scenarios are considered. Finally, these centroids resulting from the clustering are used as the training stage of SVM and/or KNN classifiers. To evaluate the performance, a single sensor -a marketed smartwatch-placed on the wrist was used. Data was gathered from the sensor when used by several members of the research team and also using an emergency rescue training manikin under different fall scenarios to simulate the falls. The diagram of the solution can be found in Fig. 1.

The structure of the paper is as follows: Next section deals with identifying fall-like events and extracting the features from the time series of the detected acceleration peaks. Section 3 explains the different approaches considered for clustering and classification modeling. Section 4 describes the dataset used and how it was validated using 10-fold cross-validation. Section 5 shows and discusses the obtained results. Finally, conclusions are drawn.

2   Peak Detection and Feature Extraction

The same peak detection and feature extraction from [1,11] is proposed in this study. A very simple finite state machine is used to detect the falls - see Fig. 1. The data gathered from a 3DACC located on the wrist is processed using a sliding window. A peak detection is performed, and if a peak for a fall-like event is found, the data within the sliding window is analyzed to extract several features which are ultimately classified as FALL or NOT_FALL. The FD block is performed with a clustering + classification approach. In [12] it was claimed that the lower the computational cost the better as it must be run in the WD.

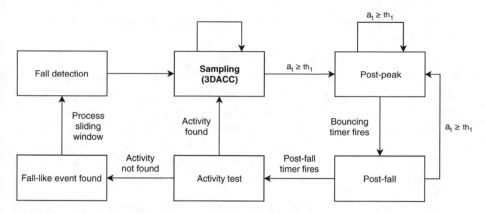

Fig. 1. Finite state machine for an acceleration magnitude a at time stamp t, using threshold $th_1 = 3 \times g$. The bouncing and post fall timers are set to 1000 and 1500 ms, respectively. For fall-like events the features are computed, and the extracted sample is classified.

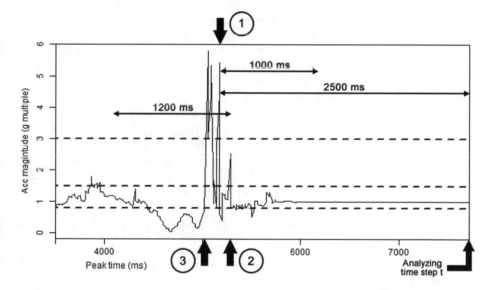

Fig. 2. Sliding window with acceleration magnitude evolution in multiples of g. Corresponds to a simulated fall with the dummy. First 3500 ms were removed for viewing purposes. **Point 1** is the peak time; **point 2** is the impact end and **point 3** is the impact start.

The feature extraction is executed whenever a peak is detected and follows the dynamics within a fall - refer to Fig. 2. Let's assume that the gravity is $g = 9.8$ m/s. Given the current timestamp t, we search a peak at **peak time** $pt = t - 2500$ ms (**point 1**). If at time pt the magnitude of the acceleration a_t - see Eq. 1 - is higher than $th_1 = 3 \times g$ and there is no other peak in the period

$(t - 2500 \, \text{ms}; t]$ (no other a_t value higher than th_1), then it is stated that a peak occurred at pt.

$$a_t = \sqrt{a_{tx}^2 + a_{ty}^2 + a_{tz}^2} \tag{1}$$

The **impact end** ie (**point 2**) denotes the end of the fall event; it is the last time for which the a_t value is higher than $th_2 = 1.5 \times g$. Finally, the **impact start** is (**point 3**) denotes the starting time of the fall event, computed as the time of the first sequence of an $a_t \leq th_3$ ($th_3 = 0.8 \times g$) followed by a value of $a_t \geq th_2$. The *impact start* must belong to the interval $[ie - 1200 \, \text{ms}, peak \, time]$. If no *impact end* is found, then it is fixed to *peak time* plus 1000 ms. If no *impact start* is found, it is fixed to *peak time*.

Whenever a fall-like peak is found, the following transformations should be computed:

- Average Absolute Acceleration Magnitude Variation, $AAMV = \sum_{t=is}^{ie} \frac{|a_{t+1} - a_t|}{N}$, with N the number of samples in the interval.
- Impact Duration Index, $IDI = impact \, end - impact \, start$.
- Maximum Peak Index, $MPI = max_{t \in [is, ie]}(a_t)$.
- Minimum Valley Index, $MVI = min_{t \in [is-500, ie]}(a_t)$.
- Peak Duration Index, $PDI = peak \, end - peak \, start$, with *peak start* defined as the time of the last magnitude sample below $th_{PDI} = 1.8 \times g$ occurred before pt, and *peak end* defined as the time of the first magnitude sample below $th_{PDI} = 1.8 \times g$ occurred after pt.
- Activity Ratio Index, ARI, calculated as the ratio between the number of samples not in $[th_{ARIlow} = 0.85 \times g, th_{ARIhigh} = 1.3 \times g]$ and the total number of samples in the 700 ms interval centered in $(is + ie)/2$.
- Free Fall Index, FFI, the average acceleration magnitude in the interval $[t_{FFI}, pt]$. The value of t_{FFI} is the time between the first acceleration magnitude below $th_{FFI} = 0.8 \times g$ occurring up to 200 ms before pt; if not found, it is set to $pt - 200 \, \text{ms}$.
- Step Count Index, SCI, measured as the number of peaks in the interval $[pt - 2200, pt]$.

The real-time fall detection is proposed as follows. The acceleration magnitude value for each instant of time t is analyzed looking for a peak that marks where a fall event candidate appears. Whenever after a peak there are 2500 ms of low activity, the entire time series for the last 7500 ms is passed to the feature extractor, which determines the *impact end*, *impact start* and all the remaining features. These features are passed to the trained clustering and classifier models, which determine whether it was a FALL or NOT_FALL.

3   Data Modeling and Classification

This study includes the use of several modelling techniques. Firstly, a clustering stage based on the K-Means Clustering Algorithm using R *stats package* [15] version v3.5.0. To perform the classification, two different methods were proposed:

(i) Support Vector Machine (SVM) classification using R *e1071 package* [13] version v1.7.0 and (ii) k-Nearest Neighbour classification using R *class package* [16] version v7.3.15.

These techniques were combined with different dataset scenarios: (i) clustering all the data together, (ii) clustering the NOT_FALL data only, (iii) clustering the NOT_FALL data only, adding the FALL data and recomputing the centroids and (iv) clustering data from each label independently and using the computed centroids. All these options are described in the next paragraphs.

Approach 1 Clustering all the data together. This scenario includes the complete dataset (both the FALL and the NOT_FALL samples). The optimal number of clusters is sought between 2 and 20. For clusters where all samples belong to the same label A then, when classifying, a new sample is labelled with this label A. In contrast, clusters in which there are a mixture of FALL and NOT_FALL classes and depending on the degree of class imbalance these clusters have, a decision has to be made between performing a SVM classification to predict the class of the test samples, or just assigning the majority class to the cluster. The best option is sought between performing SVM on clusters from 0% to 20% imbalanced, assigning the majority class if not enough imbalanced.

Approach 2 Clustering the NOT_FALL data only. In this scenario, the NOT_FALL samples are grouped, then the FALL samples are standardized to the mean and std of the NOT_FALL and assigned to the closest cluster. A SVM is trained for each cluster for which there were samples labelled for both classes.

Approach 3 Clustering the NOT_FALL data only adding the FALL data and recomputing the centroids. This option works as in the previous case but the SVMs are trained using clusters' centroids distances instead of the features themselves.

Approach 4 Clustering data from each label independently and using the computed centroids. With this approach we will use two different clustering models: one for the NOT_FALL training dataset part and another one for the FALL training dataset part. The optimal number of clusters is sought between 4 and 22 clusters for the NOT_FALLs, and between 2 and 12 clusters for the FALLs.

Clustering models' centroids are combined, in order to train a knn model (the best value of k is also sought). A new test sample is classified as FALL if the probability of being a FALL (p_F) returned by the knn model is a certain amount greater than being NOT_FALL (p_{NF}) (or classified as NOT_FALL if it is that amount greater than being a FALL). For those cases in which there is no clearly greater probability, the Eq. 2 is used.

$$pp_{CL}(sample) = (1 - \frac{\sum_{i=0}^{Q_{CL}} dist(sample, centroid_i^{CL})}{\sum_{\substack{i=0 \\ CL \in \{F, \, NF\}}}^{Q} dist(sample, centroid_i^{CL})}) * p_{CL} \quad (2)$$

pp_F (predicted probability of being a FALL) and pp_{NF} (predicted probability of being a NOT_FALL) of a *sample* are computed, using the sum of the distances of that sample to the FALL or NOT_FALL models centroids, and the obtained probabilities using the knn model (p_F and p_{NF}).

Q is the minimum centroids count in both clustering models (FALLs and NOT_FALLs). It is the same as the minimum clusters count in both models.

We look for the best parameter combination of all the above that result in less fall prediction error.

4   Experimental Design

4.1   Dataset Description

A reliable FALL/NOT_FALL dataset was needed to obtain models that were able to make a reliable fall detection. An application for accelerometry data recording was coded and installed on the WD (a marketed smartwatch).

The data saved for each detected peak consist of a 7500 ms window with the 3-axis acceleration data (and the calculated acceleration magnitude - see Eq. 1) obtained with a frequency of 100 Hz (100 samples/s). See Fig. 3.

To obtain NOT_FALL data, 3 different adult subjects have worn the WD on the wrist recording the time series of the acceleration peaks detected for 24 h. These time series correspond to activities of daily life (ADLs) like: Office work, daily household activities, driving, walking, running and other types of exercises (push-ups, etc.).

For the FALL data, an emergency rescue training dummy was used, which has the dimensions and weight equivalent to an adult person. The WD was placed on the manikin wrist, and the procedure to simulate the falls was:

Fainting: The dummy starts sitting on a chair. Then, it gets up and, when fully up in front of the chair, collapses to the ground, producing an acceleration peak.

Falls: It also starts sitting on a chair. It gets up in a normal way and starts walking, moving its arms just like walking. After 2 s, the dummy falls forward, producing an acceleration peak.

When training the algorithm there was no differentiation between falls and fainting, belonging both to the same class FALL.

After the experimentation, the obtained dataset consists of 1072 time series of NOT_FALL 3DACC values and 87 time series of FALL 3DACC values, of which 45 are fainting and 42 are falls.

4.2   Cross Validation

A 10-fold cross-validation (CV) scheme is performed, including training and testing, as follows: 10 folds are generated randomly with the NOT_FALL dataset, saving one of them for testing. In the same way, 10 folds are created randomly

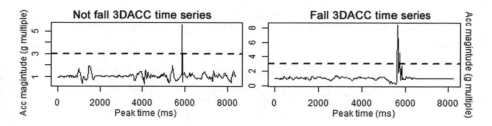

Fig. 3. FALL vs NOT_FALL time series comparison

with the FALL data, and one of them is used for testing. The remaining 18 folds are used for training. On each iteration of the CV, the two folds used for testing are interchanged with the next training folds.

5   Obtained Results and Discussion

The parameters found optimum for each classification problem (described on 3):

- **Approach 1:** Clustering with 8 clusters, and using SVM on the clusters that are more than 5% imbalanced.
- **Approach 2:** Clustering with 6 clusters, and using SVM on the clusters that are more than 10% imbalanced.
- **Approach 3:** Clustering with 6 clusters, and using SVM on the clusters that are more than 5% imbalanced.
- **Approach 4:** Clustering NOT_FALLs with 22 clusters and FALLs with 8 clusters. Using k value = 5 for the knn. Always using the Eq. 2 to calculate the predicted probability.

Table 1 shows the statistical accuracy (Acc) obtained performing the 10-fold cross-validation on the entire FALL/NOT_FALL dataset.

Because the dataset is imbalanced (7.5% FALLS - 92.5% NOT_FALLS), accuracy does not reflect prediction performance of the smaller class, and the accuracy of the FALL samples needed to be measured separately. It was then split in: Acc only for the FALL class predictions (up), which is the minor class in the dataset, and the Acc for both classes predictions (bottom). Cross-validation was also performed due to the limited FALL data samples.

From the Table 1 it can be seen that the approach that gave the best result was *Approach 4* with 83.61% accuracy for FALL samples (and 98.01% general accuracy).

Another thing that we can see in the Table 1 is that training SVM with imbalanced clusters' centroids distances instead of the features of the data series from imbalanced clusters did not improve the accuracy of the fall predictions (70.56% using clusters' centroids vs 72.92% using features).

Table 1. 10-fold CV results obtained for each one of the approaches. The class prediction accuracy for each fold, only for FALL class predictions (up), or for both classes (bottom).

| Fold | Accuracy (Acc) only for FALL class predictions | | | |
	Approach 1	Approach 2	Approach 3	**Approach 4**
1	0.8750	0.8750	0.6250	**1.0000**
2	0.5556	0.7778	0.6667	**0.7778**
3	0.7778	0.6667	0.7778	**0.8889**
4	0.5000	0.7500	0.8750	**0.8750**
5	0.5000	0.6250	0.6250	**0.7500**
6	0.6667	0.7778	0.8889	**1.0000**
7	0.7778	0.6667	0.6667	**0.6667**
8	0.7778	0.7778	0.5556	**0.7778**
9	0.5000	0.8750	0.7500	**0.8750**
10	0.5000	0.5000	0.6250	**0.7500**
mean	0.6431	0.7292	0.7056	**0.8361**
median	0.6111	0.7639	0.6667	**0.8264**
std	0.1484	0.1161	0.1128	**0.1105**

| Fold | Accuracy (Acc) including both FALLs and NOT_FALLs | | | |
	Approach 1	Approach 2	Approach 3	**Approach 4**
1	0.9826	0.9913	0.9739	**1.0000**
2	0.9573	0.9829	0.9744	**0.9829**
3	0.9828	0.9741	0.9828	**0.9741**
4	0.9652	0.9826	0.9913	**0.9826**
5	0.9655	0.9741	0.9741	**0.9741**
6	0.9741	0.9828	0.9914	**1.0000**
7	0.9828	0.9741	0.9741	**0.9741**
8	0.9828	0.9828	0.9655	**0.9655**
9	0.9652	0.9913	0.9826	**0.9913**
10	0.9652	0.9652	0.9739	**0.9565**
mean	0.9723	0.9801	0.9784	**0.9801**
median	0.9698	0.9827	0.9742	**0.9784**
std	0.0098	0.0082	0.0084	**0.0142**

6 Conclusions

This research exposes a fall detection solution based on a wearable device placed on the wrist. This solution is also based on thresholds. The basic idea is to detect high acceleration peaks and when one of them occurs, predict whether it was due to a fall or not using some clustering and classification algorithms. This study improves previous studies by using a new dataset, with more realistic and longer time series and taken with a greater frequency (100 samples/sec). This dataset allowed us different options to be considered when performing clustering and classification of the samples, that are described on this paper. One of these options resulted in a better fall detection accuracy than those reported for commercial devices, although we still have some undetected fall events.

In order to improve the fall detection accuracy, some improvements can be tried: obtaining new FALL data making use of the dummy to train with new fall scenarios; combining accelerometry data with other data collected by the same wearable device, like heart rate; or adjusting the thresholds/features used in the peak detection and feature extraction to the new dataset.

Acknowledgment. This research has been funded by the Spanish Ministry of Science and Innovation, under project MINECO-TIN2017-84804-R.

References

1. Abbate, S., Avvenuti, M., Bonatesta, F., Cola, G., Corsini, P.: AlessioVecchio: a smartphone-based fall detection system. Pervasive Mobile Comput. **8**(6), 883–899 (2012)
2. Abbate, S., Avvenuti, M., Corsini, P., Light, J., Vecchio, A.: Monitoring of human movements for fall detection and activities recognition in elderly care using wireless sensor network: a survey. In: Wireless Sensor Networks: Application - Centric Design, p. 22. Intech (2010)
3. Bourke, A., O'Brien, J., Lyons, G.: Evaluation of a threshold-based triaxial accelerometer fall detection algorithm. Gait Posture **26**, 194–199 (2007)
4. Delahoz, Y.S., Labrador, M.A.: Survey on fall detection and fall prevention using wearable and external sensors. Sensors **14**(10), 19806–19842 (2014). http://www.mdpi.com/1424-8220/14/10/19806/htm
5. Fang, Y.C., Dzeng, R.J.: A smartphone-based detection of fall portents for construction workers. Procedia Eng. **85**, 147–156 (2014)
6. Fang, Y.C., Dzeng, R.J.: Accelerometer-based fall-portent detection algorithm for construction tiling operation. Autom. Constr. **84**, 214–230 (2017)
7. Hakim, A., Huq, M.S., Shanta, S., Ibrahim, B.: Smartphone based data mining for fall detection: analysis and design. Procedia Comput. Sci. **105**, 46–51 (2017). http://www.sciencedirect.com/science/article/pii/S1877050917302065
8. Huynh, Q.T., Nguyen, U.D., Irazabal, L.B., Ghassemian, N., Tran, B.Q.: Optimization of an accelerometer and gyroscope-based fall detection algorithm. J. Sens. **2015**, 8 (2015)
9. Igual, R., Medrano, C., Plaza, I.: Challenges, issues and trends in fall detection systems. Biomed. Eng. Online **12**, 66 (2013). http://www.biomedical-engineering-online.com/content/12/1/66
10. Kangas, M., Konttila, A., Lindgren, P., Winblad, I., Jämsää, T.: Comparison of low-complexity fall detection algorithms for body attached accelerometers. Gait Posture **28**, 285–291 (2008)
11. Khojasteh, S.B., Villar, J.R., Chira, C., Gonzalez, V.M., de la Cal, E.: Improving fall detection using an on-wrist wearable accelerometer. Sensors **18**(5), 1350 (2018)
12. Khojasteh, S.B., Villar, J.R., Chira, C., González, V.M., de la Cal, E.: Improving fall detection using an on-wrist wearable accelerometer. Sensors **18**, 1–20 (2018)
13. Meyer, D., et al.: Probability Theory Group (Formerly: E1071), TU Wien - Package 'e1071' (2019). https://cran.r-project.org/web/packages/e1071/e1071.pdf

14. Purch.com: Top ten reviews for fall detection of seniors (2018). www.toptenreviews. com/health/senior-care/best-fall-detection-sensors/
15. R Core Team and contributors: K-means clustering in R stats package (2019). https://stat.ethz.ch/R-manual/R-devel/library/stats/html/kmeans.html
16. Ripley, B., Venables, W.: Functions for classification - package 'class' (2019). https://cran.r-project.org/web/packages/class/class.pdf
17. Khojasteh, S.B., Villar, J.R., de la Cal, E., González, V.M., Sedano, J., Yazğan, H.R.: Evaluation of a wrist-based wearable fall detection method. In: 13th International Conference on Soft Computing Models in Industrial and Environmental Applications, pp. 377–386 (2018)
18. Wu, F., Zhao, H., Zhao, Y., Zhong, H.: Development of a wearable-sensor-based fall detection system. Int. J. Telemed. Appl. (2015). https://www.hindawi.com/journals/ijta/2015/576364/
19. Zhang, S., Wei, Z., Nie, J., Huang, L., Wang, S., Li, Z.: A review on human activity recognition using vision-based method. J. Healthc. Eng. **2017**, 31 (2017)
20. Zhang, T., Wang, J., Xu, L., Liu, P.: Fall detection by wearable sensor and one-class svm algorithm. In: Huang DS., Li K., I.G. (ed.) Intelligent Computing in Signal Processing and Pattern Recognition, Lecture Notes in Control and Information Systems, vol. 345, pp. 858–863. Springer Berlin Heidelberg (2006)

Data Generation and Preparation

Creation of Synthetic Data with Conditional Generative Adversarial Networks

Belén Vega-Márquez$^{(\boxtimes)}$, Cristina Rubio-Escudero, José C. Riquelme,
and Isabel Nepomuceno-Chamorro

Department of Computer Languages and Systems,
University of Sevilla, Sevilla, Spain
bvega@us.es

Abstract. The generation of synthetic data is becoming a fundamental task in the daily life of any organization due to new protection data laws that are emerging. Generative Adversarial Networks (GANs) and its variants have attracted many researchers in their research work due to its elegant theoretical basis and its great performance in the generation of new data [19]. The goal of synthetic data generation is to create data that will perform similarly to the original dataset for many analysis tasks, such as classification. The problem of GANs is that in a classification problem, GANs do not take class labels into account when generating new data, they treat it as another attribute. This research work has focused on the creation of new synthetic data from the *"Default of Credit Card Clients"* dataset with a Conditional Generative Adversarial Network (CGAN). CGANs are an extension of GANs where the class label is taken into account when the new data is generated. The performance of our results has been measured by comparing the results obtained with classification algorithms, both in the original dataset and in the data generated.

Keywords: Synthetic data ·
Conditional Generative Adversarial Networks · Deep Learning ·
Credit Card Fraud Data

1 Introduction

The introduction of the new data protection law [7] has supposed that the process of sharing personal data has become increasingly tough and difficult, especially in the medical field, where data is highly personal and can be used to harm patients themselves. Because of this scientists and doctors have to establish agreements between themselves before sharing any personal data. These requirements slow down or even prevent the exchange of data between researchers [1].

Facing with this problem, several solutions have been contemplated that seek to find or simulate data that are similar to the real one without involving individuals. Among these solutions, the use of Deep Learning techniques to generate

F. Martínez Álvarez et al. (Eds.): SOCO 2019, AISC 950, pp. 231–240, 2020.
https://doi.org/10.1007/978-3-030-20055-8_22

synthetic data similar to real ones stands out [4, 20]. The purpose of this synthetic data is to be used to train machine learning models that can then be used in the real data, so that the training is done without having to make the real data public. The precision of this technique is measured by comparing the results obtained with real data and synthetic data, so that they are as similar as possible.

Generative Adversarial Networks (GANs) [8] have shown to be one of the most successful techniques in the creation of synthetic data from real data, such as generating clinical data on blood pressure [1] or even generating new magnetic resonance images for segmentation tasks [14]. Generative Adversarial Networks in which two networks are trained against each other in a zero-sum game framework. Commonly one network is known as Generator and the other as Discriminator [17].

The purpose of this article is to evaluate the utility of the samples generated by an adversarial neural network with the Credit Card Fraud Detection Data from Kaggle. To work with this dataset, we have used a Conditional Generative Adversarial Network (CGANs) [11] that takes into account the class to which the instances belong. We considered two methods to evaluate the work: the first method is to measure the correlation between the real data and synthetic data. As mentioned above, the objective of the use of these techniques is the privacy of the data, so it is advisable that the transformation process is unidirectional so that real data can not be obtained from false data. Pearson's correlation index will measure this phenomenon, so that a low correlation index would be optimal, meaning that the two sets of data are not correlated and cannot be inferred from each other. The second method is to compare the accuracy obtained with a classification algorithm, specifically the XGBoost [3] for the two sets of data. If this accuracy is similar it means that the model trained with the false set serves to reach conclusions about the real set without having to use it for training.

The article is organised as follows: Sect. 2 provides a detailed description about the methodology used in all the process. Section 3 shows the results obtained with the previous techniques previously described, and finally, Sect. 4 shows the conclusions that have been obtained after the research.

2   Methodology

Our aim in this study is to provide a Deep Learning approach to simulate new data based on the Credit Card Fraud Detection Data. We used a type of GAN known as Conditional Generative Adversarial Network (CGAN) which is the key technique in our approach. This is because this type of networks shows very good results in data sets that have a target class, since they take into account this detail to train the neural network so that the new data fits as closely as possible to the data according to which class each of the instances belongs to [13].

2.1   Generative Adversarial Networks

Generative Adversarial Networks are a deep learning model which comprise two different neural networks, a generator and a discriminator who are simultaneously trained competitively, as in a zero-sum game framework.

The generative network (G) is in charge of learning how to assign elements of a latent space (noise) to a certain data distribution, i.e., what it does is to generate new data that is as close as possible to the real data. On the other hand, the functionality of the discriminator (D) consists in differentiating between elements of the original distribution and those created by the generative network by calculating the probability of belonging to one set or another [8]. To summarise, the discriminator network is a standard convolutional network that can categorise the examples fed to it, a binomial classifier labelling instances as real or fake. The generator is an inverse convolutional network, in a sense: while a standard convolutional classifier takes an example and downsamples it to produce a probability, the generator takes a vector of random noise and upsamples it to an instance. The first throws away data through downsampling techniques like max pooling, and the second generates new data. Figure 1 shows the basic architecture of a GAN network.

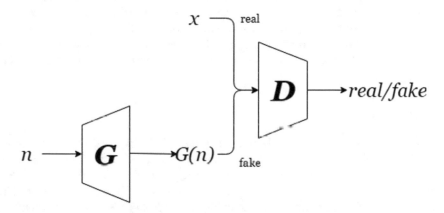

Fig. 1. GAN network architecture

2.2   Conditional Adversarial Networks

Figure 2 illustrates the basic architecture of a conditional adversarial network. It can be observed that the structure is very similar to the typical adversarial neural networks, however there is one more factor to take into account, and that is the class c to which the instance belongs.

A GAN does not take into account any type of condition with respect to the data. Usually the synthetic data to be generated has a type of property that distinguishes it, which must also be used to obtain synthetic data as close as possible to the real ones.

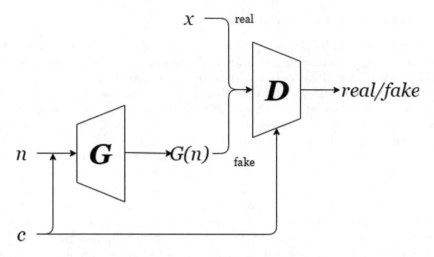

Fig. 2. CGAN network architecture

After this approach, Conditional Adversarial Networks (CGANs) arised. CGANs are an extension of GAN where some condition is taken into account. This condition implies that both the discriminator and the generator have to take into account some additional information, let's call it c, where c can be any type of additional information, such as data from another nature or some class label [11].

2.3   Dataset

The dataset chosen for this study was *"Default of Credit Card Clients Dataset"* available in [10]. The dataset contains 30000 examples and 25 variables with information about default payments, demographic factors, credit data, history of payment and bill statements of credit card clients in Taiwan from April 2005 to September 2005. The label class is called *Default payment* and it indicates if the next month the payment will be carried out or not (1 or 0).

2.4   Software and Experimental Setting

The CGAN network used in this study has been implemented with the Keras library [5]. Keras is a high-level neural networks API, written in Python and capable of running on top of Tensorflow. The classification of the data has been carried out with the scikit-learn library [12]. The executions were performed on an Intel machine, specifically Intel(R) Core(TM) i7-8700 CPU @ 3.20 GHz, with 64 GB of RAM and 12 cores.

3   Results

3.1   Generating New Credit Card Data with CGANs

First, in order to obtain better results, a preprocessing step was performed due to the fact that the data does not fit a normal distribution. This preprocessing consists of a standard normalization.

To apply CGAN architecture to the Credit Card Data Fraud dataset the GAN-Sandbox[6] package was used. GAN-Sandbox has a number of popular GAN architectures implemented in Python using the Keras library and a Tensorflow backend. All the results obtained are available as a Jupyter Notebook in [18].

As mentioned above, the neural network in turn is composed of two networks, the discriminator and the generator which have the following structure:

1. Generator Network
   - 1 Input layer: the input layer receives the real data with which the model is going to be trained.
   - 6 Dense layers with the parameters specified below:
     - First Dense layer: 30 neurons and rectified linear activation function
     - Second Dense layer: 60 neurons and rectified linear activation function
     - Third, Fourth and Fifth Dense layer: 120 neurons and rectified linear activation function
     - Sixth Dense layer: 20 neurons which correspond to the number of the columns of the dataset used to train the model

2. Discriminator Network
   - 1 Input layer: the input layer receives the fake data generated by the generator network
   - 1 Dense layer with 120 neurons and rectified linear activation function
   - 1 Dropout layer with a dropout rate of 0.1
   - 1 Dense layer with 60 neurons and rectified linear activation function
   - 1 Dropout layer with a dropout rate of 0.1
   - 1 Dense layer with 30 neurons and rectified linear activation function
   - 1 Dense layer with 1 neuron and sigmoid activation function

After CGAN training with 5000 epochs, 30000 new instances were generated in order to have the same number of real and fake examples. Figure 3 shows FOUR scatterplot of 4 variables of the dataset. To the right of the image are the graphs corresponding to the new data generated, differentiating by colors the classes to which they belong. In the left column you can see the graphs corresponding to the real data. It can be seen that a priori the data obtained is not similar to real data, but it was not a problem due the fact that one of the objectives of this work is to obtain false data that behave in the same way as real ones in classification task, but without being able to establish a relationship between these two sets of data.

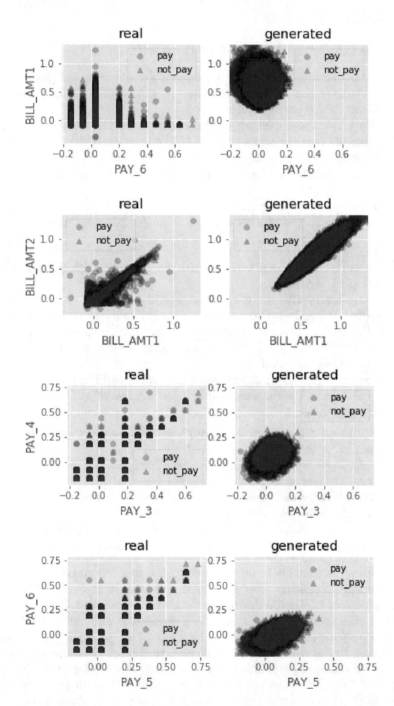

Fig. 3. Scatterplot of some attributes of dataset distinguished by real and generated data

3.2   Similarity of the Data

One of the procedures we followed to check whether or not our method was appropriate was to measure the relationship between the variables in the original dataset and the new dataset generated. The objective was to obtain values that indicated that the correlation between these sets was null or minimal in order to be able to use this technique in controversial fields such as medicine or banks.

In this section we have calculated three correlation indicators to find out whether the variables in the real and false datasets are correlated or not. These indicators are Pearson's correlation coefficient, covariance and Spearman's correlation coefficient.

Pearson's Correlation Coefficient [16] is a measure of the linear correlation between two variables X and Y. The value has a range between -1 and 1, where 1 is a total positive linear correlation, that is, when X increases, Y too, -1 indicates there is a total negative linear correlation (when X increases, Y decreases or vice versa) and finally, a zero value means there is no correlation between the two variables.

Secondly, covariance [9] is defined as the expected value of variations of two variables from their expected values, that is, covariance measures how much variables change together. The sign of the covariance can be interpreted as follows: positive sign means two variables change in the same direction, negative sign means they change in different opposite directions. A zero value indicates that both variables are completely independent. As for the magnitude of the value, at an interpretational level, a higher covariance value in absolute value will indicate a stronger linear relationship between the two variables. The disadvantage of this value is that it does not have a maximum or minimum value as it happens with the Pearson's or Spearman's coefficient.

Finally, Spearman's correlation coefficient [15] is a Pearson correlation coefficient calculated with the ranks of the values of each of the two variables instead of their actual values. It can be used to summarise the strength between the two variables when it is not supposed that the two variables are related by linear relationship. As Pearson's correlation coefficient, the measures are between -1 and 1 where -1 indicates a total negative correlation and 1 means perfectly positively correlation.

Table 1 shows the Pearson's correlation coefficient, covariance and Spearman's correlation coefficient between real and fake dataset for each of the column in the study. The table shows that all three coefficients are very close to zero or even zero when comparing variables in data sets two by two, indicating that there is no real correlation between the variables of the original dataset and the newly created dataset. This aspect is very important, since, as mentioned above, in areas such as medicine or banks, user data are affected by the new data protection law and must be carefully treated. Moreover, if these data do not present any correlation with the originals but behave in the same way, they can be used indistinctly for any of the tasks in which they are used.

Table 1. Similarity between variables

Column	Pearson index	Covariance	Spearman's index
LIMIT BAL	0.046451	0.000306	0.051436
AGE	−0.003058	0	0.000192
PAY 0	0.127477	0.000784	0.113877
PAY 2	0.086022	0.000580	0.066049
PAY 3	0.034958	0.000154	0.027432
PAY 4	0.075370	0.000449	0.056721
PAY 5	0.071400	0.000433	0.052301
PAY 6	0.062410	0.000351	0.045897
BILL AMT1	0.004100	0	0.002037
BILL AMT2	0.005826	0.000108	0.003781
BILL AMT3	0.007589	0.000175	0.005216
BILL AMT4	0.006202	0.000106	0.007436
BILL AMT5	0.004516	0	0.005910
BILL AMT6	0.001630	0	0.003899
PAY AMT1	0.017059	0.000154	0.034533
PAY AMT2	0.003157	0	0.012173
PAY AMT3	−0.001662	0	0.012225
PAY AMT4	−0.012410	0	−0.026648
PAY AMT5	0.010240	0	0.012197
PAY AMT6	−0.015963	−0.000117	−0.017851

3.3 Classification Results

After verifying that the data generated did not correlate to the real data, it was checked whether both sets of data behaved in the same way when faced with the classification task. The XGBoost [2] algorithm was used to perform this task. XGBoost is an implementation of gradient boosted decision trees designed for speed and performance. It can be seen in Table 2 that the results obtained are practically equal for the two sets of data even without having any correlation between them. These results show that the generated data can be used indistinctly for the same tasks as the real ones obtaining the same results.

Table 2. Result of classification with XGBoost

	Accuracy	F1-score	AUC
Real dataset	0.821	0.479	0.661
Generated dataset	0.826	0.509	0.676

4   Discussion

In this paper, we proposed a Conditional Generative Adversarial Network (CGAN) that generates new synthetic data from training data which can be used indistinctly for the same tasks without having to reveal the actual data. This type of network has been used and not a traditional generative adversarial network (GAN) because the data on which it has been tested had a class label that has been taken into account for the generation of new data.

The results obtained have been evaluated in two different ways: first, it has been verified that the correlation between the new data and the original data is minimal, so that they can be used in controversial fields, such as medicine or banks, in which client data must be treated with special care so as to avoid privacy problems. Secondly, since the dataset had a label that could be used for classification tasks, the same algorithm, XGBoost, has been tested with the same parameters in the two sets of data. The results have shown that the accuracy of classification is similar in both cases.

In conclusion, the research finding of this study have provided some evidence that Deep Learning methods can be used, with good performance, in synthetic data generation. For future work we will consider new variants of adversarial networks to perform this task, as well as the adjustment of parameters to get the most reliable results.

References

1. Beaulieu-Jones, B.K., Wu, Z.S., Williams, C., Lee, R., Bhavnani, S.P., Byrd, J.B., Greene, C.S.: Privacy-preserving generative deep neural networks support clinical data sharing. bioRxiv, p. 159756, Jan 2018. http://biorxiv.org/content/early/2018/12/20/159756.abstract
2. Chen, T., Guestrin, C.: XGBoost. In: Proceedings of the 22nd ACM SIGKDD International Conference on Knowledge Discovery and Data Mining - KDD 2016 (2016)
3. Chen, T., Guestrin, C.: XGBoost: a scalable tree boosting system. CoRR abs/1603.02754 (2016). http://arxiv.org/abs/1603.02754
4. Choi, E., Biswal, S., Malin, B., Duke, J., Stewart, W.F., Sun, J.: Generating multi-label discrete electronic health records using generative adversarial networks. CoRR abs/1703.06490 (2017). http://arxiv.org/abs/1703.06490
5. Chollet, F., et al.: Keras (2015). https://keras.io
6. Dietz, M.: GAN-Sandbox (2017). https://github.com/mjdietzx/GAN-Sandbox
7. Generales, C.: Ley orgánica 3/2018, de 5 de diciembre, de protección de datos personales y garantía de los derechos digitales, December 2018. https://www.boe.es/buscar/doc.php?id=BOE-A-2018-16673. Accessed 14 Feb 2019
8. Goodfellow, I.J., Pouget-abadie, J., Mirza, M., Xu, B., Warde-farley, D., Ozair, S., Courville, A., Bengio, Y.: GANs. In: NIPS (2014)
9. Kim, H.Y.: Statistical notes for clinical researchers: covariance and correlation. Restorative Dent. Endod. **43**(1), e4 (2018). http://www.ncbi.nlm.nih.gov/pubmed/29487835. http://www.pubmedcentral.nih.gov/articlerender.fcgi?artid=PMC5816993

10. Lichman, M.: UCI machine learning repository (2013). http://archive.ics.uci.edu/ml
11. Mirza, M., Osindero, S.: Conditional generative adversarial nets. CoRR abs/1411.1784 (2014). http://arxiv.org/abs/1411.1784
12. Pedregosa, F., Varoquaux, G., Gramfort, A., Michel, V., Thirion, B., Grisel, O., Blondel, M., Prettenhofer, P., Weiss, R., Dubourg, V., Vanderplas, J., Passos, A., Cournapeau, D., Brucher, M., Perrot, M., Duchesnay, E.: Scikit-learn: machine learning in Python. J. Mach. Learn. Res. **12**, 2825–2830 (2011)
13. Ramponi, G., Protopapas, P., Brambilla, M., Janssen, R.: T-CGAN: conditional generative adversarial network for data augmentation in noisy time series with irregular sampling. CoRR abs/1811.08295 (2018). http://arxiv.org/abs/1811.08295
14. Rezaei, M., Yang, H., Meinel, C.: Multi-task generative adversarial network for handling imbalanced clinical data. CoRR abs/1811.10419 (2018). http://arxiv.org/abs/1811.10419
15. Schober, P., Boer, C., Schwarte, L.A.: Correlation coefficients. Anesth. Analg. **126**(5), 1763–1768 (2018). http://insights.ovid.com/crossref?an=00000539-20180 5000-00050
16. Sedgwick, P.: Pearson's correlation coefficient. BMJ **345**, e4483 (2012). https://www.bmj.com/content/345/bmj.e4483
17. Triastcyn, A., Faltings, B.: Generating differentially private datasets using GANs (2018). https://openreview.net/forum?id=rJv4XWZA
18. Vega, B.: Syntheticdata (2019). https://github.com/bvegaus/syntheticData
19. Xie, L., Lin, K., Wang, S., Wang, F., Zhou, J.: Differentially private generative adversarial network, February 2018
20. Yoon, J., Jordon, J., van der Schaar, M.: PATE-GAN: generating synthetic data with differential privacy guarantees. In: International Conference on Learning Representations (2019). https://openreview.net/forum?id=S1zk9iRqF7

Data Selection to Improve Anomaly Detection in a Component-Based Robot

Nuño Basurto(iD) and Álvaro Herrero$^{(\boxtimes)}$(iD)

Grupo de Inteligencia Computacional Aplicada (GICAP),
Departamento de Ingeniería Civil, Escuela Politécnica Superior,
Universidad de Burgos, Av. Cantabria s/n, 09006 Burgos, Spain
{nbasurto,ahcosio}@ubu.es

Abstract. The rise in complexity of robotic systems usually leads to an increase in failures of such systems. To improve the maintenance of this type of systems and thus reducing economic costs and downtime, present paper addresses anomaly detection in a component-based robot. To do so, the problem of anomaly detection is modelled as a classification problem, being Support Vector Machine (SVM) the selected classifier. It is applied to a publicly-available and recent dataset containing useful information about the performance of the software system in a component-based robot when certain anomalies are induced. Different preprocessing strategies and data sources are compared to get the best scores for some classification metrics through cross-validation.

Keywords: Anomaly detection · Component-based robotic systems ·
Preprocessing · Missing values · Classification ·
Support vector machines

1 Introduction and Previous Work

At present time we are experiencing the fourth industrial revolution, which proposes a manufacturing interconnected process. Traditional and rising technologies, such as the Internet of Things (IoT), generate a huge amount of data in the industrial sector, among others. The large size of these volumes of data makes it difficult to analyze and deal with them correctly and to extract knowledge from them. In order to solve this problem and to support the transition to industry 4.0 there are different facilitators [14].

This advance in the industrial field evolves in parallel with the automation and digitization of the processes, supported by the incorporation of robotic systems to production chains [6]. Recently, the Industrial Federation of Robotics (IFR) has reported a 30% increase in sales of industrial robots [8], reaching 381,000 units sold worldwide in a year. It is not only the sales but also the complexity of robotic systems that increases as they evolve. In parallel, there is an increasing demand for reliability and robustness, although a complete absence of failures is impossible up to now. The maintenance of robots (both corrective

© Springer Nature Switzerland AG 2020
F. Martínez Álvarez et al. (Eds.): SOCO 2019, AISC 950, pp. 241–250, 2020.
https://doi.org/10.1007/978-3-030-20055-8_23

and predictive) still is an open challenge due to the wide variety of failures they may suffer from, like any other cyber-physical system.

The problem addressed in present work is the detection of performance anomalies in the software system of a robot. As explained in greater detail in Sect. 3.1, this work analyzes data from a robot that undertakes different actions, some of them conditioned by an anomalous behavior. The analyzed dataset [15] is publicly available [16] and anomalies are tagged, being possible to apply supervised machine learning. The classification performance obtained by applying different pre-processing strategies (managing of missing values) is analyzed in order to improve the detection of anomalies. Taking into account previous work on this same dataset, present paper is intended to improve the classification results previously obtained by using Support Vector Machines (SVM), what is discussed in Sect. 3.2.

Authors of the dataset applied One-Class SVM (OCSVM) in order to compare its performance with that obtained when applying their own model [17]. Present paper goes one step further, comparing the effect of different pre-processing strategies in order to improve the obtained results. In a sequel paper by Wienke et al. [18], an analysis of each one of the components of the robot is individually analyzed. Each one of them is compared with the rest to know the potential changes that can take place in the use of resources. That is, from the values that they collect from one component, authors try to predict the behavior of another component regarding the use of resources.

As it is well known, the detection and/or identification of anomalies does not exclusively applies to robotic systems [3,11]. In the industrial context in general [7], and for the optimization and automation of the anomaly detection in particular, machine learning techniques has been successfully applied to date in different fields [9]. More specifically, some studies have previously proposed the application of SVM to fault detection in a variety of fields. Shon et al. [13] proposed the use of SVM for Network Intrusion Detection System (NIDS). When trying to detect possible intrusions in the network, it exhibited a low false-positive rate. Banerjee et al. [1] tackled the problem of fault detection in a multi-sensor motor. They support the SVM classification with Short-Time Fourier Transform in the data pre-processing, reducing costs in the maintenance of the electro-mechanical system. Recently, Zidi et al. [19] has applied SVM for fault detection in Wireless Sensor Networks. The anomalies are caused by software, hardware, and communication failures, such as sensor or energy problems. This work obtains positive results as compared to other proposals, such as Hidden Markov Models or Naive Bayes. Taking into account the previous work, present research proposes the application of SVM to a brand-new dataset about software performance in a robotic system. Furthermore, as the importance of data pre-processing and selection is known [4,10,12], the impact of some different strategies of pre-processing when dealing with missing values and the impact of the data-sources selection are measured.

The rest of this paper is organized as follows: the methods and metrics used for anomaly detection are described in Sect. 2 while the setup of experiments, the dataset under analysis, and the results obtained are described in Sect. 3. Finally, the conclusions of present study as well as proposals for future work are stated in Sect. 4.

2   Anomaly Detection

Anomaly detection can be defined as the problem of finding certain patterns in the data that do not conform to a expected behavior [3]. To detect such unexpected behaviours, there are mainly two alternatives: the first one is to know and model the "normal" (expected) behavior, creating one or multiple groups of usual behavior and leaving out anomalies. The other alternative is to model the known anomalies and identify as anomalous the new data similar to them. Each of these two alternatives has its advantages and disadvantages, which vary significantly in terms of false positives and false negatives (See Sect. 2.2). In present paper, anomaly detection in a robot system is performed, by applying the well-known SVM classifier (See Sect. 2.1). To analyze the performance of the classifier, some different metrics have been applied (described below).

2.1   Support Vector Machines

The SVM [2,5] is a learning model, based on Statistical Learning Theory, whose purpose is to find a hyperplane that maximizes the margin of separation for the training data set, in order to universalize the archetype to be applied to new data. This is the Structural Risk Minimization perspective, as opposed to many other models (such as ANN) that are based on the Empirical Risk Minimization paradigm. SVMs can be seen as classifiers (in the case of one-class classification) where the loss function is the Hinge function, defined as:

$$L[y, f(x)] = max[0, 1 - yf(x)] \tag{1}$$

Being x an observation from input features, y the class x belongs to, and $f(x)$ the output of the classifier. Once trained, that is the support vectors are identified and the margin is maximized, a SVM can be seen as:

$$f(x) = \Sigma_{i \in s} \alpha \cdot y_i \cdot \langle x_i, x \rangle + \beta_0 \tag{2}$$

Being S the set of support vectors, α the classifier coefficients, and β the predictor coefficients.

The SVM configuration applied in present work has a sigmoidal kernel, which corresponds to the Eq. 3.

$$k(x, y) = tanh(ax^T y + c) \tag{3}$$

2.2   Metrics

As previously stated, the performance of SVM when applied to the dataset is measured according to some metrics. They are calculated from the number of data included in each one of the four possible cases, as a result of a one-class classification. In the case of the detection of anomalies, these cases are: True Positives (TP) that are these anomalies properly classified, True Negatives(TN) these normal data rightly classified, False Positives (FP) are normal data labeled as anomaly and finally False Negatives (FN) anomalies classified as normal data. Based on these figures, the following standard metrics are calculated:

Accuracy. Calculate the global hit ratio, regardless of whether the data is an anomaly or not. It is calculated as follows:

$$Accuracy = \frac{TP + TN}{TP + TN + FP + FN} \tag{4}$$

Precision. It gives an idea about the proportion of data that the model classifies as anomalous and in fact they are, with respect to the total of anomalies. It is calculated as follows:

$$Precision = \frac{TP}{TP + FP} \tag{5}$$

Recall. This indicator, also known as True Positive Rate (TPR), focuses on highlighting the relevant data in the dataset (in present case the anomalies). It is obtained in the following way:

$$Recall = \frac{TP}{TP + FN} \tag{6}$$

False Positive Rate (FPR). It is calculated as the percentage of "normal" data erroneously classified as anomalies. Together with TPR it is used to calculate the ROC Curve.

$$FPR = \frac{FP}{FP + TN} \tag{7}$$

F$_1$ Score. Given the impossibility of improving all the measures at the same time, since there are dependencies between them (if precision grows, Recall decreases), this metric is used that expresses a balance between both, as can be observed in your calculation:

$$F_1 = 2 * \frac{Precision * Recall}{Precision + Recall} \tag{8}$$

ROC Curve. ROC is a probability curve, which confronts TPR with FPR, to find the point that provides the best balance between both indicators. The larger the area under the curve (AUC), the better. AUC is the key metric in present paper as it is the one used by original authors [17] and consequently the one to get the fairest comparison.

3 Experiments and Results

This section present the analyzed dataset as well as the obtained results in the different executions and the experimental setup.

3.1 Dataset

The dataset used in the present work (A Data Set for Fault Detection Research on Component-Based Robotic System) is available in a public repository [16]. This dataset, generated by researchers from the Bielefeld University (Germany), is thoroughly explained by the authors in a paper [15]. It contains information about the problems that are addressed, details about the induced anomalies and information about the robot from which the data were collected, as well as the context.

The dataset consists of 71 trials, referring to the interaction of the robot with its environment, during a given time. Each trial data is divided into 13 series sets, according to the component of the robot where the data come from.

Each data instance from the data set is a sampling of the different sensors at a certain time. Each feature is the value of a characteristic at that time. Associated with each component of the robot for each of the trials, two data sources, with different types of data, are used: features and counters. Counters are the raw export of the performance counters for this component, whereas features are a combination of performance counters and events with the timing of the counters.

To ease comparison and for the sake of brevity, present paper analyses anomalies related to one of the components: armcontrol, which is responsible of controlling the robot arm. To move the arm, a planning of the movements and a connection to the arm interface are required. The anomaly that concerns this module is named armserverAlgo. The movements performed by the robot arm are valid, while the induced anomalies cause unnecessary, but still valid, movements.

3.2 Results on the Whole Dataset

This section shows the results obtained when trying to detect the armserver-Algo anomaly by means of SVM. For all the trials which are valid and do not comprise unexpected anomalies, features, counters and the union of both have been extracted. Finally different experiments have been carried out, as explained below (See Table 1), using a validation set of 25% of the data. One of the main

contributions of present paper is the comparison of pre-processing alternatives
when dealing with the missing values (MV). As SVM can not handle MV, they
must be removed before providing the model with the data. To remove these
values, two main alternatives have been considered: eliminating instances (rows)
with any MV or eliminating characteristics (columns) with any of these values.
In present work, both alternatives have been combined, in a variety of ways.
That is the reason why the MV ratios of 10%, 20%, and 50% have been used,
which reflect the permissiveness of MV at the time of choosing some characteristics or others. After that, as there can not be MV in the data provided to the
model, the instances (with the selected characteristics according to the selected
ratio), containing at least one of these values, are eliminated. As a results of the
different combinations of permissiveness, data sources (features, counters and
both of them), and repetitions, 6 different executions have been performed, as
can be seen in Table 1.

Table 1. Description of the executions carried out. F = *Features*, C = *Counters*, and
F+C = *Features + Counters*

Execution	MV ratio	Number of rows			Number of columns			Repetitions
		F	C	F+C	F	C	F+C	
1	0%	21887	21887	21887	33	9	42	5
2	0%	21887	21887	21887	33	9	42	10
3	0%	21887	21887	21887	33	9	42	20
4	10%	18505	18530	21887	51	9	60	5
5	20%	14429	18530	11762	56	11	67	5
6	50%	7350	18530	5974	66	11	77	5

The obtained values for the Accuracy and AUC metrics in the different experiments for the 3 different data sources are shown in the Fig. 1. It can be observed
how the execution 3 stands out for AUC when only using Counters, and obtaining
the significantly best value when using Features and Counters. On the contrary,
some of the lowest values are obtained by execution 6, for the two metrics. All in
all, the increase in the MV ratio has lead to worse results when using counters
due to the low amount of MV that can be found in it, when compared with the
results in the union of both data sources.

Complementary, the obtained values for the Precision, Recall, and F_1 Score
metrics in the different experiments for the 3 different data sources are shown in
the Fig. 2. When compared to Accuracy and AUC results, it is worth mentioning
that the execution 6 has obtained the best values for Precision, Recall and F_1
score when only using Features.

Cross-Validation. As previously mentioned, cross-validation has been applied
to exhaustively validate the generated model. In order to do it, three new executions have been carried out, varying the number of partitions (folds) from the

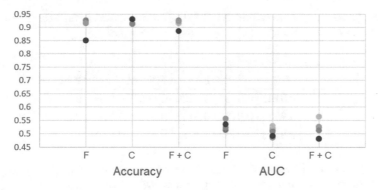

Fig. 1. Accuracy and AUC values for the different executions and data sources.

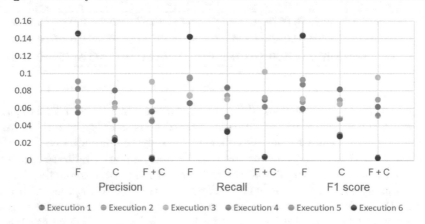

Fig. 2. Precision, recall, and F_1 score values for the different executions and data sources.

dataset in execution 3; the different values for the number of partitions are 5, 10, and 20. The obtained results for these new experiments can be seen in Tables 2 and 3. From this table, it can be said that the values obtained are similar to those have been previously found, where it can be highlighted the high values obtained for Precision, Recall, and F_1 score for 20 folds and the combination of both data sources. Meanwhile, similar values to those previously stated were obtained for the AUC and Accuracy metrics.

3.3   Results on Trial 21

Among all the trials in the dataset, one of them (the number 21) can be identified as a special one regarding the anomaly selected in the present work (armServer-Algo). Together with the trial number 23, they are the only two trials in which there are two instances of the induced anomalies while in all the others, there

Table 2. Values for accuracy and AUC using cross-validation on experiment 3. F = *Features*, C = *Counters* and F+C = *Features + Counters*.

Folds	Accuracy			AUC		
	F	C	F+C	F	C	F+C
5	0.9131	0.9107	0.9139	0.5051	0.506	0.5177
10	0.918	0.9131	0.9151	0.5196	0.5181	0.524
20	0.9154	0.9132	**0.923**	0.5141	0.5133	**0.545**

Table 3. Values for precision, recall and F_1 score using cross-validation on experiment 3. F = *Features*, C = *Counters* and F+C = *Features + Counters*.

Folds	Precision			Recall			F_1 score		
	F	C	F+C	F	C	F+C	F	C	F+C
5	0.0525	0.0578	0.0775	0.0616	0.0611	0.0857	0.0566	0.0594	0.0808
10	0.0778	0.0806	0.0889	0.0968	0.085	0.1014	0.0858	0.0825	0.0942
20	0.0697	0.0722	**0.129**	0.0784	0.0722	**0.141**	0.0737	0.0721	**0.134**

is only one instance. The main difference between trials 21 and 23 is that in the former, the anomaly instances were induced consecutively. For this reason, special attention is paid to the trial 21 that is analyzed individually. The SVM has been trained with the data of the trial 21 and has been validated with the rest of the trials in the whole dataset that contain the same anomaly. Trial 21 has been analyzed comprising both features and counters data, and with a 0% rate of MV as this was the combination that obtained the best results in the previous execution. The dimensions of both training and testing sets, have the same number of columns, but there is a big difference between the number of instances, since there are 667 for the training set and 21887 for the validation set, being a big difference.

Obtained values for the different metrics when analyzing the trial 21 are shown in the Table 4. As can be seen, positive results have been obtained in the case of Accuracy, taking into account the dimensions of the training and validation sets. A satisfactory value for F_1 score is also observed, comparing it with the previous results, as this is one of the highest values. The main reason for that is the value obtained for Precision. With regard to AUC, a similar value has been obtained when compared to those obtained in previous experiments. This is remarkable given that the training set is much smaller than the validation one.

Table 4. Values for the different metrics for the experiment with the trial 21.

Accuracy	Precision	Recall	F_1 score	AUC
0.7014	0.379	0.0638	0.109	0.5484

4   Conclusions and Future Work

In present paper, different alternatives for data selection (management of missing values and selection of data sources) have been validated for the detection of anomalies in a component-based robotic system. Obtained results for the same classifier (SVM) can be seen in the Sect. 3.2 is a key element given that depending on the metrics that is analyzing the results. That is because obtained results vary depending on the metric as none of the data-selection alternatives obtained the best scores for all the metrics.

As it has been observed in the obtained results, the different values of MV ratio means different values for the metrics under analysis. A high MV ratio provides very good values in Precision, Recall and F1 score for the Features data source. At the same time, the accuracy and AUC metrics are penalized. On the contrary, a 0% ratio in both data sources leads to the best AUC values while maintaining good stability in the rest of the metrics.

The best results obtained in present work can be compared to the results obtained by the original authors of the dataset [17]. For this comparison, the values of the AUC metric are the ones that can be compared as this is the only metric provided by these authors. Additionally, results are shown for a standard OCSVM classifier on the raw dataset and for an alternative model proposed by the authors. Execution number 3 of the present work has obtained a value of 0.563 for the AUC metric (see Fig. 1). This value is higher than that reported by Wienke et al. [17]. Complementary, the value of Accuracy obtained in execution 3 is quite positive (0.9166), so it can be concluded that a model with a good capacity for differentiation between the anomalous and normal state has been obtained when combining a 0% MV ratio and Features + Counters data sources.

As a next step in this ongoing research, it is proposed to apply the same data selection and classification models to all the anomalies in the dataset, in order to validate them. Furthermore the multi-class perspective will also be considered.

References

1. Banerjee, T.P., Das, S.: Multi-sensor data fusion using support vector machine for motor fault detection. Inf. Sci. **217**, 96–107 (2012)
2. Boser, B.E., Guyon, I.M., Vapnik, V.N.: A training algorithm for optimal margin classifiers. In: Proceedings of the Fifth Annual Workshop on Computational Learning Theory, COLT 1992, pp. 144–152. ACM, New York (1992)
3. Chandola, V., Banerjee, A., Kumar, V.: Anomaly detection: a survey. ACM Comput. Surv. **41**(3), 15:1–15:58 (2009)
4. Corchado, E., Herrero, Á., Sáiz, J.M.: Testing cab-ids through mutations: on the identification of network scans. In: Gabrys, B., Howlett, R.J., Jain, L.C. (eds.) Knowledge-Based Intelligent Information and Engineering Systems, pp. 433–441. Springer, Heidelberg (2006)
5. Cortes, C., Vapnik, V.: Support-vector networks. Mach. Learn. **20**(3), 273–297 (1995)
6. Graña, M., Alonso, M., Izaguirre, A.: A panoramic survey on grasping research trends and topics. Cybern. Syst. **50**, 40–57 (2019)

7. Herrero, Á., Jiménez, A.: Improving the management of industrial and environmental enterprises by means of soft computing. Cybern. Syst. **50**(1), 1–2 (2019)
8. IFR: International Federation of Robotics. https://ifr.org/ifr-press-releases
9. Malhotra, R.: A systematic review of machine learning techniques for software fault prediction. Appl. Soft Comput. **27**, 504–518 (2015)
10. Pérez, H., Alfonso-Cendón, J., Fernández-Robles, L., Sánchez-Gonzaález, L., Castejón-Limas, M., Corchado, E., Quintian, H.: Use of classifiers and recursive feature elimination to assess boar sperm viability. Log. J. IGPL **26**(6), 629–637 (2018)
11. Ranshous, S., Shen, S., Koutra, D., Harenberg, S., Faloutsos, C., Samatova, N.F.: Anomaly detection in dynamic networks: a survey. Wiley Interdisc. Rev. Comput. Stat. **7**(3), 223–247 (2015)
12. Sedano, J., González, S., Herrero, Á., Baruque, B., Corchado, E.: Mutating network scans for the assessment of supervised classifier ensembles. Log. J. IGPL **21**(4), 630–647 (2012)
13. Shon, T., Kim, Y., Lee, C., Moon, J.: A machine learning framework for network anomaly detection using SVM and GA. In: Proceedings from the Sixth Annual IEEE SMC Information Assurance Workshop, pp. 176–183, June 2005
14. Shrouf, F., Ordieres, J., Miragliotta, G.: Smart factories in industry 4.0: a review of the concept and of energy management approached in production based on the internet of things paradigm. In: 2014 IEEE International Conference on Industrial Engineering and Engineering Management, pp. 697–701, December 2014
15. Wienke, J., zu Borgsen, S.M., Wrede, S.: A data set for fault detection research on component-based robotic systems. In: Alboul, L., Damian, D., Aitken, J.M. (eds.) Towards Autonomous Robotic Systems, pp. 339–350. Springer, Cham (2016)
16. Wienke, J., Wrede, S.: A Fault Detection Data Set for Performance Bugs in Component-Based Robotic Systems (2016)
17. Wienke, J., Wrede, S.: Autonomous fault detection for performance bugs in component-based robotic systems. In: 2016 IEEE/RSJ International Conference on Intelligent Robots and Systems (IROS), pp. 3291–3297. IEEE (2016)
18. Wienke, J., Wrede, S.: Continuous regression testing for component resource utilization. In: IEEE International Conference on Simulation, Modeling, and Programming for Autonomous Robots (SIMPAR), pp. 273–280. IEEE (2016)
19. Zidi, S., Moulahi, T., Alaya, B.: Fault detection in wireless sensor networks through SVM classifier. IEEE Sens. J. **18**(1), 340–347 (2018)

Addressing Low Dimensionality Feature Subset Selection: ReliefF(-k) or Extended Correlation-Based Feature Selection(eCFS)?

Antonio J. Tallón-Ballesteros[1]([envelope]), Luís Cavique[2], and Simon Fong[3]

[1] Department of Electronic, Computer Systems and Automation Engineering,
University of Huelva, Huelva, Spain
`antonio.tallon@diesia.uhu.es`
[2] Universidade Aberta, Lisbon, Portugal
[3] Department of Computer and Information Science, University of Macau,
Taipa, Macau SAR, China

Abstract. This paper tackles problems where attribute selection is not only able to choose a few features but also to achieve a low performance classification in terms of accuracy compared to the full attribute set. Correlation-based feature selection (CFS) has been set as the baseline attribute subset selector due to its popularity and high performance. Around hundred data sets have been collected and submitted to CFS; then the problems fulfilling simultaneously the conditions: (a) a number of selected attributes lower than six and (b) a percentage of selected attributes lower than a forty per cent, have been tested onto two directions. Firstly, in the scope of data selection at the feature level, an advanced contemporary approach have been conducted as well as some options proposed in a prior work. Secondly, the pre-processed and initial problems have been tested with some sturdy classifiers. Moreover, this work introduces a new taxonomy of feature selection according to the solution type and the followed way to compute it. The test bed comprises seven problems featured by a low dimensionality after the CFS application, three out of them report a single selected attribute, another one with two extracted features and the three remaining data sets with four or five retained attributes; additionally, the initial feature set is between six and twenty-nine and the complexity of the problems, in terms of classes, fluctuates between two and twenty-one, throwing averages of sixteen and around five for both aforementioned properties. The contribution concluded that the advanced procedure (extended CFS) is suitable for problems where only one or two attributes are selected by CFS; for data sets with more than two selected features the baseline method is preferable to the advanced one, although the considered feature ranking method achieved intermediate results.

Keywords: Machine learning · Correlation · Feature subset selection · Feature ranking · Extended feature subset selection · CFS

© Springer Nature Switzerland AG 2020
F. Martínez Álvarez et al. (Eds.): SOCO 2019, AISC 950, pp. 251–260, 2020.
https://doi.org/10.1007/978-3-030-20055-8_24

1   Introduction

Data pre-processing [15] is a major step inside CRISP-DM (CRoss Industry Standard Process for Data Mining, [17]) which is by its part integrated into Data Engineering [4]. Pre-processing aims at transforming the raw input data into an appropriate format for subsequent analysis [2]; because of the many ways data can be collected and stored, data pre-processing is perhaps the most laborious and time-consuming [15] step in the overall CRISP-DM process.

This paper addresses the scenario reached when Correlation-based Feature Selection is able to only extract a few attributes and goals to mitigate this problem via proposing a new feature selection method as well as using contemporary strategies. The rest of this paper is organised as follows: Sect. 2 provides a brief background. Section 3 details the proposed procedure. Section 4 explains the experimentation via the algorithms, data preparation methods and their setting. Section 5 details and analyses the results. Finally, Sect. 6 states the conclusions.

2   Background

The motivation to reduce the dimensionality of the feature space is closely related to the decreased time required to double information in the world every year. Surveys on feature selection methods can be found in [9]. The aim of feature selection is to find a subset of the attributes from the original set which are representative enough for the data, and the attributions in the subset are highly relevant to the prediction.

There are three basic models in feature selection: Filter, Wrapper and Hybrid model. In the Filter model the most popular independent criteria are consistency, distance, correlation and information measures. The Wrapper model is divided into two steps, but with strong interaction between the feature selection phase and the learning phase, where the results of the prediction are used as a criterion of feature choice. Hybrid methods have been proposed to reduce features in classification by combining the advantages of the two previous methods.

3   The Proposed Approach

This contribution introduces a new possibility in the classical taxonomy of feature selection according to the generation procedure which is inherently tied to the solution type. The traditional options are feature ranking and feature subset selection. Supported by the concept of logical complement which is a very convenient operator in the branch of logic, the category of extended Feature Subset Selection (eFSS) is proposed, which comprises the combination of features that are extracted both the application of feature subset selection with the initial feature set and those features retained from the logical complement of the prior feature selection, i.e., the features which have not been selected from the first step of feature selection. Basically, the new type of method is the application of a particular method of feature subset selection more than once with feature spaces overlapped partially. Figure 1 depicts the whole taxonomy of feature selection according to the operation mode as well as the solution type.

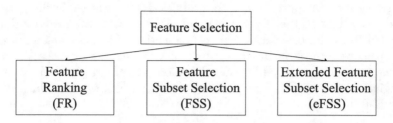

Fig. 1. Taxonomy of feature selection methods according to the operation mode and solution type.

The current paper addresses how to cope with data sets where feature subset selection based on correlation only selects a few attributes. Particularly, we focus on CFS method which is one the most outstanding procedures in the data mining research. CFS may fail to select relevant features, however, when data contains strongly interacting features or features with values predictive of a small area of the instance space. Sometimes, CFS may retain a number of attributes lower or equal than five. A problem which is described by one attribute and the label class may be thought not to be very interesting in terms of predictability; if for any unseen sample we need only one property to distinguish the class, two possible scenarios may happen: the first means that we are dealing with a very simple problem or the second represents an inaccurate application of feature selection. The reason which motivates us is the analysis of the complementary reduced set. Some extra features may be considered as relevant, after the application of the second step of feature subset selection on the negation reduced set. In this way, we are dividing the search space and separating some potential interacting attributes. Moreover, instead of keeping a stand alone in the feature subset methods, the feature ranking approaches may be very suitable for the scenarios where FSS only retains a few attributes. Concretely, under this context the method ReliefF($-k$) [13] -also named as Leave-k-out ReliefF- showed to be a good tool in the sense that some previous experiments where conducted and the behaviour was very acceptable. The values for k needed to be determined empirically, although the preliminary study is now very useful because we can compare a new method with one already published and initially tuned. The scope of this paper considers problems where feature subset selection via CFS is only able to choose between one and five features and additionally that the percentage of selected attributes is below a forty per cent. The last factor is taken into account to avoid especially data sets where the percentage of selected attributes is not in the range from almost the half feature set up to a close value to the whole feature set. Returning to the idea of applying a feature subset selection more than once, in a very recent paper we proposed a similar method to characterise the features as essential and important; nonetheless the scope is different because our purpose was to establish different levels in the pyramid of features according to the importance and also to assess the synergy that may be created from the interaction between essential and important features [12]. We deep more into eFSS in the context of CFS and we propose the method extended Correlation- based Feature Selection (eCFS) which is an extension of the original CFS method, a very prolific procedure which was introduced 20 years ago by Hall [8].

4   Experimentation

Several classifiers have been used to assess the current proposal. The machine learning algorithms, which have been chosen, belong to different typologies according to the way to represent the knowledge. As decision tree learner, the classifier is C4.5 [10] which is an extension of the popular ID3 algorithm [11], one of the classical approaches inside Artificial Intelligence. As lazy learning method, the candidate is 1NN [1] which is a subtype of the algorithm k-nearest neighbours [5]. PART (PARTial Decision Tree) [6] is the option within the methods based on decision rules. Finally, SVM (Support Vector Machines) is a classifier which creates a hyperplane or set of hyper-planes [16]. From the above classifiers, C4.5 and 1NN has been tested traditionally in the scope of feature selection in several works such as PhD Theses [8] or surveys [3]; additionally, the two aforementioned methods as well as the classifiers PART and SVM have been tested in a personal review about feature selection which was dated in 2016 [14].

Table 1. Feature selection procedures used in the experimentation

Feature selection method	Type	Parameter/Property	Value	Reference
ReliefF(−k)	*FR*	*Number of attributes to drop*	*Depends on the problem properties*	[13]
CFS	*FSS*	*Attribute evaluation measure*	*Correlation*	[8]
		Search method	*Best First*	
		Consecutive expanded nodes without improving	*5*	
		Search direction	*Forward*	
eCFS	*eFSS*	*Attribute evaluation measure*	*Correlation*	
		Search method	*Best First*	
		Consecutive expanded nodes without improving	*5*	
		Search direction	*Forward*	

Table 2. Classification data sets

Data set	Samples			Labels	Features				
	Total	Training	Testing		Number			Selected percentage	
					Original	CFS	eCFS	CFS	eCFS
Liver	345	259	86	2	6	1	2	16.7	33.3
Lenses	24	18	6	3	6	1	2	16.7	33.3
Post − op	90	67	23	3	20	1	2	5.0	10.0
Hypo	3772	2829	943	4	29	2	4	6.9	13.8
Breast	286	215	71	2	15	4	5	26.7	33.3
Smoking	2855	2141	714	3	13	5	10	38.5	76.9
Primary − tumor	339	251	88	21	23	5	6	21.7	26.1
Average	1101.6	825.7	275.9	5.4	16.0	2.7	4.4	18.9	32.4

As feature selection methods, we have considered CFS and ReliefF(-k) as the baseline procedures. ReliefF(-k) is based on the Weka implementation of ReliefF. The current proposal took CFS as the supporting and hence has been also implemented under the Weka (Waikato Environment for Knowledge Analysis) framework [7] which is an open source software created and maintained by the University of Waikato in New Zealand. For the classifiers we have conducted all the experiments with the default values since these are the recommendation of the own authors of the algorithms. Table 1 describes the setting of the attribute selection methods as well as some remarkable properties available in the implementations provided by the workbench Weka.

The test bed includes seven data sets available at the repository maintained by the University of California at Irvine which has been partitioned following a stratified hold-out procedure with three and one quarters, respectively for the training and testing sets. Table 2 gives count about the problems along with its properties; the rows are sorted in ascendant order for the number of features selected and then the number of labels.

5 Results

This section reports the results which are measured in test accuracy. From the whole test-bed two different scenarios have been created to characterise, follow and understand the proposal. Firstly, we show the results for data sets where CFS extracts one or two features. Secondly, problems, where CFS retains more than two attributes and up to five, are analysed.

Table 3 details the results in the landscape where the final feature set after CFS comprises one or a couple of attributes. We have represented the results with the full feature set, after CFS as well as some variations of the method ReliefF(-k) -or R(-k) to shorten within the table- and the proposed method eCFS. It is of particular interest to mention that the alternative/s were chosen as were proposed as the recommendation is a previous contribution published in 2017. Moreover, in specific cases an extra alternative has been considered to compare the performance of eCFS, an eFSS method, with a solution with the same complexity based on FR such as ReliefF(-k). Just to have a global appearance, we have included a pairwise comparison. We have not represented the ties since there is a good number of them; we have included instead the improvements (+) or the deterioration (-). After the individual results for each problem and classifier, the average is computed to have other kind of value to compare more concretely. As eCFS is an enhanced version of CFS, we have compared the former and the latter, in this order, and the letter B in the row entitled Ind. Avg. (individual average) means that the first method (eCFS) is better than (CFS) in terms of average. The column with the label Summ. includes the wins, ties or losses (W/T/L) and is a very convenient metric to have a qualitative value to compare different approaches. The total row shows the global summation of the below W/T/L values. Going down in the table, we see the global and the partial averages; the first value has been obtained for the methods where all data sets have been assessed and the second one has been computed for the procedures where a subset (of at least two) of the whole test bed has been considered, e.g. R(-2) has been tested with Liver, Lenses and Post-op and hence the partial averages of FULL, CFS and eCFS are useful to compare R(-2) with the competitors in the same situation. After having explained the contents of the table, it is time to provide some insights. eCFS is preferable to CFS in the situation I. By its part, conservative Relief(-k), i.e. ReliefF(-2) in the first three data sets where CFS selected one attribute or ReliefF(-3) for Hypo, is more convenient than eCFS although with a higher computation time since are discarding a couple of three attributes with Relief(-k). Additionally, eCFS gets a better than Relief(-k) with a similar complexity; clearly the advantage of eCFS is that is an almost parameter-free method instead of requiring a very specific setting for different groups of data sets as ReliefF(-k) requires. In terms of classifiers and global results, eCFS is a very convenient option for 1NN, followed by C4.5 and PART with two wins; finally, for SVM only one win is got.

Table 4 shows the assessment in data sets where CFS retains four attributes, as in Breast or Smoking, or five features for Primary-tumor. The meaning of the different columns of the table is exactly as described in the previous table and we do not repeat it due to space issues. Roughly speaking, eCFS is not able to keep, at least, same results as CFS. It represents that for problems where CFS selects more than three attributes is not undoubted that eCFS is better than CFS and a very careful analysis is hence necessary. Exceptionally, in Smoking there is one win and one loss for eCFS compared to CFS; although the average is higher for eCFS, it does not mean a better performance versus CFS as the rank-based methods assert. The good news is that eCFS performs better than Relief(-k) in global terms according to the qualitative values for the problems with four selected features by CFS. Unfortunately, for Primary-tumor the results without feature selection are better than with any dimensionality reduction procedure. It is very important to remind that there are 21 labels and 23 features in this problem which makes it very challenging. We move on to the performance of

Table 3. Situation I. Test classification results with possible approaches for data sets where CFS selects one or two attributes

Data set	Classifier	Feature set							Pairwise comparison					Alternative/s	Remark/s
		FULL	CFS	R(−2)	R(−3)	R(−4)	R(−5)	eCFS	eCFS versus CFS +/−	eCFS versus Alt.1 +/−	Alt.2 +/−	Alt.3 +/−	Summ. +/−		
Liver	C4.5	68.60	58.14	68.60		58.14		58.14	−					R(−2)	Alt. 2 : R(−4) similar complexity to eCFS
	1NN	61.63	39.53	54.65		51.16		54.65	+	+					
	PART	61.63	58.14	68.60		58.14		58.14	−						
	SVM	58.14	53.49	62.50		56.40		57.27 B							
Ind. Avg.		62.50	53.49	62.50		56.40		57.27					1/3/0		
Lenses	C4.5	66.67	50.00	66.67		66.67		66.67	+					R(−2)	Alt. 2 : R(−4) similar complexity to eCFS; R(−4) selects same features as eCFS
	1NN	16.67	50.00	83.33		66.67		66.67	+	−					
	PART	66.67	50.00	66.67		66.67		66.67	+						
	SVM	66.67	50.00	83.33		66.67		66.67 B		−					
Ind. Avg.		54.17	50.00	75.00		66.67		66.67 B					4/0/0		
Post − op	C4.5	52.17	56.52	52.17		52.17		56.52	+	+				R(−2)	Alt. 2 : R(−4) similar complexity to eCFS
	1NN	56.52	4.35	56.52		56.52		8.70	−	−					
	PART	65.22	56.52	56.52		56.52		56.52	+						
	SVM	56.52	56.52	56.52		56.52		44.57 B							
Ind. Avg.		57.61	43.48	55.43				44.57 B					1/3/0		
Hypo	C4.5	99.15	96.92		99.26	98.94		97.56	+	−	−	+		R(−3), R(−4), R(−5)	
	1NN	90.99	96.50		90.99	90.88		95.44	−	+	+	+			
	PART	98.83	96.92		98.83	98.73		97.35	+	+	−	−			
	SVM	93.85	93.32		93.85	93.85		93.32	+	−	−	−			
Ind. Avg.		95.70	95.92		95.73	95.60		95.92					2/1/1		
Total													8/7/1		
Global Avg.	C4.5	71.65	65.40					69.72	+				2/2/0		
	1NN	56.45	47.60					56.36	+				3/0/1		
	PART	73.09	65.40					69.67	+				2/2/0		
	SVM	68.79	64.50					68.66	+				1/3/0		
Partial Avg.	C4.5	62.48	54.89	62.48				60.44	+						
	1NN	44.94	31.29	64.83				43.34	+						
	PART	64.50	54.89	63.93				60.44	+						
	SVM	60.44	54.89	66.00				60.44	+						

+ means improvement and − means deterioration. B means Better

R(−k) stands for ReliefF(−k) where k is a positive value.   Alt. represents an alternative

Summ. is a summary which is expressed as W/T/L

Table 4. Situation II. Test classification results with possible approaches for data sets where CFS selects more than three attributes and less or equal than five

Data set	Classifier	Feature set							Pairwise comparison				Alternatives	Remark
		FULL	CFS	R(−2)	R(−3)	R(−4)	R(−5)	eCFS	eCFS versus CFS Summ. +/−	eCFS versus Alt. 1 +/−	Alt. 2 +/−	Alt. 3 +/−		
Breast	C4.5	70.42	69.01	70.42	70.42	70.42		64.79	−	−	−		R(−2), R(−3), R(−4)	
	1NN	64.79	70.42	67.61	69.01	69.01	70.42	70.42		+	+	+		
	PART	69.01	71.83	64.79	63.38	67.61		70.42		+	+	+		
	SVM	64.79	66.20	66.20	66.20	64.79		66.20		+				
Ind. Avg.		67.25	69.37	67.25				67.96	Worse 0/2/2					
Smoking	C4.5	68.63	69.47	67.65	69.47			69.47	+	+			R(−2), R(−3), R(−4)	
	1NN	54.76	38.52	56.86	50.28	50.14		50.28	−	−				Summation for
	PART	61.48	67.36	62.75	61.76	66.11		62.18	−		+			the alternatives
	SVM	69.47	69.47	69.47	69.47			69.47	−					
Ind. Avg.		63.59	61.20	64.18	62.75	63.80		62.85	Better 1/2/1				R(−3), R(−4), R(−5)	W/T/L : 10/8/6
Primary − tumor	C4.5	45.46	42.05	40.91	40.91	40.91	43.18	40.91	−	−	−		R(−3), R(−4), R(−5)	
	1NN	36.36	30.68	36.36	37.50	37.50		30.68	−	−	−			
	PART	43.18	42.05	40.91	39.77	38.64		38.64	−	−	−			
	SVM	47.72	42.05	48.86	48.86	47.73	40.91	40.91	−	−	−			
Ind. Avg.		43.18	39.20	41.76	41.76	43.18	41.76	37.78	Worse 0/1/3					
Total									1/5/6	3/2/3	3/4/1	4/2/2		
Global Avg.	C4.5	61.50	60.18	60.18	60.27			58.39	−	0/1/2				
	1NN	51.97	46.54	46.54	51.88	52.22		50.46	+	1/2/0				
	PART	57.89	60.41	60.41	55.35	57.83		57.08	−	0/0/3				
	SVM	60.66	59.24	59.24	61.51	61.04		58.86	−	0/2/1				
Partial Avg.	C4.5	61.50	60.18	69.03										
	1NN	51.97	46.54	62.23										
	PART	57.89	60.41	63.77										
	SVM	60.66	59.24	67.83										

R(−k) stands for ReliefF(−k) where k is a positive value.   Alt. represents an alternative [9]
Summ. is a summary which is expressed as W/T/L
+ means improvement and − means deterioration.

the different supervised machine learning algorithms. 1NN is the single classifier which takes advantage of eCFS compared to CFS; SVM keeps the same result twice although the third result is a loss. The remaining classifiers do not achieve competitive results and for these problems CFS is more convenient than eCFS.

6   Conclusions

This paper presented a new category inside the feature selection taxonomy, according to the generation procedure, which has been named eFSS which stands for extended FSS. A new eFSS method, called eCFS (extended CFS), has been introduced, as an interesting alternative to deal with problems where CFS only selects a few attributes. It is very noticeable that for data sets where CFS picks up one or two features eCFS is very recommendable. On the contrary, for problems where CFS retains more than two features and up to five, eCFS is only a more convenient approach than ReliefF(-k) even with the fine-tuning for k parameter; nonetheless in this situation CFS is preferable to eCFS since a lower feature set is achieved and the performance is not always worse.

Acknowledgments. This work has been partially subsidised by TIN2014-55894-C2-R and TIN2017-88209-C2-2-R projects of the Spanish Inter-Ministerial Commission of Science and Technology (MICYT), FEDER funds and the P11-TIC-7528 project of the "Junta de Andalucía" (Spain).

References

1. Aha, D.W., Kibler, D., Albert, M.K.: Instance-based learning algorithms. Mach. Learn. **6**(1), 37–66 (1991)
2. Bilalli, B., Abelló, A., Aluja-Banet, T., Wrembel, R.: Intelligent assistance for data pre-processing. Comput. Stand. Interfaces **57**, 101–109 (2018)
3. Chen, Y., Li, Y., Cheng, X.-Q., Guo, L.: Survey and taxonomy of feature selection algorithms in intrusion detection system. In: International Conference on Information Security and Cryptology, pp. 153–167. Springer, Heidelberg (2006)
4. Cho, S.-B., Tallón-Ballesteros, A.J.: Visual tools to lecture data analytics and engineering. In: International Work-Conference on the Interplay Between Natural and Artificial Computation, pp. 551–558. Springer, Heidelberg (2017)
5. Cover, T., Hart, P.: Nearest neighbor pattern classification. Inf. Theor. IEEE Transact. **13**(1), 21–27 (1967)
6. Frank, E., Witten, I.H.: Generating accurate rule sets without global optimization. In: Shavlik, J. (eds.) Fifteenth International Conference on Machine Learning, pp. 144–151. Morgan Kaufmann (1998)
7. Hall, M., Frank, E., Holmes, G., Pfahringer, B., Reutemann, P., Witten, I.H.: The weka data mining software: an update. ACM SIGKDD Explor. Newsl. **11**(1), 10–18 (2009)
8. Hall, M.A.: Correlation-based feature selection for machine learning. Ph.D thesis, University of Waikato, Hamilton, New Zealand (1999)
9. Miao, J., Niu, L.: A survey on feature selection. Proc. Comput. Sci. **91**, 919–926 (2016)

10. Quinlan, J.R.: C4.5: Programs for Machine Learning, vol. 1. Morgan kaufmann, Burlington (1993)
11. Shapiro, A., Niblett, T.: Automatic induction of classification rules for a chess endgame. In: Advances in Computer Chess, pp. 73–92. Elsevier (1982)
12. Tallón-Ballesteros, A.J., Correia, L., Xue, B.: Featuring the attributes in supervised machine learning. In: International Conference on Hybrid Artificial Intelligence Systems, pp. 350–362. Springer (2018)
13. Tallón-Ballesteros, A.J., Riquelme, J.C.: Low dimensionality or same subsets as a result of feature selection: an in-depth roadmap. In: International Work-Conference on the Interplay Between Natural and Artificial Computation, pp. 531–539. Springer (2017)
14. Tallón-Ballesteros, A.J., Riquelme, J.C., Ruiz, R.: Merging subsets of attributes to improve a hybrid consistency-based filter: a case of study in product unit neural networks. Connect. Sci. **28**(3), 242–257 (2016)
15. Tan, P.-N.: Introduction to Data Mining. Pearson Education, India (2018)
16. Vapnik, V.N.: The Nature of Statistical Learning Theory. Springer, Heidelberg (1995)
17. Wirth, R., Hipp, J.: CRISP-DM: towards a standard process model for data mining. In: Proceedings of the 4th International Conference on the Practical Applications of Knowledge Discovery and Data Mining, pp. 29–39. Citeseer (2000)

A Predictive Maintenance Model Using Recurrent Neural Networks

Alberto Rivas[1]([✉]), Jesús M. Fraile[1], Pablo Chamoso[1],
Alfonso González-Briones[1], Inés Sittón[1], and Juan M. Corchado[1,2,3,4]

[1] BISITE Research Group, University of Salamanca, Edificio I+D+i, Calle Espejo 2, 37007 Salamanca, Spain
{rivis,jezu1996,chamoso,alfonsogb,isittonc,corchado}@usal.es
[2] Air Institute, IoT Digital Innovation Hub (Spain), Carbajosa de la Sagrada, 37188 Salamanca, Spain
[3] Department of Electronics, Information and Communication, Faculty of Engineering, Osaka Institute of Technology, 535-8585 Osaka, Japan
[4] Pusat Komputeran dan Informatik, Universiti Malaysia Kelantan, Karung Berkunci 36, Pengkaan Chepa, 16100 Kota Bharu, Kelantan, Malaysia

Abstract. One of the main goals of Industry 4.0 is to anticipate machine breakdowns. Being able to prevent failures is important because downtime implies high cost and production loss. For this reason, the calculation of the number of remaining cycles or Remaining Useful Life (RUL) until a breakdown occurs is essential for machine maintenance. The calculation of the RUL should be based on previous observations, if possible under the same conditions. Research on RUL estimation has become central to the development of systems that monitor the current state of machines. Although this field has been studied in-depth, there is no single universal method. The lack of a universal method is the motivation behind this proposal in which the designed system uses recurrent neural networks (RNN) in a predictive maintenance problem.

Keywords: Remaining useful life · Recurrent neural network · Predictive maintenance · Industry 4.0

1 Introduction

Data analysis has become very important due to the considerable increase in computers' computing capacity. The industrial sector is one of those that has gained the greatest benefit from this development, since the large computing capacity has made it possible to analyse and study monitoring systems almost in real time [1,8].

Initially, monitoring was understood as the measurement of a physical variable representative of the condition of the machine and its subsequent comparison with known values, which indicated whether or not the machine was functioning correctly. With the advances in computing and the current automation of these techniques, the meaning of the word "monitoring" has also been

© Springer Nature Switzerland AG 2020
F. Martínez Álvarez et al. (Eds.): SOCO 2019, AISC 950, pp. 261–270, 2020.
https://doi.org/10.1007/978-3-030-20055-8_25

extended to the acquisition, processing and storage of data. Monitoring, protection, diagnosis and forecasting are the key functionalities necessary for the proper monitoring of a machine.

These studies have provided great benefits in terms of increased machine production or machine breakdown prevention [3,7,15].

This paper focuses on the creation of a model based on recurrent neural networks with the aim of identifying a possible proximate malfunction. The designed system analyzes a dataset of historical sensor data in order to perform a predictive maintenance of a set of engines [2,14].

The rest of the article is structured as follows: the next section provides background information about the different types of maintenance as well as the main machine learning models used for this purpose. The case study is described in Sect. 3. In Sect. 4 different network architectures are presented. Section 5 outlines the results and finally, Sect. 6 draws conclusions from the conducted research.

2   Background

For decades, machine maintenance has played a vital role in the industry due to the enormous economic effects that breakdowns or malfunctions have on a company. Today, maintenance remains a vital element of any industry, and there are four maintenance models that can be followed [18]:

Corrective Maintenance: this kind of maintenance consists in correcting the detected faults or problems to make the system work properly again. There are two types of corrective maintenance:

- **Unplanned Corrective Maintenance:** is the kind of maintenance that is performed when a failure occurs unexpectedly and it is necessary to repair the equipment before it can continue to be used. In this sense, unplanned corrective maintenance focuses on fixing the failures as quickly as possible in order to prevent material and human damage and reduce economic loss.
- **Scheduled Corrective Maintenance:** the objective of this type of maintenance is to anticipate possible failures or damages in the equipment that can occur in the moment. On the basis of previous experience, maintenance periods are established for parts of equipment to identify worn out parts or possible breakdowns. Hence, this type of maintenance proceeds with a general revision that diagnoses the state of the machinery. The time for the revision of equipment is scheduled beforehand so hours of inactivity or little activity can be taken advantage of.

Preventive Maintenance: intends to lessens the likelihood of an equipment failing by looking for solutions to different problems before they occur. This kind of maintenance prevents unplanned corrective maintenance. Although it is aimed at increasing reliability and reducing costs, it does not guarantee that failures will not occur in the future.

Predictive Maintenance: acts as a complement to corrective and preventive maintenance. In this type of maintenance, a series of parameters are monitored and analyzed to determine possible anomalies. In essence, the process consists in generating estimates or assumptions about the status of a particular component. When well-defined processes are predicted, especially in control theory [17], it is possible to generate a reliable mathematical model that faithfully represents the reality [13]. However, other types of processes require experimental techniques, such as classification algorithms [12] or ANN [19]. This approach attempts to extract information and model the system on the basis of historical data.

Proactive Maintenance: is a strategy used to maintain the stability and performance of a machine [10]. The useful life of the equipment is extended while avoiding errors and breakdowns. There are two types of repair [4]: perfect maintenance, when a machine returns to its ideal state (in this case the cost is usually high), and imperfect maintenance, where there is a considerable loss of the quality but the cost of repair is reduced. Another point to take into account in this type of maintenance is the performance of the machine which, due to its natural deterioration over time, should receive periodic revisions throughout its useful life. Thus, periodic maintenance can be carried out where the performance of the machine is evaluated periodically despite its correct performance. The latter solution is not optimal when the revision period is short and the machines operation is optimal. Alternatively, the condition of the machinery can be monitored and its parameters evaluated. A model that combines these two options is presented in [21], where performance loss is predicted on the basis of the failure rate and performance degradation.

One of the most common techniques for dealing with this type of problem are neural network models. Neural networks are a simplified model that emulates the way the human brain processes information. They consist of a large number of interconnected processing units that work simultaneously to perform information processing; they look like abstract versions of neurons. There are different models of neural networks which group neurons differently. Below, we describe the most prominent models.

Recurrent Neural Network (RNN): This is a type of ANN whose architecture structure incorporates memory. It is possible to implement memory because some neurons receive as input the output of one of the previous layers. There are several types of RNNs. Among them, the Long Short-Term Memory (LSTM) network.

- **Long Short-Term Memory (LSTM):** The structure of this type of network consists of memory cells as described in [6]. Their function is to store a value and determine for how long it should be stored. In addition, these cells select which entries to store and decide whether to remember them, delete them or send them as a network output. It is one of the most commonly used models when working with time-dependent data.

Regarding the evaluation of classification models, there are several types of metrics that can be used with the intention of assessing how good a classification is. Some of the most used are precision and recall. In binary classification, precision is the proportion of positive classifications that were correct, similarly, recall is the proportion of real positives that have been identified correctly. With these two metrics F1-Score is obtained which is the harmonic mean of precision and recall. With regard to regression models, the two most common methods of quantifying model performance are the mean squared error (MSE) and the mean absolute error (MAE).

3   Case Study

The aim of this study is to design a predictive maintenance model applicable to the industrial field. Due to the difficulty of obtaining real data and measurements from companies it has been decided to create a functional prototype of predictive maintenance using a public database.

To evaluate the model, a public domain dataset has been selected which can be accessed at [16]. It consists of 1414 motor operating cycles with 21 measurements from different sensors. As can be seen in the Fig. 1 there is a clear upward trend as the machine approaches the end of its useful life. The same applies to all other sensors.

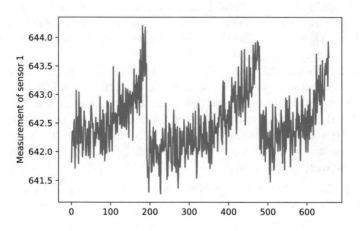

Fig. 1. Measurement of sensor 1 for the first 3 machines

The data have been preprocessed using different smoothing techniques such as moving average, central moving average and linear Fourier smoothing. These database transformations improve network performance [20]. In addition, a min-max normalization model has been applied to the data to scale the attributes to values between 0 and 1 because LSTM networks render better results with scaled data. Due to the increase in characteristics, attributes have been reduced through the application of the Principal Component Analysis (PCA) algorithm. After applying these transformations we have obtained a dataset with 45 columns.

Finally, since the data are based on time series, sliding windows of length 25 and step 3 have been used to divide the dataset. In this way, the dataset is labeled in two classes according to the number of remaining life cycles. The engines with more than 100 remaining life cycles are labelled as class 0. The motors with less than 100 remaining life cycles are labelled as class 1.

The case study has consisted in creating a predictive maintenance system. LSTM networks have been used since this type of structure is considered to be one of the most effective for the study of predictive maintenance, given that the values measured by the sensors in previous stages are indicative of the remaining useful life.

The deigned system consists of two trained neural networks. The first one is a classifying network that predicts the class of an engine according to its state. This network classifies the engines in two different classes, the engines that have many remaining life cycles (> 100) and those that have only few (≤ 100). The second neural network is regressive and allows to calculate the number of remaining life cycles of the engines that have little useful life remaining. Finally, as shown in Fig. 2, a model combining the classifying and regressive networks is evaluated to predict the RUL of new inputs.

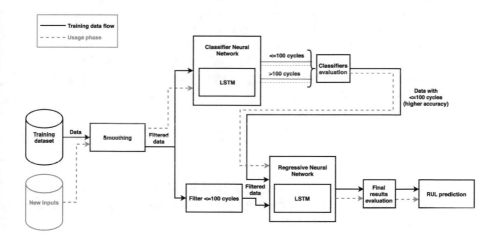

Fig. 2. System flow

The proposed model has been chosen because training a single LSTM network to predict the RUL of a machine provides a greater amount of errors, independently of the number of the machine's remaining life cycles.

4   LSTM Architecture

This section presents the architecture designed for the predictive maintenance problem. The model is based on the use of recurrent LSTM networks since this type of network is normally used to solve problems with time dependencies. A classifying network and a regressive network will be trained to later form the final model joining them together.

Neural Network Classifier. Various classification networks have been trained by searching for the best hyperparameters. For this purpose, a grid search has been used with the following hyperparameters:

$$\text{Number of hidden states} \in \{9, 10, \cdots, 14\}$$

$$\text{Batch size} \in \{200, 300, 400, 500\}$$

$$\text{Number of epochs} \in \{300, 350, 400, 450, 500\}$$

In addition, a similar procedure has been performed for networks with 2 hidden layers and the results did not improve.

All networks have been trained with an optimizer called Adam [11] using a binary crossentropy function [9] to quantify the error.

After the training of the different networks, the one with which we obtain the best results, as can be seen in Fig. 3, is formed by a hidden layer with 11 hidden states, a dropout of 0.15 to avoid overfitting and an increase in the generalization capacity, 400 epochs and a batch size of 300. The activation function used in this layer is a Hyperbolic Tangent, defined as Eq. (1). Finally, since it is intended to make a binary classification, we have 2 neurons and a Sigmoid activation function in the output layer [5] that will give us the probabilities of the equipment belonging to the different classes.

$$R(z) = \frac{e^z - e^{-z}}{e^z + e^{-z}} \tag{1}$$

Regressive Neural Network. Several regressive networks have been trained by searching for the best hyperparameters. For this purpose, a grid search with the following hyperparameters has been used:

$$\text{Number of hidden states} \in \{16, 17, \cdots, 21\}$$

$$\text{Batch size} \in \{200, 300, 400, 500\}$$

$$\text{Number of epochs} \in \{300, 350, 400, 450, 500\}$$

In addition, a similar procedure has been performed for networks with 2 hidden layers and the results were not better.

All networks have trained the network with an Adam optimizer [11] using an MSE function to quantify the error.

The best performing regressive network consists of a hidden layer with 19 hidden states. In addition, a dropout of 0.15 has been added to avoid overfitting and increase the generalization capacity, 400 times and a batch size of 200.

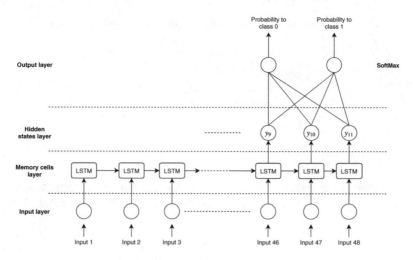

Fig. 3. LSTM classifier

Finally, in the output layer there is 1 neuron which will give us back the number of remaining life cycles. In the hidden layer, an activation function Hyperbolic Tangent (1) has been used and in the output layer, a linear activation function has been used, defined in Eq. (2).

$$R(z) = z \tag{2}$$

5   Results

Having detailed the architecture of the networks to be used, this section evaluates their operation. For this purpose, a study of the results of the regressive network and the classifying network has been performed. Finally, the results of the two networks have been analyzed individually to determine the performance of the model as a whole.

5.1   Network Classifier

Once the RNN classifier has been trained, the model is applied to the test dataset, obtaining results as shown in Table 1.

It can be seen in Table 1 that this network renders better results when classifying engines with more than 100 remaining life cycles (partly because this class has a larger number of instances). From the point of view of a company, the best approach is to correctly classify the number of life cycles of the machines with less than 100 life cycles. Therefore, when selecting the network classifier it has been taken into account that the goal is to get better results when classifying the life cycles of machines with less than 100 remaining life cycles.

Table 1. Metrics obtained after training the classifier networks

	Precision	Recall	F1-score
Class 0	0.93	0.88	0.91
Class 1	0.65	0.77	0.70
Average	0.87	0.86	0.86

5.2 Regressive Network

Having trained the regressive RNN, the test dataset has been applied. The Mean Absolute Error (MAE) defined as Eq. (3) was used to evaluate network performance.

$$MAE = \frac{\sum_{i=1}^{n} |y_i - \hat{y}_i|}{n} \tag{3}$$

After evaluating the test data, an average error of 10.4691 has been obtained. The main drawback of this network is that it is trained only for engines with less than 100 remaining life cycles, thus, it can commit considerable errors when predicting the RUL of an engine with more than 100 cycles.

5.3 Final Model

The final stage in the workflow of the proposed model consists in predicting the class to which an engine belongs. For this purpose, a series of measurements from different sensors are introduced into the network classifier. Subsequently, if the above classification is class 1, the number of remaining cycles is predicted.

To measure the performance of this model, the error function given by the algorithm (1) has been proposed.

Algorithm 1. Final model error

1: **function** MODELERROR(Ts)
2:      $predictedclass$ = ClassifierNetwork(Ts)
3:      **if** $realclass$!= 0 and $predictedclass$!= 0 **then**
4:          $predictedrul$ = RegressiveNetwork(Ts)
5:          $error = |predictedrul - realrul|$ ▷ Absolute Error
6:      **end if**
7:      **return** $error$
8: **end function**

The application of this function to the test set and the calculation of the average error has given the value of 25.6027.

6   Conclusions

When predicting the RUL of a machine the most common procedure is to perform a regression with the set of historical measurements of the different sensors of the machine. The main problem with this type of procedure is that if the range of the number of life cycles is too wide, large errors can occur when performing the regression. This is why the proposed model works better than the single network model. The main drawback of this system is that when calculating the final error of the model, the individual errors of each of the networks are included. Therefore, for this system to work correctly, a low number of errors are needed in each of the individual networks.

Acknowledgments. This research has been partially supported by the European Regional Development Fund (ERDF) under the IOTEC project grant 0123_IOTEC_3_E and by the Spanish Ministry of Economy, Industry and Competitiveness.

References

1. Briones, A.G., Chamoso, P., Rivas, A., Rodríguez, S., De La Prieta, F., Prieto, J., Corchado, J.M.: Use of gamification techniques to encourage garbage recycling. A smart city approach. In: International Conference on Knowledge Management in Organizations, pp. 674–685. Springer, Heidelberg (2018)
2. Candanedo, I.S., Nieves, E.H., González, S.R., Martín, M.T.S., Briones, A.G.: Machine learning predictive model for industry 4.0. In: International Conference on Knowledge Management in Organizations, pp. 501–510. Springer, Heidelberg (2018)
3. Chamoso, P., González-Briones, A., Rivas, A., De La Prieta, F., Corchado, J.M.: Social computing in currency exchange. Knowl. Inf. Syst., pp. 1–21 (2019)
4. Do, P., Voisin, A., Levrat, E., Iung, B.: A proactive condition-based maintenance strategy with both perfect and imperfect maintenance actions. Reliab. Eng. Syst. Saf. **133**, 22–32 (2015)
5. Funahashi, K.-I.: On the approximate realization of continuous mappings by neural networks. Neural Netw. **2**(3), 183–192 (1989)
6. Gers, F.A., Schraudolph, N.N., Schmidhuber, J.: Learning precise timing with LSTM recurrent networks. J. Mach. Learn. Res. **3**(Aug), 115–143 (2002)
7. González-Briones, A., Prieto, J., De La Prieta, F., Herrera-Viedma, E., Corchado, J.M.: Energy optimization using a case-based reasoning strategy. Sensors **18**(3), 865 (2018)
8. González-Briones, A., Rivas, A., Chamoso, P., Casado-Vara, R., Corchado, J.M.: Case-based reasoning and agent based job offer recommender system. In: The 13th International Conference on Soft Computing Models in Industrial and Environmental Applications, pp. 21–33. Springer, Heidelberg (2018)
9. Gregor, K., Danihelka, I., Graves, A., Rezende, D.J., Wierstra, D.: Draw: A recurrent neural network for image generation. arXiv preprint arXiv:1502.04623 (2015)
10. Higgins, L.R., Mobley, R.K., Smith, R., et al.: Maintenance Engineering Handbook. McGraw-Hill, New York (2002)
11. Kingma, D.P., Ba, J.: Adam: A method for stochastic optimization. arXiv preprint arXiv:1412.6980 (2014)

12. Krishnanand, K.R., Dash, P.K., Naeem, M.H.: Detection, classification, and location of faults in power transmission lines. Int. J. Electr. Power Energy Syst. **67**, 76–86 (2015)
13. Na, M.G.: Auto-tuned PID controller using a model predictive control method for the steam generator water level. IEEE Transact. Nucl. Sci. **48**(5), 1664–1671 (2001)
14. Rivas, A., Martín, L., Sittón, I., Chamoso, P., Martín-Limorti, J.J., Prieto, J., González-Briones, A.: Semantic analysis system for industry 4.0. In: International Conference on Knowledge Management in Organizations, pp. 537–548. Springer, Heidelberg (2018)
15. Rivas, A., Martín-Limorti, J.J., Chamoso, P., González-Briones, A., De La Prieta, F., Rodríguez, S.: Human-computer interaction in currency exchange. In: International Conference on Knowledge Management in Organizations, pp. 390–400. Springer, Heidelberg (2018)
16. Saxena, A., Goebel, K.: Turbofan engine degradation simulation data set. NASA Ames Prognostics Data Repository (2008)
17. Smith, C.A., Corripio, A.B., Basurto, S.D.M.: Control automático de procesos: teoría y práctica. Limusa (1991). Number 968-18-3791-6. 01-A3 LU. AL-PCS. 1
18. Swanson, L.: Linking maintenance strategies to performance. Int. J. Prod. Econ. **70**(3), 237–244 (2001)
19. Taher, S.A., Sadeghkhani, I.: Estimation of magnitude and time duration of temporary overvoltages using ann in transmission lines during power system restoration. Simul. Model. Pract. Theory **18**(6), 787–805 (2010)
20. Trinh, H.C., Kwon, Y.K.: An empirical investigation on a multiple filters-based approach for remaining useful life prediction. Machines **6**(3), 35 (2018)
21. Zhou, D., Zhang, H., Weng, S.: A novel prognostic model of performance degradation trend for power machinery maintenance. Energy **78**, 740–746 (2014)

Soft Computing Applications

Prototypical Metric Transfer Learning for Continuous Speech Keyword Spotting with Limited Training Data

Harshita Seth, Pulkit Kumar, and Muktabh Mayank Srivastava[✉]

Paralleldots, Inc., Gurgaon, India
{harshita,pulkit,muktabh}@paralleldots.com

Abstract. Continuous Speech Keyword Spotting (CSKS) is the problem of spotting keywords in recorded conversations, when a small number of instances of keywords are available in training data. Unlike the more common Keyword Spotting, where an algorithm needs to detect lone keywords or short phrases like *"Alexa"*, *"Cortana"*, *"Hi Alexa!"*, *"Whatsup Octavia?"* etc. in speech, CSKS needs to filter out embedded words from a continuous flow of speech, ie. spot *"Anna"* and *"github"* in *"I know a developer named Anna who can look into this github issue."* Apart from the issue of limited training data availability, CSKS is an extremely imbalanced classification problem. We address the limitations of simple keyword spotting baselines for both aforementioned challenges by using a novel combination of loss functions (Prototypical networks' loss and metric loss) and transfer learning. Our method improves F1 score by over 10%.

Keywords: Audio keyword detection · Prototypical metric loss · Few-shot · Transfer learning

1 Introduction

Continuous Speech Keyword Spotting (CSKS) aims to detect embedded keywords in audio recordings. These spotted keyword frequencies can then be used to analyze theme of communication, creating temporal visualizations and word clouds. Another use case is to detect domain specific keywords which ASR (Automatic Speech Recognition) systems trained on public data cannot detect. For example, to detect a TV model number *"W884"* being mentioned in a recording, we might not have a large number of training sentences containing the model number of a newly launched TV to finetune a speech recognition (ASR) algorithm. A trained CSKS algorithm can be used to quickly extract out all instances of such keywords.

We train CSKS algorithms like other Keyword Spotting algorithms by classifying small fragments of audio in running speech. This requires the classifier

---

H. Seth and P. Kumar—Authors contributed equally.

© Springer Nature Switzerland AG 2020
F. Martínez Álvarez et al. (Eds.): SOCO 2019, AISC 950, pp. 273–280, 2020.
https://doi.org/10.1007/978-3-030-20055-8_26

model to have a formalized process to reject unseen instances (everything not a keyword, henceforth referred to as background) apart from ability to differentiate between classes (keywords). Another real world constraint that needs to be addressed while training such an algorithm is the availability of small amount of labeled keyword instances. We combine practices from fields of transfer learning, few-shot learning and metric learning to get better performance on this low training data imbalanced classification task.

Our work involves :

1. Testing existing Keyword Spotting methodologies [1,2] for the task of CSKS.
2. Proposing a transfer learning based baseline for CSKS by fine tuning weights of a publicly available deep ASR [3] model.
3. Introducing changes in training methodology by combining concepts from few-shot learning [4] and metric learning [5] into the transfer learning algorithm to address both the problems which baselines have a) missing keywords and b) false positives.

Rest of the paper is orgainzed as follow. Section 2 contain past work related to keyword spotting. In Sect. 3 and Sect. 4, we have described the dataset used and our approach to solve the CSKS task. Results and Conclusion are mentioned in Sect. 5 and Sect. 6 respectively.

2 Related Work

In the past, Hidden Markov Models (HMM) [6–8] have been used to solve the CSKS problem. But since the HMM techniques use Viterbi algorithms (computationally expensive) a faster approach is required.

Owning to the popularity of deep learning, many recent works such as [10–14] have used deep learning techniques for many speech processing tasks. In tasks such as ASR, Hannun et al. [3] proposed a RNN based model to transcribe speech into text. Even for plain keyword spotting, [1,2,15–18] have proposed various deep learning architectures to solve the task. But to the best of our knowledge, no past work has deployed deep learning for spotting keywords in continuous speech.

Recently, a lot of work is being done on training deep learning models with limited training data. Out of them, few-shot techniques as proposed by [4,19] have become really popular. Pons et al. [18] proposed a few-shot technique using prototypical networks [4] and transfer leaning [20,21] to solve a different audio task.

We took inspiration from these works to design our experiments to solve the CSKS task.

3 Dataset

Our learning data, which was created in-house, has 20 keywords to be spotted about television models of a consumer electronics brand. It was collected by making 40 participants utter each keyword 3 times. Each participant recorded in normal ambient noise conditions. As a result, after collection of learning data we have 120 (3 × 40) instances of each of the 20 keywords. We split the learning data 80:20 into train and validation sets. Train/Validation split was done on speaker level, so as to make sure that all occurrences of a particular speaker is present only on either of two sets. For testing, we used 10 different 5 min long simulated conversational recordings of television salesmen and customers from a shopping mall in India. These recordings contain background noise (as is expected in a mall) and have different languages (Indians speak a mixture of English and Hindi). The CSKS algorithm trained on instances of keywords in learning data is supposed to detect keywords embedded in conversations of test set.

4   Approach

4.1   Data Preprocessing

Our dataset consisted of keyword instances but the algorithm trained using this data needs to classify keywords in fragments of running conversations. To address this, we simulate the continuous speech scenario, both for keyword containing audio and background fragments, by using publicly available audio data which consisted of podcasts audio, songs, and audio narration files. For simulating fragments with keywords, we extract two random contiguous chunks from these publicly available audio files and insert the keyword either in the beginning, in the middle or in the end of the chunks, thus creating an audio segment of 2 s. Random 2 s segments taken from publicly available audio are used to simulate segments with no keywords(also referred to as background elsewhere in the paper). These artificially simulated audio chunks from train/validation set of pure keyword utterances were used to train/validate the model. Since the test data is quite noisy, we further used various kinds of techniques such as time-shift, pitch-shift and intensity variation to augment the data. Furthermore we used the same strategy as Tang et al. [2] of caching the data while training deep neural network on batches and artificially generating only 30% data which goes into a batch. By following these techniques, we could increase the data by many folds which not only helped the model to generalise better but also helped reduce the data preparation time during every epoch.

4.2   Feature Engineering

For all the experiments using Honk architecture, MFCC features were used. To extract these features, 20 Hz/4 kHz band pass filters was used to reduce the random noise. Mel-Frequency Cepstrum Coefficient (MFCC) of forty dimension were constructed and stacked using 20 ms window size with 10 ms overlap. For all the experiments using deep speech architecture, we have extracted spectrograms of audio files using 20 ms window size with 10 ms overlap and 480 nfft value.

4.3   Deep Learning Architectures

Honk. Honk is a baseline Neural Network architecture we used to address the problem. Honk has shown good performance on normal Keyword Spotting and thus was our choice as the first baseline. The neural network is a Deep Residual Convolutional Neural Network [22] which has number of feature maps fixed for all residual blocks. The python code of the model was taken from the open source repository [23]. We tried changing training strategies of Honk architecture by the methods we will describe later for DeepSpeech, but this did not improve the accuracy.

DeepSpeech-finetune. DeepSpeech-finetune is fine tuning the weights of openly available DeepSpeech [3] model (initial feature extraction layers and not the final ASR layer) for CSKS task. The architecture consists of pretrained initial layers of DeepSpeech followed by a set of LSTM layers and a Fully Connected layer (initialized randomly) for classification. Pretrained layers taken from Deep-Speech are the initial 2D convolution layers and the GRU layers which process the output of the 2D convolutions. The output of Fully Connected layer is fed into a softmax and then a cross entropy loss for classification is used to train the algorithm. Please note that the finetune trains for 21 classes (20 keywords + 1 background) as in aforementioned Honk model. The architecture can be seen in Fig. 1.

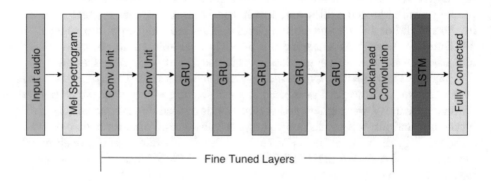

Fig. 1. Architecture for DeepSpeech-finetune

DeepSpeech-finetune-prototypical. The next model we try is fine tuning DeepSpeech model but with a different loss function. This loss function is taken from [4]. Prototypical loss works by concentrating embeddings of all data points of a class around the class prototype. This is done by putting a softmax on the negative distances from different prototypes to determine the probability to belong to corresponding classes. The architecture 2 is same as DeepSpeech-finetune, except output of pre-final layer is taken as embedding rather than

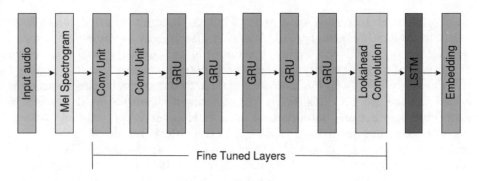

Fig. 2. Architecture for DeepSpeech-finetune-prototypical

applying a Fully Connected layer for classification. These embeddings are then used to calculate euclidean distances between datapoints and prototypes, represented as $d(embedding1, embedding2)$ in formulae. The softmax over negative distances from prototypes is used to train cross-entropy loss. During training, examples of each class are divided into support and query embeddings. The support embeddings are used to determine prototypes of the class. Equation 1 shows derivation of prototype of kth class where f_ϕ is the neural network yielding the embedding and S_k is the set of support vectors for the class. The distance of query vectors from the prototypes of the class they belong to are minimized and prototypes of other classes is maximized when training the prototypical loss. The negative distances from the prototypes of each class are passed into softmax to get the probability of belonging in a class as shown in Eq. 2. We see better results when we train the algorithm using prototypical loss than normal cross entropy. On qualitatively observing the output from DeepSpeech-finetune-prototypical we see that the mistakes involving confusion between keywords are very less compared to datapoints of the class background being classified as one of the keywords. We hypothesize that this might be due to treating the entire background data as one class. The variance of background is very high and treating it as one class (a unimodal class in case of prototypes) might not be the best approach. To address this, we propose the next method where we use prototypes for classification within keywords and an additional metric loss component to keep distances of background datapoints from each prototype high.

$$\mathbf{c}_k = \frac{1}{S_k} \sum_{(\mathbf{x}_i, y_i) \in S_k} f_\phi(\mathbf{x}_i) \tag{1}$$

$$p(y = k|\mathbf{x}) = \frac{e^{-d(f_\phi(\mathbf{x}), \mathbf{c}_k)}}{\sum_j e^{-d(f_\phi(\mathbf{x}), \mathbf{c}_j)}} \tag{2}$$

DeepSpeech-finetune-prototypical+metric. We hypothesize the components of loss function of this variant from failures of prototypical loss as stated earlier. The architecture is same as in 2, but the loss function is different from DeepSpeech-finetune-prototypical. While in DeepSpeech-finetune-prototypical, we trained prototype loss with 21 classes (20 keywords + 1 background), in DeepSpeech-finetune-prototypical+metric prototype loss is trained only amongst the 20 keywords and a new additional metric loss component inspired from [5] is added to loss function. This metric loss component aims to bring datapoints of same class together and datapoints of different class further. Datapoints belonging to background are treated as different class objects for all other datapoints in a batch. So for each object in a batch, we add a loss component like Eq. 3 to prototypical loss. $\mathbf{c}^+$ is all datapoints in the batch belonging to the same class as $\mathbf{x}$ and $\mathbf{c}^-$ is all datapoints belonging to different classes than $\mathbf{x}$ (including background). This architecture gets the best results.

$$L_{metric} = \frac{e^{average(d(f_\phi(\mathbf{x}),\mathbf{c}^+))}}{e^{average(d(f_\phi(\mathbf{x}),\mathbf{c}^+))} + e^{average(d(f_\phi(\mathbf{x}),\mathbf{c}^-))}} \tag{3}$$

5   Experiments and Results

Testing of the models is carried out on the conversational recordings where keywords are embedded. Overlapping segments of recordings are passed through models to classify them into one of the keywords or background noise. During testing of models with prototype losses, instead of taking the output with maximum probability from softmax, distance of a datapoint is checked with all the prototypes to determine its predicted class.

Table 1 lists down recall, precision and F1 scores for the different deep learning architectures mentioned in previous section on the testing data. It was observed that our baseline, DeepSpeech-finetune, had comparatively both lower recall and precision. While analysing the false positives of DeepSpeech-finetune, it was observed that the model gets confused between the keywords and also has tendency to wrongly classify background noise into one of the classes. When we tried fine tuning DeepSpeech model with prototypical loss (DeepSpeech-finetune-prototypical)), the results get better from Deepspeech-finetune architecture. When results of this model were qualitatively checked, we saw that the model confused between keywords a lot lesser, but still classifies background noise into one of the keywords. To improve this, we combined prototypical loss with a metric loss to reject background (DeepSpeech-finetune-prototypical+metric)).

As it can be seen in Table 1, DeepSpeech-finetune-prototypical+metric clearly beats the baselines in terms of both precision and recall. Honk is a respectable baseline and gets second best results after DeepSpeech-finetune-prototypical+metric, however, attempts to better Honk's performance using prototype loss and metric loss did not work at all.

Table 1. Results of all experiments

Model	Recall	Precision	F1
Honk	0.46	0.34	0.39
DeepSpeech-finetune	0.267	0.244	0.256
DeepSpeech-finetune-prototypical	0.36	0.33	0.344
DeepSpeech-finetune-prototypical+metric	**0.55**	**0.488**	**0.51**

6 Conclusion

We believe that our method to combine prototypical loss with metric learning can be used for any few-shot classification task which has a set of classes and a large background class as it significantly improves the quality of the prototypes of each class. But its effectiveness needs to be tested on other tasks.

References

1. Sainath, T.N., Parada, C.: Convolutional neural networks for small-footprint keyword spotting. In: Interspeech (2015)
2. Tang, R., Lin, J.: Honk: a PyTorch reimplementation of convolutional neural networks for keyword spotting. arXiv preprint arXiv:1710.06554, 18 October 2017
3. Hannun, A., Case, C., Casper, J., Catanzaro, B., Diamos, G., Elsen, E., Prenger, R., Satheesh, S., Sengupta, S., Coates, A., Ng, A.Y.: Deep speech: scaling up end-to-end speech recognition. arXiv preprint arXiv:1412.5567, 17 December 2014
4. Snell, J., Swersky, K., Zemel, R.: Prototypical networks for few-shot learning. In: Advances in Neural Information Processing Systems, pp. 4077–4087 (2017)
5. Hoffer, E., Ailon, N.: Deep metric learning using triplet network. In: International Workshop on Similarity-Based Pattern Recognition, vol. 12, pp. 84–92. Springer, Cham (2015)
6. Weintraub, M.: Keyword-spotting using SRI's DECIPHER large-vocabulary speech-recognition system. In: 1993 IEEE International Conference on Acoustics, Speech, and Signal Processing 1993, ICASSP 1993, vol. 2, pp. 463–466. IEEE, 27 April 1993
7. Wilpon, J.G., Rabiner, L.R., Lee, C.H., Goldman, E.R.: Automatic recognition of keywords in unconstrained speech using hidden Markov models. IEEE Transact. Acoust. Speech Sig. Proc. **38**(11), 1870–1878 (1990)
8. Rose, R.C., Paul, D.B.: A hidden Markov model based keyword recognition system. In: 1990 International Conference on Acoustics, Speech, and Signal Processing 1990, ICASSP 1990, pp. 129–132. IEEE, 3 April 1990
9. Nouza, J., Silovsky, J.: Fast keyword spotting in telephone speech. Radioengineering **18**(4), 665–70 (2009)
10. Lee, H., Pham, P., Largman, Y., Ng, A.Y.: Unsupervised feature learning for audio classification using convolutional deep belief networks. In: Advances in Neural Information Processing Systems, pp. 1096–1104 (2009)
11. Mohamed, A.R., Dahl, G.E., Hinton, G.: Acoustic modeling using deep belief networks. IEEE Trans. Audio Speech Lang. Process. **20**(1), 14–22 (2012)

12. Grosse, R., Raina, R., Kwong, H., Ng, A.Y.: Shift-invariance sparse coding for audio classification. arXiv preprint arXiv:1206.5241, 20 June 2012
13. Hinton, G., Deng, L., Yu, D., Dahl, G.E., Mohamed, A.R., Jaitly, N., Senior, A., Vanhoucke, V., Nguyen, P., Sainath, T.N., Kingsbury, B.: Deep neural networks for acoustic modeling in speech recognition: the shared views of four research groups. IEEE Signal Process. Mag. **29**(6), 82–97 (2012)
14. Dahl, G.E., Yu, D., Deng, L., Acero, A.: Context-dependent pre-trained deep neural networks for large-vocabulary speech recognition. IEEE Transact. Audio Speech Lang. Process. **20**(1), 30–42 (2012)
15. Li, K.P., Naylor, J.A., Rossen, M.L.: A whole word recurrent neural network for keyword spotting. In: 1992 IEEE International Conference on Acoustics, Speech, and Signal Processing 1992, ICASSP 1992, vol. 2, pp. 81–84. IEEE, 23 March 1992
16. Fernández, S., Graves, A., Schmidhuber, J.: An application of recurrent neural networks to discriminative keyword spotting. In: International Conference on Artificial Neural Networks, vol. 9, pp. 220–229. Springer, Heidelberg, September 2007
17. Chen, G., Parada, C., Heigold, G.: Small-footprint keyword spotting using deep neural networks. In: ICASSP, vol. 14, pp. 4087–4091, 4 May 2014
18. Pons, J., Serrá, J., Serra, X.: Training neural audio classifiers with few data. arXiv preprint arXiv:1810.10274, 24 October 2018
19. Vinyals, O., Blundell, C., Lillicrap, T., Wierstra, D.: Matching networks for one shot learning. In: Advances in Neural Information Processing Systems, pp. 3630–3638 (2016)
20. Kunze, J., Kirsch, L., Kurenkov, I., Krug, A., Johannsmeier, J., Stober, S.: Transfer learning for speech recognition on a budget. arXiv preprint arXiv:1706.00290, 1 June 2017
21. Choi, K., Fazekas, G., Sandler, M., Cho, K.: Transfer learning for music classification and regression tasks. arXiv preprint arXiv:1703.09179, 27 March 2017
22. He, K., Zhang, X., Ren, S., Sun, J.: Deep residual learning for image recognition. In: Proceedings of the IEEE Conference on Computer Vision and Pattern Recognition, pp. 770–778 (2016)
23. Honk: A PyTorch Reimplementation of Convolutional Neural Networks for Keyword Spotting. https://github.com/castorini/honk
24. Rohlicek, J.R., Russell, W., Roukos, S., Gish, H.: Continuous hidden Markov modeling for speaker-independent word spotting. In: 1989 International Conference on Acoustics, Speech, and Signal Processing 1989, ICASSP 1989, pp. 627–630. IEEE, 23 May 1989
25. Wilpon, J.G., Miller, L.G., Modi, P.: Improvements and applications for key word recognition using hidden Markov modeling techniques. In: 1991 International Conference on Acoustics, Speech, and Signal Processing 1991, ICASSP 1991, pp. 309–312. IEEE, 14 April 1991
26. Silaghi, M.C., Bourlard, H.: Iterative posterior-based keyword spotting without filler models. In: Proceedings of the IEEE Automatic Speech Recognition and Understanding Workshop, pp. 213–216, 12 December 1999

Characteristic of WiFi Network Based on Space Model with Using Turning Bands Co-simulation Method

Anna Kamińska-Chuchmała[(✉)]

Faculty of Computer Science and Management,
Wrocław University of Science and Technology,
Wybrzeże Wyspiańskiego 27, 50-370 Wrocław, Poland
anna.kaminska-chuchmala@pwr.edu.pl

Abstract. Recently, mostly users prefer an access to the wireless network than wired. Due to the fact of mobility of users and choosing of mobile devices such as smartphone, tablet, smartwatch etc. Extensively growth of wireless network users affect on reliable connection to the network. In this research parameters from Access Points (APs) contained in PWR-WiFi an open WiFi network belonging to Wrocław University of Science and Technology (WUST) in Poland are investigated. A central issue in this paper is to create space models prediction of WiFi network efficiency by Turning Bands Method (TBM). Statistical analysis of considered WiFi daily data were conducted. Acquired results were discussed and conclusions with future research directions to WiFi network efficiency predictions were drawn.

Keywords: WiFi · Wireless network efficiency · Space models ·
Turning bands method · Co-simulation method

1 Introduction

In pursuance of Cisco Visual Networking Index (Cisco VNI) document [1] about global IP traffic forecast develops the claim that in 2016, wired devices accounted for the majority of IP traffic at 51%, moreover traffic from wireless and mobile devices will account for more than 63% of total IP traffic by 2021 and wired devices will account only 37% of IP traffic. Growing demand on wireless network highlight that it should be focus on investigation of WiFi models of efficiency and reliable connection to the Internet. Research in this paper support the above view of that situation. Namely, open access WiFi network is analysed in the context of performance connection to the Internet through APs. 3D models of performance wireless network by TBM are prepared.

Paper consist of section with related works. Subsequently, simulation TBM are described and then preliminary and structural analysis are presented. Next section with space models of PWR-WiFi network are discussed. Conclusions and plan of future research ending the paper.

F. Martínez Álvarez et al. (Eds.): SOCO 2019, AISC 950, pp. 281–290, 2020.
https://doi.org/10.1007/978-3-030-20055-8_27

2 Related Work

This section studies briefly related researches in context of wireless network efficiency.

In literature [2] is proposed WiFi performance estimation model based on Machine Learning method. The main issue of this model is allows service providers to predict WiFi saturated throughput from easy measurement on the AP. Author based their research on a SVM-based classification model to work as a prediction function.

Different approach to estimation of throughput in wireless networks is shown in [3]. Researchers investigated in their work data from an active measurements and using statistical learning tools. The idea of their studies was the system which was trained during short periods with application flows and probe packets bursts. Continuous non intrusive methodology was obtained as result which allows to determine the maximum throughput of a wireless connection only knowing some characteristics of the network.

Authors in [4] focus on green efficiency of wireless links and carry out a comprehensive analysis of the efficiency metrics like: spectrum efficiency, energy efficiency, power efficiency and proposed by Authors green efficiency of a wireless link. They prove that green efficiency is a strict convex function with respect to transmission power and distance.

Studies [5] draw attention on performance of expected transmission count (ETX) under different traffic loads. Authors presents the design and implementation of a variation of ETX called high load - ETX (HETX), which reduces the impact of route request broadcast packets to link metric values under high load.

Summarizing the review of actual literature, to the best of knowledge no work exists using TBM to prepare space model of wireless networks efficiency expect Author's paper [6]. In subsequent section, Turning Bands algorithm is introduced and the method is explained.

3 Turning Bands Method

One of the most classical approach among simulation methods in applied spatial statistics domain is Turning Bands method (TBM). The oldest mention about TBM were in [7,8] by Matheron and subsequently by Lantuejoul [9]. Generally, the main idea of TBM is reduction of multidimensional simulation to a one-dimensional one. TBM rest on the reduction of a Gaussian random function of covariance C to the simulation of an independent stochastic process of covariance C_θ. Let $(\theta_n, n \in \mathbf{N})$ be a sequence of directions $\mathbf{S}_d^+$, and let $(X_n, n \in \mathbf{N})$ be a sequence of independent stochastic processes of covariance C_{θ_n}. Random function:

$$Y^{(n)}(x) = \frac{1}{\sqrt{n}} \sum_{k=1}^{n} X_k(<x, \theta_k>), \qquad x \in \mathbb{R}^d \tag{1}$$

assumes covariance equal to:

$$C^{(n)}(h) = \frac{1}{n} \sum_{k=1}^{n} C_{\theta_k}(<h, \theta_k>).\qquad(2)$$

Algorithm of TBM

1. Transform input data using Gaussian anamorphosis.
2. Choose a set of directions $\theta_1, ..., \theta_n$. Simulate of covariance C_3 by summing, the projection of the simulation for a given number of *turning bands* lines of covariance C_1.
3. Generate independent standard stochastic processes $X_1, ..., X_n$ with covariance functions $C_{\theta_1}, ..., C_{\theta_n}$.
4. Compute $Y^{(n)}(x)$ for any $x \in D$.
5. Make kriged estimate $y^*(x)$ for each $x \in D$.
6. Simulate a Gaussian random function with mean 0 and covariance C in domain D on condition points.
7. Make kriged estimate $z^*(x)$.
8. Calculate final equation: $W(x) = y^*(x) + z(x) - z^*(x)$, where $x \in D$.
9. Perform a Gaussian back transformation to return to the original data.

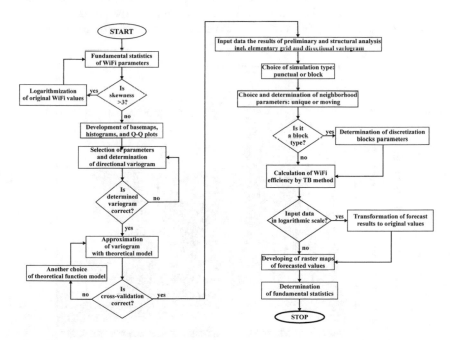

Fig. 1. Block schema of analysis and creating the space model of PWR-WiFi efficiency calculated by TBM method

To draw attention to particularly steps of preparing elementary and structural statistics rest on WiFi parameters before run TBM algorithm and to final analysis of results after calculation, the schema block in Fig. 1 is presented with step by step Author's approach to methodology. The aim of this research is to apply TBM to create space model of PWR-WiFi efficiency based on measurements obtained from APs. The purpose of this investigation are considered four specific main parameters from APs and prepare co-model by co-simulation TBM.

4   Characteristic of PWR-WiFi Network

PWR-WiFi is a WiFi network with free access located at Wrocław University of Science and Technology (WUST) in Poland. All APs use LWAPP (Light Weight Access Point Protocol) communication protocol as form of communication between them and controllers. APs are wirelessly connected to switch and configured to get IP address from network. APs are connecting to WiFi controller in star topology. PWR-WiFi network using frequency 2.4 GHz in IEEE 802.11b/g/n standards and APs with using 5 GHz frequency in IEEE 802.11a/n standards.

The data gathered in this research are obtained from nine Access Points (APs) given in five-storey building named B4 (5 floors plus ground floor) located in main campus at WUST. Indicated APs in B4 building are presented in Fig. 6a. Investigated data for analysis of the network are acquired from 4th April 2016.

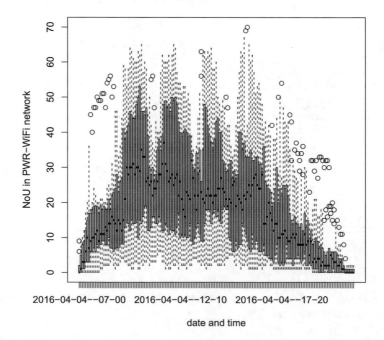

Fig. 2. NoU connected to PWR-WiFi network in 4th March 2016 derived from 9 APs

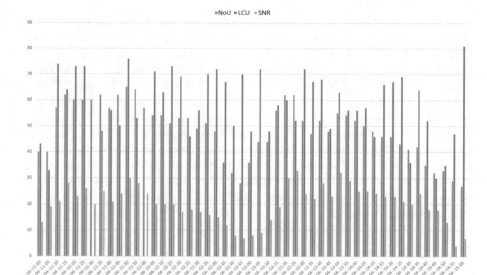

Fig. 3. Measurements from one AP between 11:00 AM and 15:00 PM on 4th April 2016

Author have chosen this day because of this is a typical day of work - Monday, without any holidays thus PWR-WiFi network was functional normally and all students and employees had an access to the network. Data were gathered with five minutes intervals for whole day. The database consists of: date and time of taking the measurement; name of AP; X, Y and Z coordinates; and parameters: Number of Users (NoU), Multiple Retry Client (MRC), Load Channel Utilization (LCU), and Signal to Noise Ratio (SNR). The graph (Fig. 2) presents data relating to parameter of NoU connected to the PWR-WiFi network in 4th April 2016 aggregate data from all 9 APs. NoU parameter represents number of users connected to the access point in the last measurement interval. On basis Fig. 2 we could notice that main traffic is between 9:00 AM and 5:00 PM. Moreover uneven shape of the graph would be response to regular mode of classes. Predominantly, the first classes are started at 7:30 AM and the last one finished at 8:30 PM. During the day, classes usually are started quarter after an odd hour. Students coming on new classes wants to connect to WiFi network (visible in Fig. 2 increase of NoU at the begging of odd hour) and at the end of course (decrease just before odd hour) they turn off their mobile devices. PWR-WiFi is an open university network thus most of users are students, albeit users contains also employees of WUST like lecturers or administrative workers. The B4 building is closed by night, hence main traffic in network is during office hours and lectures. In regard of above mentioned facts as exemplary graph of three parameters from one AP is taken a period between 11:00 AM and 15:00 PM and is presented in Fig. 3.

The horizontal axis indicates date and time of measurement and vertical axis shows in the same scale depend on parameter different units: NoU [number], LCU [%], SNR [number]. LCU means channel utilization, i. e. the percentage of the communication channel used for data transmission and SNR describes number of users with a low SNR (signal-to-noise ratio) connected to the AP in the last measurement interval. Considered AP are located on 3rd floor, where also are two libraries, lecture hall and many scientist and administrative employee's offices. The relationships between NoU, LCU and SNR show in this figure typical correlation in functionality of WiFi network. If more users are connected to the PWR-WiFi network then utilization of channel is higher and signal noise to ratio is lower. Moreover, in some cases LCU is high whilst NoU is small. It is caused by interference phenomenon, because of in B4 building are also other WiFi networks available.

5   Preliminary and Structural Analysis of Data

All research presented in this paper were performed in R language under RStudio in version 1.1.447 which is available as Free Software under GNU license [10]. Additionally package under RStudio named RGeostats was used in version 11.2.1 [11] to made geostatistical co-simulation space model with TBM. Fundamental statistics are presented in Table 1. Both of high variance of NoU and LCU parameters indicate on significant data differentiation. Furthermore dispersion of data characterized by standard deviation is also substantial for these two parameters on contrary to mean value. Maximum value equals 70 of NoU is high relative to mean value equals 18. It means that predominantly frequencies is focus on values less than 10. MRC parameter is a counter shall increment when the transmission is successful after more than one retransmission, what means on basis high values of MRC that quality of transmission was on relatively low level. Generally the maximum value of SNR is equal 34 what means that less than 50% users have a problem with good connection to APs. The highest frequencies of users connected to the PWR-WiFi are in between 0 to 5 and 5 to 10 users which confirm histogram presented in Fig. 4a. Additionally, in Fig. 4b MRC parameter is presented with three distinguished bins. This histogram is asymmetric. Histogram of LCU presented in Fig. 5a shows that the highest frequencies of channel utilization is between 10 to 20 %. Moreover overloaded channel is rather rare situation because of frequency between 90% to 120% is less than 10. Parameter

Table 1. Basics statistics of network parameters for all considered APs located in B4

Network parameter	Number of users NoU	Multiple retry count MRC	Load channel utilization LCU	Signal to noise ratio SNR
Minimum value Z_{min}	0	4907132	0	0
Maximum value Z_{max}	70	155247323	112	34
Mean value Z	18	48583053	35.89	4.94
Standard deviation S	16.93	44615087	23.25	6.32
Variance V	286.59	1990506014	540.72	39.95

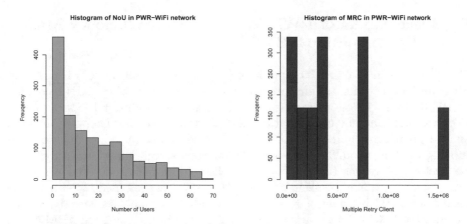

Fig. 4. Histogram of: (a) NoU and (b) MRC in PWR-WiFi network in 4th March 2016

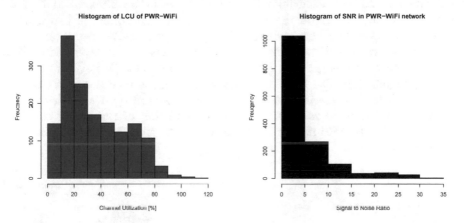

Fig. 5. Histogram of: (a) LCU and (b) SNR in PWR-WiFi network in 4th March 2016

of SNR in Fig. 5b shows that less than 5 users have often problems with SNR. Subsequent step in conducted research is structural analysis. Firstly, it have to calculation of variogram. Considered variogram for all APs in B4 building in 4 April 2016 was calculated in four different directions: 0, 45, 90, 135 degrees. In the next step, variogram function was approximated by theoretical functions: nugget effect, exponential, and spherical.

6   Space Model of PWR-WiFi Network Efficiency Made by Co-simulation TBM

Created space model has 3 dimensions (x and y geographical coordinates and z the altitude coordinate) and prediction cover the space, where APs belonging to PWR-WiFi wireless network spread the signal in building B4. Space prediction was performed by co-simulation TBM described in third chapter. Model of spatial prediction contains theoretical model of variogram approximation and moving neighborhood. The moving neighborhood search is performed by angular sectors. Moreover the neighborhood ellipsoid is anisotropic and has three dimensions. As a final result in Figs. 6b, 7a and 7b are presented raster maps for each floor in B4 building of NoU, LCU and SNR parameters respectively during 4th April 2016. These space models conducted by co-simulated TBM shows efficiency of PWR-WiFi network depend on different parameters. Depending on localization of APs on each floor the spread of users are different. The main concentration of users is on the first two floors in surrounding of two deaneries and on the last two floors where are several lecture halls, two libraries and many administrative and scientist offices. The consensus view seems to be that are clearly correlations between considered parameters of PWR-WiFi network. If NoU is higher than also LCU and SNR is growing up. It means that quantity of users indicated on loads of channel utilization and also on number of users with low signal-to-noise ratio. According to final result of research it could be claimed that PWR-WiFi network, during considered time, has a good efficiency as a wireless network.

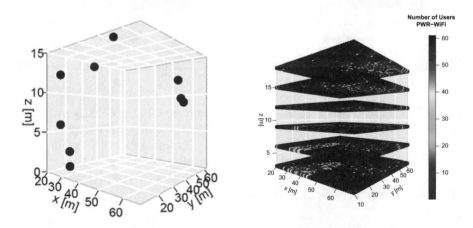

Fig. 6. Scatter 3D plot of: (a) APs localization and (b) NoU parameter of PWR-WiFi network on each floor in B4 building at WUST in 4th April 2016.

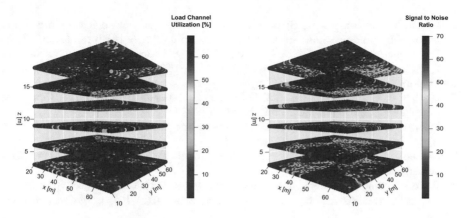

Fig. 7. Scatter 3D plot of: (a) LCU and (b) SNR parameters of PWR-WiFi network on each floor in B4 building at WUST in 4th April 2016

7 Conclusions

The aim of this research was presented space models of PWR-WiFi wireless network efficiency with using co-simulation TBM. Before prediction, preliminary and structural analysis of data had been discussed. Results indicate on high correlation between investigated PWR-WiFi network parameters and good efficiency of the wireless network. Generally, this kind of spatial prediction could be very helpful for administrators of networks. Space models show the good spread of signal from Access Points, because there are more users connected to the wireless network. It could be valuable to see efficiency and capacity of network on this kind of scatter 3D plot.

As a future plan, perform prediction of PWR-WiFi wireless network in 4D (space-time) model of wireless network is considered.

References

1. Cisco Visual Networking Index: Forecast and Methodology, 2016–2021, White paper, Cisco public, 6 June 2017
2. Pan, D.: Analysis of Wi-Fi performance data for a Wi-Fi throughput prediction approach. M.Sc. Thesis, KTH Royal Institute of Technology, School of Information and Communications Technology (ICT), Stockholm, Sweden, June 2017
3. Rattaro, C., Belzarena, P.: Throughput prediction in wireless networks using statistical learning. In: LAWDN - Latin-American Workshop on Dynamic Networks, Buenos Aires, Argentina, November 2010, p. 4 (2010). (inria-00531743)
4. Zhao, L., Zhao, G., O'Farrell, T.: Efficiency metrics for wireless communications. In: IEEE 24th Annual International Symposium on Personal, Indoor, and Mobile Radio Communications (PIMRC) (2013). https://doi.org/10.1109/PIMRC.2013.6666628
5. Tran, A.T., Mai, D.D., Kim, M.K.: Link quality estimation in static wireless networks with high traffic load. J. Commun. Netw. **17**(4), 370–383 (2015)

6. Kamińska-Chuchmała, A.: Spatial models of wireless network efficiency prediction by turning bands co-simulation method. In: Graña, M. (eds.) International Joint Conference SOCO 2018-CISIS 2018-ICEUTE 2018, Proceedings, Advances in Intelligent Systems and Computing, vol. 771, pp. 155–164. Springer, Cham (2019). ISSN 2194-5357

7. Matheron, G.: Quelques aspects de la montée. Internal Report N-271, Centre de Morphologie Mathematique, Fontainebleau (1972)

8. Matheron, G.: The intrinsic random functions and their applications. JSTOR Adv. Appl. Probab. **5**, 439–468 (1973)

9. Lantuejoul, C.: Geostatistical Simulation: Models and Algorithms. Springer, Heidelberg (2002)

10. R Core Team: R: a language and environment for statistical computing. R Foundation for Statistical Computing, Vienna, Austria (2017). https://www.R-project.org

11. Renard, D., Bez, N., Desassis, N., Beucher, H., Ors, F., Freulon, X.: RGeostats: The Geostatistical R package 11.2.1 MINES ParisTech/ARMINES. http://cg.ensmp.fr/rgeostats

Inconsistency Detection on Data Communication Standards Using Information Extraction Techniques: The ABP Case

Sonia León[1(✉)] ⓘ, José Antonio Rodríguez-Mondéjar[2] ⓘ,
and Cristina Puente[2] ⓘ

[1] University of Las Palmas de Gran Canaria, Las Palmas, Spain
sonia.leon@ulpgc.es
[2] Comillas Pontifical University, Madrid, Spain
{mondejar,cristina.puente}@icai.comillas.edu

Abstract. The present research aims mainly, at establishing an error tolerant procedure that extracts information from Natural Language (NL) Communication Standard Documents along with storing error knowledge. The error knowledge will contain information about the detected errors and inconsistencies as well as the actions taken to solve them. It will act as a key tool for solving the detected errors at various levels of the procedure. As a particular scope, the searching of errors and inconsistencies will be based on comparing results from two NLP tools, parsing and chunking. Information Extraction (IE) technics, aided by some specific-developed heuristic algorithms, are used. The approach has been applied to two different-written texts describing the Alternating Bit Protocol (ABP). A Semantic Net is automatically extracted. The error knowledge provides information to the user about what fragments of the text contained inconsistent structures or words and how they were or not solved. The implemented algorithm solved inconsistencies related to words tagged differently by the NLP tools and showed other errors due to the use of complex syntactic structures. Specific metrics were extracted that permitted identify some features of the texts.

Keywords: Natural language processing · Information Extraction ·
Heuristic algorithm · Industrial communication standard · Syntactical patterns ·
Setting chunking · Semantic Network · NLP tools · Error tolerant process

1 Introduction

The research method has been applied to two texts that explain the same communication protocol, the Alternating Bit Protocol (ABP). The first text (*ABP_Standard_Text*) is taken from the IEC 60870-5-2 industrial standard and is used for protection against duplication of message transmission and, the second text (*ABP_NO_Standard_Text*) from a publication where it was first explained by its author, [1]. The two texts are supposed to explain the same exchange rules and be based on the idea of including a bit into the sent or received frames, whose value is alternated depending on a set of studied cases that may occur during the handshaking.

© Springer Nature Switzerland AG 2020
F. Martínez Álvarez et al. (Eds.): SOCO 2019, AISC 950, pp. 291–300, 2020.
https://doi.org/10.1007/978-3-030-20055-8_28

These texts are chosen since they are quite short ones and initially, explain a not much complex protocol. With this research we do not intend to assess the eventual completeness, consistency or reliability of the ABP protocol; contributes of this kind can be found in the literature; as an example see [2, 3].

The texts are written with a very different style: the first one uses few words and all the verbs are in passive mode; the second one uses more words and more linguistic resources. After having studied the two texts, these listed central ideas have come up to drive the present approach:

1. Find and apply a method to find contextual relationships that permits to extract the entities and the main of-interest relations between them.
2. Implement a top-down approach: identify packs of words that perform a main syntactical function inside the sentences, divide the sentences into significant segments and finally, detect the relationships between those packs. The result will be a Semantic Network showing entities and their main relationships.
3. Search for a feasible and generic method to produce reliable information, that is, an error tolerant method that stores information about every found error so that next errors in the next levels can be solved or at least identified, and that, ultimately, shows the reliability grade of the final data.
4. Rather than obtain a 'perfect' result, the target is to produce an annotated result that informs when the text is likely to contain ambiguities.
5. As NL is used, the text may contain vagueness and incompleteness. The use of NLP may produce errors as well. Therefore, the method applies the idea of finding errors from two sources of NLP-generated information, from a parsing and from a chunking tool and their subsequent comparison.

2 Background and Related Works

The present approach is oriented to find inconsistences in communication standard documents, usually written in NL. Known NLP mistakes [4], the large use of NL to write standard documents [5, 6] and the problematics they produce [7, 8], currently cause great and severe problematics in industrial communication, due to concept and expression ambiguity, getting the point of producing real difficulties of interoperability or non-conformant protocol implementations [9, 10].

There can be found many works in the literature that study software requirement validation from NL texts and their eventual ambiguity or incompleteness at various levels, as these two surveys show [11, 12]. Nevertheless, the application over standard texts of protocol description are neither so abundant nor developed [13, 14].

The procedure for applying Information Extraction (IE) from a NL text, in most cases, always starts applying some NLP tool [11]; so this background is made over this kind of studies found in the literature. The algorithms applied afterwards, or the logic used to extract information depend largely on the pursuit objectives.

Some works have created a complex environment, as [15, 16], aided with previous text annotation and knowledge bases or ontologies previously populated with same patterns or models or lexical databases or rules to identify requirements. They use either a proprietary parsing or tagging tool [15, 17] or a commercial one. The inconsistency matter is limited to the field of requirement semantics. The main algorithms are constructed based on comparing the parsing trees with the patterns or models, and applying the rules. In [18], the results of the parsing are compared with the data in the ontology and the parsing errors are semantically corrected. The text is annotated in order to extract a semantic model. On the contrary, we only want to use context relations to extract information so that the process can be controlled since it begins, using NLP tools and studying their results. We do not use any semantic relation at this first stage.

Both, text chunking and parsing tools, are also used to extract information from NL texts to find certain known pre-defined patterns [19]. The algorithms are based on comparing the searched patterns with the results of the NLP tools. The idea of a sentence division into primary units is also used. Inconsistencies of the NLP tools are not treated. We have also adopted the idea of dividing but applying primary divisions using a chunking and a heuristic grammar, that searches for the main constituents of the sentences, (NP, V, VG, adding nexuses as key particles) and then, secondary divisions of the sentences, into heuristic patterns.

In [16] a standard text is studied, aided only by a tagger and by the writing structure of the whole document. Algorithms are not deeply showed. A certain state model is constructed. In [14], some sentences are selected from of the OSI session protocol specification written in the form of sequence description; only a NLP tool is used, producing a parsing of the sentences. Information is extracted only from the parsed sentences; no inconsistencies from the NLP tool are taken into account. The algorithms only deal with few syntactic inconsistencies or rules to find a state model. Instead, we have not used a full standard, but two complete and short descriptions of the same communication protocol and our objective model is a sort of conceptual graph.

The matter of inconsistencies or ambiguities found in a free-written text is not fully addressed in the literature. We agree with [12] that NLP tools and techniques of EI themselves have to be evaluated, the use of inappropriate vocabulary must be addressed [16] and that the inconsistencies have to be avoided o resolved [12]. Our closer primary objective is to automatically construct a conceptual graph, initially based on a Semantic Network as a main model, that may contain the information the text holds along with its owns inconsistencies.

3 Research Method

3.1 Initial Works on the Text Sentences: Tagging, Setting Chunking, Heuristic Split and Parsing

Before the text can be studied, some previous steps, though not trivial ones, has been taken. Here is a short explanation of them and a diagram, Fig. 1, to help follow their sequence: (1) The text was introduced in the process in the form of plain text and in unstructured NL. (2) Then the tagger tool of Python was applied. (3) Later on, a special setting chunking was implemented. The grammar used for applying the chunking was created for finding the sets of words that had a main syntactic role in the text. Each pattern was constructed following some previously studied heuristics rules, around a main word, that is, around a noun (obtaining a NP set), or a verb (obtaining a V or a VG set).

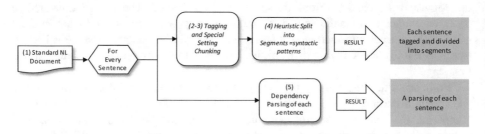

Fig. 1. Initial works on the texts: Tagging, Setting Chunking, Heuristic split and Parsing.

The rest of words were labeled as *nexus* (Nx) *or punctuation marks* (punct). Therefore, the result was a net formed by *sets of words*, already marked, joint by nexuses and some punctuation marks. (4) After that, each sentence was divided into previously studied *syntactic patterns* that we called *segments*, based on the net elements of the previous step. (5) Concurrently, for each sentence, a Dependency Parser (Stanford Parser) was applied over the same sentence, obtaining each word tagged and labeled after the syntactic dependency relationships between them. (4)(5) Once arrived to this point, we had: (a) each sentence, as a net of *word sets*, and divided into segments and (b) a parsing of each.

Below, an example is shown consisting of a compound sentence: sentence 1 from *ABP_Standard_Text (ABP_ST)*, Fig. 2. The *syntactic patterns (4)*, in Fig. 1, are found and the sentence is divided into *segments*. The words in blue background are the main words of each syntactic set:

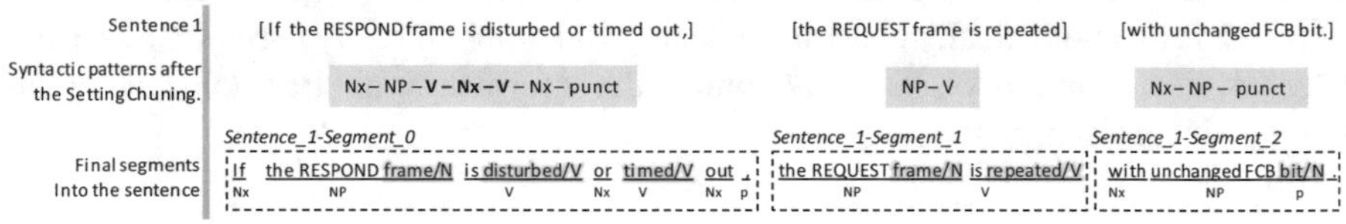

Sentence 1	[If the RESPOND frame is disturbed or timed out,]		[the REQUEST frame is repeated]	[with unchanged FCB bit.]
Syntactic patterns after the Setting Chuning.	Nx – NP – V – Nx – V – Nx – punct		NP – V	Nx – NP – punct
	Sentence_1-Segment_0		Sentence_1-Segment_1	Sentence_1-Segment_2
Final segments Into the sentence	If the RESPOND frame/N is disturbed/V or timed/V out ,		the REQUEST frame/N is repeated/V	with unchanged FCB bit/N .
	Nx NP V Nx V Nx p		NP V	Nx NP p

Fig. 2. Syntactic patterns and segments of sentence 1 from *ABP_ST*

Steps three and four are not deeper described because they are not the focus for this document. Let us just say that these two operations over the texts turned to be very important to achieve an easier management of the text, and to help the effectiveness of the algorithms. The algorithms will work always over the main word into the syntactic set. This way, the sets of words are consider as a whole till they arrive to the last level of the algorithms, where each set will be deeper studied.

3.2   Searching and Extraction Algorithms

The following applied algorithms are divided into three logical levels, showed in Fig. 3:

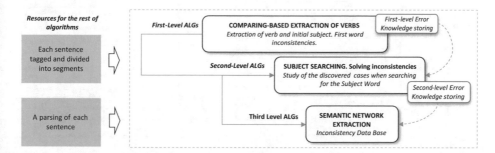

Fig. 3. General logic levels of the core *Searching and Extraction Algorithms.*

First-Level Comparing-Based Verb Extraction Algorithms. <u>Basis</u>: If the *segment* has a verb, it is searched in the chunking and in its entirely parsed sentence, and the various results are compared. The gerund verbs are also treated. <u>Procedure</u>: The verb and its tag in the chunking are compared to the same word/s and tag/s found in the parsed sentence, Fig. 4. <u>Result</u>: Some kinds of inconsistencies are determined, Fig. 5.

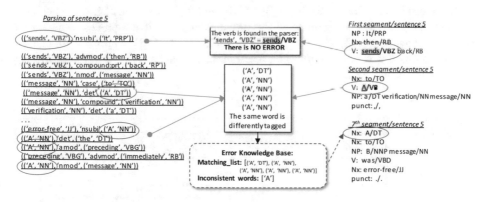

Fig. 4. Verb extraction algorithms applied to a sentence

At this level, the errors or eventual inconsistences are generated owing to a non-concordance between the TAG attached to identical words, found in the results of the parsing and chunking processes. The words under study are stored, creating the Error Knowledge Base, as a *matching list* or as *inconsistent words*, Fig. 4, that will be used in the next algorithm level.

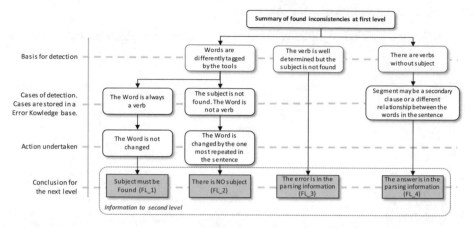

Fig. 5. Inconsistencies found at first-level *Comparing-Based Verb Extraction Algorithms.*

Second-Level Subject Searching Algorithms. <u>Basis:</u> For each first-level case, the process of searching for the subject continues, new cases are explored.

<u>Procedure:</u> The subject is searched in the parsing information for each preceding first-level case. The searching can be successful or instead, the information from the parser could not be enough to solve it. This is expressed by *Logic Stage A* in Fig. 6. If this occurs, the Error Knowledge Base (EKB) is consulted, the information is crossed with the parsing, and the searching is repeated, *Logic Stage B* in Fig. 6. <u>Result:</u> normally, the subject is found. 2_SL, 3_SL, 5_SL and 7_SL cases in Fig. 6 are achieved whether the verb was determined in the previous stage or if the process was able to overcame *Logic Stage A*, with the help of the EKB. 1_SL, 4_SL and 6_SL are reached if *Logic Stage B* was not able to find a solution. If this happens, the error/inconsistency is annotated in the EKB.

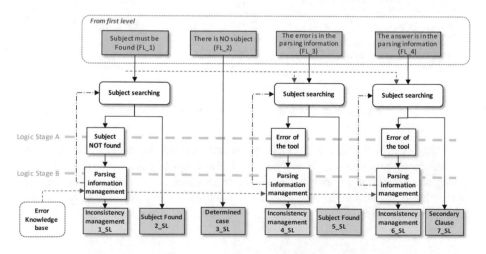

Fig. 6. Simplified logic of second-level *Subject Searching Algorithms*

Third-Level Semantic Network Extraction Algorithms. Based on the EKB and relation between nexuses and nouns, a Semantic Network is constructed. A short outline of the approached process is shown in Fig. 7.

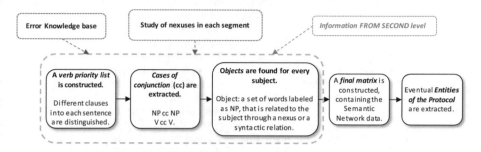

Fig. 7. Summarized logic of third-level *Semantic Network Extraction Algorithms*.

4   Results

4.1   Results Obtained from the *ABP_Standard_Text*

This text is formed by six sentences. The 18 segments derived from them could be directly solved due to the sober writing style used, just proper of a standard text. The only inconsistence found is at first level (Fig. 5), FL_1; the word '*received*' is found differently tagged in the parser, as a verb and as an adjective, but in this case, the fact is normal and the subject is faultlessly found. The information is stored in the EKB.

About the third-level algorithms, here it worth say, that the idea of an automatic extraction of a Semantic Network could be achieved; the entity objects could be extracted from the main nouns of the NP sets that constituted the subjects and objects in the sentences; an early list of attributes of the entities and their classification could be extracted as well.

4.2   Results Obtained from the *ABP_NO_Standard_Text*

The second text contained 20 sentences that in turn provided 76 segments. For this text, the results were more interesting. They showed that the previous classified errors, helped solve others found in next stages. From 45 treated segments with a verb, 15 (33,3%) had errors, mainly because of not finding the subject.

From these latter:

- Twelve of them (80% of the total error-containing segments) appeared owing to the presence of the stored inconsistent words, the word 'A', Table 1. 'A' as verb, was changed by 'A' as noun and thereby, this 12 segments were finally solved, the subject could be found. The word 'A' turns to be an inappropriate word.

Table 1. List of eventual inconsistent words in *ABP_NST*

n_sent	n_seg	Value in Chunking	Value in parser	Selected word	Annotation	Error knowledge	Result
0	4	('receiver', 'VB')	('receiver', 'NN')	('receiver', 'NN')	ERROR of tagging in the chunking, the word and tag is changed in the segment. The final word is not a verb	FL_2. Figure 5	Solves one segment
5	1	('A', 'VB')	('A', 'DT'), ('A', 'NN')	('A', 'NN')	The word is a verb and in the chunking is tagged differently every time it appears in the parser	FL_2 Fig. 5 2_SL. Figure 6	Solves ten segments
18	1	('fetched', 'VBN')	('fetched', 'VBD')	Taken as verb	The word in the chunking is a verb tagged differently in the parser	FL_1. Figure 5	Solves one segments
18	3	('accepted', 'VBN')	('acepted', 'VBD')	Taken as verb	The word in the chunking is a verb tagged differently in the parser	FL_1. Figure 5 6_SL. Figure 6	Annotated Not solved

- In two of them, there was not any inconsistency. Using the *Logic Stage B* in Fig. 6, studying the dependencies in the parsing, these cases could be solved.
- The last case is a 4_SL case (Fig. 6). The parser wrongly tagged the words in the sentence, since it was a quite syntactically complex one; so the error was stored but not solved.

The other inconsistencies from 0 and 18 sentences solved, each in turn, one segment at the first level of algorithms.

5 Conclusions

As a first conclusion, it can be said that the results directly depend, obviously, on the way the studied text is written: (a) the kind of used words related to the importance of the concepts they are supposed to express (these kind of words are likely to be the main entities), and (b) the level of complexity of the syntactic structure (how the main constituents, NP, VP, GV, are arranged.)

The first used text, *ABP_ST,* has very few varied words and is written in an extremely concise, obscure and succinct way. No inconsistencies are detected.

With this first trial, we conclude that: (a) the basic structure and initial basis used to elaborate the algorithms work, and that (b) an automatically constructed Semantic Network can be achieved. That means that the subjects, verbs and objects, and its relationships can be extracted from the text. The constructed SN was manually validated as well. On the other hand, based on the constructed SN, some other inconsistencies would be found, such as the entities that produced the actions over the objects cannot be directly determined since all verbs were passive.

The second text, *ABP_NST*, is more verbose and uses more linguistic structures, and this makes it easier to understand when read. Nevertheless, studying this text, we have discovered: (a) some inconsistencies due to tagging and parsing errors of the NLP tools (words differently tagged by the NLP tools, which means an inappropriate used vocabulary, and syntactic structures that yield an erroneous parsing), (b) the iterative process of the designed algorithm structure permits to solve inconsistencies based on the EKB or, even more important, detect new ones, (c) that the whole process permits as main objective, to detect, store and annotate the text with the inconsistencies found, regardless of whether certain syntactic solution is achieved or not.

6 Future Works

We want to exploit the described approach to apply it as an analysis tool on communication standards: to recognize when a communication protocol is well described, what kinds of inconsistencies it contains, how much understandable it is, that is, to determine the degree of reliability and consistency of the standard text. So far, part of the main skeleton of the process has been designed.

On this basis, the next steps are to explore more protocol description texts and so, find more patterns and metrics to improve the texts in an automatic way.

The algorithms of the third level, used to construct the SN, has to be still deeper validated in order to extract, to the most extent, all the information (types of inconsistencies and EKB).

Second-level algorithms have to be optimized and test their generalization, specially the *Logic Stage B that* is, working with the parsing information to solve and/or annotate the rest of cases that can be found.

Study and optimize the first treatments over the text (Setting Chunking and Heuristic split into segments) is another objective, since the study of certain syntactic patterns could help simplify some parts of the algorithms and serve as a source for finding more inconsistencies or metrics.

The final and key point is to focus the efforts on the finding of protocol requirement descriptions patterns and find specific metrics to evaluate their linguistic descriptions, (correctness, structuration, completeness, ambiguousness, consistency). The approach has just started and there are still many matters and subtle aspects to adjust to contribute to help implementers to more clearly understand standard requirements and easier synthesize communication protocols.

References

1. Lynch, W.C.: Reliable full-duplex file transmission over half-duplex telephone lines. Commun. ACM **11**(6), 407–410 (1968)
2. Holzmann, G.J., Smith, M.H.: An automated verification method for distributed systems software based on model extraction. IEEE Trans. Softw. Eng. **28**(4), 364–377 (2002)

3. Holzmann, G.J., Smith, M.H.: A practical method for verifying event-driven software, pp. 597–607 (1999)
4. Berry, D., Gacitua, R., Sawyer, P., Tjong, S.F.: The case for dumb requirements engineering tools. Lecture Notes Computer Science (including Subser. Lecture Notes Artificial Intelligence Lecture Notes Bioinformatics), vol. 7195, pp. 211–217 (2012)
5. Ryan, K.: The role of natural language in requirements engineering. In: Proceedings of IEEE Interantional Symposium on Requirements Engineering, pp. 240–242 (1993)
6. Fabbrini, F., Fusani, M., Gnesi, S., Lami, G.: The linguistic approach to the natural language requirements quality: benefit of the use of an automatic tool. In: 26th Annual NASA Goddard Software Engineering Workshop, IEEE/NASA SEW 2001, pp. 97–105 (2001)
7. Leveraging Natural Language Processing in Requirements Analysis: How to Eliminate over Half of All Design Errors Before they Occur, pp. 1–11 (2016)
8. Barlas, K., Koletsos, G., Stefaneas, P.: Extending standards with formal methods: open document architecture. In: 2012 International Symposium on Innovations in Intelligent Systems and Applications, pp. 1–5 (2012)
9. Bruns, G., Staskauskas, M.: Applying formal methods to a protocol standard and its implementations. In: Proceedings of International Symposium on Software Engineering for Parallel and Distributed Systems, pp. 198–205 (1998)
10. Drechsler, R., Harris, I.G., Wille, R.: Generating formal system models from natural language descriptions. In: 2012 IEEE International High Level Design Validation and Test Workshop (HLDVT), pp. 164–165 (2012)
11. Yue, T., Briand, L.C., Labiche, Y.: A systematic review of transformation approaches between user requirements and analysis models. Requirments Eng. 16(2), 75–99 (2011)
12. Bano, M., Addressing the challenges of requirements ambiguity: a review of empirical literature. In: 2015 IEEE Fifth International Workshop on Empirical Requirements Engineering (EmpiRE), pp. 21–24 (2015)
13. Kobayashi, Y., Enoki, H., Ohta, T.: Understanding natural language requirement descriptions for telecommunication services. In: Proceedings of 7th IEEE International Conference on Tools with Artificial Intelligence, pp. 295–302 (1995)
14. Ishihara, Y., Seki, H., Kasami, T.: A translation method from natural language specifications into formal specifications using contextual dependencies. pp. 232–239 (1993)
15. Ambriola, V., Gervasi, V.: On the systematic analysis of natural language requirements with c irce. Autom. Softw. Eng. 13(1), 107–167 (2006)
16. Greghi, J.G., Martins, E., Carvalho, A.M.B.R.: Semi-automatic generation of extended finite state machines from natural language standard documents. In: Proceedings - 2015 45th Annual IEEE/IFIP International Conference on Dependable Systems and Networks Workshops, DSN-W 2015, pp. 45–50 (2015)
17. Wang, Y., Zhang, J.: Experiment on automatic functional requirements analysis with the EFRFs semantic cases. In: PIC 2016 - Proceedings of the 2016 IEEE International Conference on Progress in Informatics and Computing, pp. 636–642 (2017)
18. Wang, Y.: Semantic information extraction for software requirements using semantic role labeling. In: Proceedings of 2015 IEEE International Conference on Progress in Informatics and Computing, PIC 2015, pp. 332–337 (2016)
19. Arora, C., Sabetzadeh, M., Briand, L., Zimmer, F.: Automated checking of conformance to requirements templates using natural language processing. IEEE Trans. Softw. Eng. 41(10), 944–968 (2015)

Mobile Architecture for Forest Fire Simulation Using PhyFire-HDWind Model

Alejandro Hérnández[1], David Álvarez[2], M. Isabel Asensio[2],
and Sara Rodríguez[1(✉)]

[1] BISITE Digital Innovation Hub, University of Salamanca, Salamanca, Spain
{ahg, srg}@usal.es
[2] SINUMCC Numerical Simulation and Scientific Computation,
University of Salamanca, Salamanca, Spain
{daalle, mas}@usal.es

Abstract. This article presents the design and implementation of a new visualization system for mobile platforms for the PhyFire-HDWind fire simulation model, called AppPhyFire. It proposes a mobile computing infrastructure, based on ArcGIS Server and REST architecture, which improves the user experience in actions associated with the fire simulation process. The PhyFire-HDWind model, of which the system presented here forms part, is a forest fire propagation simulation tool developed by the SINUMCC research group of the University of Salamanca, based on two own simplified physical models, the PhyFire physical fire propagation model, and the HDWind high definition wind field model, resolved using efficient numerical and computational tools and parallel computing, allowing simulation times shorter than the real time fire propagation, integrated into a Geographical Information System, and accessible through a server by the AppPhyFire. The system presented in this article allows a quick visualization of simulations results in mobile devices. This work presents the detailed operation of the system and its phases of operation.

Keywords: Simulation · Mobile architectures · Prediction models · PhyFire · HDWind

1 Introduction

Indonesia in 2015, Canada and Spain in 2016, Chile and Portugal in 2017, Greek and California in 2018, are just some examples of how forest fire is a global problem that is getting worse due to several factors. One of these factors, is the devastating impact of climate change. Wildfires are more severe, more intense and longer lasting, causing more loss of life and property.

The scientific community makes efforts to seek solutions for this growing problem, relying on emerging technology, seeking ways of adapting, preventing and combating forest fires. An example of these efforts are the numerous forest fires spread models existing (FARSITE [1], Prometheus [2], WRF-Fire [3], etc.), and how current technologies (Geographical Information Systems, Remote Sensing, Supercomputing) are improving the applicability and efficiency of these models. The introduction of

© Springer Nature Switzerland AG 2020
F. Martínez Álvarez et al. (Eds.): SOCO 2019, AISC 950, pp. 301–310, 2020.
https://doi.org/10.1007/978-3-030-20055-8_29

programming techniques for the use of multiprocessor systems has allowed a significant increase in the computing capacity of the simulations tools based on these models. Moreover, current computational technology offers great potential for the effective modeling of wildfire behaviour through more complex models.

Mobile computing is also being using as a useful tool in firefighting, such as an early warning system for wildfire alerts [4] or to view active fire maps on mobile devices [5]. But predicting fire spread and behaviour on a wildfire, in the field through a mobile device, can be determinant in order to minimize many risks for firefighters, to improve the success of their work and prevent the loss of human life. Some experimental wildfire spread systems have recently developed mobile applications [6]

We present here a system that allows a quick visualization of the simulations performed in SINUMCC Server thanks to the fusion of physical simulation models (Phyfire and HDWind server models) and current communication and data processing technologies (such as Api REST, JSON and ArcGIS Server). This article presents the detailed communication with the system and its operation phases for a mobile app: AppPhyFire.

The rest of the paper is structured as follows: Sect. 2 presents a background about the prediction models and technology used. The proposed system is described in Sect. 3. Finally, results and conclusions are summarized in Sect. 4.

2 Background

Wildfires are a growing problem and the climate change with higher temperatures and drier landscape increase their frequency and severity. Wildfires are not only due to the outcome of climate change, they also emit greenhouse gases and therefore also contribute to global warming.

In recent years, thanks to the technological advances, there has been an increasing development of wildfire spread models. The simulation of wildfire spread has direct applications in prevention and firefighting, from risk mapping, reforestation policies, to evacuation planning or resource optimization. There are several types of wildfire spread models, that range from the purely physical (based on physical and chemical principles involved in the combustion of biomass fuel and behaviour of wildfire) to purely empirical (based on phenomenological description or statistical regression of observed fire behaviour) passing through approaches from one end of the spectrum to the other [7, 8]. All kind of models should try to strike a balance between fidelity and fast execution to be operational. The use of empirical models, most of them based on Rothermel's model [9], is widespread due to its simplicity and computational efficiency, but their applicability is limited. Physical or semi-physical models are more complex and has high computational costs, but the current computational technology offers great potential for the effective modeling of wildfire behaviour through more complex models.

The PhyFire-HDWind model, upon which the mobile architecture presented in this paper has been developed, is a wildfire spread simulation tool developed by the research group SINUMCC at the University of Salamanca, based on two own simplified physical models, the physical fire spread model PhyFire, and the high definition wind field model HDWind, solved making use of efficient numerical and computational tools and parallel computation, allowing simulation times shorter than the real time of the fire spread.

2.1 PhyFire-HDWind Models

The PhyFire model is the current version of a simple 2-D one-phase physical fire spread model first published in [10], based on the energy and mass conservation equations. This model considers convection and radiation as dominant thermal transfer mechanisms, and mainly depends on metheorological data (direction and intensity of the wind, ambient temperature and humidity), orography, and fuel type and load. The influence of fuel moisture content and heat absorbtion by pyrolysis is considered by means of a multivalued operator representing the enthalpy [11]. Radiation is represented by a non-local radiation term that allows the modelling of the radiation from the flame above the fuel layer, enabling to cope with the effect of wind and slope over the flame tilt [12]. Efforts have been made to improve the feasibility of the PhyFire model with simulations of real fires [13], and experimental fires [14]. The simulation of experimental examples has been used to perform a global sensitivity analysis of the model allowing to conclude that the model properly reflects the most important factors affecting a wildland fire spread, to configure the design of model parameter adjustment, and to improve and update the model itself.

The numerical methods used to solve the non-dimensional equations derived from the model, include finite element method combined with different finite difference schemes. In addition, the computational cost has been reduced by means of definition of active nodes and making use of parallel computation techniques in order to become competitive compared with some other simpler models. The model code is programmed in C++, using the API OpenMP in order to take advantage of the multiprocessor platforms to reduce the computational time [15]. It was compiled using the GNU Compiler (GCC) version 4.6.3.

PhyFire can operate with constant wind or with wind data provided by a wind model, particularly with wind data provided by the HDWind model. HDWind is a mass consistent vertical diffusion wind field model. The idea behind this model is to adapt the principles of the shallow water models, where the horizontal dimensions are much larger than the vertical one, to the forest fire convective phenomena. Through an asymptotic approximation of the primitive Navier-Stokes equations, this model provides a 3D velocity wind field (which satisfies the incompressibility condition) in the air layer under the influence of the fire, by solving only 2D linear equations, so that it can be coupled with the 2D PhyFire model. The 2D equations of the HDWind model depend on the temperature surface distribution, the surface height and the meteorological wind flow on the surface boundary. The model depends on a single parameter, the air friction coefficient which is related with the roughness length of the surface. The nonlinear terms are neglected and it is assumed that the air temperature decreases linearly with the height, however, the model takes into account buoyancy forces, slope effects, and mass conservation [16]. Last version of HDWind [17] allows to obtained a wind field that fits to several punctual wind velocity measurements at different points in the 3D domain (the air layer) by an optimal control problem in which the wind flow on the surface boundary is the control.

Both models, PhyFire and HDWind, can be compiled for any platform, and can operate either together or separately. In order to automate the processes of input data capture and output data visualization during the simulating process, PhyFire and HDWind have been integrated into a Geographic Information System [15]. This GIS-based interface has a dual purpose, on the one hand, this provides a more accessible tool to a broader audience that might not be familiar with the models; and on the other hand, this facilitates the testing and validation process. The PhyFire integrated in the GIS tool uses the following input data: topography, fuel load and type, weather conditions, ignition location and fire suppression tactics; and predicts the fire spread for the established time period. The outputs provided at each time steps are: the burnt area perimeter and the fire front position and depth. Likewise, the HDWind integrated in the GIS tool also uses topography, surface roughness and weather conditions, and provides a wind velocity field that is well adapted to the domain studied. Both have been developed for its use throughout Spain, so the scope of the spatial information currently used is limited to that area. For this purpose, a geodatabase has been developed containing the three maps needed for extracting the spatial information our models use: a first map containing the height of the surface, a second map gathering all the information related to fuel type, both maps used by the two models, and a third map collecting all the elements involving the function of either artificial or natural fuel-breaks that affect the fire spread, only used by the PhyFire model [18]. This geodatabase has been generated from several public map service as the Spanish National Geographic Institute or the Ministry of Agriculture, Food and Environment of the Spanish Government.

The GIS tool chosen for this integration was ArcMap 10.4 of Esri's ArcGIS Desktop suite and the interface was developed as a Python add-in for ArcMap. The functionality of each tool was implemented as a script using the Python programming language and the ArcPy geoprocessing library.

Mobile technology is having a very strong growth at present. Since computing power is limited in these devices, an architecture for offering the PhyFire and HDWind models has been designed via web services, so these devices can easily consume the model operations over the Api REST deployed in a server, https://sinumcc.usal.es/.

3 Proposed System

The PhyFire and HDWind server models have a loose coupling approach, where the linkage between the GIS and the modelling system is made through the import–export of data, specifically ASCII grid text files as inputs and outputs.

Thus, it is key to understand the phases that the tool performs in each simulation: Input spatial data, pre-processing, processing and visualization (Fig. 1).

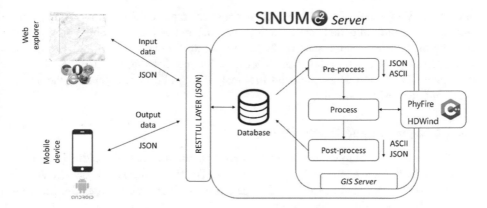

Fig. 1. SINUMCC server architecture overview

3.1 Input Spatial Data

The PhyFire-HDWind tool displays to the user a base map, where the simulation area must be selected. In addition, the user must enter other data such as fire source, lines of defense (optionally), ambient temperature, humidity and wind (speed and direction). Wind data can be constant for the whole domain or can be a collection of punctual data (direction and intensity) in several points of the domain. If constant wind is established, only the PhyFire model is executed, whereas when wind point data are entered, the HDWind model calculates a wind field over the simulation area that best fits these punctual data, so the simulated wind field will be used as wind data in the PhyFire model. The simulation time can also be set by the user. All these data are collected and saved in a database server in JSON format identified by an ID of the simulation

3.2 Pre-process

Once desired data has been stored in the server, the PhyFire tool takes JSON saved data and start making transformation for obtaining three files corresponding to the topography, fuel type and fuel load of the study domain, in order to get the georeferenced data necessary for the simulation [18]. Therefore, it is enough to clip these data corresponding to the study domain from the geodatabase developed for this purpose, described above.

3.3 Process

Once all the input data for the selected area are pre-processed, clipped, checked to avoid errors, converted to raster and exported to ASCII grid text files, the PhyFire and eventually the HDWind model read the files they need, and the simulation is run providing the corresponding output ASCII grid text files.

The PhFire model provides two types of output data: the solid fuel mass fraction and the non-dimensional solid fuel temperature. These data are also stored in ASCII files that collect the rasters of these variables on each point of the simulation area mesh, and at every moment of time in which a graphical representation is required. Comparing the output solid fuel mass fraction with the initial fuel mass fraction inputted into the model provides the state of the landscape. So for each point of the domain we can determine whether or not that specific point has been burnt, defining the fire perimeter. The non-dimensional solid fuel temperature allows to determine the active fire front, its position and depth.

3.4   Post-process and Results Display

ASCII files with the output data from the simulation require a transformation in order to be displayed in their corresponding cartography. Therefore, the ASCII data generated by the models are transformed by GIS operations to finally obtain a JSON file containing a representation of the results.

The data are stored in the database as a result of the simulation using the previously indicated identifier. These data stored in JSON will be used to visualize the results on the map (Fig. 2).

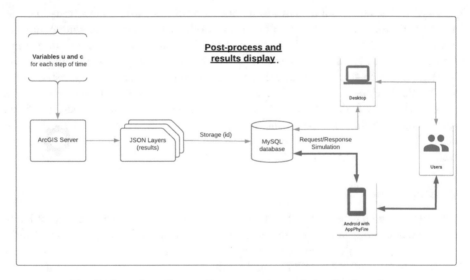

Fig. 2. Data flow diagram. Post-processing and visualization phase.

Due to ArcGIS Server's ability to generate the simulation layers in JSON format, we can use its Android API to collect and show them to users on their terminals with the application installed, thus leaving the framework of traditional devices. AppPhyFire is thus integrated with the rest of the PhyFire system and manages to show users any simulation they have made, giving numerous additional advantages such as the ability to view it from the action area itself and thus improve data capture or validation of existing data.

4   Results and Conclusions

The final system can represent both demonstrations of how the physical models work and the simulations performed by the user, for example, the one shown below (Fig. 3):

(a)

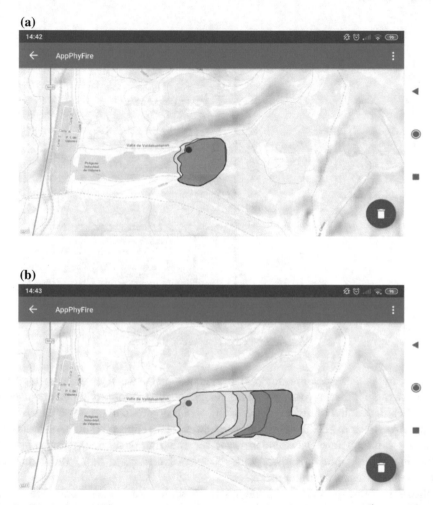

(b)

Fig. 3. Displaying PhyFire spread simulation results in the AppPhyFire (a) 1st step. Simulates 0:30 h of fire spread. (b) 4th step. Simulates 2 h of fire spread. It's also showed the intermediate steps (0:30 h, 1 h, 1:30 h, 2 h).

This is a simulation that has four half-hour time steps each and, in each step, there are two layers, a layer corresponding to the position of the fire front (showed in light red on previous pictures) and another layer corresponding to the burned area (showed in light gray). The focus (or focuses) identifies the ignition point of the fire begins, and it is possible to add defense lines that have been also included as an input of the simulation.

The part of hardware or physics necessary for the start-up of the application, as well as all the existing relationships between them, is shown in the following deployment diagram (Fig. 4):

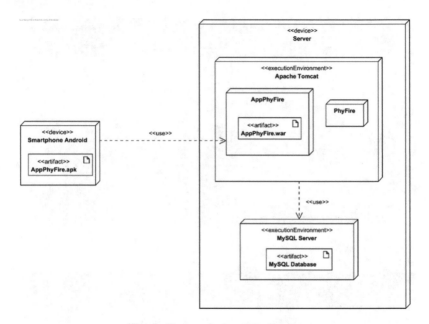

Fig. 4. System deployment diagram

This article describes the different techniques that have been used to create an innovative mobile display system for fire simulations that involve a high computational cost. After a detailed study of the existing technologies our proposal consisted in joining ArcGIS and REST architecture for the visualization of the simulations. The necessary infrastructure was implemented for the introduction, preprocessing and processing of the simulation data, with the aim of creating a light, well-structured, scalable system adapted to the user.

In order to test the stability of the system, visualization requests have been simulated from a mobile terminal to the server. For this purpose, 1000 requests have been made, corresponding to a login request and the loading of a simulation with 4 time steps, with a delay of 1 s between each of them, resulting in an average of 542 ms in obtaining the response and with no loss. The server specifications are: Windows Server 2016, Intel Xeon E5-2660v4 256 GB RAM; The mobile device is a Pixel 2XL with Android 9 over a 4G network.

In a future work, the data that is collected and fed to the server that carries out the computation of the models, could be modified in real time, with what entails a visualization of the simulations in a dynamic way. Given the processing capacities of the current light devices, it is possible to unload the weight of the computation and do it in a more distributed way. The professionals dedicated to fire prevention constantly deal with simulations information and prediction models knowledge. Because the proposed system can be executed on mobile devices, simulation data are always available, and can provide ubiquitous access for the users. These kinds of devices allow the system the use of several context-aware technologies [19, 20] to acquire information from users and their environment. For instance, in the future it would be possible to add mobile alerts, identification or positioning services. Moreover, it is also expected to validate the system with new sources of simulation data.

Acknowledgments. This work has been partially supported by the Consejería de Educación of the regional government, Junta de Castilla y León (SA020U16) and by the University of Salamanca General Foundation (PROTOTIPOS TCUE 2017-18), both with the participation of FEDER funds.

This work was also developed as part of "Virtual-Ledgers-Tecnologías DLT/Blockchain y Cripto-IOT sobre organizaciones virtuales de agentes ligeros y su aplicación en la eficiencia en el transporte de última milla", ID SA267P18, project cofinanced by Junta Castilla y León, Consejería de Educación, and FEDER funds.

References

1. Finney, M.A.: FARSITE: fire area simulator-model development and evaluation. United States Department of Agriculture Forest Service, Rocky Mountain Research Station, Research Paper RMRS-RP-4, March 1998. Revised February 2004
2. Tymstra, C., Bryce, R.W., Wotton, B.M., Taylor, S.W., Armitage, O.B.: Development and structure of prometheus: the Canadian wildland fire growth simulation. Information Report NOR-X-417 Northern Forestry Centre Canadian Forest Service (2010)
3. Mandel, J., Beezley, J.D., Kochanski, A.K.: Coupled atmosphere-wildland fire modeling with WRF 3.3 and SFIRE 2011. Geosci. Model Dev. **4**(3), 591–610 (2011)
4. CalFire. http://www.readyforwildfire.org/Ready-for-Wildfire-App/. Accessed 11 Mar 2019
5. Firemap. https://play.google.com/store/apps/details?id=com.mapadebolsillo.firemap&rdid=com.mapadebolsillo.firemap. Accessed 11 Mar 2019
6. Monedero, S., Ramirez, J., Cardil, A.: Predicting fire spread and behaviour on the fireline. Wildfire analyst pocket: a mobile app for wildland fire prediction. Ecol. Model. **392**, 103–107 (2019)
7. Sullivan, A.L.: Wildland surface fire spread modelling, 19902007. 1: physical and quasi-physical models. Int. J. Wildl. Fire **18**(4), 349–368 (2009)
8. Sullivan, A.L.: Wildland surface fire spread modelling, 19902007. 2: empirical and quasi-empirical models. Int. J. Wildl. Fire **18**(4), 369–386 (2009)
9. Rothermel, R.C.: A mathematical model for predicting fire spread in wildland fuels. USDA Forest Service Research Paper INT USA, no. INT-115, p. 40 (1972)
10. Asensio, M.I., Ferragut, L.: On a wildland fire model with radiation. Int. J. Numer. Methods Eng. **54**(1), 137–157 (2002)
11. Ferragut, L., Asensio, M.I., Monedero, S.: A numerical method for solving convection-reaction-diffusion multivalued equations in fire spread modelling. Adv. Eng. Softw. **38**(6), 366–371 (2007)
12. Ferragut, L., Asensio, M.I., Monedero, S.: Modelling radiation and moisture content in fire spread. Commun. Numer. Methods Eng. **23**(9), 819–833 (2006)

13. Ferragut, L., Asensio, M.I.: A simplified wildland fire model applied to a real case. In: Casas, F., Martínez, V. (eds.) Advances in Differential Equations and Applications. SEMA SIMAI Springer Series, vol. 4, pp. 155–167 (2014
14. Prieto, D., Asensio, M.I., Ferragut, L., Cascón, J.M.: Sensitivity analysis and parameter adjustment in a simplified physical wildland fire model. Adv. Eng. Softw. **90**, 98–106 (2015)
15. Álvarez, D., Prieto, D., Asensio, M.I., Cascón, J.M., Ferragut, L.: Parallel implementation of a simplified semi-physical wildland fire spread model using OpenMP. In: de Pisón, F.M., Urraca, R., Quintián, H., Corchado, E. (eds.) Hybrid Artificial Intelligent Systems: HAIS 2017. LNCS, vol. 10334, pp. 256–267. Springer, Cham (2017)
16. Asensio, M.I., Ferragut, L., Simon, J.: A convection model for fire spread simulation. Appl. Math. Lett. **18**(6) Spec. Iss. 673–677 (2005)
17. Ferragut, L., Asensio, M.I., Simon, J.: High definition local adjustment model of 3D wind fields performing only 2D computations. Int. J. Numer. Method. Biomed. Eng. **27**(4), 510–523 (2011)
18. Herráez, D.P., Sevilla, M.I.A., Canals, L.F., Barbero, J.M.C., Rodríguez, A.M.: A GIS-based fire spread simulator integrating a simplified physical wildland fire model and a wind field model. Int. J. Geogr. Inf. Sci. **31**(11), 2142–2163 (2017)
19. Chamoso, P., González-Briones, A., Rodríguez, S., Corchado, J.M.: Tendencies of technologies and platforms in smart cities: a state-of-the-art review. Wirel. Commun. Mob. Comput. (2018)
20. Casado-Vara, R., Chamoso, P., De la Prieta, F., Prieto, J., Corchado, J.M.: Non-linear adaptive closed-loop control system for improved efficiency in IoT-blockchain management. Inf. Fusion **49**, 227–239 (2019)

A Proposal of Robust Leak Localization in Water Distribution Networks Using Differential Evolution

Maibeth Sánchez-Rivero[1], Marcos Quiñones-Grueiro[1], Carlos Cruz Corona[2],
Antônio J. Silva Neto[3], and Orestes Llanes-Santiago[1(✉)]

[1] Universidad Tecnológica de La Habana José Antonio Echeverría (Cujae),
calle 114, No. 11901 e/Ciclovía y Rotonda, Marianao, La Habana, Cuba
orestes@tesla.cujae.edu.cu
[2] Universidad de Granada, Granada, Spain
[3] IP-UERJ, Rua Bonfim, 25, Vila Amélia, Nova Friburgo, RJ, Brazil

Abstract. The performance of popular leak location strategies that depend on the use of a leak sensitivity matrix (LSM) is highly dependent on the conditions used for generating such matrix. Since noise is present in real Water Distribution Networks (WDNs), the location results can deteriorate seriously when using the former methods. A robust model-based leak localization approach by using Differential Evolution (DE) algorithm is presented in this article. The proposal uses the topological information of the water distribution network in the location of the leak and it does not depend on the sensitivity matrix. The proposed method demonstrates its robustness compared with the LSM approach by using the Hanoi network as case study. The performance achieved by the proposal improves the leak location accuracy by 20% with respect to LSM-based leak location method in the presence of different noise magnitudes.

Keywords: Differential Evolution · Measurement's noise ·
Model-based leak localization · Robustness

1 Introduction

Water loss in Water Distribution Networks (WDNs) is an issue of great concern for water utilities, strongly linked with operational costs and water resource savings. For this reason, great efforts are made in the development of methods for detection and localization of leakages [11].

A method based on pressure measurements and analysis of the leakage sensitivity is proposed in [2,8,9]. Such methodology compares the residuals (difference between the measured values and those estimated from the hydraulic model) with respect to a threshold and the sensitivity matrix (SM) is used to determine the leak location. The sensitivity matrix evaluates the effect of a leak in the pressure head at the nodes of the network.

© Springer Nature Switzerland AG 2020
F. Martínez Álvarez et al. (Eds.): SOCO 2019, AISC 950, pp. 311–320, 2020.
https://doi.org/10.1007/978-3-030-20055-8_30

Casillas et al. proposed a Leak Signature Space (LSS) location method that associates every possible leak within the network with a specific signature independently of the leak magnitude [4]. LSS considers the relationship between pressure residuals and leakage as an approximate linear model that is projected onto a hyperplane.

The performance of the above-mentioned leak location methods depends on the sensitivity matrix. Hence, the best performance for the localization task is achieved when the leakages sizes are close to those used to build the sensitivity matrix. Nonetheless, if the leakage outflow is different from the nominal value, the performance deteriorates.

Although such strategy showed good results under ideal conditions, its performance decreases when there are uncertainties in the model and noise in the measurements.

The leak localization in WDN, formulated like an inverse problem, was proposed by Pudar in 1992 [10]. Recently, Steffelbauer et al. [14] also presented the leak localization as the solution of an inverse problem that is formulated as an optimization problem and solved by using the metaheuristic algorithm Differential Evolution [15]. The main conclusion of their research is that the leakage localization depends on the distance metric used to compute the objective function, as well as the labeling of the leakage positions in the parameter space (node of the network).

Pressure sensor devices are effective and less costly for the leak location task, because they are easier to install and maintain. Nonetheless, pressure sensors are affected by measurement's noise. Therefore, it is necessary to have leak location methods robust against noise in the measurements [4].

A novel robust approach for leak localization in WDNs, which does not depend on the sensitivity matrix is presented in this paper. The proposal considers the leak localization as the solution of an inverse problem using metaheuristic optimization algorithms. The novelty lies in the leak search strategy, which is based on the topological configuration of the network such that it does not depend on the labeling of the nodes. Therefore, the main contributions of this paper are: a robust search strategy for leak localization by using metaheuristic optimization algorithms that improves the performance and the efficiency of the traditional approach. Besides, the proposal does not depend on the sensitivity matrix and present a better performance in presence of noise in the measurements than other methods reported in literature. The Hanoi Water Distribution Network is used to validate the proposed strategy where the metaheuristic algorithm Differential Evolution is used as optimization tool.

2   Model-Based Leak Localization Methods in WDN

Most model-based leak detection and localization methods for water distribution networks assume the occurrence of leaks only in nodes of the network, where they can be described as an extra demand or a demand dependent of the pressure [10].

A realistic approach used for simulating the latter behavior is to set an emitter coefficient EC in a node that will generate a leakage outflow of magnitude f given by

$$f = (EC)h^\gamma \tag{1}$$

where h is the pressure head at the node, and $\gamma = 0.5$ [12].

The model-based leak localization minimizes the residuals obtained from the difference between m_i measurements and the corresponding $\hat{m}_i$ estimated values found due to potential leaks obtained by using the network hydraulic model.

$$r_i = m_i - \hat{m}_i \qquad i = 1, \ldots, n \tag{2}$$

where n is the number of pressure sensor installed in the network. Then, these residuals are analyzed in order to locate the leak.

Leak location in water distribution networks based on models has been achieved by using the Leak Sensitivity Matrix method (LSM), proposed by Pudar in 1992 [10]. Such matrix measures the effect of all possible leaks considered in the WDN at every one of the available pressure sensors. Initially, the sensitivity matrix must be analytically obtained, but this is very complex because a water network is a large scale problem described by a multivariable non-linear system of equations which may also be non-explicit. For that reason, it is obtained by simulation as follows:

$$S = \begin{pmatrix} \frac{\partial m_1}{\partial f_1} & \cdots & \frac{\partial m_1}{\partial f_{n_n}} \\ \vdots & \ddots & \vdots \\ \frac{\partial m_{n_n}}{\partial f_1} & \cdots & \frac{\partial m_{n_n}}{\partial f_{n_n}} \end{pmatrix} \tag{3}$$

where every element S_{ij} of the sensitivity matrix S measures the effect of the leak f_j, known as nominal leak, in the pressure sensor m_i. Every element is normalized according to the magnitude of the leak. The sensitivity matrix has as many rows as sensors in the WDN and as many columns as considered leaks.

2.1   Leak Detection and Localization Methods Based in LSM

In [8,9], a method based on pressure measurements and analysis of the leakage sensitivity is proposed. This methodology compares the residuals with respect to a threshold obtained according to uncertainties in the parameters of the model and noise. When any of these residuals exceed the threshold, a sensitivity matrix is used to determine the leak location. Different metrics such as the Pearson correlation, the angle between vectors and the Euclidean distance have been evaluated for the previous methods. Among them, the angle method achieved the best performance.

The previous works are extended [2] and a comparison of the pressure vectors measured with the sensitivity matrix is proposed by using different metrics such as the Pearson correlation, the angle between vectors and the Euclidean distance. Moreover, an optimization method based on LSM is added, which lies in finding a proper leak size for every node in the network that explains the present pressure measurements. The node affected by the leak is selected depending on the similarity with respect to each column of the sensitivity matrix. This method allows to obtain information about the leak initially, because it allows to estimate the size of the leak that produces the observed pressure data.

2.2   Leak Localization Methods Using Optimization Algorithms

The method proposed by Steffelbauer in 2017 [14] defines an optimization problem where an arbitrary metric can be used to determine the difference $d(m, \hat{m}(x))$ between the measurements of the real world m and the corresponding values of hydraulic simulations $\hat{m}(x)$ to find the localization of the leak.

$$f(x) = d(m, \hat{m}(x)) \tag{4}$$

The proposed problem, where $f(x)$ is a an one-dimensional function called the fitness or objective function, and $x \in R^2$ is defined as:

$$\min_{x} f(x) \quad s.a. \quad 0 < L_L < L_{Lmax} \quad 0 < L_P < n_n \tag{5}$$

$$x = \begin{pmatrix} L_L \\ L_p \end{pmatrix} \tag{6}$$

where $L_L \in R$ is the leak value, L_{Lmax} is the maximum outflow of leak to localize, and $L_P \in N$, is the leak position (the node where the leak occurs) and n_n is the number of nodes of the network.

The leak position and size can be estimated by looking for a vector which minimizes the objective function in the $L_L - L_P$. The objective function will have different local minimums. Thus, an algorithm based on the descending gradient fails in finding the global minimum. This issue can be solved in a satisfactory way by using optimization methods based on the use of metaheuristics, in this paper the Differential Evolution method is used.

Figure 1 shows the localization method proposed in this research, which is based in the use of the pressure residuals combined with optimization tools. The optimization algorithm to determine the leak causes those differences with major probability.

If L_P is introduced as an optimizing parameter and the nominal value of the node label (where the leak takes place) is taken, an in-depth search is required in order to find the global optimum. The results obtained from an in-depth search depend on the amount of information collected on the network configuration introduced in the node labeling. Besides, the labeling strategies do not take into account the specific relationship between the network topological configuration and the node label. Hence, the localization performance is sensitive to the labeling method [14].

The proposal states that the leak magnitude coding and the leak position should be dealt with in different ways for every localization algorithm. The leak magnitude localization should be developed in a continuous space while the leak position localization should take into account the network topological configuration instead of the node label.

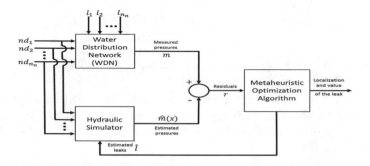

Fig. 1. Methodology for localization and size determination of leaks in a WDN.

2.3   Topological Differential Evolution (T-DE)

The Differential Evolution (DE) algorithm [1,15] is applied because it is an evolutionary method based on populations with very good results in several applications [1,14]. However, the proposed strategy can be combined with other metaheuristic algorithms applied for optimization purposes.

DE is based on three operators: mutation, crossing and selection of vector operations. The general idea behind DE is to provide a new solution by varying the solutions relative to the population up to that moment. The solution model can be expressed by using the notation $DE/x/y/z$, where x denotes the vector to disturb (also called the base vector), y indicates the number of pair of vectors for disturbing and z represents the distribution function, which will be used during the crossing.

In this paper, consideration has been given to $DE/rand/1/bin$ where a random vector (x_{rand}) and the difference of a vector pair $(y = 1)$ were applied in order to generate a mutation. This can be obtained through a binomial crossing model $(z = bin)$. The mutation operator is expressed as:

$$\bar{X}_i^{(j,t)} = X_i^{(\alpha,t)} + F_S(X_i^{(\beta,t)} - X_i^{(\gamma,t)}) \qquad j = 1, \ldots, N; i = 1, \ldots, D \qquad (7)$$

where $X_i^{(\alpha,t)}$, $X_i^{(\beta,t)}$, $X_i^{(\gamma,t)} \in \Re^D$ are elements of the population generated in the previous iteration, and α, β and γ are chosen randomly in the interval $[0, N]$ where N is the population size and D is the localization space dimension.

F_S is a real and constant value known as scaling factor which determines the influence of the vector pair used in the mutation. The crossing operator for each component of the solution vector, is defined as:

$$\bar{\bar{X}}_i^{(j,t)} = \begin{cases} \bar{X}_i^{(j,t)} & if \quad q_{rand} < C_R \quad i = 1, \ldots, D \\ X_i^{(j,t)} & otherwise \end{cases} \qquad (8)$$

where $0 < C_R < 1$ is the crossing constant and q_{rand} is a random number generated when using the z distribution function; in this case the binomial distribution function. Finally, the selection operator is:

$$X^{(j,t+1)} = \begin{cases} \bar{\bar{X}}^{(j,t)} & if \quad F(\bar{\bar{X}}^{(j,t)}) < F(X^{j,t}) \\ X^{(j,t)} & otherwise \end{cases} \tag{9}$$

where $F(\bar{\bar{X}}^{(j,t)})$ and $F(X^{(j,t)})$ are the objective values for the $\bar{\bar{X}}^{(j,t)}$ test vector and the $X^{(j,t)}$ current vector respectively.

In this problem, the search vector is made up by a real value (L_L), whose optimal value can be found by applying the traditional DE method previously analyzed and a discrete value (L_P) is determined by the node label where the leak takes place [14].

In this article, DE is reformulated, called Topological-Differential Evolution (T-DE), such that the mutation operator for L_P will select one of the $X^{(j,t)}$, neighboring nodes (connected to it through a pipe) in the WDN defined as

$$\bar{X}_i^{(j,t)} = X_i^{(\theta)} \tag{10}$$

where $X_i^{(\theta)} \in P = v_1, v_2, \ldots, v_l$ such that v_l is $X_i^{(j,t)}$ neighbor, θ is randomly selected in the $[0, L]$ interval and L is, in general, the number of neighbors to the node $X_i^{(j,t)}$. Figure 2 shows the pseudo-code for the T-DE metaheuristic algorithm.

3 Hanoi Network

The water distribution trunk network in Hanoi, Vietnam, introduced by Fujiwara and Khang in 1990 [7], is the model of hydraulic network selected to validate the proposed strategy and to compare its performance with the method reported in the literature [2,3,5,6,13]. The network topology is shown in Fig. 3.

The network consists of 34 pipes, 32 nodes, and 3 loops. It is fed by gravity from a single fixed head source and it is designed to satisfy the customers demand at specific minimum pressures. The network configuration is the one proposed in [16] for a total demand of 2770 lps.

In this article, the considerations set forth by Soldevila et al. [13] were applied. The experiments were conducted by measuring pressures in nodes 12 and 21 when considering two sensors.

4 Experiments and Results

The goal of this section is to demonstrate the robustness of the proposed method in presence of noise when compared with one of the most popular model-based leak localization methods. EPANET software package is used for hydraulic simulations [12].

A comparison will be performed with respect to the proposal presented in [2,8,9] due to its good findings. Such approach will be identified as LSM method. It is worthwhile to remark that these location strategies depend on the sensitivity matrix and the leak nominal value. In this case, the angle between vectors is used as a metric.

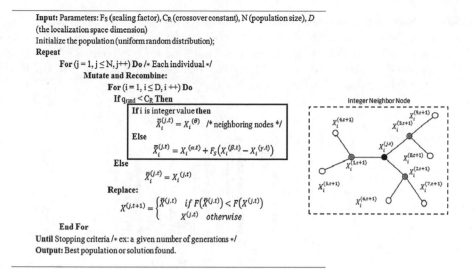

Fig. 2. Pseudo-code for T-DE algorithm.

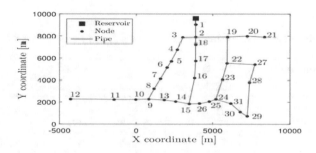

Fig. 3. Hanoi network topology.

The parameters used in the T-DE algorithm are the $CR = 0,9$ to ensure a low influence of the population solution obtained up to that moment and $Fs = 0.6$, which determines the influence of the mutation operator. The population is formed by 20 individuals ($N = 20$). The stopping criteria constitutes an error below 10^{-4} and, in this case, the highest number of generations is set to 100. The metric used to formulate the objective function is the Euclidean Distance because the best localization results have been achieved with it.

A set of 1000 leak scenarios randomly generated for an $EC \in [10, 28]$ (leakage outflows from 26 to 78 lps that represent from 1,6 to 4,7% of the total demand of the network in a 30% load) are used to compare T-DE and the method based on LSM.

Figure 4 shows the performance of both methods by using the accuracy in the leak localization as performance measure (defined as the percentage of leaks accurately localized in relation to the total number of simulated leaks). The figure shows the performance for different percentages of Gaussian noise chosen according to typical noise features in WDN pressure measurements. Whenever the noise level increases the performance of both methods decreases. Nonetheless, the proposed method is always better than the LSM approach, and in the worst case, the achieved performance is higher than 70%.

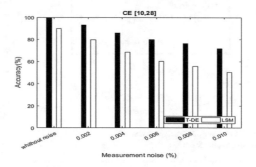

Fig. 4. Performance comparison of proposed method (T-DE) and LSM with angle (LSM).

Table 1. Number of incorrectly located leaks by T-DE and LSM

Measurement noise (%)	T-DE			LSM		
	Total errors	NN	no NN	Total errors	NN	no NN
Without noise	4	4	0	104	72	32
0.002	77	50	22	205	98	107
0.004	141	91	50	317	140	177
0.006	270	118	82	399	146	253
0.008	238	122	116	445	144	301
0.010	284	146	138	496	171	325

In order to compare the performance of both methods the topological distance (defined as the longitude of the shortest path from the node i to the node j) is used. This performance measure allows to consider a soft threshold for the error when the leak is located in a neighbor node of the obtained solution.

Table 1 shows the number of incorrectly located leaks out of 1000 scenarios. Moreover, it specifies which of them were located in a neighbor node (NN)- or in a more distant node (no NN) with respect to the place where the leak really occurs. As it can be observed, the number of errors in the leak location task is always smaller for the T-DE method. The proposed method locates the majority of leaks in the neighbor node of the real leak independently of the noise level. For the LSM method, most incorrectly located leaks are found far from one-node distance to the real leak. Such problem gets worse when the level of noise increases, which demonstrates that the proposed approach is more robust.

5   Conclusions

A novel robust leak localization approach based on the solution of an inverse problem by using metaheuristic tools has been presented in this paper. This approach attempts to adapt optimization tools so that the topological configuration of the network can be used in leak localization. Two advantages of the proposed approach are that it does not depend on the use of a sensitivity matrix, neither it depends on the node labeling method. The proposal was applied by using Differential Evolution (T-DE) for the Hanoi network benchmark. The satisfactory results demonstrate the robustness of the proposed approach in comparison with the popular LSM-based method for different levels of measurement noise.

References

1. Camps Echevarría, L., Llanes-Santiago, O., da Silva Neto, A.J.: An approach for fault diagnosis based on bio- inspired strategies. In: Proceedings of the IEEE Congress on Evolutionary Computation, CEC 2010. No. September 2015 (2010). https://doi.org/10.1109/CEC.2010.5586357
2. Casillas Ponce, M.V., Castañón, L.E.G., Puig, V.: Model-based Leak Detection and Location in Water Distribution Networks considering an Extended-horizon Analysis of Pressure Sensitivities. Journal of Hydroinformatics 16(3), 649–670 (2014). https://doi.org/10.2166/hydro.2013.019
3. Casillas Ponce, M.V., Garza-Castañón, L.E., Puig, V.: Optimal Sensor Placement for Leak Location in Water Distribution Networks using Evolutionary Algorithms. Water pp. 6496–6515 (2015). https://doi.org/10.3390/w7116496
4. Casillas Ponce, M.V., Garza-Castañón, L.E., Puig, V., Vargas-Martinez, A.: Leak Signature Space: An Original Representation for Robust Leak Location in Water Distribution Networks. Water pp. 1129–1148 (2015). https://doi.org/10.3390/w7031129
5. Cugueró-Escofet, M.A., Puig, V., Quevedo, J.: Optimal pressure sensor placement and assessment for leak location using a relaxed isolation index: application to the barcelona water network. Control Eng. Pract. 63(June), 1–12 (2017). https://doi.org/10.1016/j.conengprac.2017.03.003
6. Ferrandez-Gamot, L., Busson, P., Blesa, J., Tornil-Sin, S., Puig, V., Duviella,E., Soldevila, A.: Leak Localization in Water Distribution Networks using Pressure Residuals and Classifiers. IFAC-PapersOnLine 48(21),220–225 (2015). https://doi.org/10.1016/j.ifacol.2015.09.531, http://dx.doi.org/10.1016/j.ifacol.2015.09.531
7. Fujiwara, O., Khang, D.B.: A two-phase decomposition method for optimal design of looped water distribution networks. Water Resour. Res. 26(4), 539–549 (1990)
8. Pérez, R., Puig, V., Pascual, J., Quevedo, J., Landeros, E., Peralta, A.: Methodology for leakage isolation using pressure sensitivity analysis in water distribution networks. Control Eng. Pract. 19(10), 1157–1167 (2011). https://doi.org/10.1016/j.conengprac.2011.06.004
9. Pérez, R., Puig, V., Peralta, A., Landeros, E., Jordanas, L.: Pressure sensor distribution for leak detection in Barcelona water distribution network. Water Sci. Technol. Water Supply 9(6), 715–721 (2009). https://doi.org/10.2166/ws.2009.372

10. Pudar, R.S., Liggett, J.A.: Leaks in pipe networks. J. Hydraul. Eng. **118**(7), 1031–1046 (1992)
11. Puust, R., Kapelan, Z.S., Savic, D., Koopel, T.: A review of methods for leakage management in pipe networks. J. Urban Water **7**(1), 25–45 (2010). https://doi.org/10.1080/15730621003610878
12. Rossman, L.A.: Epanet 2 Users Manual (2000)
13. Soldevila, A., Blesa, J., Tornil-Sin, S., Duviella, E., Fernandez-Canti, R.M., Puig, V.: Leak localization in water distribution networks using a mixed model-based/data-driven approach. Control Eng. Pract. **55**, 162–173 (2016). https://doi.org/10.1016/j.conengprac.2016.07.006
14. Steffelbauer, D.B., Günther, M., Fuchs-Hanusch, D.: Leakage localization with differential evolution: a closer look on distance metrics. Procedia Eng. **186**, 444–451 (2017). https://doi.org/10.1016/j.proeng.2017.03.251
15. Storn, R., Price, K.: Differential evolution a simple and efficient adaptive scheme for global optimization over continuous spaces. Technical Report TR-95-012, Int CS Institute, University of California (1995)
16. Tospornsampan, J., Kita, I., Ishii, M., Kitamura, Y.: Split-pipe design of water distribution network using simulated annealing. Int. J. Comput. Inf. Syst. Sci. **1**(4), 28–38 (2007)

Neural Model of a Specific Single Proton Exchange Membrane PEM Fuel Cell

Jose Manuel Lopez-Guede[1,4(✉)], Manuel Graña[2,4], and Julian Estevez[3,4]

[1] Department of Systems Engineering and Automatic Control,
Faculty of Engineering of Vitoria, Basque Country University (UPV/EHU),
Nieves Cano 12, 01006 Vitoria-Gasteiz, Spain
jm.lopez@ehu.es
[2] Department of Computer Science and Artificial Intelligence,
Faculty of Informatics, Basque Country University (UPV/EHU),
Paseo Manuel de Lardizabal 1, 20018 Donostia-San Sebastian, Spain
[3] Department of Mechanics, Faculty of Engineering of Gipuzkoa,
Basque Country Univesrsity (UPV/EHU),
Plaza Europa 1, 20018 Donostia-San Sebastian, Spain
[4] Computational Intelligence Group,
Basque Country Univesrsity (UPV/EHU), Donostia-San Sebastian, Spain

Abstract. Hydrogen based technologies are a growing field among the renewable energies options, being the single proton exchange membrane (PEM) fuel cells (FC) one of the most promising. There are a number of possibilities to model the electrical behavior of this kind of devices. Among them, in this paper authors have tackled the problem using an Artificial Neural Network based approach. As a proof of concept, a specific 2 KW PEM FC device has been used to obtain the data needed to carry out the process. Using these data, a model of only one signal at both the input and the output has been designed, i.e., the current I_{FC} at the input and the voltage V_{FC} at the output. As result of the modeling process that is explained along the paper, a relatively small model showing a medium squared error of $0.012\,V^2$ with test data has been obtained.

1 Introduction

Fuel cell generation systems have been recognized as new-energy technology to reduce environment pollution and also with big contribution in solving the petrol energy crisis. Among these, proton exchange membrane (PEM) fuel cells seem to be the best alternative due to its simple structure, high power density, quick start, no moving parts and superior reliability and durability, low operating temperature and environmental aspects. Taking into account the more and more increasing interest in PEM fuel cell research and development, its operational performances are focused on various mathematical models, expressing both the dynamic and steady-state conditions. Thus the PEM fuel cell system behavior can be analyzed starting even with the design step by different pathways with

© Springer Nature Switzerland AG 2020
F. Martínez Álvarez et al. (Eds.): SOCO 2019, AISC 950, pp. 321–329, 2020.
https://doi.org/10.1007/978-3-030-20055-8_31

respect to computer simulations in different operational conditions [1,2]. In the state of the art there are a number of successful efforts to model the electrical behavior of PEM fuel cells obtaining the voltage provided, but those works usually rely on complex models using many input variables. In [3] three input variables are used, namely, partial pressures of hydrogen and oxygen, and cell operating current. A model of similar complexity is obtained in [4], where the temperature of the fuel cell and one sample time delayed values of the current and voltage are used. Slightly more complex models of four inputs are also used. In [5] authors consider the stack current, stack temperature, hydrogen flow and oxygen flow as inputs, while in [6] the hydrogen and oxygen pressures, room temperature and stack current are taken. A model with five inputs is used in [7], where hydrogen and oxygen inlet pressures, stack temperature, oxygen relative humidity and the stack current are taken as inputs. Finally, six inputs are taken in the most complex models as in [8] with current density, hydrogen and air flow rates, hydrogen and air humidification temperatures, and cell temperature. In [9] authors consider anode and cathode relative humidity, anode and cathode flow rate temperatures, cell temperature and current density.

The objective of this paper is to obtain an efficient model of a commercial PEM fuel cell, i.e., it should be simple but accurate enough. Besides, it should use as little calculation capacity as possible. Following these ideas, we have obtained a very simple model that obtains the voltage V_{FC} from the current I_{FC} using a very small neural network, showing an MSE of $0.012\,\mathrm{V}^2$ and a mean error test of $0.0766\,\mathrm{V}$. The remaining of the paper is structured as follows. Section 2 gives a short background on the basic concepts this paper rely on, Sect. 3 describes the experimental design while Sect. 4 discusses the obtained results. Finally, Sect. 5 enumerates our main conclusions.

2 Background

This section provides background information about PEM fuel cells in the first subsection, and about artificial neural networks in the second one.

2.1 PEM Fuel Cells

Their distinguishing features are based on low temperatures (50 to 80 °C) low pressures ranges (0–2 bar) and a special polymer electrolyte membrane. PEM fuel cells are developed mainly for transport applications, as well as for stationary and portable FC applications. The structure of PEM fuel cells includes several parts, namely: (i) membrane electrode assemblies (MEA) consisting in electrodes, solid membrane electrolyte, catalyst, and gas diffusion layers; (ii) bipolar and electrical plates. The reactants are hydrogen (H_2) supplied at anode and oxygen (O_2) supplied at cathode. The chemical mechanism reaction assumes the dissociation of hydrogen molecules into electrons and ionized atoms in form H+. The oxygen picks up the electrons and completes the external circuit. The electrons go through this circuit thus providing the electrical current. The oxygen then combines with the H^+ hydrogen, and water (H_2O) is formed as the waste product leaving the fuel cell. The electrolyte allows the appropriate ions to pass between the electrodes [10]. The most important feature in a single fuel cell is that it typically produces about 0.5–1 volts, depending on operational conditions. Therefore, fuel cells are usually connected in series in order to form a fuel cell assembled system. The stack is the system used in cars, generators, or other electrochemical devices that produce power.

2.2 Artificial Neural Networks

Artificial Neural Networks (ANNs) [11] are computational models inspired by a rough simplification of the human brain structures. They are able to learn complex (even nonlinear) relationships between a set of inputs and outputs of a given system without analytical methods. They are very useful in problems where the dynamics of the system is complex, nonlinear, distributed in nature, and particularly vague or totally unknown with uncertain parameters. ANNs have found to be relevant solving difficult problems in a number of areas of engineering including signal processing, pattern recognition, power electronics, fault detection, adaptive control, expert systems, function approximation, forecasting and control of electrical machinery.

3 Experimental Design

This section gives the experimental design that has been carried out in the paper. The first subsection describes the commercial PEM fuel cell that has been used in the paper; the second one provides the data gathering process and finally, the last subsection explains how the modeling task has been accomplished.

3.1 Commercial PEM Fuel Cell Description

The experimental investigation was performed using a 2 kW PEM fuel cell stack with 16 cells of a 230 cm^2 surface area. The fuel cell stack was manufactured by Nedstack (Holland). The PEMFC stack was connected to a test bench of Arbin Instruments Fuel Cell Test Station (Fig. 1) that was run by a MITS Pro-FCTS software. FCTS is located at National Center for Hydrogen and Fuel Cell (Romania) and operated taking into account the expertise of our group in dynamic and steady-state conditions. FCTS is composed of five basic subsystems: gas input and control module, humidifier module, cooling water module, power and control module. The gas input and control module included mass flow controls (MFC), installed into both fuel and oxidant lines sensors (in order to assure wide power range capacities from very small to 20 kW power output), temperature and pressure regulators. The temperature was maintained through a specific cooling water system. The humidifier module allows the required humidity in both fuel and oxidant channels. The power and control module assures mainly the monitoring of environmental and experimental conditions, electronic programmable load, emergency and main power switches, and heat coil control.

Fig. 1. PEM fuel cell test station (Arbine instruments FCTS-20 kW)

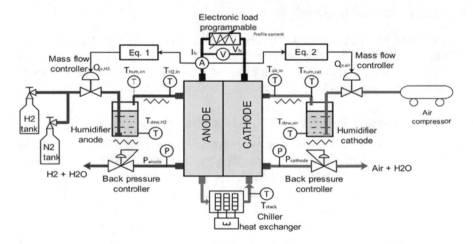

Fig. 2. Schematic of the PEM fuel cell testing system

3.2   Data Gathering Process

The experiments and data gathering process was carried out following the process described by the schematic of Fig. 2. During studied experiments, the fuel hydrogen with 99.998% purity was supplied at anode and the inlet pressure was maintained at 60 psi. Air was supplied at cathode. Both fuel and air were humidified using de-ionized water before entering into the test fuel cell. The excess of hydrogen and air were exhausted to the atmosphere. The reactants flow rates were calculated using the Faraday's law of Eqs. (1) and (2) taking into consideration mainly the expected drawn current and also several parameters such as stack temperature, reactants pressure and reaction stoichiometry.

$$Q_{v,H_2} = \frac{60,000 \cdot R \cdot T_{stack} \cdot n \cdot i_{fc} \cdot \lambda_{H_2}}{2 \cdot F \cdot P_{anode} \cdot x\,(\%)}; \; Q_{v,H_2} > 5, \tag{1}$$

$$Q_{v,Air} = \frac{60,000 \cdot R \cdot T_{stack} \cdot n \cdot i_{fc} \cdot \lambda_{air}}{2 \cdot F \cdot P_{cathode} \cdot y\,(\%)}; \; Q_{v,Air} > 10, \tag{2}$$

The tests were performed using the test schedule using the MITS PRO software. A typical fuel cell polarization of current and voltage curve was obtained in following conditions: stack temperature 338 K, anode hydrogen stoichiometry 1.25, cathode air stoichiometry 2.0, anode compartment pressure 18 psi and cathode compartment pressure 14.7 psi.

3.3 Modeling Process

As indicated in the Introduction section, the aim of the paper is to obtain a model as simple as possible, but meeting the accuracy conditions that are expected from a model. That model should obtain the voltage V_{FC} given by the PEM fuel cell from the current I_{FC}. We have chosen the fixed feed-forward structure of the ANN that will be trained. The data obtained following the process described in the previous subsection have been used for the training and test steps, and for that purpose have been divided them into three subsets, i.e., 60% are used for training, 20% are used for validating and the last 20% is used as a independent test of network generalization. Given the input/output specification of the model, the trained ANN has only one neuron at both the input and output layers, and in order to maintain the simplicity requested by the specifications, it is a single hidden layer ANN containing only two neurons with the tangent sigmoid activation function in that hidden layer.

4 Experimental Results

In this subsection we discuss the results achieved following the experimental design described in the previous section. Due to space limitations only the final

Table 1. Results of the trained model (test data)

Magnitude	Value	Percentage of the minimum voltage	Percentage of the mean voltage
MSE	0.012 V^2	0.129%	0.105%
Maximum error	0.490 V	5.468%	4.434%
Mean error	0.077 V	0.855%	0.694%

326   J. M. Lopez-Guede et al.

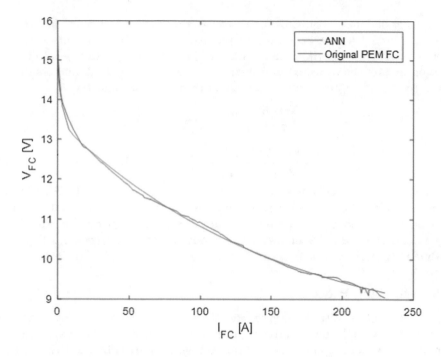

Fig. 3. Original PEM fuel cell behavior vs ANN output (test data)

achieved results are given, omitting intentionally the information regarding the training process itself. Since the experimental design takes into account the training of a very simple ANN, it is obvious that the simplicity condition imposed by the specifications is met. With regard to the accuracy requirements, Table 1 shows the results achieved after the training process with the test dataset. More specifically, it shows that the trained ANN has reached a MSE of $0.012\,\mathrm{V}^2$, a maximum error of $0.490\,\mathrm{V}$ and a mean error or $0.077\,\mathrm{V}$. In that table there are also these absolute values expressed as the percentage of the minimum and mean voltage V_{FC}. In order to understand the meaning of these values, several figures have been obtained. On one hand, Fig. 3 shows the polarization curve of the PEM fuel cell, comparing the original test data (as were gathered from the device) to the values obtained by the ANN after the training process, where it is possible to see that the model prediction of V_{FC} is accurate along all range of I_{FC} values. On the other hand, as the curves of Fig. 3 are very close, Fig. 4 shows augmented the difference between them, i.e., the test error. The first figure shows the original value of the error, while the second one gives the absolute value of that error. It is obvious that only for two values of I_{FC} the absolute error is larger than $0.2\,\mathrm{V}$, and for the most part of the range of I_{FC} values, the absolute error is less than $0.1\,\mathrm{V}$, meeting a mean error of $0.077\,\mathrm{V}$.

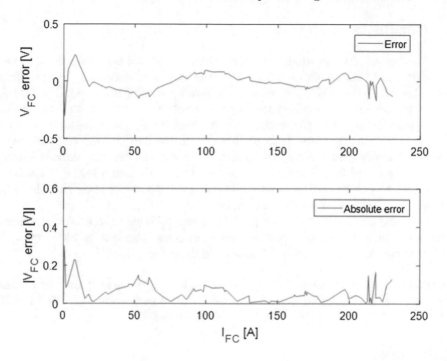

Fig. 4. ANN errors (test data)

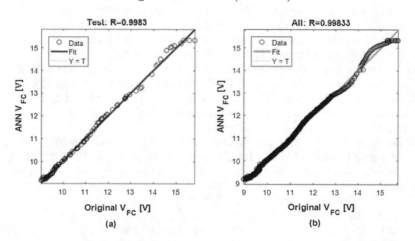

Fig. 5. Correlation values

Finally, Fig. 5 shows the correlation values obtained when the trained ANN is evaluated. There is possible to verify that the quality of obtained ANN is very high, given the correlation coefficients for both the test and complete datasets.

5   Conclusions

In this paper an efficient model of a commercial PEM fuel cell has been obtained. In order to get that result we have used artificial neural networks, profiting from their characteristics. We have started the paper giving a background of its core points, then we have described the experimental design paying attention to the data gathering process, the design process and finally, we have discussed the promising obtained results. It is possible to notice that despite of being being the model a very simple ANN, the results obtained with the test dataset reach a maximum error of 5.47% and 4.43% regarding the minimum and mean voltage respectively. Is more clarifying that the mean squared error regarding the same voltages are 0.13% and 0.11% respectively.

As future work, once that the promising results have been analyzed, we will deal with the problem of obtaining an even more accurate model, trying to maintain the ANN that holds the model as simple as possible.

Acknowledgments. The work in this paper has been partially supported by FEDER funds for the MINECO project TIN2017-85827-P, and project KK-2018/00071 of the Elkartek 2018 funding program of the Basque Government.

References

1. Asl, S.S., Rowshanzamir, S., Eikani, M.: Modelling and simulation of the steady-state and dynamic behaviour of a PEM fuel cell. Energy **35**(4), 1633–1646 (2010). Demand Response Resources: The US and International Experience: http://www.sciencedirect.com/science/article/pii/S0360544209005301
2. Daud, W.W., Rosli, R., Herianto, E., Hamid, S., Mohamed, R., Husaini, T.: PEM fuel cell system control: a review. English Renew. Energy **113**, 620–638 (2017)
3. Mammar, K., Chaker, A.: Neural network-based modeling of PEM fuel cell and controller synthesis of a stand-alone system for residential application. Int. J. Comput. Sci. Issues (IJCSI) **9**, 244–253 (2012)
4. Belmokhtar, K., Doumbia, M., Agboussou, K.: PEM fuel cell modelling using artificial neural networks (ANN). Int. J. Renew. Energy Res. **4**, 725–730 (2014)
5. Hatti, M., Tioursi, M., Nouibat, W.: Static modelling by neural networks of a PEM fuel cell. In: IECON 2006 - 32nd Annual Conference on IEEE Industrial Electronics, November 2006, pp. 2121–2126 (2006)
6. Puranik, S.V., Keyhani, A., Khorrami, F.: Neural network modeling of proton exchange membrane fuel cell. IEEE Trans. Energy Convers. **25**(2), 474–483 (2010)
7. Han, I.-S., Chung, C.-B.: Performance prediction and analysis of a PEM fuel cell operating on pure oxygen using data-driven models: a comparison of artificial neural network and support vector machine. Int. J. Hydrogen Energy **41**(24), 10:202–10:211 (2016). http://www.sciencedirect.com/science/article/pii/S036031991530389X
8. Bhagavatula, Y.S., Bhagavatula, M.T., Dhathathreyan, K.S.: Application of artificial neural network in performance prediction of PEM fuel cell. Int. J. Energy Res. **36**(13), 1215–1225 (2012). https://doi.org/10.1002/er.1870
9. Falcão, D., Pires, J., Pinho, C., Pinto, A., Martins, F.: Artificial neural network model applied to a PEM fuel cell, pp. 435–439 (2009)

10. Jayakumar, A., Sethu, S.P., Ramos, M., Robertson, J., Al-Jumaily, A.: A technical review on gas diffusion, mechanism and medium of PEM fuel cell. Ionics **21**(1), 1–18 (2015). https://doi.org/10.1007/s11581-014-1322-x
11. Widrow, B., Lehr, M.A.: 30 years of adaptive neural networks: perceptron, madaline, and backpropagation. Proc. IEEE **78**(9), 1415–1442 (1990)

Special Session - Soft Computing Methods in Manufacturing and Management Systems

A Hybrid Heuristic Algorithm for Multi-manned Assembly Line Balancing Problem with Location Constraints

Damian Krenczyk$^{(\boxtimes)}$ and Karol Dziki

Faculty of Mechanical Engineering, Silesian University of Technology,
Gliwice, Poland
{damian.krenczyk,karol.dziki}@polsl.pl

Abstract. Mass customization production is the next stage in the development of production systems that combines an individual approach to the client needs and benefits of mass production. This approach forces manufacturers to seek new, more effective methods of production flow planning, in particular methods for solving the assembly line balancing problem. The traditional approaches and methods proposed for solving balancing problems require adaptation to new constraints associated with the increasingly widespread introduction of multi-manned and spatially divided assembly workstations. This requires considering additional location restrictions and a more complex allocation of tasks in contrast to restricted only by technological precedencies and time constraints for Simple Assembly Line Balancing Problem. The paper presents a proposal for solving the problem of line balancing with location constraints using new hybrid heuristic algorithm, which is a combination of a modified RPW algorithm and a local search of task sequence on assembly stations zones. Moreover, the concepts of smoothness and efficiency is referred to two separate areas: stations and employees. Experimental results for the literature case of a 30 tasks problem indicate the effectiveness of the proposed approach in practice.

Keywords: Heuristics · Assembly line balancing ·
Multi-manned assembly line · Space and location constraints

1 Introduction

The main goal of modern production line designing and planning is to produce as many products as possible, at the lowest unit cost while maintaining the assumed quality level. Most effective way to achieve this goal is to produce standardized goods in large batches that provide low fixed cost per unit. On the other hand, modern market is focused on client needs and industry is forced to accept that they need to redesign assembly lines for customized products – to meet certain requirements of clients groups. There are two approaches for this problem. First is production of small batches of certain product option – multi-model. The quantity and repeatability of each batch is determined to obtain optimal line efficiency based on the cycle time and production plan. Second is designing the assembly line in such a way that it is possible to produce various product options – mixed-model. The only condition is that different options

© Springer Nature Switzerland AG 2020
F. Martínez Álvarez et al. (Eds.): SOCO 2019, AISC 950, pp. 333–343, 2020.
https://doi.org/10.1007/978-3-030-20055-8_32

must have similar processing time and core, base production path must be the same for each product type. To solve balancing problem of such lines type is possible, depending of tasks sequence and processing time variability, to use methods commonly used to simple assembly line balancing problem (SALBP) or general assembly line balancing problem (GABLP). The SALBP and GABLP, due to the its nature, is NP-hard class problem and therefore, optimal solutions to real problems is unfeasible to obtain in a reasonable amount of time using deterministic algorithms [1–4]. According to [5], genetic algorithm and swarm optimization are the most frequently used meta-heuristic algorithms in SALBP, but the frequency usage of other heuristics algorithms are relatively similar [6], as the most commonly used soft computing methods.

Due to the increasing complexity of products, in order to maintain a certain (as small as possible) number of stations, it is necessary to assign workers to workstations in a way that allows to perform as many as possible assembly tasks in same time, on same workstation [2, 3, 7–9]. This problem is not related only with maximal usage of available time of workstation, but also available time of worker and available space, for certain operations, around workstation. The traditional approaches and methods proposed for solving balancing problems require adaptation to new constraints that appear in response to above-mentioned changes. In the paper a model of multi-manned assembly line balancing problem with location constraints is presented.

The paper is organized as follows: In Sect. 2, the problem of multi-manned assembly line is presented. In Sect. 3, a new model of multi-manned location constrained assembly line balancing problem (mLALBP) is defined. The proposed hybrid heuristic method is descripted in Sect. 4. Section 5 presents an computational example, which is solved by the hybrid heuristic method. Finally, Sect. 6 concludes the research with directions for future work.

2 Multi-manned Assembly Line's Workstations with Location Constraints

Multi-manned assembly line balancing problems (mALBP) are typical for assembly lines in industries used to produce large-size products [10], in particular automotive industry. In this type of assembly line, the product stops during the cycle time at each workstation, where there are several workers simultaneously performing different operations on the same product [11]. In traditional approach to ALBP only restriction are relations that derive from association graph. mLALBP make it necessary to search of new solutions for additional problems. Those problems, among others, are finding how many tasks can complete worker on workstation, what is optimal allocation of minimal number of necessary line workers and they allocation on workstations. In ALBP, such problems do not occur because one operator is assigned one task at a given position. It should also be emphasized that multi-manned assembly stations are stations where operators can perform various tasks at the same station during one cycle. A similar solution is to develop an assembly line divided into assembly sections. Then teams of employees are created, whose task is to mount a set of parts before the product passes to the next station [2, 3, 7–10]. Regardless of the scope, balancing takes place in relation to two typical problems. The first one determines the number of stations, having a given cycle time (mLALBP-1), while the second one for a fixed number of workstations determines the cycle time (mLALBP-2). In this study the problem of the first type is considered, assuming a permanent division of the assembly station into zones in which technological operations are carried out by independent operators or teams (Fig. 1).

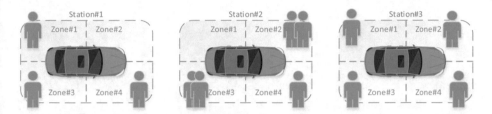

Fig. 1. Example of area constrained multi-manned assembly line

The issues, which is also connected with the combination of task relationship constraints and spatial limitations that occur in standard multi-station assembly lines, are the disturbance of the balanced line's smoothness index and line efficiency. However, in the case of mLALBP, where different divisions are possible in the implementation of tasks for individual stations, as well as the implementation of tasks by the workers in different location at the same station, the concepts of smoothness and efficiency should refer to two separate areas – stations and employees. In contrast to typical mALBP models [10–13], such an expansion takes into account the model we have proposed in the next section. The methods of resolving the mLALBP also needed to be extended, so that additional constraints related to the variable number of workers and space assignment that are not present in the classical approach were taken into account. For this purpose, it the extension of the known heuristic methods and combine them with local search methods is proposed. The most commonly used classic heuristics is Ranked Positional Weight Algorithm (RPW), Killbridge-Wester or IUFF [14]. The first of these involves the allocation of positional scales to the operation. The weight represents the sum of task execution times and all tasks performed on it. On the basis of weights, the longest path is determined and assigning tasks begins. The second groups operations for operations that precede them - relations graphs in such a way that in the first part there are operations that do not have a predecessor, and in each subsequent operation to which it is necessary to end operations from the previous part. The third one takes into account the operation times, the number of operations before and after and the path topography.

3 mLALBP Model

3.1 Problem Definition

In this paper, a multi-manned location constrained assembly line is considered as a series of multi-manned location constrained workstations denoted by W on which a set of technological tasks denoted by *A* must be performed with a predetermined cycle time *T*. Each workstation is divided into a fixed number of work zones (locations) determined by the system designer (Fig. 2). In each workstation there is the possibility of assigning more than one operator to each location in accordance with technological requirements.

Fig. 2. Example of dividing areas on workstations

3.2 Problem Assumptions

Some specific constraints of mLALBP are considered, such as zoning constraints and workers allocation. The general characteristics and basic assumptions of the mLALBP considered in this paper are given as follows:

- the precedence relationships among the tasks are known,
- a single product is assembled on a multi-manned workstations,
- an assembly task times are deterministic, independent of the assigned station and assigned to a particular zone.
- an assembly task is performed by a given number of workers,
- an assembly task cannot be divided among several workstations,
- only one task can be performed by a worker(s) on workstation's zone at a given time,
- parallel assembly tasks can be performed, each in a separate zone at the workstation,
- zones can be connected at a given station, if it results from technological requirements,
- the transportation times are ignored,
- the goal is to minimize the total number of station, and in turn to maximize the line efficiency.

3.3 Model Notation

In order to formally describe the mLALBP, the following notation to define the mathematical model is employed [14]:

$$W = \{W_1, W_2, \ldots, W_N\} \tag{1}$$

where:

$W_j = \{\{w_{j,\ 1},\ w_{j,\ 2}, ...,\ w_{j,\ k}\},\ ws_j\}$ – j-th workstation, $j = 1, ..., N$,

w_{jk} – k^{th} zone on the jth workstation, $k = 1, ..., K$,

ws_j – the number of workers designated for the service of the jth workstation,

N – number of workstations,

K – number of zones within each station.

$$A = \{a_1 = \{at_1, ap_1, as_1\}, a_2 = \{at_2, ap_2, as_2\}, ..., a_I = \{at_I, ap_I, as_I\}\} \tag{2}$$

where:

a_i – i^{th} task, $i = 1, ..., I$,

at_i – processing time of the ith task,

ap_i – the zone # or set of zones of the ith task,

as_i – number of workers required for the implementation of the ith task,

I – the number of tasks to be performed on the assembly line.

E – workers matrix, where:

$e_{ijh} = 1$, if i^{th} task is performed by the hth worker on j^{th} station; 0, otherwise;

Z – assigning matrix, where:

$z_{ij} = 1$, if i^{th} task is assigned to j^{th} station; 0, otherwise;

T – cycle time;

As each task can be assigned to one or several zones on a station, the ap, value is an binary-coded decimal numeral representation of set of zones. This encoding corresponds to the binary number represented by a sequence of bits with the number of positions (bits) corresponding to the number of zones K. 1 or 0 in a given position means that the given (i^{th}) task is performed or not in the #*position* zone. For example, for five zones on a workstation if task no. 2 is performed on the fourth one (01000), the value of ap_2 is 8; if the execution of the task no. 2 requires the simultaneous use of zones 2 and 3 (00110) the value of ap_2 is 6.

The value of the assembly line cycle T is also given, for which minimum number of stations N and the assignment of tasks to individual zones should be determined.

The most commonly used objective functions are the minimization of the smoothness index (SI and SIs) and maximization of the line efficiency (e and s), which allows for evenly and efficient distributing the workload amongst the stations. [10–12, 15]. As mentioned earlier, the smoothness and efficiency formulas for two separate areas of stations (3) and (5), and employees (4) and (7) are defined as follows:

$$e = \frac{\sum_{i=1}^{I}\left(at_i \sum_{p=0}^{\log_2 ap_i}\left(\left\lfloor \frac{ap_i}{2^p}\right\rfloor \bmod 2\right)\right)}{KNT} \cdot 100\% \tag{3}$$

$$s = \frac{\sum_{i=1}^{I} at_i as_i}{\sum_{j=1}^{N} ws_j T} \cdot 100\% \tag{4}$$

$$SI = \sqrt{\frac{\sum_{j=1}^{N}\left(e_{max} - e_j\right)^2}{KN}}, \tag{5}$$

where:

e_j - efficiency indicator of the jth station:

$$e_j = \frac{\sum_{i=1}^{I}\left[z_{ij}at_i \sum_{p=0}^{\log_2 ap_i}\left(\lfloor\frac{ap_i}{2^p}\rfloor \bmod 2\right)\right]}{KT} \quad (6)$$

$e_{max} = max_j\,(e_j)$.

$$SIs = \sqrt{\frac{\sum_{j=1}^{N}\left[\sum_{k=1}^{ws_j}\left(s_{max}-s_{jk}\right)^2\right],}{\sum_{j=1}^{N}ws_j}} \quad (7)$$

s_{jk} - efficiency indicator of the kth workers on j^{th} station:

$$s_{jk} = \frac{\sum_{i=1}^{I}e_{ijk}at_i}{T} \quad (8)$$

$s_{max} = max\,(e_{jk})$.

4   The Proposed Hybrid Heuristic Method

The mLALBP model presented in this paper requires proposing new algorithms for solving this class of problems. We have proposed a new hybrid heuristic algorithm for this purpose, which is a combination of a modified RPW algorithm and a local search of task sequence on assembly stations zones.

RPW is one of the most frequently used method of solving simple assemble balancing line problem. In this method, based on total processing times and precedence relationships, the positional weight of task is calculated as the total time on the longest path needed by the task and its successors to the last operation of the diagram. The choice of this method was dictated by properties of method, especially its precedence and connections constraints. This gives possibility of using the local search for a preselected set of tasks, which is the sum of subsets for specific zones, and for large number of tasks from which it could be chosen, it is more likely to find the combination that will maximize the line efficiency. A simplified diagram of the implementation of the proposed hybrid algorithm is shown in Fig. 3.

5   An Illustrative Example

Proposed hybrid, heuristic method of task assignment for multi-manned workstations with positional restrains is illustrated below, based on modified, known from the literature on the subject set of comparative data [16, 17]. The sequence relation graph shown in Fig. 4, is modified for the purpose of division of tasks into zones, proposed

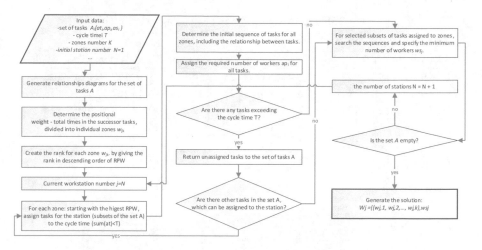

Fig. 3. A hybrid RPW heuristic & local search algorithm

by Sawyer, consisting of 30 tasks. All task parameters are given in the figure, as indicated in the Sect. 3.3. It was assumed that the assembly will take place at stations with three working zones (ap_i = 1, 2 or 4), and the cycle time T is 26.

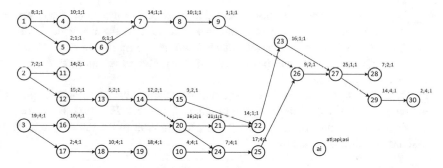

Fig. 4. Precedence relationships diagram

For given input data, a solution is sought in the form of assignment of tasks to stations, which will allow to determine its number and indicators of balance assessment.

1st Step: For solving featured balancing problem using RPW method preparing data for assignment of weights for tasks is needed. The weight values calculated as the total time on the longest path needed by the task and its successors to the last operation on each zone and sorted from highest to lowest is presented in Table 1.

2nd Step: Tasks, according to the order resulting from RPW sorting, are assigned to the workstation. For example, on Fig. 5 the initial subset of the tasks from the set of remaining tasks A for Workstation 4 are shown.

Table 1. Assign RPW in descending order

Zone#1				Zone#2				Zone#3			
Task	Weight	Task	Weight	Task	Weight	Task	Weight	Task	Weight	Task	Weight
1	93	22	71	2	147	26	50	3	137	18	28
21	92	8	61	12	140	11	14	16	127	19	18
4	85	23	57	13	125	28	7	10	78	29	16
5	83	9	51	20	108			24	74	30	2
6	81	27	41	14	92			25	67		
7	75			15	80			17	30		

WORKSTATION 4

AREA 1   Task 21 (21) 1 OP

AREA 2   Task 26 (9) 1 OP   Task 11 (14) 1 OP

AREA 3   Task 25 (17) 1 OP

Fig. 5. Initial assignment of available task to workstation 4

3th Step: All tasks whose execution exceeds the available time have been returned to the *A* set.

4th Step: The sequence relations between particular zones and operators are checked, resulting in the ordering of tasks in zones. Due to the occurring sequence relations, tasks that have remained from selected subsets are returned to the set of tasks waiting for assignment and the possibility of assigning other alternative tasks is checked, according to Table 1.

5th Step: For tasks assigned to workstations, using the local search possible sequencing is checked for minimizing the number of workers. Figure 6 shows the best solution for workstation 4. In the previous case, the number of operators was 3, in the new assignment number it is reduced to 2.

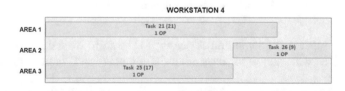

WORKSTATION 4

AREA 1   Task 21 (21) 1 OP

AREA 2   Task 26 (9) 1 OP

AREA 3   Task 25 (17) 1 OP

Fig. 6. Workstation with rearranged tasks for the best workers number

6th Step: Next workstation is created and remaining task are assigned according to RPW order. Steps 2–5 are looped until task collection will be empty.

The steps for the remaining stations have been omitted due to the limited capacity of the paper and only the results of task assignment are given in Fig. 7.

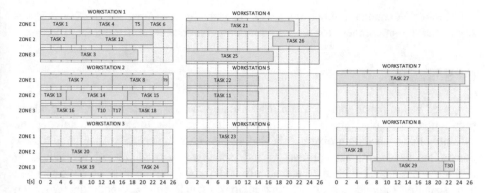

Fig. 7. Final assignment of tasks to workstations

For evaluation of assembly line balancing indicators mentioned in Sect. 3.3 were used. In addition number of operators (Num. of op.) and assembly workstations (Num. of WS). Table 2 contains results for each indicator for cycle time $T = 26$ and 36.

Table 2. mLALBP results - different cycle time

Indicator		
T	26	36
e	52%	50%
s	78%	75%
SI	0,01	0,014
SIs	0,09	0,429
Num. of WS	8	6
Num. of op.	16	12

A quick analysis of the results could indicate that a total line efficiency of 52% is not a good result, but given that each workstation is divided into 3 independent zones, it gives better results (T = 26 for 8 stations) than the same assembly line without dividing the workstations into zones (T = 41 [17]). Rearrangement of assigned tasks and switching some parallel to each other tasks, results high workers time efficiency, ~ 80% is very good outcome, especially that smoothness for assembly line and workers saturation is on very low level. It should be emphasized that if in this case, use a classic assembly line, with one assembly area, the higher efficiency result obtained, would be occupied by a significant increase in the number of assembly stations, with a slight change in the resource utilization rate.

For checking impact of giving workstations tasks with location constrains to assembly line cycle time was increased by almost 40%. Despite a small decrease in the efficiency indicators, the number of required workers and workstations was reduced, which would probably reduce production costs using the line referred to in the example. However, for the increased cycle time the smoothness indicators have deteriorated. A smaller number of stations suggests that the impact on the SI index was small, but the smoothness for workers saturation has been almost threefold increased.

6 Summary

Proposed in the paper approach is a response to seek new, more effective methods of production flow planning by manufacturer, in particular methods for solving the assembly line balancing problem. It is also an answer to the need for adaptation to new constraints the traditional approaches and methods proposed for solving balancing problems, associated with the increasingly widespread introduction of multi-manned and spatially divided assembly work-stations, which enforces additional location restrictions. An extension of known heuristic method and its connection with local search methods was proposed in order to determine both the allocation of tasks to independent zones within the assembly station and the assignment of workers to the station. The subject of further research in the discussed area will be the development of methods for assigning variable task times for the mix- and multi-model lines, which will enable the development of a comprehensive tool supporting the planning of modern assembly lines.

References

1. Pape, T.: Heuristics and lower bounds for the simple assembly line balancing problem type 1: Overview, computational tests and improvements. Eur. J. Oper. Res. **240**(1), 32–42 (2015). https://doi.org/10.1016/j.ejor.2014.06.023
2. Gansterer, M., Hartl, R.F.: One- and two-sided assembly line balancing problems with real-world constraints. Int. J. Prod. Res. **56**(8), 3025–3042 (2018). https://doi.org/10.1080/00207543.2017.1394599
3. Make, M.R.A., Rashid, M.F.F.A., Razali, M.M.: A review of two-sided assembly line balancing problem. Int. J. Adv. Manuf. Technol. **89**, 1743 (2017). https://doi.org/10.1007/s00170-016-9158-3
4. Zemczak, M., Skolud, B., Krenczyk, D.: Two-stage orders sequencing system for mixed-model assembly. In: IOP Conference Series: Materials Science and Engineering, vol. 95, p. 012130 (2015). https://doi.org/10.1088/1757-899x/95/1/012130
5. Rashid, M.F.F., Hutabarat, W., Tiwari, A.: A review on assembly sequence planning and assembly line balancing optimisation using soft computing approaches. Int. J. Adv. Manuf. Technol. **59**, 335 (2012). https://doi.org/10.1007/s00170-011-3499-8
6. Fathi, M., Ghobakhloo, M.: A technical comment on "a review on assembly sequence planning and assembly line balancing optimisation using soft computing approaches". Int. J. Adv. Manuf. Technol. **71**, 2033–2042 (2014). https://doi.org/10.1007/s00170-014-5613-1
7. Tuncel, G., Topaloglu, S.: Assembly line balancing with positional constraints, task assignment restrictions and station paralleling: a case in an electronics company. Comput. Ind. Eng. **64**(2), 602–609 (2012). https://doi.org/10.1016/j.cie.2012.11.006
8. Cheng, Y., Sun, F., Zhang, Y., Tao, F.: Task allocation in manufacturing: a review. J. Ind. Inf. Integr. (2018, in press). https://doi.org/10.1016/j.jii.2018.08.001

9. Krenczyk, D., Skolud, B., Herok, A.: A heuristic and simulation hybrid approach for mixed and multi model assembly line balancing. Adv. Intell. Syst. Comput. **637**, 99–108 (2018). https://doi.org/10.1007/978-3-319-64465-3_10

10. Hamid, Y., Mustafa, Y.: Multi-manned assembly line balancing problem with balanced load density. Assem. Autom. **35**(1), 137–142 (2015). https://doi.org/10.1108/AA-05-2014-041

11. Dimitriadis, S.G.: Assembly line balancing and group working: a heuristic procedure for workers' groups operating on the same product and workstation. Comput. Oper. Res. **33**(9), 2757–2774 (2006). https://doi.org/10.1016/j.cor.2005.02.027

12. Roshani, A., Roshani, A., Roshani, A., Salehi, M., Esfandyari, A.: A simulated annealing algorithm for multi-manned assembly line balancing problem. J. Manuf. Syst. **32**(1), 238–247 (2013). https://doi.org/10.1016/j.jmsy.2012.11.003

13. Prasad, M.M., Ganesan, K., Suresh, R.K.: An optimal balancing of multiple assembly line for a batch production unit. Int. J. Lean Think. **4**(2), 22–32 (2013)

14. Krenczyk, D., Dziki, K.: A multi-manned assembly line balancing in spatial restrictions. In: Knosala, R. (ed.) Management Engineering. Digitalization of Production. Research news, PWE Warszawa (2019, in press). (in polish)

15. Grzechca, W.: Estimation of time and cost oriented assembly line balancing problem. In: 19th International Conference on Systems Engineering, Las Vegas, NV, pp. 248–253 (2008). https://doi.org/10.1109/icseng.2008.48

16. Assembly Line Balancing Data sets & Research topics (2019). https://assembly-line-balancing.de/. Accessed 01 Feb 2019

17. Scholl, A.: Data of assembly line balancing problem. Darmstadt Technical University (1993)

A Comparison Analysis of the Computer Simulation Results of a Real Production System

Production System Modelling with FlexSim and Plant Simulation Software

Cezary Grabowik[✉], Grzegorz Ćwikła, Krzysztof Kalinowski,
and Magdalena Kuc

Faculty of Mechanical Engineering, Silesian University of Technology,
Gliwice, Poland
{cezary.grabowik,grzegrz.cwikla}@polsl.pl

Abstract. Use of the production systems modelling and simulation tools allows integrating both technical and organisational production preparation areas. The higher integration level the better utilisation of enterprise resources is. In the paper a comparison analysis of simulations results got for the model of a real production line is performed. The main motivation of the paper was to compare simulation results obtained in the two simulation tools it is FlexSim and Tecnomatix Plant Simulation in order to state if the simulation quality depends on the choice the simulation tool.

Keywords: System modelling and simulation · FlexSim ·
Tecnomatix plant simulation · Comparison analysis

1 Introduction

The simulation technology is an important tool in a complex system design, implementing and operating [1–4]. Computer simulation is a method which tries to imitate processes that proceed in the real world by means of mathematical models with computer programs. Such kind of simulation is commonly used for inferencing about processes flow for which the direct observation is impossible or too expensive. The biggest advantage of computer simulation is possibility of replacing of expensive industrial test. Moreover, computer simulation are repeatable and their results are easy to interpretation [5–7].

2 The Model of the Production System

The modelled production system is a part of the enterprise production system. The modelled production line is used in the production process of a car bumper; it is in assembly process of the bumper mounting components. The considered production

F. Martínez Álvarez et al. (Eds.): SOCO 2019, AISC 950, pp. 344–354, 2020.
https://doi.org/10.1007/978-3-030-20055-8_33

process starts with the bumper initial quality control operation; a worker takes a product from a transport rack and checks organoleptic its quality. The damages and faults which undergo the quality check procedure are as follows: (i) particle inclusions, painted over hair, chips and underpaintings. If the product quality fulfils quality parameters, it is next send to a saw station where its outer edges are cut to proper dimensions. Next, the product is put into interop buffer where it waits for further processing. During the second assembly operation a worker takes it from interop buffer and attaches all screws to the bumper assembly, the worker performs quality check operation simultaneously. In the third operation a worker welds handles to the bumper. So, at first the bumper is attached to a forming die, handles are set in the right position relative to the bumper and next pressed in order to weld them. The final assembly operation is being done on the two parallel work stations; workers alternately take products from the appropriate interop buffer and install seals and screws, after then they send bumpers to the final quality control operation. The last operation of the production process is packaging. The schema of the bumper production process in the Fig. 1 is shown.

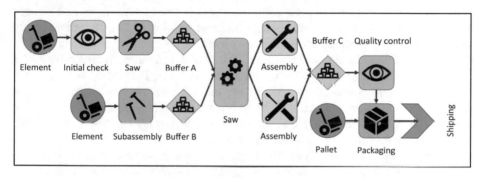

Fig. 1. The flow of the bumper production process

The process and set-up times were taken during direct observations of the production process performed in the enterprise of the automotive industry. In Table 1 the average operation times and the type of activities a worker take part during the production process are shown.

Table 1. The operation and set-up times and distribution of the workers duties

Worker	Station	Process times	Worker's presence	Set-up times	Worker's presence
OP1	Initial check	15 s	✓	–	✗
	Saw	30 s	✗	5 s	✓
OP2	Element subassembly	67 s	✓	–	✗
OP3	Welder	100 s	✗	27 s	✓
OP4 OP5	Element assembly	140 s	✓	8 s	✓
OP6	Final check	41 s	✓	–	✗
	Packaging	60 s	✓	30 s	✓

2.1   The Simulation Model of the Production System Built with FlexSim

The production line model was made with the following FlexSim objects: (i) *source*, responsible for creating and sending flow elements – products, (ii) *queue*, used for storing elements that wait for further processing on the next work station, (iii) *processor*, an object which processes elements, a single operation is done, (iv) *multiprocessor* is an object with the same characteristic like processor one, it additionally allows performing several operations on the product in earlier predetermined way, (v) *combiner* used for connection flow elements, (vi) operator a worker responsible for doing certain tasks, (vii) *sink* which takes and remove finished products from the simulation model. At start, in the simulation model the first object of source type was defined. This object, a painted bumper, is an input to the production process. Next, the objects responsible for representing of production system workstations, it is the initial quality control and saw stations were defined. Having this done, the next object of the *source* type and *multiprocessor* were generated. The *source* object represents a component which is attached to the bumper body, whilst *multiprocessor* represents operations performed at this stage of the production process; it is screws subassembly and interop quality control. The next modelling stage enclosed creating of interop buffers for the saw and subassembly workstations, the two objects of the *queue* type were applied for this purpose. These objects were called as the buffer_A and buffer_B respectively. The capacity of the buffer_A was set to 20, and 25 for the buffer_B. Next the two objects of the *combiner* type were introduced. In the model these objects represent welder workstations. On these workstations the welding operations of bumper mounting elements to the bumper body are carried out. In the following step the two assembly workstations were added to the simulation model, it was done by introducing of the two objects of the *multiprocessor* type. The sequence of operation was set in the following way: (i) set-up, (ii) screws assembly, (iii) gaskets assembly. Then, the interop buffer was defined; the buffer name was set to buffer-C, and its capacity to 25 (Fig. 2).

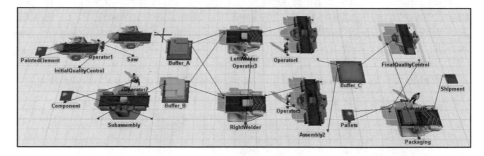

Fig. 2. The simulation model of the production line made with FlexSim

The Production Process Indexes

The two types of simulation were carried out, one for the 1 h production span time, and for 8 h long production shift. It should be noticed that the results got during the simulation were rounded up; the fraction part of the number was in every case omitted. This remark relates both to manufacturing resources utilisation level and workers utilisation level. At first, the report that shows manufacturing resources utilisation level was created. The manufacturing resources state pie in the Fig. 1 is presented. The particular labels denote: (i) *idle*, resource is idle, not working, (ii) *processing*, a task allocated on a resource is being processed, (iii) *collecting*, a resource waits until the number of products reached the given number, (iv) *waiting for operator*, a resource waits for an operator, (v) *setup*, the percentage rate of the set-up times operations (Fig. 3).

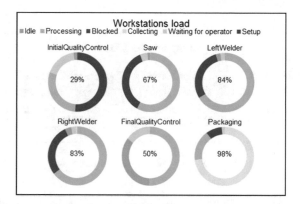

Fig. 3. The chart pie of manufacturing resources utilisation

In the modelled production line the most loaded resources are subassembly, assembly1 and packing stations. The most unloaded resource is the initial quality control station. It comes from fact that this station is blocked for more than its half directive working time waiting for the saw to become idle. The Fig. 4 shows the workers utilisation level. In this chart the labels means as follows: (i) *idle*, a worker is idle, does not work, (ii) *travel* empty, a worker travel back after performing a task, (iii) *utilize*, a worker works or do some setups.

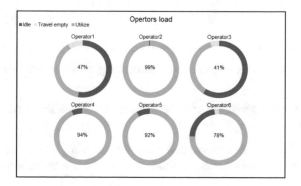

Fig. 4. The chart pie of workers utilisation

The most loaded workers are those who serve of components subassembly and assembly workstations; it is operator1, operator2 and operator5 respectively. The low utilisation level for the opertor1 comes from fact that the operator1 waits for the following station in the production line (the saw station) become idle. For the operator3 the situation is different. The operator3 duty is to perform setup operations on the welder workstation, when the welder station welds components together the operator3 waits for the welding cycle ends. In order to show the level of buffers utilisation the state bar object was used (see: the Fig. 5).

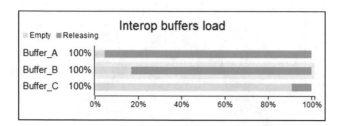

Fig. 5. The report of the interop buffer usage

Basis on this report it can be said that utilisation of the buffer_A and B, located in the areas of the saw and subassembly workstations, is very high, so they are necessary for the production to be flow without blocks and delays. The next production index taken into account was workstations throughput (see: the Fig. 6). This diagram shows the number of products that leave the particular workstation.

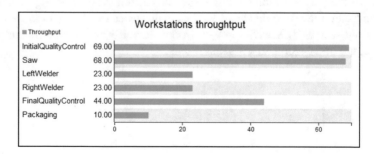

Fig. 6. The production line manufacturing resources throughput

2.2 The Simulation Model of the Production System Built with Plant Simulation

The production line model was created with the following Plant Simulation objects: (i) *source*, responsible for creating mobile units *mu*, it can create either one or many model objects in any order, it may represents both an input store or a workstation that delivers products to the further processing, (ii) *entity*, the smallest unit which can be processed or moved through the modelled process, (iii) *buffer*, is an simulation object, which placed between another the two objects serves for storing a part, till the moment it can transferred to next model object, it is a compensation station which compensate variable processing and transport times, it prevents the system blocking, (iv) *assemblyStation*, add components to a main part or simulates the assembly process by removing a single parts and creating a new part which makes the assembled part, (v) *drain*, a drain is an active flow object, it has a single place and destroys *mu's* as soon as they are processed, (v) *broker*, mediates between suppliers and demanders of services in the simulation model, (vi) *workerPool*, an object that creates the number of workers according to a creation table and makes them available for registered model services after a request, in case workers are idle the Plans Simulation sends them to *workerPool*, (vii) *workplace*, only a worker stays on a workplace whilst performing a job, with workplace you connect an event of the machine with a request of a service, (viii) *worker*, instantiation of an worker object is realized by the *workerPool* object. At the beginning of the modelling process, like in case of FlexSim simulation model, the *source* object that represents the painted bumper and the two object of *singleProc* type were introduced. The *singleProc* objects were used for modelling of the initial quality control and the saw stations. Next, the *broker* and *workplace* objects were created. The number of workers was set to 6, and to each worker a single job was assigned. Taking into consideration that in Plant Simulation there is no corresponding object to the *multiprocessor* FlexSim object it was necessary for the subassembly workstation model to apply the two *singleProc* Plant Simulation objects. These objects represent the screws subassembly and quality control operations. Having done this, the two interop buffers were generated; like in the FlexSim model they were called buffer_A and buffer_B respectively, and their capacity was set as follows 20 for the buffer_A and 25 for the buffer_B. In the following step the two welder workstation were modelled with *assemblyStation* object. Next modelling step enclosed activities connected with modelling of assembly workstations. Like in case of the subassembly workstation due to lack of the *multiprocessor* object it was required to use the *singleProc* objects in order to model assembly workstation. The elements that leave assembly stations are stored in the buffer_C object. At the end of the modelling process the following objects were modelled: the final quality control (*singleProc* object), pallets (*source* object) and dispatching (*drain* object). The Fig. 7 shows the simulation model made in Plant Simulation.

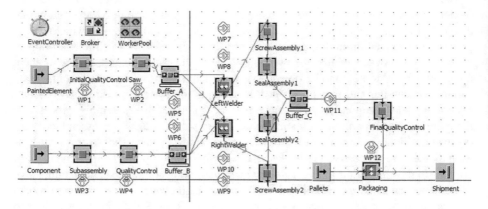

Fig. 7. The simulation model of the production line made with Plant Simulation

The Production Process Indexes

At first, a report that shows resources utilisation level was generated (see: the Fig. 8). The particular labels denote: (i) *working*, a task allocated on a resource is being processed, (ii) *set-up*, a resource is served by an operator, some set-up actions are performed, (iii) *waiting*, a workstation is idle, it is waiting for an operator or for collecting the suitable number of products, (iv) *blocked*, the workstation is blocked because the next station is not able to take products from the predecessor.

Object	Working	Set-up	Waiting	Blocked	Powering up/down	Failed	Stopped	Paused	Unplanned	Portion
InitialQualityControl	29.44%	0.00%	9.72%	60.83%	0.00%	0.00%	0.00%	0.00%	0.00%	
Saw	57.78%	9.72%	0.42%	32.08%	0.00%	0.00%	0.00%	0.00%	0.00%	
QualityControl	22.08%	0.00%	77.92%	0.00%	0.00%	0.00%	0.00%	0.00%	0.00%	
LeftWelder	66.67%	18.42%	2.08%	12.83%	0.00%	0.00%	0.00%	0.00%	0.00%	
SealAssembly1	48.97%	5.11%	45.92%	0.00%	0.00%	0.00%	0.00%	0.00%	0.00%	
SealAssembly2	50.06%	2.56%	47.39%	0.00%	0.00%	0.00%	0.00%	0.00%	0.00%	
FinalQualityControl	50.11%	0.00%	49.89%	0.00%	0.00%	0.00%	0.00%	0.00%	0.00%	
Packaging	17.44%	9.17%	73.39%	0.00%	0.00%	0.00%	0.00%	0.00%	0.00%	
Subassembly	77.92%	0.00%	22.08%	0.00%	0.00%	0.00%	0.00%	0.00%	0.00%	
ScrewAssembly1	38.33%	2.67%	59.00%	0.00%	0.00%	0.00%	0.00%	0.00%	0.00%	
ScrewAssembly2	38.33%	2.56%	59.11%	0.00%	0.00%	0.00%	0.00%	0.00%	0.00%	
RightWelder	66.67%	18.00%	12.25%	3.08%	0.00%	0.00%	0.00%	0.00%	0.00%	

Fig. 8. The report of workstations utilisation level

Basis on this report it can be said the highest utilisation level is for subassembly and welders workstations and the worst for packing workstation. The next report was made for workers utilisation (see: the Fig. 9). For every worker the following states can be discussed: (i) *working*, workers perform tasks that were assigned to them, (ii) *setting-up*, a worker does a set-ups actions, (iii) *en-route to job*, a worker comes back to a workplace after performing certain operation e.g. placing a part in an interop buffer, (iv) *waiting for importers*, waits for the following station to become idle.

	Working	Setting-up	Repairing	Transporting	En-route to job	Waiting for Importers	Waiting for MUs	Failed
Worker1	28.78%	9.58%	0.00%	0.00%	7.06%	54.58%	0.00%	0.00%
Worker2	94.72%	0.00%	0.00%	0.00%	5.28%	0.00%	0.00%	0.00%
Worker3	0.00%	35.61%	0.00%	0.00%	1.81%	62.58%	0.00%	0.00%
Worker4	83.33%	7.41%	0.00%	0.00%	4.05%	5.21%	0.00%	0.00%
Worker5	84.26%	5.00%	0.00%	0.00%	3.90%	6.84%	0.00%	0.00%
Worker6	64.50%	9.17%	0.00%	0.00%	1.37%	24.96%	0.00%	0.00%

Fig. 9. The report of workstations utilisation level

It appeared that the most loaded operator is the worker2 which is responsible for performing subassembly and quality control operations. His engagement is at almost 95% level. The least loaded operator is worker1. The highest rates of operators blocking were achieved for worker1 and worker3, initial quality control and welding operations. The last generated report concerned the utilisation level of interop buffers. Like in FlexSim model, the least load buffer is buffer_C which is used for only 10% of its capacity (Fig. 10).

Object	Number of Entries	Number of Exits	Minimum Contents	Maximum Contents	Relative Empty	Relative Full	Relative Occupation without Interruptions	Relative Occupation with Interruptions
Bufor_A	68	48	0	20	3.59%	49.27%	76.32%	76.32%
Bufor_B	50	48	0	4	52.30%	0.00%	3.56%	3.56%
Bufor_C	42	42	0	1	84.10%	0.00%	0.80%	0.80%

Fig. 10. The report of interop buffers utilisation level

3 The Comparative Analysis

The comparative analysis of the results achieved in FlexSim and Plant Simulation was bases on the production process indexes. The production process indexes taken into account were as follows: workstations utilisation, buffers utilisation, operator utilisation and the production line throughoutput. The production simulation time span was set to 1 h, and 8 h long working shift. The workstation utilisation level was the first considered production index. It was not possible to compare all workstation utilisation levels directly, because the models of the considered production line were slightly different. The main differences are observed in the areas of subassembly workstation. It comes from fact that in Plan Simulation there is no multiprocessor object, so it was necessary to introduce into Plant Simulation model the two equivalent *singleProc* objects. Taking into account this for further analysis were taken workstations utilisation levels for those workstations which had the full reflection in the both models. In case of the 1 h production span time the difference in workstation utilisation levels was in range 0–1,58%. The highest deviation rate was observed for waiting and blocking parameters for welder workstations. The smallest deviations were observed for set-ups in case of the saw and the packaging. As the duration of the simulation increases the differences for workstations become slightly bigger and lower in case of setups operations and waiting times. In the Table 2 simulation results of workstations workload for 1 h time span are shown (Table 3).

Table 2. The simulation results for 1 h long simulation time

Workstation	Processing/Working		
	FlexSim 2018	Plant Simulation 13	Difference
	Processing	Working	
InitialQualityControl	29,17%	29,44%	0,27%
Saw	57,50%	57,78%	0,28%
LeftWelder	66,67%	66,67%	0%
RightWelder	65,09%	66,67%	1,58%
FinalQualityControl	50,11%	50,11%	0%
Packaging	16,67%	17,44%	0,77%
Workstation	Setup		
	FlexSim 2018	Plant Simulation 13	Difference
	Setup	Setting–up	
InitialQualityControl	0%	0%	0%
Saw	9,58%	9,72%	0,14%
LeftWelder	18,00%	18,42%	0,42%
RightWelder	18,00%	18,00%	0,00%
FinalQualityControl	0%	0%	0%
Packaging	9,09%	9,17%	0,08%
Workstation	Blocked		
	FlexSim 2018	Plant Simulation 13	Difference
	Blocked	Blocked	
InitialQualityControl	52,93%	60,83%	7,9%
Saw	28,15%	32,8%	3,93%
LeftWelder	12,97%	12,83%	0,14%
RightWelder	12,68%	3,08%	9,6%
FinalQualityControl	0,00%	0,00%	0,00%
Packaging	0,00%	0,00%	0,00%

Table 3. The simulation results for 8 h long simulation time

Workstation	Processing/Working		
	FlexSim 2018	Plant Simulation 13	Difference
	Processing	Working	
InitialQualityControl	21,41%	20,94%	0,47%
Saw	42,68%	41,77%	0,91%
LeftWelder	67,36%	65,97%	1,66%
RightWelder	67,30%	65,91%	1,39%
FinalQualityControl	54,67%	53,53%	1,14%
Packaging	19,92%	19,38%	0,54%

(*continued*)

Table 3. (*continued*)

Workstation	SETUP		
	FlexSim 2018	Plant Simulation 13	Difference
	Setup	Setting–up	
InitialQualityControl	0%	0%	0%
Saw	7,12%	6,96%	0,16%
LeftWelder	18,28%	17,81%	0,47%
RightWelder	18,19%	17,81%	0,38%
FinalQualityControl	0%	0%	0%
Packaging	10,00%	9,79%	0,21%
Workstation	Blocked		
	FlexSim 2018	Plant Simulation 13	Difference
	Blocked	Blocked	
InitialQualityControl	65,54%	72,10%	6,56%
Saw	47,16%	51,22%	4,06%
LeftWelder	14,06%	15,95%	1,89%
RightWelder	13,98%	15,90%	1,92%
FinalQualityControl	0,00%	0,00%	0,00%
Packaging	0,00%	0,00%	0,00%

4 Summary

In the paper a comparative analysis of simulation results got with FlexSim and Plant Simulation modelling packages were presented. The main motivation of this work was to check if the selection of the simulation software has an influence on simulation results quality. In order to make the results more reliable the data used for simulation originated from a real production line. The simulation experiments were carried out both for 1 and 8 h long working shifts. This assumption was made in order to check if simulation time has influence on the results got in FlexSim and Plant Simulation packages. The main production indexes taken into consideration were workstation utilisation level, buffers and operators utilisation and production line throughput. As a result of simulation it appeared that difference in the simulation results got for both packages are slight. For workstations utilisation levels the difference between simulations models were less than 2%, and they become lower as the simulation time grows. The bigger differences other production indexes results from different distances between particular stations it the both models. Summarising up, there is in fact very little difference in simulation results. So, during the selection of simulation software other factors like software price, software maintenance and user-friendly interface could be taken into account during simulation packages selection.

References

1. Bangsow, S.: Manufacturing Simulation with Plant Simulation and SimTalk. Springer, Berlin (2010)
2. Krenczyk, D., Skolud, B., Herok, A.: A heuristic and simulation hybrid approach for mixed and multi model assembly line balancing. In: Advances in Intelligent Systems and Computing, vol. 637, pp. 99–108 (2018)
3. Siderska, J., et al.: Application of tecnomatix plant simulation for modeling production and logistics processes. Bus. Manag. Educ. 14(1), 64–73 (2016)
4. Law, A.M., et al.: How to build valid and credible simulation models. In: Proceedings of the 40th Conference on Winter Simulation, pp. 39–47 (2008)
5. Banks, J., Carson, J.S., Nelson, B.L., Nicol, D.: Discrete-Event System Simulation, 5th edn. Prentice Hall, Upper Saddle River (2010)
6. Krenczyk, D., Olender, M.: Production planning and control using advanced simulation systems. Int. J. Mod. Manuf. Technol. 6, 38–43 (2014)
7. Burduk, A.: Stability analysis of the production system using simulation models. In: Process Simulation and Optimization in Sustainable Logistics and Manufacturing, pp. 69–83. Springer, Heidelberg (2014)

Multiple Fault Diagnosis in Manufacturing Processes and Machines Using Probabilistic Boolean Networks

Pedro J. Rivera Torres[1]([✉]) (iD), Antônio José Silva Neto[2] (iD), and Orestes Llanes Santiago[3] (iD)

[1] Department of Computer Science - School of Natural Sciences, University of Puerto Rico-Río Piedras, 17 Ave Universidad STE 1701, San Juan, PR 00925-2537, Puerto Rico
pedro.riveral@upr.edu

[2] Instituto Politécnico da Universidade do Estado do Rio de Janeiro (IPRJ/UERJ), Rua Bonfim 25 - Campus UERJ, Vila Amélia, Nova Friburgo, RJ 28625-570, Brazil
ajsneto@iprj.uerj.br

[3] Dpto. de Automática y Computación, Universidad Tecnológica de la Habana José Antonio Echeverría (CUJAE), Calle 114, no. 11901, 19390 Marianao, La Habana, Cuba
orestes@tesla.cujae.edu.cu

Abstract. Developing methodologies for fault diagnosis in industrial/manufacturing systems is an active area of research. In this paper, a fault diagnosis scheme based on the Probabilistic Boolean Networks (PBN) model is proposed for a group of machines in a manufacturing process. The proposal takes into account the failure modes which affect the function and performance of the system. Firstly, the modes are identified and divided into two groups: faults and failures. The former implies detectable degradation of system function until the threshold for fault, which is eventual catastrophic loss of system, is surpassed. The latter leads to catastrophic fault. Then, using PBN, both classifications can be diagnosed and actions to mitigate them can be taken. The proposal also allows to forecast a time in hours by which the fault or failure will be imminent. The method herein discussed was applied to a ultrasound welding cycle, and a PBN model was created, simulated and verified through by means of model checking in PRISM. Results obtained show the validity of this methodology.

Keywords: Fault Detection and Isolation · Probabilistic Boolean Networks · Multiple faults · Reliability

1 Introduction

Compliance with regulations such as environmental, and safety, while producing high-quality goods requires constant process improvement. Any equipment fault impacts negatively system availability, operational and environmental safety, and the productivity of the business. Proper operation of these systems imply that faults have to be detected,

© Springer Nature Switzerland AG 2020
F. Martínez Álvarez et al. (Eds.): SOCO 2019, AISC 950, pp. 355–365, 2020.
https://doi.org/10.1007/978-3-030-20055-8_34

isolated and mitigated, tasks related to Fault Detection and Isolation (FDI) [4]. FDI methods are mainly divided in two categories, model-based and process-history-based [16, 17].

Model-based methods make use of either an analytical or computational model of the systems. A varied spectrum of the proposed model-based methods are supported by basic concepts such as the parity space, observer approach and the parameter identification or estimation approach [3]. In [19] is showed that the observers and parity space methods do not always permit the isolation of actuator faults. For models that are non-linear in nature, the complexity on the observer design method increases, whereas a precise system model is needed for the parity space method. To overcome these problems, a more recent approach based in the solution of an inverse problem using computational intelligence tools has been presented [2, 3]. In general, research has been limited to diagnosing independently occurring faults. Diagnosing simultaneous faults has not sufficiently addressed in scientific literature. Multiple fault detection in dynamic systems can be challenging, because the effects of a fault may hide or be compensated with the effects of different type of fault, and because equal types of multiple faults can manifest themselves in different forms, considering their order of occurrence. Computational Intelligence tools have been the most used to address this area [12]. In this sense, research has focused on static systems [15], and solutions to the multiple faults problem through observations on imperfect tests as in [13], to determine the closest evolution relative to the state of the fault. The authors of [18] postulate an algorithm-based pattern recognition method for diagnosis, which resulted in high efficiency and precision, but with cases in experimental data where particular fault tests didn't have a solution. Other developments include SLAT patterns for multiple fault diagnosis, and model-based methods for describing multiple faults in rotor systems [1]. However, multiple fault diagnosis is a current research area which demands the development of novel strategies for improving the performances of the fault diagnosis systems. This paper's main goal is to present a new approach of multiple faults diagnosis in industrial systems by using Probabilistic Boolean Networks (PBN).

PBNs have been used to model Gene Regulatory Networks (GRN) and to understand the general rules that govern gene regulation in genomic DNA [14]. PBNs are transition systems that satisfy the Markov Property, (memoryless, not dependent on the history of the system). Proposed in [14] by extending Kauffman's Boolean Network (BN) concept, they combine the rule-based modeling of BNs with uncertainty principles. These PBNs consist of a series of constituent BNs that have assigned selection probabilities, where each BN may be considered a context. Data for each cell comes from dissimilar sources; where each represents a cell context. In each given point in time t, a system can be governed by one of these constituent BNs, and the system switches to another constituent BN at another time, with a given switching probability. PBNs for manufacturing systems were introduced in [10] and further developed [8, 9, 11].

When compared to other biomimetic modeling and analysis methods, PBNs have the advantage of providing models that require less complexity in their definition, while still providing an adequate level of representation, based mainly in the logical relationships of their components, and their ability to accurately predict system behavior. Neural networks are structurally more complex than PBNs, because PBNs have no threshold, do not require training, or defining complex relationships between inputs and outputs, or layers. Also, the time that can take the computation of the next state of the network can be lower. PBNs, as Multi-Agent Systems (MAS), can model complex behavior of manufacturing systems, and can, similar to Holonic MAS, self-organize, because each context of a PBN is a constituent BN that has transitions that may lead to attractor states. The study of those attractors, contexts and networks yields information about the steady states of the system. They also, as Holonic MAS, can explain/exhibit emergent characteristics. Unlike MAS, Holonic MAS, and swarms, PBNs do not require the definition of agents and behaviors.

This article expands the use of PBNs in manufacturing systems to allow the consideration of faults that may lead to catastrophic failure, being this the first contribution of this research. The proposed model allows detecting and classifying single and multiple faults, which constitutes another contribution of the proposal. It allows identification of fault states in which it might possible to continue operation, and those where it is not possible to continue (failure). It also allows to forecast a time in hours by which the fault or failure will be imminent. As a final contribution, the system provides information about the maximum probability of fault and failure occurrence, which allows better maintenance planning. This paper is organized in the following manner: Sect. 2 discusses PBNs and their use in manufacturing systems modeling, Sect. 3 presents how these PBNs can be used for FDI in these systems. Section 4 discusses the experimental results. Finally, the conclusions of this research and future work are presented.

2 Probabilistic Boolean Networks and Their Use in Manufacturing Systems

Boolean Networks (BN) [5] and PBNs [14] have been proposed as a way of modeling manufacturing systems and process' dynamics (validated through model checking), and predict their future behaviors with statistical analysis and discrete event simulation [8–11]. Intervention [14] is used to steer the evolution of the network and guide it away from undesired states (disease).

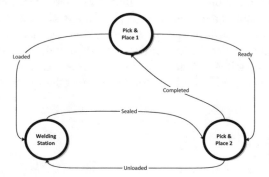

Fig. 1. Ultrasonic welding process from [11].

BNs are a finite set of Boolean variables (nodes), with states approximated to 0 or 1, for which, their state is determined by the current state of other nodes in the BN. It has a set of input nodes called regulatory nodes and a set of Boolean functions (predictors) that regulate the value of a target node. If the set of nodes and their corresponding functions is defined, the BN is defined. PBNs are basically a collection of BNs for which at any discrete time interval, the node state vector transitions are based on one of the rules of the constituent BNs. These context-sensitive, dynamical and probabilistic BN satisfy the Markov property. In [10] the authors demonstrated that PBNs are valid for modeling manufacturing systems, by establishing the method, validating it through model checking, and comparing the results obtained through simulation with actual machine data. In [8, 11], the authors used the same methodology applied to a manufacturing process to obtain quantitative occurrence data for DFMEA. In [9], the authors expands the application of PBNs in industrial manufacturing systems by incorporating the intervention mechanism to guide a modeled manufacturing system away from possible failure modes, thus delaying eventual failure of the system. For a detailed description of PBNs, see [14].

3   PBNs for FDI in Manufacturing Systems

To present the proposed method, the system introduced in [10, 11], consisting of three elements, an off-the-shelf Ultrasonic Welding Station (WS), and two off-the-shelf robotic hands, or Pick and Place (PP) machines, will be modeled. The WS system has a 2.5 KW power supply unit, an actuator housing a 3-inch air cylinder, a 20-micron converter, a 1:2.0 gain booster and a 20 kHz, 1:1 gain horn. This station will join two rigid parts. The PP has movement in both the x and y axes, and using a gripper holds, places and removes parts to and from the WS into an assembly line. The PP loads the parts into the WS. Once these parts are welded, a second PP will remove the welded parts. Figure 1 presents a finite-state machine of the above described welding system.

The method proposed in this paper adapts the FDI scheme described in [7], and shown in Fig. 2, where a model is used for normal operation of the process and another model is used for each one of the different faults. PBNs self-organize into attractor states, which relate to the different failure modes that the system experiences. Model construction and semantics are identical to [11]. Through characterization of the failure modes, the models can, with the property verification, characterize the state of their relevant components to determine which component failures correlate to machine and/or system fault conditions. A notable difference from past efforts is that this system is modeled as a PBN of PBNs. Each one of the nodes of the systems' PBN is in itself a PBN.

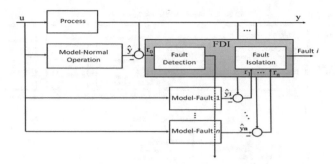

Fig. 2. Fault detection and isolation method from [7].

The PBN for the PP machines and the WS are detailed in [10] with its components, predictors and selection probabilities for each of the functions, in addition to its BN realizations, vector functions, attractors and the selection probabilities for each realization. The method is very flexible, and the design of the PBN and its state transitions depends on the amount of resolution that the experts need, based on design specifications. The system can grow in complexity and expression depending on the needs of the experts. Normal operation is modeled through simulation of the system's machines, based on the reliability analysis performed in [10, 11]. This can be modeled for the system as a whole, or for each of the machines that compose it, through simulation of their relevant components, based on each of the component's Mean Time Between Failures (MTBF) data. Each of the system's faults are modeled based of the Design Failure Mode and Effects Analysis (DFMEA) conducted in [11], and similarly for each of the possible faults for each machine. Therefore, the model is able to detect and isolate single machine and multiple machine faults for the system, and also single and multiple component faults on the individual machines.

With this structure, it is possible to classify faults and failure modes per machine (through the individual machine's PBN) and system faults and failure modes (through the system's PBN). The authors propose the establishment of the model using the PRISM model checker [6], in order to validate its use and check its formal correctness using Probabilistic Computational Tree Logic (PCTL). The model is composed of an input module, which uses PRISM's local non-determinism to provide the input to the PBNs. Three modules for each PBN model the machines involved in the process, and a fourth system PBN module models the behavior of the whole process. An output module produces the system state based on the state of the individual modules. This way, given the different faults and failure modes of the individual machines (which are based on the possible fault conditions of their components) the model produces the failure modes corresponding to the system. The failure modes for each machine were discussed in [11], and are based of DFMEAs conducted on each of the machines involved in the process. The process of producing the PBN for the system is presented graphically in Fig. 3.

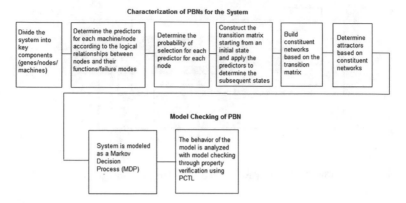

Fig. 3. PBN for FDI. Manufacturing system modeling process from [11].

4 Experimental Results

In this section, the experimental results are presented. PRISM was employed to validate the model quantitatively and to produce data required for statistical tests, used to determine the level of correspondence. Experiments were conducted using three PBN models: a model for the PP robots, one for the WS, and a one for the complete process. The models presented in [8–11] were expanded to include fault conditions that may lead to faults/failures on the individual machines and system. This allows predicting conditions that do not cause total failure, but rather failure modes that may lead unhealthy states where the system continues its operation, but cannot perform the required task to specifications, and if untreated, lead to failure. For each machine, and complete process, FMEAs were performed as per [11], and a determination was made of which system components and failure modes can produce a failure or a fault. As an example, the relevant nodes of the PP PBN are the gripper, a rotary axis, a fixed axis, motors for the rotary and fixed axes, and a power supply for the machine. On the PP machines, failure of the gripper and the axes will cause a fault on the PP as a whole. Failure of the motors or the power supply will cause a failure on the PP. Three modules constitute the complete models in PRISM, an Input module, a module for the PBN, and an Output module. The current state of the PBN's components is in module Input. The PBN module uses the state of the input variables and applies the corresponding predictors to transition to the next state. Based on the values of these variables, and the fault conditions, the state of a global PP variable is changed, giving us the current state of the machine. In these experiments, time is expressed in hours (h). Control groups were created through modeling and simulation of the systems' relevant components, with the components' corresponding MTBF obtained from real technical data sheets. These groups were established for the PP, the WS, and the complete system. Control data was used to compare against the PBN models, representing expected values.

Property verification in PRISM was employed for determining the maximum probability of occurrence of any of the failure modes that could lead to fault, for each of the presented models. From an initial state for each of the machines, a determination is made about the maximum probability of reaching one of the different identified fault conditions. Statistically significant differences between both the control and experimental groups (PBN models) were checked. Property verification in PRISM not only allows us to verify the models; they also allow, through experiments, to reach an estimate in time about when fault occurrence is certain.

Detection: The models can detect faults and failures, based on the application of the PBN. Given the current state of the network genes, the PBN selects an appropriate context and self-organize into one of the attractor states of their constituent BNs. In [11], context was equated to the different failure modes that can occur. The input module randomizes the current state of the machine, and based on it, the PBN module will apply the predictors and select a BN. The output module contains all of the identified fault conditions/failure modes of a machine, and after applying the predictors, evaluates the state of the machine's components, and makes a determination of the state of the machine as a whole, which can be in a complete failure condition, or in a fault condition, that can be specifically described based on the condition of the components, thus allowing detection and isolation of individual or multiple faults.

The first test determines the maximum probability of reaching any of the failure modes leading to fault of the PP through verification of the Probabilistic real time Computation Tree Logic (PCTL) property. This property was tested for the PP's PBN model, and the control group. Two sample T-tests were performed using Minitab 16 to look for statistically significant differences among the group means. The null hypothesis states that there is no difference between the Control and PBN groups, or H_0: μControl = μPBN. The alternative hypothesis would be finding differences between the Control and PBN Model groups, or H_1: μControl $\neq$ μPBN. For an α-level of 0.05 for the test, the conclusion is that for the PP, there are no statistically significant differences between the groups (p − value > 0.05). This means that there is no difference between both groups. Results of the two-sample T test are presented in Figs. 4 and 5.

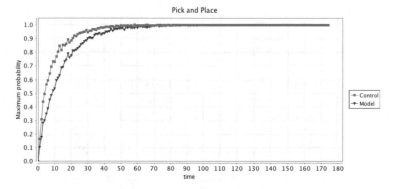

Fig. 4. Maximum probability: PP PBN vs. Control group

```
Two-Sample T-Test and CI: Control-PP, Model-PP

Two-sample T for Control-PP vs Model-PP

              N    Mean   StDev   SE Mean
Control-PP   175   0.953  0.139    0.011
Model-PP     175   0.923  0.178    0.013

Difference = μ (Control-PP) - μ (Model-PP)
Estimate for difference:  0.0292
95% CI for difference:  (-0.0043, 0.0628)
T-Test of difference = 0 (vs ≠): T-Value = 1.71   P-Value = 0.088   DF = 328
```

Fig. 5. Two-sample T test: Pick and Place PBN vs Control group

Diagnosis and Prognosis: Labels in PRISM can be used to highlight specific states, or sets of states, and as such, used to highlight single or multiple faults. When the PBN is applied and a constituent BN is selected, the labels filter out the fault conditions, or the correct machine operation. Within the output module, all of the possible failure and fault conditions on the machine caused by the identified components are expressed, allowing determination of its future state. This allows not only to discern which specific fault or combination of faults is occurring, but through property verification labels can be used to produce a *prognosis*, an estimate in time of when the fault is expected to occur. Knowing the probability of failure occurrence allows the system designers to make decisions about the downtime needed for maintenance. Experiments were performed to verify the capability of detecting multiple faults of the system, using the System's PBN model, specifically simultaneous faults detected on both PP machines. Figure 6 shows a simultaneous fault on PP1 and PP2, and the time estimate (700 h) at which it will manifest.

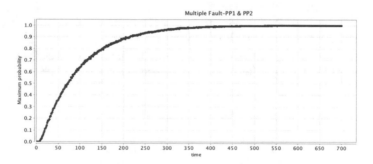

Fig. 6. Simultaneous faults of Pick and Place 1 and Pick and Place 2.

With PRISM, the states of variables in a simulation can be plotted to track their changes. In the case of PP system, there are 64 states. Table 1 presents a subset of fault, failure and operation states in which the PP model can be.

Table 1. Some states of the Pick and Place PBN.

Machine state	Description
47	Failure due to motor 1a and power
59	Multiple fault due to fixed axis and power
62	Single fault due to rotary axis
64	Normal operation

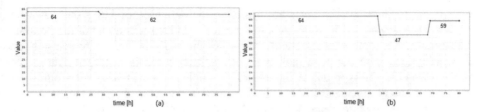

Fig. 7. Fault diagnosis using the Pick and Place's PBN model

In Fig. 7, diagnosis of faults and failures is illustrated. States of the PP machine have been labeled, and each state of the machine can be individually identified. This means that all faults, single or multiple, can be singled out specifically. State 64 is the normal operating state of the machine. In Fig. 7(a), after 29 h of normal operation, the model and simulation identifies a single fault of the rotary axis. If the machine continues to operate without intervention, this fault may develop into a failure. The initial state of the system is presumed to be the normal operating state. Figure 7(b) shows another simulation of 80 h of continuous operation, where after 48 h of normal operation, the system detects and diagnoses a failure of motor1a at 49 h, and a fault of the fixed axis at 69 h.

This novel method of modeling in manufacturing systems provides a simple, straightforward mechanism of detecting, isolating and classifying single and multiple faults. Through the obtained experimental results, it is statistically demonstrated that model-based PBN FDI performs satisfactorily when examining a machine's possible behavior, for detection, classification and isolation of single and multiple faults. The system will experience a fault condition on any of its components at about 60 h of operation. Given an operational state, and a set of possible fault states, the model is able to characterize possible failure modes in which the machine or system may be, given the state of its variables. Model-based approaches depend upon the knowledge of experts that classify the different fault conditions and failure modes. In [11], a table with possible failure modes is given that define the failure modes of the WS. The model in [11] is able to predict these failure modes, but the table does not contemplate all the possible combinations that can occur. One of the models presented here is an evolution of that model, including faults of each of the component, therefore, new fault conditions and failure mode may occur that are not covered in the table. Experts would have to examine these and determine the corresponding new failure modes for each system.

5 Conclusions

A bioinspired method that allows modeling single and multiple faults on manufacturing systems using PBNs was presented. The modifications herein proposed to the aforementioned architecture allowed the classification of single and multiple failures. These permit the FDI scheme shown in Fig. 2, detection and isolation of single and multiple faults, along with an estimate of when these faults will present themselves. Statistical tests performed of this data validate the proposed approach for future use and further development. Since these models are based on the definition of PBNs derived from regulating genes/nodes, this discretization creates a limitation in terms of the possible states that it can represent. but greatly simplifies the analysis. For future research, an interesting idea is to design a fault diagnosis system based in historical data of the process with the ability to detect and classify multiple and novel faults. Expanding the use of non-binary quantized PBNs will also allow in the future a richer mechanism of expressing fault conditions and failure modes. Another possible avenue of development is the use of Intervention in FDI-enabled PBN models.

Acknowledgements. The authors acknowledge the financial support provided by FAPERJ, Fundação Carlos Chagas Filho de Amparo à Pesquisa do Estado do Rio de Janeiro; CNPq, Conselho Nacional de Desenvolvimento Científico e Tecnológico; CAPES, Coordenação de Aperfeiçoamento de Pessoal de Nível Superior, research supporting agencies from Brazil; UPR-RP, University of Puerto Rico-Río Piedras; UERJ, Universidade do Estado do Rio de Janeiro and CUJAE, Universidad Tecnológica de La Habana José Antonio Echeverría.

References

1. Bachschmid, N., Pennacchi, P., Vania, A.: Identification of multiple faults in rotor systems. J. Sound Vib. **254**, 327–366 (2002)
2. Camps Echevarría, L., Campos Velho, H.F., Becceneri, J.C., Silva Neto, A.J., Llanes-Santiago, O.: The fault diagnosis inverse problem with ant colony optimization and ant colony optimization with dispersion. Appl. Math. Comput. **227**(15), 687–700 (2014)
3. Camps Echevarría, L., Llanes-Santiago, O., Campos Velho, H.F., Silva Neto, A.J.: Fault Diagnosis Inverse Problems: Solution with Metaheuristics. Springer, Heidelberg (2019)
4. Isermann, R.: Fault-Diagnosis Applications: Model-Based Condition Monitoring: Actuators, Drives, Machinery, Plants, Sensors, and Fault-tolerant Systems, vol. 24. Springer, Heidelberg (2011)
5. Kauffman, S.A.: Homeostasis and differentiation in random genetic control networks. Nature **224**, 177–178 (1969)
6. Kwiatkowska, M.Z., Norman, G., Parker, D.: Prism 4.0: verification of probabilistic real-time systems. In: Gopalakrishnan, G., Qadeer, S. (eds.) Computer Aided Verification. Lecture Notes in Computer Science, vol. 6806, pp. 585–591. Springer, Heidelberg (2011)
7. Mendoça, L., Sousa, J., da Costa, J.S.: An architecture for fault detection and isolation based on fuzzy methods. Expert Syst. Appl. **36**, 1092–1104 (2009)
8. Rivera Torres, P., Serrano Mercado, E.: Probabilistic Boolean network modeling as an aid for DFMEA in manufacturing systems. In: Proceedings of 18th Scientific Convention in Engineering and Architecture (CCIA 2016), La Habana, Cuba (2016)
9. Rivera Torres, P., Serrano Mercado, E., Llanes-Santiago, O., Anido Rifón, L.: Modeling preventive maintenance of manufacturing processes with Probabilistic Boolean Networks with Interventions. J. Intell. Manuf. (2015)
10. Rivera Torres, P.J., Serrano Mercado, E., Anido, R.L.: Probabilistic Boolean Network modeling of an industrial machine. J. Intell. Manuf. **29**, 875–890 (2015)
11. Rivera Torres, P.J., Serrano Mercado, E., Anido Rifón, L.: Probabilistic Boolean Network modeling and model checking as an approach for DFMEA for manufacturing systems. J. Intell. Manuf. **29**, 1393–1413 (2015)

12. Rodríguez Ramos, A., Domínguez Acosta, C., Rivera Torres, P.J., Serrano Mercado, E.I., Beauchamp Báez, G., Anido Rifón, L., Llanes-Santiago, O.: An approach to multiple fault diagnosis using fuzzy logic. J. Intell. Manuf. (2016)
13. Ruan, S., Zhou, Y., Feili, Y., Pattipati, K., Willett, P., Patterson-Hine, A.: Dynamic multiple-fault diagnosis with imperfect tests. IEEE Trans. Syst. Man Cybern. Part A: Syst. Humans **39**, 1224–1236 (2009)
14. Shmulevich, I., Dougherty, E., Kim, S.: Probabilistic Boolean Networks: a rule-based uncertainty model for gene regulatory networks. Bioinformatics **18**(2), 261–274 (2002)
15. Sobhani-Tehrani, E., Talebi, H., Khorasani, K.: Hybrid fault diagnosis of nonlinear systems using neural parameter estimators. Neural **50**, 12–32 (2014)
16. Venkatasubramanian, V., Rengaswamy, R., Yin, K., Kavuri, S.N.: A review of process fault detection and diagnosis-Part I: quantitative model-based methods. Comput. Chem. Eng. **27** (3), 293–311 (2003)
17. Venkatasubramanian, V., Rengaswamy, R., Yin, K., Kavuri, S.N.: A review of process fault detection and diagnosis-Part III: process history based methods. Comput. Chem. Eng. **27**(3), 327–346 (2003)
18. Wang, Z., Marek-Sadowska, M., Tsai, K., Rajski, J.: Analysis and methodology for multiple-fault diagnosis. IEEE Trans. Comput.-Aided Design Integr. Circuits Syst. **25**, 558–575 (2006)
19. Witczak, M.: Modelling and Estimation Strategies for Fault Diagnosis of Non-Linear Systems From Analytical to Soft Computing Approaches, vol. 354. Springer, Heidelberg (2007)

Concurrent Planning and Scheduling of Heterogeneous Production System. Case Study

Bożena Skołud[1]([⊠]), Agnieszka Szopa[2], and Krzysztof Kalinowski[1]

[1] Institute of Engineering Processes Automation and Integrated Manufacturing Systems, Faculty of Mechanical Engineering, Silesian University of Technology, Konarskiego 18A, 44-100 Gliwice, Poland
{bozena.skolud,krzysztof.kalinowski}@polsl.pl
[2] The Institute of Ceramics and Building Materials, Refractory Materials Division in Gliwice, Toszecka 99, 44-100 Gliwice, Poland
a.szopa@icimb.pl

Abstract. The problem of heterogeneous production planning and scheduling is presented in this paper. The basic assumption of the study is the "methodology of the practice", i.e. simple, fast and flexible, enabling to determine whether the given order can be implemented in the system on the basis of meeting or not specified conditions. The case study presented scheduling of the production of refractory materials with KbRS software is presented.

Keywords: Scheduling · Production planning ·
Heterogeneous production systems

1 Introduction

Heterogeneous production systems which combines three types of production: make to stock (MTS), make to order (MTO) and also engineer to order (ETO) (production which is project oriented) is the most common in small and medium enterprises existing in competitive condition. In the 90s, production changed its character to process-oriented and market-oriented. The product life time has been significantly reduced and the product quality has improved. The plants have various locations around the world, which created a network of production plants. Nowadays, production is more and more often oriented to individual customer expectations. The time of product life has rapidly decreased, and thus the time of introducing new products to the market. For this reason, enterprises increase flexibility, which will enable them to change the configuration of production to new customer expectations. Challenges and changes in the industry require a new product development strategy that combines different types of activities and planning methods, and different levels of complexity and flexibility. Numerous companies create heterogeneous systems that offer products that are at different stages of product life, products with varying volatility and repeatability. This means that one system produces typical products for which the market demand is predictable and determined based on marketing researches. Current

© Springer Nature Switzerland AG 2020
F. Martínez Álvarez et al. (Eds.): SOCO 2019, AISC 950, pp. 366–375, 2020.
https://doi.org/10.1007/978-3-030-20055-8_35

production is make to stock (MTS) [6]. At the same time, make to order (MTO) production is started based on the customer's order. These products are characterized by a certain degree of individualization under strictly defined customer requirements. In addition, the company complements its offer with the new product introduced to the market. Production of these products requires a multi-stage implementation, the project oriented production called engineer to order (ETO) [2, 4]. Planning and scheduling of such complex systems belong to the NP-hard class and it is not possible to find optimal solution in disposal time. Despite the difficulty of managing such a complex system, enterprises decide to make the best use of resources at the disposal of the enterprise, in particular workstations, machines, equipment and human resources [5, 7].

Each enterprise make production in constraints conditions, some of them are deterministic (they are known and permanent), i.e. the number of resources and their availability, other constraints are accidental, they are related to special situations such as device failure, employee absence. The company can only affect them through predictive actions that reduce the probability of their occurrence or minimize their effects. In these paper deterministic resources constraints are considered. In this context, the precondition is to plan the activity providing access to limited resources (i.e.: machines/human resources) over time. Each decision about orders execution is related to the analysis of the resources availability in the period in which the order (set of orders) should be performed. In the context of the project, it is necessary to verify constraints resulting from the scope of the project, defined start and end, and costs, which depends on each other.

The implementation of the production plan will balance the tasks in the system, i.e. the production plan will answer the question: Is it possible to perform all tasks within the period specified by the customer (usually several months) and within limited resources? Due to the complexity of the problem, considering the amount of data in the non-homogeneous system, an optimal solution is not available.

This paper outlines the methodology of production planning in a heterogeneous system. The essence of this methodology is to develop a sequence of proceedings (making decisions) and to determine the principles of so-called dispatching (priority) rules that prioritize the allocation of tasks to the resources [1]. This methodology and the software 'Knowledge base rescheduling system' (KbRS) elaborated by one from authors [3, 7] can be included to the class of soft computing methods. The proposed approach guarantees the achievement of an acceptable solution (preparation of the plan and schedule) in a short time.

One must remember that the basic assumption of the study is the "methodology of the practice", i.e. simple, fast and flexible, enabling to determine whether the given order can be implemented in the system on the basis of meeting certain conditions.

2 Methodology

A heterogeneous system of multi-assortment production is considered, including discrete production processes running simultaneously. Heterogeneous systems are dedicated to entrepreneurs who want to meet the client's expectations. The system characterize [2]:

- the resources of the system are renewable, i.e. that after the process is completed, the resource is ready for the next process;
- each process is a sequence of a finite number of operations. Operations are non-preemptive, i.e. the operation cannot be interrupted before its completion;
- production orders are defined by: the size of the order, completion date, technological route, lot size;
- processes compete for access to limited resources in the mutual exclusion mode (asynchronous), i.e. only one process can be carried out at a time;
- constraints in the system are deterministic;
- the next operation starts immediately after the completion of the preceding one, under the condition that the resource is available (order restriction);
- times of transport operations and setups are included in the duration of technological operations,

The original methodology for determining the production plan and the acceptable schedule (i.e. allocating resources to tasks and determining the sequence of tasks and providing deadlines on resources) consists of the following steps:

Step 1 Planning - Initial estimation of the availability of machines and devices.
For the assumed time period, a resource availability calendar is defined. The deterministic system constraints, such as working days, shift system, planned shutdowns in connection with repairs, modernization, etc. are taken into account. A general plan can be created every year (or every month) and is a tool for the management level in the company. The created plan is the basis for performing a new schedule.

Step 2 Planning of MTS orders.
MTS orders are planned "forward" in a given period. The deadlines for individual orders are not determined. It is assumed that unused capacity will be left to execute orders for which arise urgent demand.

Step 3 Analysis of the system's production capacity.
For some production capacity part of it is reserved for ETO and MTO orders. It is determined on the base of historical data. It should be at least the minimum production capacity that ensures the execution of urgent orders. The indicator (of free capacity) may change during the planning period, which results from periodic increased market demand for a given product. In the case of unsatisfactory schedule, the planner performs rescheduling.

Step 4 ETO Scheduling.
The ETO production orders (project oriented) are planned considering the resource availability calendar taking into account the implementation of MTS orders. ETOs constitute one of the milestones of the external project being implemented. The planner fulfils of this type of order, it is necessary to preserve the availability of needed resources in the range from the fastest possible date of execution (ES) to the latest possible deadline (LF). In the case when the resource availability is not enough the MTS schedule is changed.

Step 5 MTO Scheduling.

Orders of this type often appear in the system. Unlike step 2, MTO orders are planned "backwards", because for them a directive ending time of the order is defined. In the case when the implementation of MTO is not possible in needed period the planner assesses the validity of the order as compared to the previously planned. If the new order is more important, then changes to the current schedule are allowed (e.g. transfer of some tasks MTS to subsequent planning periods)

Step 6 The schedule evaluation.

From the solutions obtained, the best plan is chosen in the light of the adopted criteria or in the absence of a satisfactory plan, the assumptions are verified and re-scheduling is run.

3   Case Study

Given is production system of refractory materials. Basic production is carried out according to MTS and MTO. However in recent years, the market environment has changed fundamentally, because thanks to the support of EU funds, enterprises have invested in R & D departments within their structures. Therefore, the Institute was forced to change its production strategy and, apart from MTS and MTO production types, develops innovative materials in accordance with ETO principles (project-oriented activity). Such orders have the highest priority. Their timely implementation is important due to the need to preserve the schedule of the project, but also or primarily there is a strictly defined time in which this material can be verified in the investor's installation. When there is a delay in implementation, the implementation of the entire project can be postponed by up to 12 months. The effects of untimely implementation concern both the negative assessment of the producer. Additionally it will not be fulfilled by the research project on time. It draws its financial consequences but also delays the introduction of innovations to the market, which causes very measurable financial losses. Therefore extremely important is to protect the availability of resources for the duration of the project.

The system consists of 12 stations where operations are curried out successively (Fig. 1). The list of workstations is presented in Table 1. It is visible that some stations are doubled. They are: raw material weighing system, mixer, packing machine, wooden mold. The stands work in a one-shift system, excluding Saturdays, Sundays and public holidays, according 8-hour calendar (Ka_1). Resource Z8, work in continuous system (24 h/day, 7 days in week - Ka_2). The period of 30 calendar days was assumed.

Table 1. Resources

Symbol	Name	Ka_b	C_1
Z1	Raw material weighing system	Ka_1	1
Z1'	Raw material weighing system	Ka_1	1
Z2	Feeder of chemical additives	Ka_1	1
Z3	High-speed mixer	Ka_1	1
Z4	Packing machine	Ka_1	1
Z4'	Packing machine	Ka_1	1
Z6	Stirrer	Ka_1	1
Z6'	Stirrer	Ka_1	1
Z7	Wooden mold	Ka_1	1
Z7'	Wooden mold	Ka_1	1
Z8	Roasting oven	Ka_2	1
Z9	Packing machine	Ka_1	1

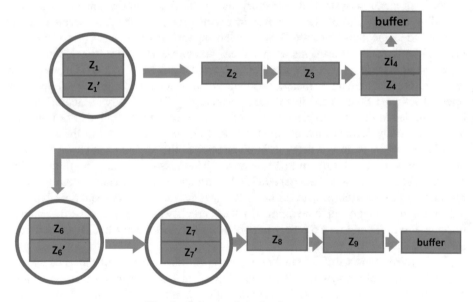

Fig. 1. Scheme of production system

The system performs many different types of monolithic materials run according to the same technological route described by the matrix:

$$M_p = \begin{bmatrix} 1 & 2 & 3 & 4 & 6 & 7 & 8 & 9 \\ 45\,\text{min} & 20\,\text{min} & 15\,\text{min} & 180\,\text{min} & 60\,\text{min} & 120\,\text{min} & 2880\,\text{min} & 60\,\text{min} \end{bmatrix}$$

$$(1)$$

One production part equals production of 3 tons of materials. The discretization cause from workstations capacities. Processes differs one from the other one. Three types of processes are distinguished:

- Pr_1 – process of monolithic materials runs from operation No. 1 to operation No. 4.
- Pr_2 – prefabrication process of monoliths, runs from operation No. 1 to No. 9.
- Pr_3 – MTS processes, runs from the operation No. 6 to No. 9.

3.1   Production Order Description

This heterogeneous systems, which offer products at various stages of product life-cycle, products is characterized by various changeability and reproducibility. Make-to-stock (MTS) and make-to-order (MTO) products, including implementation of project tasks (ETO) run in the system.

MTS is described by 5 elements:

$$M_P^{MTS}\left\{M_P, L_M, t_r, \Delta t_r, w^{MTS}\right\} \tag{2}$$

where:

M_p – matrix of process,
L_M – series size (number of elements for completion of production order),
t_r – the earliest possible time to start the production process,
Δt_r – difference between the end time of the MTS order and its re-start,
w^{MTS} – the priority rule specifying the validity of the order being processed

MTS orders are described by the earliest possible start time. It depends on the state of the rotation rate of a given product in stock of finished products and marketing forecasts. On this basis it is planned dates for commencement of production and the batch size for a given product. And thus determines the load of resources resulting from the implementation of orders.

Orders MTO, are described by:

$$M_P^{MTO}\left\{M_P, L_M, t_D, w^{MTO}\right\} \tag{3}$$

where:

M_p – matrix of process,
L_M – series size (number of elements for completion of production order),
t_D – deadline,
w^{MTO} – the priority rule specifying the validity of the order being processed.

In the case of the MTO, the key is the limitation given by the client regarding the deadline for implementation. Using the backward planning method, the time of order fulfilment and resource load is determined.

Orders ETO are specific to project-type tasks. Due to the strictly defined time frame in which the order is to be processed (production part of the project) it is characterized by:

$$M_p^{ETO}\left\{M_p, L_M, t_r t_D, w^{ETO}\right\} \tag{4}$$

where:

M_p – matrix of process,
L_M – series size (number of elements for completion of production order),

t_r – the earliest possible time to start the production process,
t_D –deadline,
w^{ETO} – the priority rule specifying the validity of the project being processed.

The subject of planning is therefore set of orders described by:

$$M = \{M_l^{MTS}, M_k^{MTO}, M_m^{ETO}, l = 1, \ldots, L, k = 1, \ldots, K, m = 1, \ldots, M) \qquad (5)$$

where:
l, k, m – the total number of orders accepted in a given production system.
In a heterogeneous system, the orders are described by:

$$M = \left\{ \sum_l [M_p, L_M, t_r, \Delta t_r, w^{MTS}], \sum_k [M_p, L_M, t_D, w^{MTO}], \sum_m [M_P, L_M, t_r, t_D, w^{ETO}] \right\} \qquad (6)$$

It is assumed that MTO orders have a higher priority than orders MTS, however have a lower priority than ETO. ETO > MTO > MTS.

In the Tables 2, 3 and 4 details of the input data and basic assumptions for scheduling are presented.

Table 2. The set of production orders

No.	Production order	Type	Quantity	Processing time	Release date	Due date	
1	ZAM/G/18/1	Pr_1	18000 kg	1560 min.	2.07.18	-	MTS
2	ZAM/G/18/2	Pr_1	35000 kg	3120 min.	2.07.18	-	MTS
3	ZAM/G/18/3	Pr_1	20000 kg	1760 min.	2.07.18	-	MTS
4	ZAM/G/B23	Pr_1	9000 kg	780 min.	09.07.18	16.07.18	ETO
5	ZAM/G/18/54-1	Pr_1	20000 kg	1760 min.	19.07.18	27.07.18	MTO
6	ZAM/G/18/54-2	Pr_1	57000 kg	4940 min.	19.07.18	27.07.18	MTO
7	ZAM/G/18/52		1000 kg	140 min.	13.07.18	20.07.18	MTO
8	ZAM/G/531	Pr_1	10000 kg	920 min.	12.07.18	20.07.18	MTO
9	ZAM/G/18/555	Pr_1	5000 kg	460 min.	23.07.18	27.07.18	MTO
10	ZAM/G/18/541	Pr_1	1000 kg	140 min.	17.07.18	31.07.18	MTO
11	ZAM/G/539-1	Pr_1	24000 kg	2080 min.	17.07.18	31.07.18	MTO
12	ZAM/G/539-2	Pr_1	16000 kg	1440 min.	17.07.18	31.07.18	MTO
13	ZAM/G/537	Pr_1	200 kg	120 min.	17.07.18	18.07.18	MTO
14	ZAM/G/18/536	Pr_1	10000 kg	920 min.	25.07.18	30.07.18	MTO
15	ZAM/G/18/587	Pr_3	300 kg	3120 min.	2.07.18	17.07.18	MTO
16	ZAM/G/18/591	Pr_3	826 kg	3120 min.	2.07.18	20.07.18	MTO
17	ZAM/G/18/594	Pr_2	1398 kg	3200	2.07.18	31.07.18	MTO
18	ZAM/18/595	Pr_2	580 kg	3200	2.07.18	25.07.18	MTO
19	ZAM/18/598	Pr_1	4000 kg	140 min.	09.07.18	16.07.18	MTO
20	ZAM/18/520	Pr_1	2000 kg	200 min.	10.07.18	25.07.2018	MTO

Table 3. Priority (sequencing) rules

Name	Rule
LIST	Priority is given to the task by given task list (with ETO > MTO > MTS order)
EDD	Priority is given to the task with the earliest due date time
LPT	Priority is given to the task with the longest processing time
SPT	Priority is given to the task with the shortest processing time

Table 4. Performance measures

Criterion	Description
Cmax	Makespan
mC	Mean completion time in the set of tasks
Fmax	Max. flow time
mF	Mean flow time
Tmax	Max tardiness
mT	Mean tardiness
Emax	Max earliness
mE	Mean earliness
U	The number of tardy tasks

The KbRS software [8] was used for creating schedules according to proposed methodology. One of the obtained solutions, created by the List rule, is shown in Fig. 2. At first step calendars were generated for determining availability of machines. Except of Z8 which works in continuous mode (24/7), all other resources work in a one-shift system excluding weekends (8/5). The white bars on the chart represent breaks from calendar generated for a given resource. In the step 2 the MTS orders are planned by forward strategy (red bars in the chart, the connections show the flow of the selected lot). Analysis of the system's production capacity in Step 3 confirms the possibility of scheduling next orders. Step 4 is intended for planning ETO orders (blue bars in the chart); resource availability does not require changes in MTS orders. In the 5 step the orders No. 5–20 are scheduled, according to the backward strategy (green bars in the chart). The list of performance measures of individual solutions, calculated at step 6 of proposed methodology, are presented in Table 5. In all solutions, a schedule was obtained without delays. Differences can be noticed in the earliness and flow time indicators – for the EDD and SPT rules. They are important when the supply of materials is planning - they can affect the reduction of production costs.

Table 5. Performance measures in the set of obtained schedules

Rule	Cmax	mC	Fmax	mF	Fsum	Tmax	mT	Emax	mE	U
List	41220	22202,5	41220	7311,59	643420	0	0	41080	14713,86	0
EDD	41220	22491,59	41220	7600,68	668860	0	0	41080	14424,77	0
LPT	41220	21413,64	41220	6522,73	574000	0	0	41080	15502,73	0
SPT	41220	22523,64	41220	7632,73	671680	0	0	41080	14392,73	0

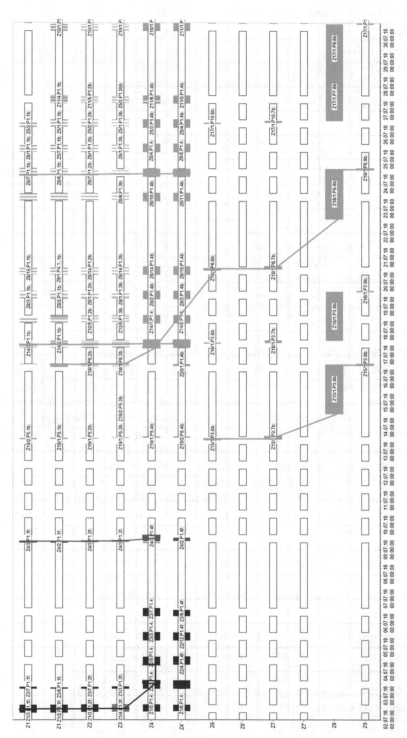

Fig. 2. The schedule obtained using the list rule

4 Summary

This article presents the original planning and scheduling methodology for heterogeneous systems. In the subsequent stages the proposed method enables planning of production orders performed in make to stock (MTS), make to order (MTO) and engineer to order (ETO) conditions. Each of them requires a slightly different orders treatment. The interaction of orders from individual groups is also considered – the iterations are permitted in which the previously scheduled order are corrected.

The procedure was presented in the KbRS software and it was shown that the procedure guarantees acceptable schedules. An example of application for data from a company is presented. The analysed example is quite complex but does not fully cover the problems of concurrent planning of various types of orders. The planning process in such conditions requires a significant participation of the decision maker in the scheduling process. This applies in particular to situations where there is a need for partially rearrange orders already placed in the schedule.

The next step will be to compare the results obtained from this method with the timely planning and scheduling methods. The issue of multi-criteria evaluation method in relation to forward and backward scheduling strategies for selected groups of orders should be also considered.

References

1. Browning, T.R., Ali, A.Y.: Resource-constrained multi-project scheduling: priority rule performance revisited. Int. J. Prod. Econ. **126**, 212–228 (2010)
2. Egri, P., Kovacs, A., Markus, A., Vancza, J.: Project-oriented approach to production planning and scheduling in make-to-order manufacturing. Prod. Syst. Inf. Eng. **2**, 23–36 (2004)
3. Kalinowski, K.: Scheduling of Discrete Production Processes. Wyd. Pol.Sl, Gliwice (2013). (in Polish)
4. Markus, A., et al.: Project scheduling approach to production planning. CIRP Ann.-Manuf. Technol. **52**(1), 359–362 (2003)
5. Marvelias, C.T., Sung, C.: Integration of production planning and scheduling overview, challenges and opportunities. Comput. Chem. Eng. **33**, 1919–1930 (2009)
6. Rafiei, H., Rabbani, M., Alimardani, M.: Novi Bi-level hierarchical production planning in hybrid MTO/MTS production contexts. Int. J. Prod. Res. **5**(2012), 1–16 (2012)
7. Skołud, B., Szopa, A.: Production planning methodology in hybrid system. In: ModTech2018 IOP Conference Series: Materials Science and Engineering, vol. 400, p. 062027 (2018)
8. http://imms.home.pl/kbrs - the kbrs scheduling software web page

Multi-domain, Advisory Computing System in Continuous Manufacturing Processes

Krzysztof Niemiec[1,2]([envelope]) and Damian Krenczyk[1]

[1] Faculty of Mechanical Engineering, Silesian University of Technology,
Gliwice, Poland
{krzysztof.niemiec,damian.krenczyk}@polsl.pl
[2] Valmet Automation Polska Sp. z.o.o, Gliwice, Poland

Abstract. Many decisions that must be made during the production process mean that limited human perception is not able to meet the growing requirements of keeping the parameters and constantly striving to increase the efficiency of current production lines. The main challenge is also the continuous increase of awareness about the process and the possibility of its modernization. This forces the expansion of the production with new elements, which are not directly related to the production line itself. And this in turn forces the expansion of knowledge, for example, cooperation with new elements. The theoretical knowledge that each employee must have from every issue going to be very general without going deeper into details. Simultaneous control of all mutual elements with the same coincidence becomes impossible. Traditional methods of failure analysis and finding reasons for its occurrence are inefficient and ineffective. This paper is attempting to create a system topology for all subsystems. A comprehensive production management system, its efficiency and failure predictive system will be discussed. The system should integrate and correlate many different databases, which are conducted according to different standards. This causes a necessity of choosing a method for seeking solutions for problems in such a large stored database. "Big data" which is popular today, is using neural networks which not always is the best choice. Especially when we don't have enough knowledge about technology and connections inside the process. Maximum use of expert knowledge, experience of employees, data acquisition and usage of unfiltered data will be presented in this paper.

1 Introduction

The modern factory with continuous production is based on computer production control systems. However, to ensure the continuity of the process, which should be efficient and stable in terms of the quality of the final production, there is a need to seek for new methods of control. The production line, depending on the complexity, consists of a larger or smaller number of subsystems, components, package deliveries and auxiliary devices. The more important or larger parts of the production line usually are equipped with their own systems or controllers, which are responsible for the basic implementation of activities provided by the device manufacturer. Often, manufacturers include their own know-how in local controllers and provide them in the form of so-called "Black box" in order to do not deliver the specific algorithm invented by them in open form to the final client. Larger machines or more responsible machines can be equipped with several supervisors and control systems. One of the systems may be responsible for controlling and executing functions. The second system can be responsible for measuring and analyzing vibrations and shaft displacements for rotating machines. The third system may be a system responsible for emergency shutdown of the device. It does not implement the logic associated with control only it implements logic with the safety of people and the process. Signals from subsystems or from other databases participating in the system are always pre-filtered. This operation allows taking these measurements to this control system. Unfortunately, it can "hide" certain phenomena appearing on the raw signal [1]. Ripples or abnormal behaviors appearing on the baseline measurement may indicate deteriorating process conditions or an upcoming failure. Unfortunately, modern control systems are varied in terms of hardware and software. The supply of equipment by various vendors forces the use of the solution of communication links and many intermedium bases. Many valuable information is lost forever. The best solution would be to connect all cooperating devices to one central control system. The result is that many variables are not scaled according to one standard and ideology. This forces the need to rank these variables and unify them using various computer methods [2].

Currently, there is such an integration trend. Connect many sub vendors in one control system eq. DCS class system [3] or similar high-level control system. However, such a necessity also exists in already started units, where deep integration of systems should be also carried out. Technological changes or the expansion of the factory or increasing the efficiency [4] always entails the need to expand the control system or add a new "control island". Repeated transmission of the signal from the system to the system and its multiple conversion distorts it and applies subsequent filters through the subsequent processing of the signal. Examples might be the obvious Quantization error, Scaling error, processing time and many more [5]. Attached filters on the input card or introduced delays all have an impact on the quality of the input data that is passed to the control and analysis system. Finding and catching very subtle symptoms, micro ripple or fast pulses may be impossible through the initial signal filtered through input circuits [6]. A typical computer control system with digitized potential degradation sites of the measurement signal is shown in Fig. 1.

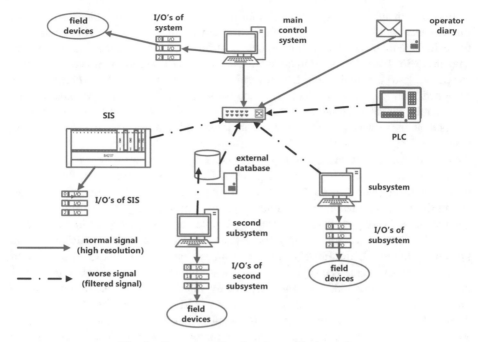

Fig. 1. Computer control system with signal flow

2 Analysis of Loss of Signal Quality at Selected Points of the Computer System

Time Synchronization
In addition to loss of signal resolution, the time stamp of a signal is often lost. Because sending a time stamp together with an event requires at least twice as much data transmission, it is a big problem with poorly developed communication solutions. Moreover, if a time stamp is already exchanged it must ensure this synchronization of all systems use a single source of time. In addition, various systems collect data at different frequencies. And some systems, despite downloading data with high resolution and time stamp, do not transfer them with the same tag and resolutions. Data from various subsystems are difficult to analyze and transfer to one common platform [7]. The shift of measurement signals or sequence of events even by a few tens of milliseconds may make it difficult to clearly indicate the cause or the starting point triggering the production discontinuing sequence. Some program cycles in industrial installations have a response time of a few milliseconds, so the shutdown sequence of the entire line production takes less than a second. Therefore, these few milliseconds of displacement with the simultaneous interpretation of signals from several systems are important in determining the cause of the failure. It has less meaning when we are analyzing a trend or try to tune a loop control. But it is a big issue when we are analyzing a "post mortem" situation. During analysis of this kind of situation it is very important to catch the first reason of stopping and to avoid it in the future.

I/O Cards

It is also shown in Fig. 2 that the signals from the very beginning, that is from the field device are sent to the control system input cards. In addition to the current or voltage in the measuring loop, which reflects the size of the physical measured value, the system can also read additional information from the intelligent transducers.

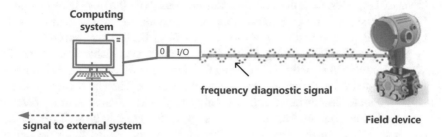

Fig. 2. Signal exchange from field device

The main measured value from the field device transmitted to the third part system is already devoid of these diagnostic values. Additional information from smart field device equipment is usually used by services responsible for the care of services field equipment. However, these are dedicated systems for this task. However, they are not always subject to the same good supervision as the main control system.

Vibration Monitoring

Another important external system that is connected to the superior control system is the vibration analysis system. This is a very important system especially for a rotating machine. Such systems have a special algorithm built in to determine many characteristics of the rotating system [8]. Vibroacoustic provides many methods and tools to determine the technical condition of the machine. These characteristics are recorded especially when starting or stopping the installation or when changing a load. An example of those characteristic trajectory of the shaft spigot is shown in Fig. 3.

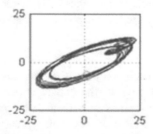

Fig. 3. Typical shaft trajectory

Usually, this check takes place periodically without observing the trend. Responsible machines such as turbines or large compressors are equipped with continuous vibration monitoring. However, even with continuous measurements, only vibration levels are transmitted to the system. Again, any analysis of the raw signal is lost by specifying only the vibration level. On the other hand, less responsive drives are periodically tested using portable measuring stations. If the condition of the drive is correct at the moment of the measurement and the installation works stable, the measurement signal may have the appropriate shape of the trajectory. In most cases, the system is responsible for generating an alarm if the vibration level rises beyond a certain value. However, with this solution, information about the raw shape of signal is lost. The alarm signal tells us nothing about the situation before the alarm was triggered. Moreover, in a situation where the signal is in the limit of acceptance, it is usually not observed and the way which alarm was made is invisible to the user. That situation is moreover not good because when the vibrations will appear, it is too late for any preventive actions and attempts to regulate the process. The alarm signal indicates a threat for people or machine, and the only thing that remains is the operator's decision whether he continues the process or stops it. It's good to be well before these phenomena and pick out these symptoms that inform much earlier about the deteriorating technical condition of the device or the incorrect course of the process. Because also the process parameters have an impact on the durability of the devices themselves. For example, a sudden change in load (unloading or overloading, sudden stop or speed change) all affect the condition of the device and affect the earlier or later the failure will occur.

Drive System
In many cases, the computer control system works with the MCC, its mean with a system of interconnected electric drives. The superordinate system connected with then, for example, sets the speed required for the entire line and all supervision over the drive unit is on inside of the MCC cabinets. Individual motor control is located on the Control Center side. Typically, information about the set speed, current speed and possible errors in the MCC is exchanged. We lose a lot of detailed information on the basic parameters of electric drives. The individual values of voltages and currents are analyzed only in the event of a necessary situation, e.q. an emergency. In real time, these data are in most cases not transmitted to the main computer control system. Correlation of these values with process values increases the scope of analysis and provides ways to create new interdependencies. We can lose a lot energy or lose energy by trying to align situation by increase a power instead of focusing on places that are really an issue [9].

Operator's Log
Beside of the control system, each continuous production plant has a shift operator log. Currently, the journal is not only in the classic paper form, but it can also be an electronic diary of the operator. It is important that it provides information on each daily activity of the service and should cover all typical and non-typical behaviors of the machines, devices and objects that occurred on the given change [10]. Depending on the service, the journal can be run more or less accurately. But almost always there is a problem in its interpretation and conversion into machine language. If we could process a written language into the language of the algorithm, then the possibilities of analyzing would increase significantly. Also, there is a problem with each other interpretation of notes making by one man and reading by other. Another thing is that

for one man occurrence of the issue can be very important and should be in diary. And for another this situation can be irrelevant and even will not mention about it in diary. But this knowledge is very important for rest of plant's crew. General knowledge sharing between people is not measurable and that reason is so hard to transfer to a computer system.

3 Concept of a Multi-domain System

A few threads were taken towards but only the one can be the main. They are: process performance optimization, quality targets, minimalization of raw material usage, chemicals and energy consumption and planning and implementation of annual shut-downs. But first issue is to recognize a big difference between a level of vibration drive rollers comparing different production batches formally of the same product. Trend was shown on Fig. 4.

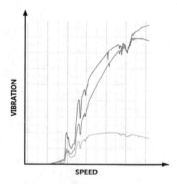

Fig. 4. Various level of vibrations

There are many numerical methods for optimizing production [11] and predicting failure. However, if we are talking about the integration of numerical methods in one central point of the factory, there must be several ways to find mutual correlations in a series of data. To integrate data from many different systems and databases, consider the numerical methods that will be used. According to this, in Fig. 5 KDD principle is shown.

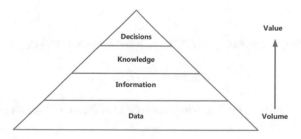

Fig. 5. Dependence of the amount of data to the value (based on [1])

For this reason, it will be necessary to use at least several data sources. For sure the vibration measurement and analysis system, data from the basic control system, data from MCC and laboratory data from the physic-chemical analysis of the raw material. As a complement to the data, the operator's diary will be used to compare the shifts that conduct the production process in analysed period of time. To solve this problem, it is necessary to combine information from several subsystems and databases. Firstly, the data must be prepared for analysis. Then the noise and doubles must be filtered out and the resolution must be unified, and the time stamp synchronized.

The method of discovering sequence patterns can be used as first method. Because is good for wide database searching method. Generally, we are looking for occurring sequences of events. We are looking for simple depending If a then b. Which means that if event A has occurred, there was also an event B and so on. The web structure is used to calculate a list of all possible items [9, 12–14]. On Fig. 6 is shows the web of events for J = {a, b, c, d}. A database that contains n items can theoretical generate up to 2n − 1 number of sequence itemset.

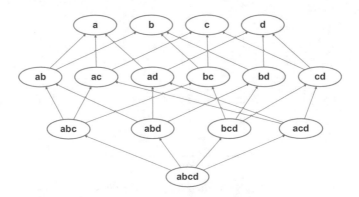

Fig. 6. Web of sequence

Therefore, for the continuous process under consideration, we will accept the state of the process in which it was located when creating a pair of connections. adopted
A   – speed of line
B   – the value of vibration
C   – assortment
D   – shift

Principal Component Analysis (PCA) will be use as method of preparation of data.

$$w_{(k)} = \left(\omega_1, \ldots, \omega_p\right)_{(k)} \tag{1}$$

Because of vibration and frequency analytic also Fourier Transform should be use. Transform is defined as:

$$f(x) = \int_{-\infty}^{\infty} \hat{f}(\varepsilon)e^{2\pi i x \varepsilon} d\varepsilon \tag{2}$$

Fast Fourier Transform in this case can serve as source of typical mechanical analysis. Still can be used for rotating elements to predict the failure [15]. Simple and fast calculation on not big amount of data can online give a briefly result of mechanical state of machine. transforming can be made by very simple CPU without necessary having to go through the calculation multiple times. The potential places of abnormalities are presented on trend as is show on Fig. 7. Another thing is verification of showed places. Next the result of recalculation can be related with calculation coming from another method [16].

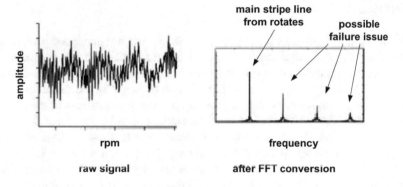

Fig. 7. Result of FFT conversion

This will be supplemented by the FFNN method, because neural networks can find non-linear connections between many inputs and outputs. This is a good method for starting data analysis without a specific model and without preliminary knowledge what is looking for [5, 17]. The results obtained cannot be treated uncritically. The results should be subjected to a verification and verification treatment. With this issue, a survey addressed to the crew of production line will be used as the verification method to exclude obvious dependencies. Because system can return a very simple connection and relationship which are knowing from long period of time and this information will be useless for that purpose. And, to eliminate completely impossible to exist. Experience and knowledge of employee in particular plant is invaluable source for verification of the result of analysis. The questionnaire will be simple questions regarding the situation and the results obtained. The questions will be:

- Do you think that the source of the situation is: bearings, drives, material or service?
- Is the situation inconvenient for smooth running of the process?
- Whether a given situation repeats itself under certain conditions or unpredictable?
- Do you know a way to correct this situation?

Again, there will be a need to compare the results obtained from various sources and their mutual verification. Then the results obtained should be presented in a univocal way to the relevant departments of the plant for their analysis.

This approach allows for earlier detection of failure situation in the process and faster reactions in the certain area of its detection. The operators will have time to prepare preventive actions or change the control of process to avoid of failure at the indicated process point. Eventually if an emergency situation occurs, then it can be possible to prepared for the repair of a specific technological area or particular device. This will enable the preparation of appropriate specialists in advance and the delivery of spare parts for it. Then the shutdown can be controlled, and the stop production is made for a specific activity and the startup is much faster.

4 Summary

In line production there are a lot of places where the data are generated. Proper way of samples acquiring and preparation for analysis should be planned and custom-made. In spite of modern numerical methods, the expert knowledge is still a source of valuable decision and help to make the final steps regarding the obtained calculation results. Currently, there is no implementation enabling automatic regulations of the parameters which will make it possible to safely continue production until the date of planning shut down. Presented approach also allows to increase the effectiveness. This will be done by reducing the number of unforeseen failures. To shorten the time of downtimes through precisely targeted corrective actions is also significant. The two results mentioned above, besides increased production efficiency, give the possibility of quality assurance over a long period of time.

References

1. Wang, X.: Data mining and knowledge discovery for process monitoring and control (1999). Springer, London. https://doi.org/10.1007/978-1-4471-0421-6
2. Vazan, P., Janikova, D., Tanuska, P., Kebisek, M., Cervenanska, Z.: Using data mining methods for manufacturing process control. IFAC PapersOnLine **50**(1), 6178–6183 (2017). https://doi.org/10.1016/j.ifacol.2017.08.986
3. Bolte, C., Kurbel, K., Rautenstrauch, C.: Integration of knowledge-based modules into a distributed production planning and control system. In: Tjoa, A.M., Wagner, R. (eds.) Database and Expert Systems Applications. Springer, Vienna (1990). https://doi.org/10.1007/978-3-7091-7553-8_16
4. Hashemian, H., Bean, W.: State of the art predictive maintenance techniques. IEEE Trans. Instrum. Meas. **60**(10) (2011). https://doi.org/10.1109/tim.2009.2036347
5. Reuter, Ch., Brambring, F., Weirich, J., Kleines, A.: Improving data consistency in production control by adaptation of data mining algorithms. Proc. CIRP **56**, 545–550 (2016). https://doi.org/10.1016/j.procir.2016.10.107
6. Cheng, Y., Chen, K., Sun, H., Zhang, Y., Tao, F.: Data and knowledge mining with big data towards smart production. J. Ind. Inf. Integr. **9** (2018), https://doi.org/10.1016/j.jii.2017.08.001

7. Cwikła, G., Grabowik, C., Kalinowski, K., Paprocka, I., Banas, W.: The initial considerations and tests on the use of real time locating system in manufacturing processes improvement. In: IOP Conference Series: Materials Science and Engineering, vol. 400, p. 042013 (2018). https://doi.org/10.1088/1757-899x/400/4/042013
8. Thornhill, N., Higglund, T.: Detection and diagnosis of oscillation in control loops. Control Eng. Pract. **5**(10), 1343–1354 (1997). https://doi.org/10.1016/S0967-0661(97)00131-7
9. Gökan, M., Barlettab, I., Stahla, B., Taisch, M.: Energy management in production: a novel method to develop key performance indicators for improving energy efficiency. Appl. Energy **149**(1), 46–61 (2015). https://doi.org/10.1016/j.apenergy.2015.03.065
10. Uraikul, V., Chan, Ch., Tontiwachwuthikul, P.: Artificial intelligence for monitoring and supervisory control of process systems. Eng. Appl. Artif. Intell. **20**(2), 115–131 (2006). https://doi.org/10.1016/j.engappai.2006.07.002
11. Bottou, L., Curtis, F., Nocedal, J.: Optimization Methods for Large-Scale Machine Learning (2018). arXiv:1606.04838 [stat.ML], https://doi.org/10.1137/16m1080173
12. Kaur, A., Kaur, K., Malhotra, R.: Soft computing approaches for prediction of software maintenance effort. Int. J. Comput. Appl. **1**(16), 69–75 (2010). https://doi.org/10.5120/339-515
13. Mitrea, C., Lee, C., Wu, Z.: A comparison between neural networks and traditional forecasting methods: a case study. Int. J. Eng. Bus. Manag. **1**(2), 19–24 (2009). https://doi.org/10.5772/6777
14. Harańczyk, G.: Prediction of failures and quality problems. StatSoft Polska (2013, in Polish)
15. Heikkinen, J., Ghalamchi, B., Viitala, R., Sopanen, J., Juhanko, J., Mikkola, A., Kuosmanen, P.: Vibration analysis of paper machine's asymmetric tube roll supported by spherical roller bearings. Mech. Syst. Signal Process. **104**, 688–704 (2018). https://doi.org/10.1016/j.ymssp.2017.11.030
16. Liuab, R., Yangab, B., Ziocd, E., Chenab, X.: Artificial intelligence for fault diagnosis of rotating machinery: a review. Mech. Syst. Signal Process. **108**, 33–47 (2018). https://doi.org/10.1016/j.ymssp.2018.02.016
17. Han, J., Kamber, M., Pei, J.: Data Mining Concepts and Techniques. The Morgan Kaufmann Series in Data Management Systems (2012). https://doi.org/10.1016/c2009-0-61819-5

Assessment of Similarity of Elements as a Basis for Production Costs Estimation

Grzegorz Ćwikła$^{(\boxtimes)}$, Cezary Grabowik, Krzysztof Bańczyk,
and Łukasz Wiecha

Faculty of Mechanical Engineering, Silesian University of Technology,
Gliwice, Poland
grzegorz.cwikla@polsl.pl

Abstract. The paper presents the method of fast production cost estimation based on similarities of elements. In order to describe elements, the shape and characteristic features of elements are encoded into a semantic network. Elements are decomposed into functional surfaces, which are characterised by structural and technological features. Similarities between elements are assessed by comparing element's semantic nets. It is assumed that elements with a high similarity factor also have similar production costs. Results are presented on the example of shafts, numerous runs of c&t similarities assessment at various settings of the weights of the network branches were carried out in order to obtain convergence of the results of assessing similarity using semantic networks with similarities in the actual production costs, calculated by standard methods.

Keywords: Cost estimation · Similarity · Semantic net · Functional surfaces

1 Introduction

Companies forced to compete in a globalized market are forced to look for methods of rapid response to the changing market situation [5, 6]. One of the elements to maintain competitiveness is the speed of response to the offer query from potential customers [12]. The main factor allowing for quick preparation of the offer of production of mechanical parts is the ability to quickly determine the cost of manufacturing, that can be done, based on the experience in the production of similar components [8], whose manufacturing costs have already been determined by precise methods [7, 13].

The method of rapid estimation of manufacturing costs of mechanical components presented in the article is based on the following assumptions [1]:

- The sought-after production cost of a new element is similar to the cost of other, previously manufactured components, assuming that these elements are similar enough, and the structural or technological features of the elements, affecting the production costs have been successfully identified.
- At the initial stage of preparing an offer, we do not always have CAD models of elements to be manufactured, so the method of estimating similarity of elements should not require the CAD model to be prepared only for cost estimation, which is time-consuming and leads to financial losses if the offer is rejected.

© Springer Nature Switzerland AG 2020
F. Martínez Álvarez et al. (Eds.): SOCO 2019, AISC 950, pp. 386–395, 2020.
https://doi.org/10.1007/978-3-030-20055-8_37

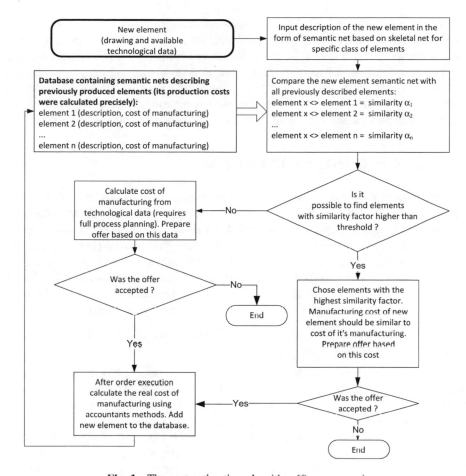

Fig. 1. The cost estimation algorithm [Source: own]

- The company has experience in the production of a specific class of components, and therefore has an extensive database of previously manufactured components, including their precisely calculated production costs.
- The database containing manufacturing costs of elements produced in the past should be updated on a regular basis (taking into account changing prices of materials, labour costs, etc.), so that it is possible to obtain sufficiently accurate data.

According to the proposed algorithm (Fig. 1), the element whose production (machining) cost is to be estimated, should first be described in a specific way (coded) and then compared with elements in the database, whose cost of manufacturing is known by the each-to-each method. If it is possible to find in the database an element whose similarity is high enough, one can assume that the cost of producing a new element will be comparable to the costs of previously produced elements [1]. Therefore, there is a need to develop a set of methods that allow to store the characteristics of the elements and to determine the similarity of the features of the elements [10].

Neural networks are probably the most commonly used method of knowledge representation in artificial intelligence methods, but in this case it was decided, due to the limitations and the higher difficulty of data preparation, to use semantic networks [3].

The assumption of this paper is that a pair of elements having high structural and technological similarity (s&t similarity) has also similar cost of manufacturing. In that case the new element's process planning and calculation of machining times is not required. Features having highest influence on production costs have to be identified.

2 Method of Description of Elements

Semantic network is one of the knowledge representation methods in artificial intelligence applications. This method is applied for determination relations between facts. A fact could be defined as one part of coded element. Semantic network lets create an object oriented description as well as support of the analysis process [11]. Semantic net could be described as a graph. Nodes of the graph correspond to facts, connections between nodes describe relations between them [9]. Semantic network lets to create representation of characteristic features of elements and estimate the similarity factor of these elements. Two ways to describe elementary features of elements are possible:

- technology - oriented features (e.g. step, thread, taper, splines, …),
- geometry - oriented features (e.g. cylinder, cone, …).

Technology-oriented description is more useful, because it lets avoid the loss of information about the structure of an element and functions of surfaces. In geometry oriented description it is possible to split element on elementary surfaces in many different ways. Therefore this method of description is ambiguous [2].

Taking into account the variety of parts that can be produced, elements have to be initially divided on some classes: e.g. shafts, sleeves, disks, etc. The method of describing the features of the element will be presented on the example of shafts.

2.1 Description of Shaft's Technological and Structural Features

Proposed features describing outer form of an element are presented below:

- type of outer form (plain, one side stepped, two sides stepped),
- total length of an element (L_{tot}),
- max. diameter of an element (D_{max}),
- type of a stock (bar, forging, cast),
- expected machining (turning only, or turning and milling, grinding or drilling),
- heat treatment (necessary or not, type of heat treatment),
- grade of material (common steel, high - grade steel, other), workability,
- planned volume of production (single-piece, series or mass production).

In the case of shafts, description of the inner features is very brief, it includes only relative diameter (if exists) of a hole. Shaft's end faces are also described.

2.2 Description of Shaft's Elementary Functional Surfaces

To define elementary surfaces of the shaft the user has to describe every main (1-st level) functional surface, starting from the left (according to specific rule) side of a shaft. The 1-st level elementary surfaces can be as follows: fine-quality step (cylindrical part, requiring turning and grinding), ordinary step (cylindrical part, only turning), thread, splines, gear, taper, non-rotational surface, spherical surface, cam, eccentric. Each of them is defined by the following relatives: length, position and diameter (calculated in relation to the L_{tot} or the D_{max}, accordingly), as well as technological features (e.g. dimensional accuracy, roughness, method of machining).

It is also possible to attach 2-nd level elementary surfaces (e.g. undercut for Seger's ring, splineway, groove, radial hole, one- or two-sided relief) that can belong to the 1-st level surfaces. If the user don't know the exact value of some parameters, it is possible to specify the range of values. The structure of a configurable skeletal net describing shaft is presented in Fig. 2. When describing elements, each of them is saved in a form of a modified and value-filled skeletal semantic network (instance).

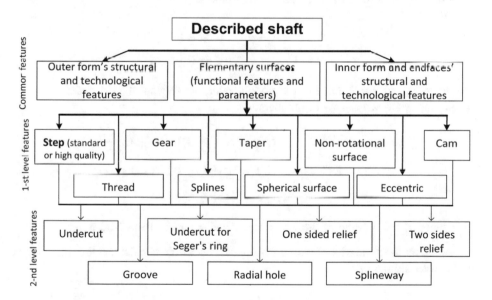

Fig. 2. The skeletal network structure describing a shaft [Source: own]

3 Assessment of Element's Similarity

The determination of the similarity of elements is carried out by comparing the semantic networks (instances) describing these elements in pairs, each to each. The corresponding branches of networks are compared. The comparison process of two nets starts from lowest level and steps up to the top of the network. Then the similarity factor of the compared networks is calculated. If one of the corresponding nodes does not exist, the similarity of these nodes equals zero. Total similarity factor of networks is

based on partial ones, calculated as an average of node similarities. The final result of
the comparison is a number called similarity factor, within [0, 1] range, where 1 means
that elements are identical [4]. Calculation method of the corresponding node similarities depends on the type of node. The value of a node can be a string, a number or a
range of numbers. The formulas (1–4) show the method of calculating the similarity of
selected types of features.

- Similarity of strings:

$$(\alpha = 1 - \text{for identical strings}, \ \alpha = 0 - \text{for different strings}) \tag{1}$$

- Similarity of numbers:

$$\alpha = 1 - \frac{|\alpha_2 - \alpha_1|}{d}, \tag{2}$$

where: α_1, α_2 - values of compared features, d - range of this value.
- Similarity of a number and a range of numbers:

$$
\begin{aligned}
\alpha &= 1 - \frac{|a_2 - a_1''|}{|a_2 - a_1'|} &&, \text{if } \alpha_2 \notin \left[a_1', a_1''\right] \\
\alpha &= 1 &&, \text{if } \alpha_2 \in \left[a_1', a_1''\right]
\end{aligned} \tag{3}
$$

where: α_1', α_1'' - range of the 1-st node's value, α_2 - value of 2-nd node.
- Similarity of two ranges of numbers:

$$
\begin{aligned}
\alpha &= 1 - \frac{|a_2' - a_1''|}{|a_2'' - a_1'|} &&, \text{if } \left[a_1', a_1''\right] \cap \left[a_2', a_2''\right] = 0 \text{ and } \alpha_1' < \alpha_2', \\
\alpha &= 1 - \frac{|a_1' - a_2''|}{|a_1'' - a_2'|} &&, \text{if } \left[a_1', a_1''\right] \cap \left[a_2', a_2''\right] = 0 \text{ and } \alpha_1' < \alpha_2', \\
\alpha &= 1 &&, \text{if } \left[a_1', a_1''\right] \cap \left[a_2', a_2''\right] \neq 0.
\end{aligned} \tag{4}
$$

where: α_1', α_1'' - range of the 1-st node, α_2', α_2'' - range of the 2-nd node.

Based on the similarity of partial features, it is possible to calculate the similarity of
nodes (c&t similarity):

$$\left(\frac{\omega_i}{\omega_k}\right) = \frac{\sum\limits_{j=1}^{n} w_j \alpha_j}{\sum\limits_{j=1}^{n} w_j} \tag{5}$$

where: ω_i, ω_k - compared nodes, n - number of features describing nodes at the specific
level, α - similarity of feature values, w - weight factor.

Calculations are carried out for all nodes, at all levels of the network. Additionally,
every branch of a skeletal network has its weight factor, which allows to decrease or
increase its influence on the overall elements similarity. Thanks to it, this method can
be tuned to the needs of different companies, because technological features influence

on the production cost depends on the organisation of the company. Every company using this method has to tune weights of semantic network branches to get the best correlation between s&t similarity and real production cost similarity. The s&t similarity values obtained in different companies can't be directly compared because the algorithm is tuned to a specific user.

4 Example of Method's Application

The described algorithm was implemented in the Prolog language. The program allows user to describe specific element by asking questions about structural and technological features of an element and letting to chose linguistic or numerical (exact or approximate) values. Decomposition of an element into functional surfaces is similar, requiring user to chose type of surface and then define specific values of surface's structural and technological parameters.

Descriptions of a set of example elements (Fig. 3) available as technical drawings, were entered into the program's database. The real manufacturing costs of these elements were known, they were calculated using the typical methods used in the company from which the data originated.

The real production cost similarity factor was defined to compare cost estimation results with real production costs. This factor is calculated from the Eqs. (6):

$$c = 1 - \frac{|c_1 - c_2|}{c_1}, \quad c_1 > c_2$$
$$c = 1 - \frac{|c_1 - c_2|}{c_2}, \quad c_2 > c_1$$

(6)

where c_1, c_2 − costs of 1-st and 2-nd element.

S&t similarities between all elements and similarities of cost were calculated. Resultant s&t and cost similarities were compared. High similarity of both results means that system of cost estimation works properly. Calculations of s&t similarities were carried out a few times with different values of weight factors. In the first set of calculations (Table 1) all weights were set to 1 (each feature has the same influence on the overall s&t similarity). In next sets of calculations weights of selected technological or structural features were increased or decreased.

The high convergence of s&t similarity factors indicates the possibility of applying the cost of a previously produced item as a basis for determining the cost of manufacturing a new item.

The best results of cost estimation were obtained in fourth set of calculations (Table 2). If similarity factor is near 1, accuracy of this method is satisfactory.

If database does not contain elements with similarity factor higher than threshold factor (in 4-th set of calculations − 0.9), production cost of the new element can't be estimated - accuracy is too low, error of estimation can be greater than ±10%. In this case it is necessary to calculate production cost with conventional methods. After this, new element can be added to database to improve abilities of the whole cost estimation system by enriching the range of produced elements types.

392   G. Ćwikła et al.

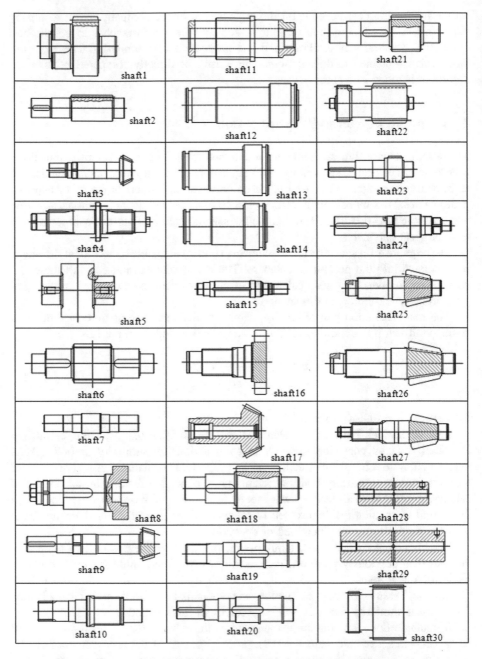

Fig. 3. Shapes of example elements entered into the database [Source: own]

Tables 1 and 2 show the pairs of elements for which the highest s&t similarity factors values were obtained, together with the similarity of the actual production costs, calculated using precise (traditional) methods.

Table 1. Pairs of elements having highest c&t similarity factor in 1-st set of calculations

Element 1	Element 2	C&t similarity factor	Cost similarity factor	Element 1 cost	Element 2 cost	Difference between similarity factors [%]
shaft25	shaft26	0.9930	0.9950	160.62	159.94	0.20
shaft12	shaft13	0.9900	0.9640	189.69	182.87	2.62
shaft13	shaft14	0.9870	0.9370	182.87	171.38	5.07
shaft12	shaft14	0.9810	0.9030	189.69	171.38	7.95
shaft18	shaft21	0.9730	0.9980	83.36	83.21	2.57
shaft1	shaft2	0.9290	0,9760	92.62	87.87	5.06
shaft29	shaft28	0.9140	0,8290	102.18	43.82	9.30
shaft1	shaft21	0.8960	0,925	92.62	83.21	3.24
shaft1	shaft18	0.8930	0,926	92.62	83.36	3.70
shaft25	shaft27	0.8610	0.8690	160.62	139.69	0.92
shaft26	shaft27	0.8600	0.8730	159.94	139.69	1.51
shaft2	shaft21	0.8500	0.9470	87.87	83.21	11.41
shaft2	shaft18	0.8470	0.9480	87.87	83.36	11.92

Table 2. Pairs of elements having highest c&t similarity factor in 4-th set of calculations

Element 1	Element 2	C&t similarity factor	Cost similarity factor	Element 1 cost	Element 2 cost	Difference between similarity factors [%]
shaft25	shaft26	0.9930	0.9950	160.62	159.94	0.20
shaft12	shaft13	0.9800	0.9640	189.69	182.87	1.63
shaft13	shaft14	0.9700	0.9370	182.87	171.38	3.40
shaft18	shaft21	0.9690	0.9980	83.36	83.21	2.99
shaft12	shaft14	0.9590	0.9030	189.69	171.38	5.84
shaft26	shaft27	0.9100	0.8730	159.94	139.69	4.06
shaft25	shaft27	0.9060	0.8690	160.62	139.69	4.08
shaft1	shaft21	0.8930	0,9250	92.62	83.21	3.58
shaft1	shaft18	0.8920	0,9260	92.62	83.36	3.81
shaft19	shaft20	0.8860	0.6820	257.69	377.54	23.02
shaft21	shaft25	0.8700	0.5180	83.21	160.62	40.46
shaft21	shaft26	0.8680	0.5200	83.21	159.94	40.09
shaft21	shaft27	0.8640	0.5950	83.21	139.69	31.13

In the carried out experiments, all tested items were treated as previously manufactured, for many of them it was still impossible to find any elements similar enough to draw conclusions about their production costs, which indicates the need to have a large database of previously manufactured elements in order to obtain satisfactory system performance.

5 Summary

Presented method allows to estimate element's production costs without the need to create it's exact CAD model, what is costly and time-consuming, and there is no guarantee, that this effort will gain any profit in the future.

Element description is simple but a lot of data (linguistic and numerical) have to be entered into the program. Still, the time needed to create a semantic network instance is much shorter than the CAD model, especially for more complex elements. Proposed method lets to avoid the costly development of the CAD model and the complete process planning for the new element. The CAD model (and technological process plan) will only have to be created when the production order is accepted, based on the offer prepared using simplified model in the form of a semantic network instance.

Any technical drawing, sketch or even an existing element or its physical model (e.g. an old element disassembled from a machine) can be the basis for entering a description into the database, which is a big advantage when there is no access to the 3D CAD model. On the other hand, if a 3D CAD model already exists, entering the element into the database requires exactly the same steps as in the case of its absence, resulting in additional work. It seems reasonable to create additional software that allows for quick translation from the 3D CAD model to the form of the skeletal semantic network instance.

The proposed method takes into account the quantitative and qualitative features of elements. It is possible to compare elements belonging to selected constructional class such as shafts using the semantic net. In the developed method, comparing elements belonging to different classes is not possible, a separate skeletal semantic network should be created for each element class, e.g. disks, levers and sleeves. The network structure and weights must be closely matched to the needs and characteristics of the company by which the method is used. In the case of classes with very different elements, it may be necessary to develop separate networks for individual subclasses, in order to get more detailed results.

The time of obtaining a comparison results depends strongly on the number of elements in the database, in this case (30 elements) it was about 1–3 s. This allows us to name the proposed method of estimating the cost as "fast".

The identification of the factors that have the strongest influence on the production costs in the company should be carried out in order to fine tune the weight system, reinforcing the impact of selected, significant features on the final result and weakening the impact of other features. The conducted research, mainly comparing the s&t similarity and real costs similarity, showed that it is possible to obtain, after tuning of the weight system, satisfactory convergence of these factors, and, consequently, correct results of production costs estimation. Ability of changing the semantic network weights factors allows to fit the system to the needs of specific user. Extensive database with descriptions of elements is necessary for proper system operation. Accuracy of cost estimation is proportional to s&t similarity factor, the more elements are in the database, the bigger is probability that base contains element similar to the new, that accurate estimation of cost will be possible.

References

1. Ćwikła, G., Knosala, R.: The cost estimation method basing on similarities of elements. In: CO–MAT–TECH 1997, Trnava, pp. 195–200 (1997)
2. Davidrajuh, R., Skolud, B., Krenczyk, D.: Performance evaluation of discrete event systems with GPenSIM. Computers **7**(1), 8 (2018). https://doi.org/10.3390/computers7010008
3. Gwiazda, A., Ćwikła, G.: Qualitative methods of elements description for classification systems. Proc. Int. Conf. Comput. Integr. Manuf. Zakop. **1996**, 147–154 (1996)
4. Knosala, R.: Methoden zur Bewertung von Bauelementen als Voraussetzung für die Entwicklung von Baukastensystemen. Dissertation B, TU Dresden (1989)
5. Paprocka, I.: The model of maintenance planning and production scheduling for maximizing robustness. Int. J. Prod. Res. (2018). https://doi.org/10.1080/00207543.2018.1492752
6. Paprocka, I.: Evaluation of the effects of a machine failure on the robustness of a job shop system - proactive approaches. Sustainability **11**(1), 65 (2019). https://doi.org/10.3390/su11010065
7. Roy, R., Souchoroukov, P., Shehab, E.: Detailed cost estimating in the automotive industry: data and information requirements. Int. J. Prod. Econ. **133**, 694–707 (2011)
8. Salmi, A., David, P., Blanco, E., Summers, J.D.: A review of cost estimation models for determining assembly automation level. Comput. Ind. Eng. **98**, 246–259 (2016)
9. Song, S., Lin, Y., Guo, B., Di, Q., Lv, R.: Scalable distributed semantic network for knowledge management in cyber physical system. J. Parallel Distrib. Comput. **118**(Part 1), 22–33 (2018)
10. Skolud, B., Krenczyk, D., Davidrajuh, R.: Solving repetitive production planning problems. An approach based on activity-oriented Petri nets. In: Graña, M., López-Guede, J., Etxaniz, O., Herrero, Á., Quintián, H., Corchado, E. (eds.) International Joint Conference SOCO'16-CISIS'16-ICEUTE'16, SOCO 2016, ICEUTE 2016, CISIS 2016. Advances in Intelligent Systems and Computing, vol. 527, pp. 397–407 (2017), https://doi.org/10.1007/978-3-319-47364-2_38
11. Sowa, J.F.: Principles of Semantic Networks: Explorations in the Representation of Knowledge. Elsevier, Amsterdam (1991)
12. Więcek, D., Więcek, D.: Production costs of machine elements estimated in the design phase. In: Burduk, A., Mazurkiewicz, D. (eds.) Intelligent Systems in Production Engineering and Maintenance – ISPEM 2017, ISPEM 2017. Advances in Intelligent Systems and Computing, vol 637. Springer (2018)
13. Więcek, D., Więcek, D.: The influence of the methods of determining cost drivers values on the accuracy of costs estimation of the designed machine elements. In: Wilimowska, Z., Borzemski, L., Świątek, J. (eds.) Information Systems Architecture and Technology: Proceedings of 38th International Conference on Information Systems Architecture and Technology – ISAT 2017, ISAT 2017. Advances in Intelligent Systems and Computing, vol 657. Springer (2018)

Special Session - Soft Computing Applications in the Field of Industrial and Environmental Enterprises

Outlier Generation and Anomaly Detection Based on Intelligent One-Class Techniques over a Bicomponent Mixing System

Esteban Jove[1,2(✉)], José-Luis Casteleiro-Roca[1], Héctor Quintián[1],
Juan Albino Méndez-Pérez[2], and José Luis Calvo-Rolle[1]

[1] Department of Industrial Engineering, University of A Coruña,
Avda. 19 de febrero s/n, 15405 Ferrol, A Coruña, Spain
esteban.jove@udc.es
[2] Department of Computer Science and System Engineering,
Universidad de La Laguna, Avda. Astrof. Francisco Sánchez s/n,
38200 S/C de Tenerife, Spain
jamendez@ull.edu.es

Abstract. One of the most important points to improve the profits in an industrial process lies on the fact of achieving a good optimisation and applying a smart maintenance plan. Under this circumstances an early anomaly plays an important role. Then, the implementation of classifiers for anomaly detection is an important challenge. As many of the anomalies that can occur in a plant have an unknown behaviour, it is necessary to generate artificial outliers to check these classifiers. This work presents different one-class intelligent techniques to perform anomaly detection in an industrial facility, used to obtain the main material for wind generator blades production. Furthermore, artificial anomaly data are generated to check the performance of each technique. The final results achieved are successful in general terms.

Keywords: Anomaly detection · Control system · Outlier generation

1 Introduction

In general terms, a high percentage of enterprises presents complex and expensive processes whose operation can be optimized [1,18]. The technological advances achieved in many different industrial fields, such as instrumentation or automation has led to the optimisation and development of industrial processes [15]. In addition, factors like of the promotion of energy efficiency policies or the increased competitiveness must be taken into consideration in any plant optimisation [20,40]. Then, a good operation in a specific process requires an early detection of any deviation from its normal operation of actuators, sensors, and so on [12,24]. This is specially important in high cost and safe-critical processes [9,28]. From an economic point of view, avoiding a wrong performance of any

© Springer Nature Switzerland AG 2020
F. Martínez Álvarez et al. (Eds.): SOCO 2019, AISC 950, pp. 399–410, 2020.
https://doi.org/10.1007/978-3-030-20055-8_38

part of a facility, can lead to economic savings in terms of energy consumption, raw material waste and corrective maintenance.

From a theoretical point of view, a data pattern that represents an unexpected behaviour is defined as an anomaly [7]. In a specific application, the anomaly detection task must tackle some issues like selecting a limit between normal and anomaly data, the availability of data during anomaly operation or the appearance of noise [7].

The previous information about the initial dataset can lead to three different cases of anomaly detection [7,16,38]:

- Type A: in this case, the anomaly detection classifier is obtained using only data from normal operation. In this case, semi-supervised techniques are applied since data is pre-classified as normal. This kind of classification is known by one-class.
- Type B: the nature of the initial dataset is not known. Then, unsupervised techniques are used to classify data between normal and anomaly data.
- Type C: the initial dataset is composed by pre-labelled normal and anomalous data, so supervised algorithms are used in this case to model the system [6,30]

The use of anomaly detection techniques is widely spread in the field of fault detection in industrial processes, fraud detection of credit cards, intrusion detection in surveillance systems or even in medicine diagnosis, where it could represent a helpful tool to, for instance, reduce the workload of the medical staff [13,22,22].

Given the difficulty of obtaining data from faulty situations, checking the performance of a one classifier may require the generation of artificial outliers. Previous works related to this task deals the outlier generation by using linear transformation between variables [29] or setting normal data geometrically out of the target set [37].

This works presents the anomaly generation and detection of an industrial plant, consisting of a bicomponent mixing system used on manufacturing process of wind generator blades. In this case, different intelligent one-class classifiers are implemented and assessed in the anomaly detection task using real data, obtained from the right plant operation is used. The techniques used are: the Approximate Convex Hull (ACH), Autoencoder and Support Vector Machine (SVM). Then, the proposed classifiers are assessed and validated using generated artificially anomalies using the Boundary Value Method [37].

This paper is structured as follows. After the present section, The case of study is introduced. Then, Sect. 3 explains the used techniques to generate the outliers and obtain the classifiers. Section 4 describes the experiments and results and finally, the conclusions and future works are presented in Sect. 5.

2   Case of Study

2.1   Bicomponent Mixing System

The installation used to develop the present work aims at mixing two different fluids to obtain a bicomponent material, whose features are suitable for wind generator blades. The primary fluids, the epoxy resin and the catalyst, are stored in two different tanks. They are boosted by two independent pumps, supplied by variable frequency drives, and mixed in a valve in charge of delivering the proper amount of bicomponent at the output [17].

Figure 1 shows a simplified scheme of the above explanation with the actuators, valves, pipes and sensors of the mixing system.

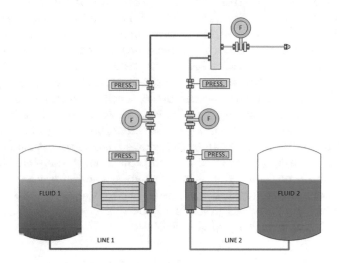

Fig. 1. Process scheme

The primary fluids to be pumped have Non-Newtonian characteristics, which means that their properties vary depending on the mechanic efforts applied on them [10]. Factors like the viscosity variation with the pump speed or the pump efficiency, make this system more sophisticated, and hence, in this context, the anomaly detection is a challenge.

2.2   Dataset Description

During the right operation of this industrial plant, the values of ten variables were registered and used as the initial dataset. This variables are: the real value and set point of the bicomponent flow rate at the output, the flow rate of fluids 1 and 2, the pressure and speeds of pumps 1 and 2 and the pressure after the flowmeter 1 and 2. After inspecting the initial dataset, the null measurements were removed from the initial 9511 samples, and finally, 8549 are considered.

3   Techniques Applied to Validate the Proposed Model

This study assesses three different one-class classification techniques, that are briefly described in Subsect. 3.1. Then, to validate the anomaly detection, outliers are artificially generated according to the procedure shown on Subsect. 3.2. To assess the performance of each classifier, each one was checked using a $k - fold$ cross-validation with $k = 10$ that was repeated 3 times.

3.1   One-Class Techniques

Approximate Convex Hull. The Approximate Convex Hull is a one-class classification technique, that offered successful results in previous works [4,11]. The basic idea of this method is to approximate the limits of a dataset $S \in \mathbb{R}^n$ using its convex hull. Due to the fact that the convex limits of a dataset with N samples and d variables has a computational cost of $O(N^{(d/2)+1})$ [4], the convex hull is approximated using p random projections on $2D$ planes and determine their convex limits on that plane. Then, after modelling the convex hull of the training data making p projections, when a new test sample arrives, it is considered as an anomaly if it is out of the convex hull in any of the projections.

In addition, a parameter λ can be defined as a factor that expands of reduces the convex limits from the centroid of each projection. A value of λ higher than 1 expands the limits, while if it is lower than 1, the limits are contracted. A proper value of the parameter λ depends on the nature of the dataset.

An example of an anomaly point in $\mathbb{R}^3$ space is shown in Fig. 2. In this case, the anomaly is out of just one of the projections.

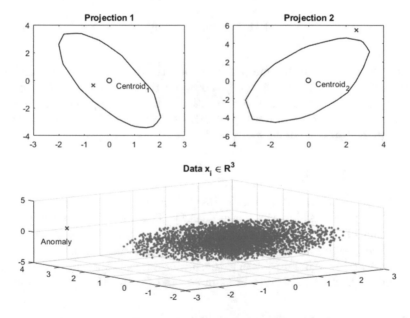

Fig. 2. Anomaly point in $\mathbb{R}^3$

Artificial Neural Networks. Autoencoder. The use of Artificial Neural Networks (ANN) for one-class using an Autoencoder configuration, has obtained very good results in different applications, such as anomaly detection or data noise reduction [14,33,36].

The most common ANN for this task is the Multilayer Perceptron (MLP) [39], which has an input layer, an output layer and a hidden layer. Each layer is composed by neurons connected with weighted links between adjacent layers, which have nonlinear activation functions [3,5,19]. The approach of Autoencoder technique is to reconstruct the patterns from the input to the output using an intermediate nonlinear dimensional reduction in the hidden layer. Then, the number of input neurons and output neurons are the same as the variables of the dataset, and the hidden layer must have at least one less neuron.

The intermediate nonlinear reduction acts like a filter, that eliminates the data that is not consistent with the dataset. Then, anomaly data must lead to high reconstruction error, which is the difference between the input and the output estimated by the MLP.

Support Vector Machine. The Support Vector Machine (SVM) is used in regression and classification tasks [8,21,23]. In one class classifications, this technique maps the dataset into a high dimensional space by means of a kernel function. In this space, a hyper-plane that maximises the distance between the data and the origin is implemented [32].

When the SVM is trained, the criteria to detect if a test data is an anomaly, is based on the distance of that point to the hyper-plane. If the distance is negative, the data is inside the target class and is considered as an anomaly otherwise.

3.2   Artificial Outlier Generation

This paper assesses the performance of three different one class techniques on the task of anomaly detection of an industrial plant, used to obtain the wind generator blades material. Then, it is necessary to generate artificial outliers, that simulate deviations on the plant operation. For this purpose, the Boundary Value Method [37] is applied in this case. For instance, from a dataset with M samples and n attributes, this technique generates anomalous points shifting the samples out of the training dataset boundary. The process followed to generate an artificial outlier from a specific data point $a \in \mathbb{R}^n$ is described next:

1. Find out the maximum and minimum of each attribute in the initial dataset and save these values in two vectors $V_{max}, V_{min} \in \mathbb{R}^n$.
2. Select randomly 2 dimensions u and v of the n-dimensional dataset.
3. Replace the value of $a(u)$ randomly by $V_{max}(u)$ or $V_{min}(u)$.
4. Replace the value of $a(v)$ randomly by $V_{max}(v)$ or $V_{min}(v)$.

The example of Fig. 3 shows the transformation of a point inside the target class (green point) in $\mathbb{R}^3$ into an anomaly. The transformation Tx replaces the x coordinate by the maximum registered (yellow dot), and the transformation

Ty changes the y value by its minimum (red dot). Then, it can be noticed that the data generated is out of the initial set.

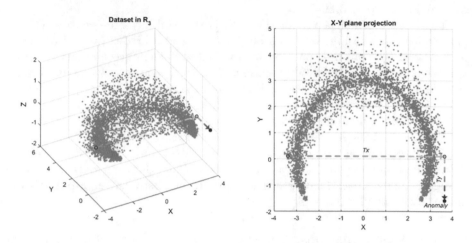

Fig. 3. Anomaly point in $\mathbb{R}^3$

4   Experiments and Results

From the initial dataset, corresponding to the right operation of the plant, 5 % of the samples were artificially modified to transform them into anomalies, following the process described in Sect. 3.2. For each sample, two dimensions are randomly selected and replaced by its maximum or minimum values. The one class techniques, used to assess the anomaly generation, were validated using a 3 times nested $k - fold$ with $k = 10$. The performance of each classifier is assessed using the Area Under Curve (AUC) parameter, that relates the true positives and false positives [2]. In addition, the standard deviation (SD) of the AUC obtained for each repetition and the training time is registered.

4.1   Approximate Convex Hull Classifier

The performance of this technique was evaluated using 5, 10, 50, 100, 500 and 1000 $2D$ projections with a λ value of 0.9, 1 and 1.1. Given the geometric nature of this technique, the dataset was not normalised. The AUC, the SD and the training time for each configuration can be seen in Table 1. The obtained results with $\lambda = 0.9$ were not successful and were omitted due to the wrong performance of this configuration.

Table 1. Results obtained with ACH classifier

Parameter λ	Projections	AUC (%)	STD (%)	Time (min)
1	5	97,09	0,13	0,005
	10	98,34	0,16	0,013
	50	98,90	0,21	0,058
	100	98,74	0,32	0,109
	500	98,35	0,34	0,560
	1000	98,40	0,42	1,114
1,1	5	93,51	0,02	0,004
	10	96,19	0,00	0,007
	50	98,67	0,07	0,065
	100	99,00	0,08	0,103
	500	99,25	0,08	0,563
	1000	99,20	0,08	1,116

4.2 Artificial Neural Network Autoencoder Classifier

The MLP Autoencoder was obtained using the Matlab function *train Autoencoder* [25]. As mentioned in Sect. 3.1, the number of neurons in the hidden layer must be less than the number of inputs. Then, to check the importance of this parameter, it is varied from 1 to 9 with a step of 2 neurons. Different configuration were tried regarding the normalisation, first with a 0 to 1 normalisation, then with z-score [35] and the results without normalisation were also checked. The criteria to select if a test data is an anomaly, is based on the reconstruction error. If the reconstruction error is higher than the one obtained with 99% of the training set, it is considered as an anomaly. The results in terms of AUC, SD and training cost for each configuration is shown in Table 2.

4.3 SVM Classifier

The SVM one-class classifier was obtained using the Matlab function *fitcsvm* [26]. The kernel function was set as Gaussian, the percentage of outlier fraction of the training data was varied from 0 to 3. The influence of the normalisation methods was assessed as for Autoencoder. As mentioned in Sect. 3.1, the criteria to decide if a test data is an anomaly is based on its distance to the decision plane. This is evaluated with the Matlab function *predict* [27]. The results are presented in Table 3

Table 2. Results obtained with autoencoder classifier

Normalisation	Number of neurons	AUC (%)	STD (%)	Time (min)
NoNorm	1	80,82	0,60	124,050
	3	96,13	0,32	210,016
	5	97,20	0,85	225,571
	7	97,49	0,25	224,424
	9	97,56	0,25	259,265
0 to 1	1	87,27	0,20	0,341
	3	88,50	0,21	1,520
	5	88,34	0,16	1,234
	7	88,49	0,28	1,428
	9	88,35	0,18	1,261
Zscore	1	91,42	0,24	1,168
	3	93,28	0,44	10,527
	5	94,55	0,82	8,285
	7	95,03	0,29	10,175
	9	95,03	0,22	11,772

Table 3. Results obtained with SVM

Normalisation	Outlier frac. (%)	AUC (%)	STD (%)	Time (min)
NoNorm	0	95,82	0,14	2,755
	1	95,32	0,35	2,580
	2	94,72	0,54	2,631
	3	94,53	0,54	2,621
0 to 1	0	95,79	0,15	2,571
	1	95,39	0,33	2,613
	2	94,83	0,60	2,582
	3	94,53	0,60	2,598
Zscore	0	95,83	0,13	2,598
	1	95,42	0,36	2,616
	2	94,90	0,53	2,586
	3	94,52	0,56	2,662

5   Conclusions and Future Works

This work assesses the artificial anomaly generation in an industrial plant used to obtain wind generator blades material with the aim of check fault detection methods. The proposal to generate anomalies shows very good results. This is shown as an extremely useful method to generate different anomaly situations that help to check the performance of different one-class techniques. The implementation of the classification techniques proposed in an industrial process, would allow its manager to detect slight variations in it. Hence, this information can be used for different tasks, such as predictive maintenance planning or corrective maintenance management, among others.

The outlier data generated were used to train different one-class classifiers. The best AUC (99.25%) value is achieved using Approximate Convex Hull with $\lambda = 1.1$ and 500 projections. However, the computational cost must be taken into consideration. In that case, this technique achieves almost the same AUC (98.90%) with $\lambda = 1$ and 50 projections with a significantly lower computational cost (0.058 min). Autoencoder offers good AUC results in general terms. However, the kind of normalisation has a high influence over the training time. In addition, an increase of the number of neurons in the hidden layer leads to a better performance in AUC and higher computational cost. Finally, the results obtained with SVM shows good performance regardless the kind of normalisation, or outlier fraction.

Taking into account the results of the three techniques, the Approximate Convex Hull with 50 projections and $\lambda = 1$ is chosen as the optimal solution.

As future works, the use of different techniques to generate anomalies can be considered. Due to the fact that the plant evolves during right operation, a re-train process can be performed to update the classifiers. The possibility of applying clustering algorithms to the initial dataset could be taken into consideration. To reduce training time, the correlation between variables can be assessed using dimensional reduction [31,34]. The one-class classifiers proposed detect anomalies in the system with good performance. However, the source of the anomaly is not identified concretely. Then, in future works, a specific one-class classifier could be implemented for each facility point.

References

1. Alaiz Moretón, H., Calvo Rolle, J., García, I., Alonso Alvarez, A.: Formalization and practical implementation of a conceptual model for PID controller tuning. Asian J. Control **13**(6), 773–784 (2011)
2. Bradley, A.P.: The use of the area under the ROC curve in the evaluation of machine learning algorithms. Pattern Recognit. **30**(7), 1145–1159 (1997)
3. Calvo-Rolle, J.L., Quintian-Pardo, H., Corchado, E., del Carmen Meizoso-López, M., García, R.F.: Simplified method based on an intelligent model to obtain the extinction angle of the current for a single-phase half wave controlled rectifier with resistive and inductive load. J. Appl. Log. **13**(1), 37–47 (2015)

4. Casale, P., Pujol, O., Radeva, P.: Approximate convex hulls family for one-class classification. In: International Workshop on Multiple Classifier Systems, pp. 106–115. Springer (2011)
5. Casteleiro-Roca, J.L., Barragán, A.J., Segura, F., Calvo-Rolle, J.L., Andújar, J.M.: Fuel cell output current prediction with a hybrid intelligent system. Complexity (2019)
6. Casteleiro-Roca, J.L., Jove, E., Sánchez-Lasheras, F., Méndez-Pérez, J.A., Calvo-Rolle, J.L., de Cos Juez, F.J.: Power cell soc modelling for intelligent virtual sensor implementation. J. Sens. (2017)
7. Chandola, V., Banerjee, A., Kumar, V.: Anomaly detection: a survey. ACM Comput. Surv. (CSUR) **41**(3), 15 (2009)
8. Chen, Y., Zhou, X.S., Huang, T.S.: One-class SVM for learning in image retrieval. In: Proceedings of 2001 International Conference on Image Processing, vol. 1, pp. 34–37. IEEE (2001)
9. Chiang, L.H., Russell, E.L., Braatz, R.D.: Fault Detection and Diagnosis in Industrial Systems. Springer, Heidelberg (2000)
10. Fan, H., Wong, C., Yuen, M.F.: Prediction of material properties of epoxy materials using molecular dynamic simulation. In: 7th International Conference on Thermal, Mechanical and Multiphysics Simulation and Experiments in Micro-Electronics and Micro-Systems, EuroSime 2006, pp. 1–4, April 2006
11. Fernández-Francos, D., Fontenla-Romero, Ó., Alonso-Betanzos, A.: One-class convex hull-based algorithm for classification in distributed environments. IEEE Trans. Syst. Man Cybern.: Syst. **99**, 1–11 (2018)
12. Garcia, R.F., Rolle, J.L.C., Castelo, J.P., Gomez, M.R.: On the monitoring task of solar thermal fluid transfer systems using NN based models and rule based techniques. Eng. Appl. Artif. Intell. **27**, 129 – 136 (2014). http://www.sciencedirect.com/science/article/pii/S0952197613001127
13. González, G., Angelo, C.D., Forchetti, D., Aligia, D.: Diagnóstico de fallas en el convertidor del rotor en generadores de inducción con rotor bobinado. Revista Iberoamericana de Automática e Informática industrial **15**(3), 297–308 (2018). https://polipapers.upv.es/index.php/RIAI/article/view/9042
14. Goodfellow, I., Bengio, Y., Courville, A., Bengio, Y.: Deep Learning, vol. 1. MIT Press, Cambridge (2016)
15. Hobday, M.: Product complexity, innovation and industrial organisation. Res. Policy **26**(6), 689–710 (1998)
16. Hodge, V., Austin, J.: A survey of outlier detection methodologies. Artif. Intell. Rev. **22**(2), 85–126 (2004)
17. Jove, E., Aláiz-Moretón, H., Casteleiro-Roca, J.L., Corchado, E., Calvo-Rolle, J.L.: Modeling of bicomponent mixing system used in the manufacture of wind generator blades. In: Corchado, E., Lozano, J.A., Quintián, H., Yin, H. (eds.) Intelligent Data Engineering and Automated Learning - IDEAL 2014, pp. 275–285. Springer, Cham (2014)
18. Jove, E., Alaiz-Moretón, H., García-Rodríguez, I., Benavides-Cuellar, C., Casteleiro-Roca, J.L., Calvo-Rolle, J.L.: PID-ITS: an intelligent tutoring system for PID tuning learning process. In: International Joint Conference SOCO 2017-CISIS 2017-ICEUTE 2017, León, Spain, 6–8 September 2017, pp. 726–735. Springer (2017)
19. Jove, E., Antonio Lopez-Vazquez, J., Isabel Fernandez-Ibanez, M., Casteleiro-Roca, J.L., Luis Calvo-Rolle, J.: Hybrid intelligent system to predict the individual academic performance of engineering students. Int. J. Eng. Educ. **34**(3), 895–904 (2018)

20. Jove, E., Casteleiro-Roca, J.L., Quintián, H., Méndez-Pérez, J.A., Calvo-Rolle, J.L.: A new approach for system malfunctioning over an industrial system control loop based on unsupervised techniques. In: Graña, M., López-Guede, J.M., Etxaniz, O., Herrero, Á., Sáez, J.A., Quintián, H., Corchado, E. (eds.) International Joint Conference SOCO'18-CISIS'18-ICEUTE'18, pp. 415–425. Springer, Cham (2018)
21. Jove, E., Gonzalez-Cava, J.M., Casteleiro-Roca, J.L., Méndez-Pérez, J.A., Antonio Reboso-Morales, J., Javier Pérez-Castelo, F., Javier de Cos Juez, F., Luis Calvo-Rolle, J.: Modelling the hypnotic patient response in general anaesthesia using intelligent models. Log. J. IGPL **27**, 189–201 (2018)
22. Moreno-Fernandez-de Leceta, A., Lopez-Guede, J.M., Ezquerro Insagurbe, L., Ruiz de Arbulo, N., Graña, M.: A novel methodology for clinical semantic annotations assessment. Log. J. IGPL **26**(6), 569–580 (2018). https://doi.org/10.1093/jigpal/jzy021
23. Li, K.L., Huang, H.K., Tian, S.F., Xu, W.: Improving one-class SVM for anomaly detection. In: 2003 International Conference on Machine Learning and Cybernetics, vol. 5, pp. 3077–3081. IEEE (2003)
24. Manuel Vilar-Martinez, X., Aurelio Montero-Sousa, J., Luis Calvo-Rolle, J., Luis Casteleiro-Roca, J.: Expert system development to assist on the verification of "TACAN" system performance. Dyna **89**(1), 112–121 (2014)
25. MathWorks: Autoencoder. https://es.mathworks.com/help/deeplearning/ref/trainautoencoder.html. Accessed 29 Jan 2019
26. MathWorks: fitcsvm. https://es.mathworks.com/help/stats/fitcsvm.html. Accessed 29 Jan 2019
27. MathWorks: predict. https://es.mathworks.com/help/stats/classreg.learning.classif.compactclassificationsvm.predict.html. Accessed 29 Jan 2019
28. Miljković, D.: Fault detection methods: a literature survey. In: 2011 Proceedings of the 34th International Convention on MIPRO, pp. 750–755. IEEE (2011)
29. Pei, Y., Zaïane, O.: A synthetic data generator for clustering and outlier analysis. University of Alberta, edmonton, AB, Canada, Department of Computing science (2006)
30. Quintián, H., Casteleiro-Roca, J.L., Perez-Castelo, F.J., Calvo-Rolle, J.L., Corchado, E.: Hybrid intelligent model for fault detection of a lithium iron phosphate power cell used in electric vehicles. In: International Conference on Hybrid Artificial Intelligence Systems, pp. 751–762. Springer (2016)
31. Quintián, H., Corchado, E.: Beta scale invariant map. Eng. Appl. Artif. Intell. **59**, 218–235 (2017)
32. Rebentrost, P., Mohseni, M., Lloyd, S.: Quantum support vector machine for big data classification. Phys. Rev. Lett. **113**, 130503 (2014). https://doi.org/10.1103/PhysRevLett.113.130503
33. Sakurada, M., Yairi, T.: Anomaly detection using autoencoders with nonlinear dimensionality reduction. In: Proceedings of the MLSDA 2014 2nd Workshop on Machine Learning for Sensory Data Analysis, p. 4. ACM (2014)
34. Segovia, F., Górriz, J.M., Ramírez, J., Martinez-Murcia, F.J., García-Pérez, M.: Using deep neural networks along with dimensionality reduction techniques to assist the diagnosis of neurodegenerative disorders. Log. J. IGPL **26**(6), 618–628 (2018). https://doi.org/10.1093/jigpal/jzy026
35. Shalabi, L.A., Shaaban, Z.: Normalization as a preprocessing engine for data mining and the approach of preference matrix. In: 2006 International Conference on Dependability of Computer Systems, pp. 207–214, May 2006

36. Vincent, P., Larochelle, H., Lajoie, I., Bengio, Y., Manzagol, P.A.: Stacked denoising autoencoders: learning useful representations in a deep network with a local denoising criterion. J. Mach. Learn. Res. **11**(Dec), 3371–3408 (2010)
37. Wang, C.K., Ting, Y., Liu, Y.H., Hariyanto, G.: A novel approach to generate artificial outliers for support vector data description. In: IEEE International Symposium on Industrial Electronics, ISIE 2009, pp. 2202–2207. IEEE (2009)
38. Wojciechowski, S.: A comparison of classification strategies in rule-based classifiers. Log. J. IGPL **26**(1), 29–46 (2018). https://doi.org/10.1093/jigpal/jzx05
39. Zeng, Z., Wang, J.: Advances in Neural Network Research and Applications, 1st edn. Springer Publishing Company, Heidelberg (2010). Incorporated
40. Zotes, F.A., Peñas, M.S.: Heuristic optimization of interplanetary trajectories in aerospace missions. Revista Iberoamericana de Automática e Informática Industrial RIAI **14**(1), 1–15 (2017). http://www.sciencedirect.com/science/article/pii/S1697791216300486

Material Flow Optimization Using Milk Run System in Automotive Industry

Dragan Simić[1]([⊠]), Vasa Svirčević[2], Vladimir Ilin[1],
Svetislav D. Simić[1], and Svetlana Simić[3]

[1] Faculty of Technical Sciences, University of Novi Sad,
Trg Dositeja Obradovića 6, 21000 Novi Sad, Serbia
dsimic@eunet.rs,
{dsimic,v.ilin,simicsvetislav}@uns.ac.rs
[2] Lear d.o.o, 21000 Novi Sad, Serbia
vasasv@hotmail.com
[3] Faculty of Medicine, University of Novi Sad,
Hajduk Veljkova 1-9, 21000 Novi Sad, Serbia
svetlana.simic@mf.uns.ac.rs

Abstract. Material flow can be characterized as an organized flow of material in the production process with the required sequence determined by the technological procedure. This paper presents biological swarm intelligence in general, and, particle swarm optimization for modelling material flow optimization using milk run system in production system of automotive industry. The aim of this research is to create model to optimize route period and number of trails for one train considering layout and space constraints.

Keywords: Milk run · Material flow · Particle swarm optimization

1 Introduction

Material flow can be characterized as an organized flow of material in the production process with the required sequence determined by the technological procedure. It is a summary of operations presented by material conveying, storage, packaging and weighing, and technological manipulations and works directly related to the production process. It also includes all kinds of auxiliary materials required for production process realization. Planning and dimensioning material flow challenges, which are difficult to overcome, especially in scenarios characterized by many hard constraints and by well-established processes [1].

Milk run, as one of the most efficient and popular models to improve logistic operations in automotive industry, food distribution, and military. It can effectively reduce inventory, while increasing the material flow between warehouse and working places and the usage rate of the vehicles, therefore decreasing the cost and improving the productivity. The use of milk-run system is based on tugger trains for industrial material transportation and has shown three important tendencies: (1) tugger train systems are being mainly adopted for material supply; (2) the adoption of tugger train

© Springer Nature Switzerland AG 2020
F. Martínez Álvarez et al. (Eds.): SOCO 2019, AISC 950, pp. 411–421, 2020.
https://doi.org/10.1007/978-3-030-20055-8_39

systems has recently increased following a linear trend and (3) the use of tugger train systems beyond the automotive sector is consistently growing [2].

This paper presents biological swarm intelligence in general, and particularly model: particle swarm optimization (PSO) for modelling material flow using milk run system in automotive industry. The aim of this research is to create model to minimize production cost, when production is given on demand while considering the price of items and inventory keeping cost. This research continuous the authors' previous researches in supply chain management, and inventory management system presented in [3–6].

The rest of the paper is organized in the following way: Sect. 2 overviews the material flow management and related work selection. Section 3 shows PSO implemented in material flow. This section also describes used dataset. Experimental results and discussion are presented in Sect. 4 and finally, Sect. 5 gives concluding remarks.

2 Material Flow and Related Work

The term material flow covers a broad spectrum of methods and approaches in the literature. In general, material flow refers to the analysis and specific optimisation of material, energy and information flows that arise during the manufacturing of row materials to final products and provision of services. Most contemporary organisations are facing many dilemmas, the essence of which is the determination of the ways of acquisition or maintenance of their position in the competitive environment.

The strategic aim of all production enterprises is to achieve the highest possible profit, for which they use attributes of lean production that are related to and support the reduction of production costs. By using theoretical knowledge and a detailed analysis of a material flow including mapping of individual production processes it is possible to obtain the necessary data sets identifying savings options by increasing efficiency, including cost savings. Methods of modelling and simulation used for the optimization of material flow are presented in [7].

One of the ways to get competitive prevalence is by analysing and optimizing material flow which contributes to the reduction of the company costs. Material flow in a production organization shows how the system is organised. In [8] the current state of material flow in a production company of the automotive branch is analysed. The material flow optimization has been proposed, and it consists of creation of a program of internal transport and optimization of the location of the working stands. That paper describes the current problems of the company's internal transport and indicates solutions to those problems.

Material handling - flow is one of the most crucial issues that should be considered for eliminating waste, reducing the cost and just in time-based delivery of the product. An effective transportation management system, optimized with milk-run concept, must be implemented to control the cost of transportation and inventory. The milk-run material supply system consists of cyclic trips, where goods are collected from one supplier and delivered to several customers. The objective of this optimization is the minimization of the total material handling and inventory holding cost, which will boost up the sales or lift the profit margin of the organization [9].

Paper [10] proposes the systematic approach which consists of the milk-run route optimization with mixed integer linear programming model; container loading heuristic to increase the utilization of milk-run vehicles and time interval assignment heuristic to avoid coincidence of milk-run vehicles at the depots of a manufacturer.

3   Modelling in Material Flow

The basic planning process of a milk-run system is divided in 4 phases, as is shown in Fig. 1. Every choice made in one phase impacts the next one. This impact is not easily assessable until all the planning has been completed. Therefore, the planning process is based on a loop system, so that in every phase of the planning, decisions made in the previous phases can be called into question and relevant parameters may be adjusted. This planning approach requires a high level of expertise and competence in order to make the right adjustments towards a consistent and useful result.

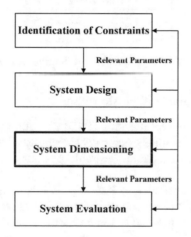

Fig. 1. Planning phases of a milk run system

The *system dimensioning* phase will be tackled, and a new method will be introduced. Therefore, it will be possible to follow from the early stage of the dimensioning phase both, the material-flow and the information-flow, together with other requirements characterizing a typical scenario in the enterprises where the freedom of the logistic planner is often restricted. Accordingly, the development of a homogeneous perception of material flows and information flows is one of the central tasks of material flow management (Fig. 2).

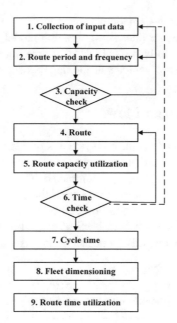

Fig. 2. Milk run system, steps of the *dimensioning system*

3.1 Particle Swarm Optimization Algorithm

The original Particle Swarm Optimization (PSO) algorithm was inspired by the social behavior of biological organisms, specifically the ability of groups of some species of animals to work, as one, in locating desirable positions in a given area, e.g. birds flocking to a food source.

The PSO is originally attributed to Kennedy, Eberhart, and Shi [11, 12]. The PSO algorithm is comprised of a collection of particles that move around the search space. An individual particle i is composed of three vectors: its position in the N-dimensional search space of solutions $\vec{x_i} = (x_{i1}, x_{i2}, \ldots, x_{iN})$, the best position that it has individually found $\vec{p_i} = (p_{i1}, p_{i2}, \ldots, p_{iN})$, and its velocity $\vec{v_i} = (v_{i1}, v_{i2}, \ldots, v_{iN})$. Particles were originally initialized in a uniform random manner throughout the search space; velocity is also randomly initialized. These are also weighted matrixes r_p, r_g dimensions - *search space* x *time horizon* - where items are random numbers of unimodal distribution in the range U <0, 1>. These collections of particles then move throughout the search space influenced by their own best past location and the best past location of the whole swarm or a close neighbor.

The Euclidian neighborhood model was abandoned in favor of less computationally intensive models for mathematical optimization. Each iteration particle's velocity is updated using:

$$v_i(k+1) \leftarrow \omega v_i(k) + c_1 r_p \left(b_i - x_i(k)\right) + c_2 r_g \left(h_i(k) - x_i(k)\right) \qquad (1)$$

where $\omega = 1$ is *Inertia weight*; $c1 = 2$ is *Personal Learning Coefficient*; $c2 = 2$ is *Global Learning Coefficient*; $v_i(k + 1)$ is the new velocity for the ith particle, $x_i(k)$ is the ith particle's position at time k, b_i is the ith particle's best-known position, and $h_i(k)$ is the best position known to the swarm. Also, a random matrix $r_p(k) \sim U <0, 1>$ for $k \in [1, N]$, is generated, as well as a random matrix $r_g(k) \sim U <0, 1>$ for $k \in [1, N]$. A particle's position is updated using: $x_i(k+1) \leftarrow x_i(k) + v_i(k)$. Repeating for all particles (individuals) in all next generations presents PSO cycle.

Algorithm 1 *The **Particle Swarm Optimization** algorithm*

Begin

Step 1: **Initialization.** A random population of individuals $\{x_i\}, i \in [1,N]$

Each individual's n-element velocity vector $\{v_i\}, i \in [1,N]$

The best-so-far position of each individuals: $b_i \leftarrow x_i, i \in [1,N]$

Neighbourhood size $\sigma < N$; $nPop = 150$; Pop. Size; Max. velocity, v_{max}; *MaxIt* $= 1000$; Max. Num. of Iter. $w = 1$; Inertia weight $c1 = 2$; Personal Learning Coeff.; $c2 - 2$; Global Learning Coeff.

Step 2: *While* (the termination criteria is not satisfied) **or** (*MaxIt* $= 1000$)

Step 3:   *for each* individual $x_i, i \in [1,N]$

$H_i \leftarrow \{ \sigma$ nearest neighbors of $x_i \}$

$h_i \leftarrow \arg \min_x \{f(x) : x \in H_i \}$

Generate a random matrix $r_p(k) \sim U <0, 1>$ for $k \in [1, N]$

Generate a random matrix $r_g(k) \sim U <0, 1>$ for $k \in [1, N]$

$$v_i(k) \leftarrow \omega v_i(k) + c_1 r_p (b_i - x_i(k)) + c_2 r_g (h_i(k) - x_i(k))$$

if $|v_i(k)| > v_{max}$ *then*

$v_i(k) \leftarrow v_i(k) v_{max} / |v_i(k)|$

end if

$$x_i(k) \leftarrow x_i(k) + v_i(k)$$

$$b_i \leftarrow \arg \min \{f(x_i(k)), f(b_i)\}$$

end for i Next individual

Step 4: *end while*   Next generation

Step 5: *Post-processing the results and visualization;*

End.

The *Algorithm 1* presents PSO used algorithm where the entire swarm is updated at each step by updating the velocity and position of each particle in every dimension when the termination criteria is not satisfied or *MaxIt* maximal number of iterations is not satisfied.

3.2   Collection of Input Data

The first step of the method is the collection of input data. Generally, the input data come from the decisions the planner has made in the previous phases of identification of constraints and system design. The data concern *Layout* and *space constraints* for instance, the number of buffer places available at each stop, which is presented in

Fig. 3. *Milk run area* is often called supermarket (SM) This set of data is fundamental because it has direct influence on the planning results and on the system evaluation, last phase of a complete system planning.

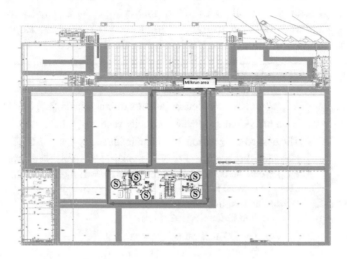

Fig. 3. Milk run system – *Layout* and *space constraints*

Punctual item-related data are usually collected in a document, which has a significant value: *Plan For Every Part* (PFEP) [13]. In this document, which is often in a form of a table, information about every item or part needed for production, logistics and procurement can be found. Typically, the PFEP is an exhaustive document. This means that for every item an extensive set of information is available, most of which are not directly related to the dimensioning phase of the tugger train planning (Table 1).

Time need to produce one final good (FG) is 1 min, that means 60 peaces per hour.

Table 1. Plan *For Every Part* - PFEP - number of peaces in one finish good

Material description	*Bill of Material* for one FG	Production needs per hour	Peaces per container	Number of containers per hour	Container type
Material 1	10	600	2500	0.2400	Small box
Material 2	10	600	600	1.0000	Small box
Material 3	2	120	1000	0.1200	Small box
Material 4	5	300	6500	0.0462	Small box
Material 5	5	300	5000	0.0600	Small box
Material 6	1	60	168	0.3571	Medium box
Material 7	1	60	68	0.8824	Medium box
Material 8	5	300	1000	0.3000	Small box

(continued)

Table 1. (*continued*)

Material description	Bill of Material for one FG	Production needs per hour	Peaces per container	Number of containers per hour	Container type
Material 9	32	1920	576	3.3333	Small box
Material 10	1	60	60	1.0000	Medium box
Material 11	1	60	52	1.1538	Medium box
Material 12	1	60	20	3.0000	Big box
Material 13	1	60	48	1.2500	Small box
Material 14	1	60	25	2.4000	Medium box
Material 15	1	60	20	3.0000	Big box
Material 16	1	60	54	1.1111	Small box
Material 17	24	1440	500	2.8800	Spool
Material 18	20	1200	1000	1.2000	Small box
Material 19	20	1200	1000	1.2000	Small box

Milk run system: (1) Average speed 4 [km/h]; (2) Length of route 265 [m]; (3) *Time for loading* in central point 4 [min].

4   Experimental Results and Discussion

According to the nature of PSO, the experiment is repeated 100 times with the data collection and experimental results for minimizing *Train sequence* with the optimized *number of trailers*. Train sequences, Containers need, trailer per hour and trailers per train are calculated for numerical iterations.

The aim of this research is the need to optimize *Route Period and Route Frequency*. Three different train sequences are shown here: 20 min, 30 min and 40 min. A number of *possible number of trailers is calculated* for that route period, There are, respectively: 4, 6 and 8 trailers for defined train sequence (Tables 2, 3 and 4). The analysis shows that optimal route period is 20 min.

Table 2. Containers need - calculation for train parameters for train sequence 20 min

Train sequence **20** (min)	Needs per hour	Number of containers per trailer	Needed trailers per hour	Needed trailers per sequence	% of train space
Small box	9.86	25	0.3944	0.131466667	4%
Medium box	5.79	2	2.895	0.965	27%
Big box	6	1	6	2	56%
Spool	2.88	2	1.44	0.48	13%
			10.7294	3.576466667	
		by the 11 trailer per hour		**must have 4 trailers per train**	

Table 3. Containers need - calculation for train parameters for train sequence 30 min

Train sequence 30 (min)	Needs per hour	Number of containers per trailer	Needed trailers per hour	Needed trailers per sequence	% of train space
Small box	9.86	25	0.3944	0.1972	4%
Medium box	5.79	2	2.895	1.4475	27%
Big box	6	1	6	3	56%
Spool	2.88	2	1.44	0.72	13%
			10.7294	5.3647	
		by the 11 trailer per hour		**must have 6 trailers per train**	

Table 4. Containers need - calculation for train parameters for train sequence 40 min

Train sequence 40 (min)	Needs per hour	Number of containers per trailer	Needed trailers per hour	Needed trailers per sequence	% of train space
Small box	9.86	25	0.3944	0.262933333	4%
Medium box	5.79	2	2.895	1.93	27%
Big box	6	1	6	4	56%
Spool	2.88	2	1.44	0.96	13%
			10.7294	7.152933333	
		by the 11 trailer per hour		**must have 8 trailers per train**	

The *Time analysis* for optimal defined *Milk Run* tact per train is 20 [min], *Total number of trains* per hour: is 3; which is shown in Tables 5 and 6.

Table 5. Time analysis for optimal sequence 20 min

Time distance [min]	4.0
Time total container change [min]	2.9
Time loading at central supermarket [min]	4.0
Time total route [min]	**10.8**

Table 6. *Milk Run tact per train* for optimal 20 min

Defined Milk Run tact per train [min]	**20**
Total number of trains per hour	3.0

4.1   Milk Run System Implementation

A typical scenario in which the newly developed method could be applied is characterized by the following distinguishing features: (1) the milk-run system is set to accomplish the task of material supply; (2) the logistic system follows the pull principle: material order triggered by the actual line consumption; (3) the material delivery is triggered by means of reordering techniques such as Kanban, e-Kanban or direct order commissioning and delivery processes are decoupled [13]. In the *traditional Kanban* approach, where the information travels together with the unit load in form of a label, the commissioning process cannot start until the tugger train has returned to the SM, because of lack of information. In *E-Kanban* approach the tugger train driver recognizes the demand (empty container) at the point of use and scans the relative barcode. The information is immediately available at the SM and the commissioning process could start right away, before the tugger train has returned to the SM. In *Direct order* system the reordering process is independent of the tugger train position along the route. The material is directly ordered by the operator who is currently working on the production or assembly line and detects an empty container. In the exact moment the container is empty, the operator presses the relative button for ordering or scans the related barcode. Alternatively, the buffer could also be equipped with sensors able to identify the material need.

4.2   Discussion, Benefits and Limitation of Milk Run System

There are numerous benefits for companies that implement milk run as a concept of internal material flow. This system can be implemented in every serial production environment such as: electrical industry, electronic industry, mechanical industry. It helps to reduce or eliminate many different wastes in accordance to the principles of lean production. Milk run optimizes transport capacities usage because of multiple transport of both full and empty containers not only for material but also for finished goods. It assures continuous production eliminating production stoppages due to missing material. Additionally, milk run is reducing material handling space in production giving more free space for production activities.

Milk run has positive economic effect for the company because of material flow optimization. Material lots are smaller and there is smaller quantity of material in work in progress that gives significant inventory reduction. Furthermore, optimized transport requires less human and transport resources. By optimizing material flow, milk run concept reduces material lots and overall inventory levels creating positive economic effect on inventories and optimization of material handling space in production. Important thing to have in mind is that the inventory is a burden to a company because inventory is trapped capital and it is therefore desirable for inventory levels to be as low as possible.

But *milk run system* has some limitations. It must be implemented with other material flow systems such as: traditional Kanban system; *E-kanban* system; some sets of *Constant-Work-in Progress* mechanism; or software for material needs control. On the other hand, it is not suitable for extremely big or extremely small production companies. In general, well implemented is relatively long. The practice has shown that good implementation of the system takes approximately from one to three years.

5   Conclusion and Future Work

This paper presents biological swarm intelligence model: particle swarm optimization for modelling material flow optimization using milk run system in automotive industry. The proposed models optimize *Route* Period and number of trails for one train considering *Layout* and *space constraints*. It should be combined with Kanban system because Kanban is a trigger when something is needed in the production and milk run assures that it is transported in the right time and in the most optimal way.

Experimental results encourage further research. The optimization method, PSO has several parameters that determine its behaviour and efficacy. The future work could focus on extending research on good choice of parameters for various optimization scenarios which should help the production manager to achieve different solutions and make better operational solutions and better production results with little effort. Then this model will be tested with original real-world dataset obtained from existing different manufacturing companies.

References

1. Wagner, B., Enzler, S.: Material Flow Management. Physica – Verlag, Heidelberg (2006)
2. Urru, A., Bonini, M, Echelmeyer, W.: Planning of a milk-run systems in high constrained industrial scenarios. In: Proceeding of 22nd IEEE International Conference on Intelligent Engineering Systems, pp. 231–238 (2018)
3. Simić, D., Simić, S.: Hybrid artificial intelligence approaches on vehicle routing problem in logistics distribution. In: Hybrid Artificial Intelligence Systems. LNCS, vol. 7208, pp. 208–220. Springer, Heidelberg (2012)
4. Simić, D., Kovačević, I., Svirčević, V., Simić, S.: Hybrid firefly model in routing heterogeneous fleet of vehicles in logistics distribution. Log. J. IGPL 23(3), 521–532 (2015)
5. Ilin, V., Ivetić, J., Simić, D.: Understanding the determinants of e-business adoption in ERP-enabled firms and non-ERP-enabled firms: a case study of the Western Balkan Peninsula. Technol. Forecast. Soc. Chang. 125, 206–223 (2017)
6. Simić, D., Svirčević, V., Ilin, V., Simić, S.D., Simić, S.: Particle swarm optimization and pure adaptive search in finish goods' inventory management. Cybern. Syst. 50(1), 58–77 (2019)
7. Kodym, O., Čujan, Z., Turek, M., Mikušová, N.: Optimization of material flow using simulation. MATEC Web of Conferences, VVaPOL 2018, vol. 263, p. 01007 (2019). https://doi.org/10.1051/matecconf/201926301007
8. Krolczyk, J.B., Krolczyk, G.M., Legutko, S., Napiorkowski, J., Hloch, S., Foltys, J., Tama, E.: Material flow optimization – a case study in automotive industry. Tehnički vjesnik – Technical Gazette 22(6), 1447–1456 (2015)
9. Patel, D.R.: Design and optimization of milk-run material supply system with simultaneous pickups and deliveries in time windows. Gujarat Technological University, Ahmedabad, India, Ph.D. thesis (2017)
10. Aksoy, A., Öztürk, N.: A two-stage method to optimize the milk-run system. Eur. J. Eng. Res. Sci. 1(3), 7–11 (2016)
11. Kennedy, J., Eberhart, R.: Particle swarm optimization. IEEE Int. Conf. Neural Netw. 4, 1942–1948 (1995)

12. Shi, Y., Eberhart, R.: A Modified particle swarm optimizer. In: IEEE International Conference on Evolutionary Computation, pp. 69–73 (1998)
13. Harris, R., Harris, C., Wilson, E.: Making Materials Flow. Lean Enterprise Institute, US (2003)

Smart PPE and CPE Platform
for Electric Industry Workforce

Sergio Márquez Sánchez[1(✉)], Roberto Casado Vara[1],
Francisco Javier García Criado[5], Sara Rodríguez González[1],
Javier Prieto Tejedor[1], and Juan Manuel Corchado[1,2,3,4(✉)]

[1] Bisite Research Group, University of Salamanca,
Calle Espejo 2, 37007 Salamanca, Spain
smarquez@usal.es
[2] Air Institute, IoT Digital Innovation Hub (Spain), 37188 Salamanca, Spain
[3] Department of Electronics, Information and Communication,
Faculty of Engineering, Osaka Institute of Technology, 535-8585 Osaka, Japan
[4] Pusat Komputeran dan Informatik, Universiti Malaysia Kelantan,
Karung Berkunci 36, Pengkaan Chepa, 16100 Kota Bharu, Kelantan, Malaysia
[5] Departamento de Cirugía, Universidad de Salamanca, Salamanca, Spain

Abstract. Smart textiles can monitor the state of people and their
environment, and react appropriately according to the parameters they
capture. They are considered ideal for personal protection equipment.
Currently, the rate of accidents among workers is high in all types of
industries. Most of those accidents are caused by the lack of prevention
measures, poor safety training and obsolete safety systems which do no
adapt technologically to the needs of today's workplace. In this work is
presented a intelligent system providing workers the tools they need to
prevent a large percentage of accidents and create safe working environments.

Keywords: Smart textile · Condition monitoring ·
Intelligent environment

1 Introduction

The technological transformation within the working environment has made it
necessary to improve the field of JSA (Job Safety Analysis). Research in intelligent PPE (Personal protective equipment) can contribute to the development
of JSA methods. In addition, it is necessary to adapt PPE to the needs of each
industry; the risks involved in specific tasks can vary considerably.

The main objective of this work is the creation of an intelligent interconnected system consisting of personal protective equipment (PPE) and collective
protective equipment (CPE). This system is going to be targeted at the electric industry, designed for particular tasks and manoeuvres that require greater
safety for both, the installation and the workforce. The aim is to integrate all

© Springer Nature Switzerland AG 2020
F. Martínez Álvarez et al. (Eds.): SOCO 2019, AISC 950, pp. 422–431, 2020.
https://doi.org/10.1007/978-3-030-20055-8_40

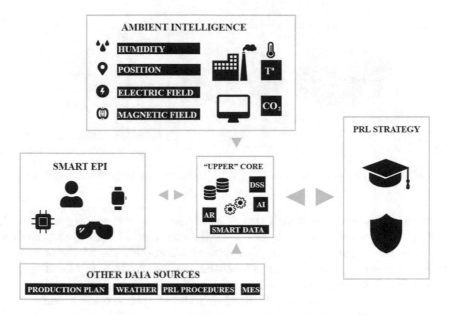

Fig. 1. Summary diagram of system operation

electronic elements (some of them wearable) in an application that collects and transmits data in real-time regarding the worker's physical condition and the working environment. A schema of the system is shown in Fig. 1. The expected results are increased safety of the workforce and the work environment, as well as increased quality of service.

The types of unexpected risks that may emerge can be: technical, environmental, chemical, physical, intrusive, etc. For example, some of the risks that might appear would be: Installations and/or equipment in poor condition, malfunction, use of inadequate PPE, slips and trips, interior lighting, chemical agents, thermal overload, falling objects, variation in the anthropometric constants of the worker, shocks and blows. Previous studies have allowed us to evaluate the different levels of risks arising from the appearance of emerging situations that distort the worker's situation [12]. These data have been taken from different sources (Ministry of Labour, prevention associations, autonomous communities, etc.). Below, Figs. 2A and 2B outline some of the studies that described common risks in the electric industry.

Different electrical installations have been studied and the elements that they have in common have been identified in order to design a comprehensive system and suitable tests. Moreover, different work practices have been considered and classified according to the common risks associated with them, also the tasks were divided into those that involve voltage and those that do not. It is important to note that maneuvers and measurements, tests and verification are not to be considered as work with voltage. All the tasks require some type

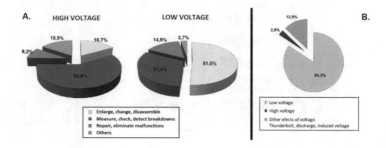

Fig. 2. A. Accidents according to work B. Accidents according to voltage

of preventive measure although the more dangerous they are, the stricter the preventive measures (electrical hazards, trips and slips). In addition, some new risks have been identified during their classification (risk of suffocation). When workers perform the tasks, sensors are going to collect data and send them to the Machine Learning system to prevent possible incidents.

The rest of this paper is organized as follows: Sect. 2 reviews the state of the art. The hardware architecture and system connections are described in Sect. 3. Section 4 outlines the conducted case study in which two sensors collected measurements and its results. Finally, conclusions are drawn from the results in Sect. 5.

2 Related Work

Different studies and statistics have already ensured us of the benefits of incorporating technology in the workplace and have demonstrated that the number of accidents among industry workers can be reduced significantly. Some tasks are particularly dangerous due to the conditions of the working environment; the intelligent protection equipment. Smart textiles have a wide range of applications and can collect different parameters, being able to either reflect the value of an input or capture a stimulus and transform it into a signal. The signals include temperature, humidity, electromagnetic field, radiation, movements, forces, biological activity, etc.

In the early 1990s, the first steps were taken towards designing intelligent textiles, such as PPE. Research focused on designing protective equipment for the military, aerospace and wearable technology in general. The discreet nature of intelligent clothing and the possibility of interacting with computers made it possible to measure the user's physical and physiological parameters. Many lines of research have been and continue to be focused on this area. Research and development are progressively leading to the commercialization of intelligent textile solutions for firefighters and the military, however, there is still a gap between prototype development and commercialization [1].

In the mid-1990s, a team of MIT researchers led by Steve Mann, Thad Starner and Sandy Pentland began developing what he called "wearable computing.

These devices consisted of traditional computer equipment attached to and carried by the body. Another group at MIT, formed by Maggie Orth and Rehmi Post, began to explore how such devices could be more functionally integrated into clothing and other soft substrates. Among other developments, this team explored the integration of digital electronics with conductive fabrics and developed a method for embroidering electronic circuits [1,2].

"Tech Wearable Motherboard" was a pioneering project that developed a garment with sensors and communication capacity. It was designed to monitor in real time the health condition of the military participating in military activities. [4] The multinational European "MyHeart" project, developed and evaluated underwear for the prevention of cardiovascular disease. This underwear was capable of performing an electrocardiogram (ECG) and measuring respiratory and heart rate and other parameters [5]. The SimpleSkin shirt combines physiological and physical detection [6]. Choi and Jiang presented a system for control of cardio-respiratory measurements during sleep [7]. In addition, to the implicit detection and analysis activities, the garment can also be used for explicit input, e.g. through tactile input [3].

Personal protection equipment proposals are designed to protect against a particular level of risk, depending on the task they are intended for. For example, intelligent protective suits for firefighters [14], rescue personnel or chemical protection in projects such as PROeTEX [9], Prospie or I-Protect31. Others are designed for protection against multiple hazards. In some professions, the type and nature of risk is variable and unpredictable. The military sector is a typical example. Intelligent textiles for the military may require a high level of self-adaptability, so that effective protection is only offered when necessary [8]. Compatibility, interoperability, modularity, ergonomics, etc. are important factors that must be considered. They collect a variety of information; advanced processing allows to evaluate complex situations [11].

3   Proposed Method

Finally, the proposed objectives have been achieved with the creation of a Hardware KIT and a software package called GESTOR Fig. 3. At first, the sensors were placed inside the installation to perform maintenance and initial operations. If an incident occurs, the GESTOR application is going to send the relevant orders to the operator or Control Center to act in the proper and known way. All the elements are connected wirelessly via Bluetooth, radio frequency and Wi-Fi for communication of the worker with the web server. In addition, we intended to design a social machine, where data is going to be transferred through the Cloud platform, to give warnings and recommendations in real time [13].

It was intended to create temporary or permanent sensor environments, so that all work areas ensure workers with equivalent levels of security, establishing an internal wireless network that connects all the elements of the installation with the workers. Our system is implemented in the PPE and CPE hardware elements. Below we will describe the characteristics of each of them:

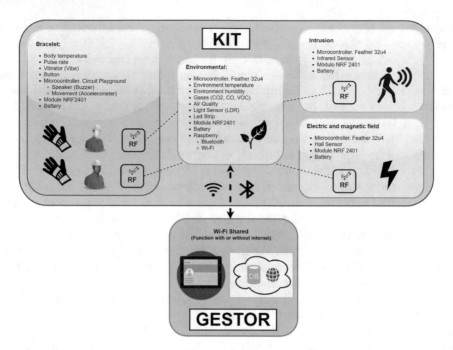

Fig. 3. Accidents according to voltage

There are several types of PPE developments worthy of mention. Wearable electronic components and electronic textiles has been used to monitor the vital signs of workers. Its features make it possible for the user to wash their hands with no impediments. Moreover, they are characterized by being small, flexible and the distribution on their pins makes them ideal for sewing on fabrics [10]. Specifically, a bracelet has been developed with which body temperature, ambient temperature, noise level, heart rate and falls can be measured Fig. 4A. In addition, the user can control the bracelet through the resistive force sensor and in this way to cancel alarms in case they are false. The alarms are received by means of a buzzer and a vibrator there would be the possibility of receiving alarms and giving warnings. In this case, the bracelet is designed so that it does not disturb the wearer and adapts to the wearer's movement. This can be done by using wire instead of cable and designing the electronics for the required use.

CPE measures environmental parameters:

- The Environmental module has the function of measuring environmental parameters such as temperature, gases, humidity, pressure or light. In addition, this module has a button to cancel the alarms and a loudspeaker and a strip of LED to notify of the possible alarms that occur Fig. 5.
- The Intrusion module is composed of a pyroelectric sensor and a loudspeaker that alerts when a person enters the working environment Fig. 4B.

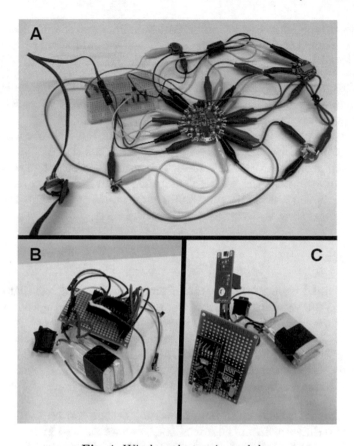

Fig. 4. Wireless electronic modules

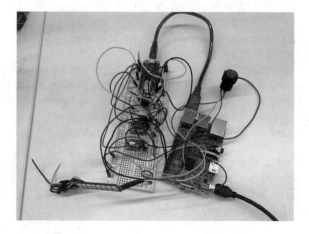

Fig. 5. Electronic components of the environment module

– The Electromagnetic Field module will be composed of a Hall effect sensor and a loudspeaker that alerts when high voltage and strong magnetic field are detected inside the transformation centre Fig. 4C.

GESTOR will be used to refer to all internal processes and intelligence that allows teams to communicate with each other, to evaluate received data and execute actions according to the evaluation, and so on.

The following sequence of actions was performed: the different wireless components were placed in suitable positions and then the different elements were activated: the sensors capture the data which are then sent to the GESTOR. The GESTOR evaluates the received information. In case the data exceed certain thresholds, the GESTOR will send a request/alarm through the bracelet, either to the team leader or to the workers [16,21–26].

4   Results

We have obtained experimental results from three different installations where we had tested our system. The outcome reflects that safety against risk is significantly improved.

Based on the model proposed in the paper [20], Fig. 6 shows the changes in speed during an accidental fall. We have identified three critical differences that characterize a fall and can serve as the criteria detecting a worker who fell. They are marked in the red boxes and explained in detail as follows:

1. **Before the fall:** Before the fall the static acceleration on the three axes is going to be different from the initial status as we can see in Fig. 6.
2. **Beginning of the fall and Impact:** Regardless of the type of fall, there is a significant increase in the acceleration on all axes when a fall occurs. Figure 6 shows the peak on the X-AXIS and Y-AXIS, indicating that a potential incident is taking place, our algorithm detects this rise and continues to observe the pattern to see if it corresponds to a fall.
3. **After the fall:** Generally, after a fall, the worker remains on the ground. This is reflected in Fig. 6 which shows that the worker's usual pattern of movement has been interrupted and at this stage we see the greatest variation in the X-Axis varying from 0 to -0.4 g and on the Z-AXIS the usual value of 1 g to 0.8 g, consequence of the fall lying on the ground after an accident.

In addition, we can see the changes in body temperature over time Fig. 7. When the temperature exceeds or falls below certain values, it means that the worker's conditions is too poor for the to continue working or that they have had an accident if the temperature increases rapidly.

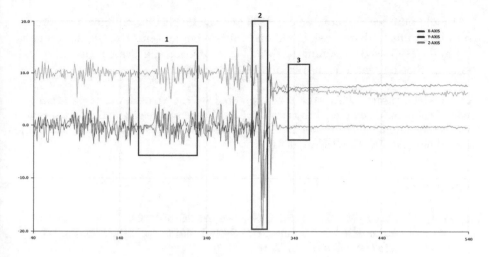

Fig. 6. Simulation of a fall

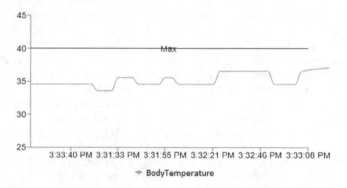

Fig. 7. Body temperature measurement

5   Conclusion and Future Work

This work has presented a novel system based on the creation of a PRL - KIT commanded by GESTOR. Our system combines wireless communication, Smart Textiles and electronic devices. Moreover, the data collected by the system can be visualized by the user on a tablet or a mobile phone. There's currently no such thing as this system.

It has been proven that the proposed system significantly improves the safety conditions of the workforce. In addition, the collected data collected are used to train the system and make possible the early detection of patterns indicative of risk.

Thanks to our knowledge of technology in the fields of circuit printing and flexible PCBs, in a future work we are going to include and improved version of communications model.

In addition, we will create new PPE and the GESTOR will have new predictive algorithms to detect risk based on Multi-Agent systems [16,19], data quality [18] and Case based reasoning (CBR) [17]. Moreover, our system provides real-time monitoring, valid for use in different industries.

Acknowledgments. This work was supported by the Spanish Junta de Castilla y León, Consejería de empleo. Project: UPPER, aUgmented reality and smart personal protective equipment (PPE) for intelligent pRevention of occupational hazards and accessibility INVESTUN/18/SA/0001.

References

1. Post, E.R., Orth, M., Russo, P.R., Gershenfeld, N.: E-broidery: design and fabrication of textile-based computing. IBM Syst. J. **39**(3.4), 840–860 (2000). https://doi.org/10.1147/sj.393.0840. ISSN 0018-8670
2. US 6210771 - Electrically active textiles and articles made therefrom
3. Schneegass, S., Voit, A.: GestureSleeve: using touch sensitive fabrics for gestural input on the forearm for controlling smartwatches. In: Proceedings of the 2016 ACM International Symposium on Wearable Computers, ISWC 2016, pp. 108–115. ACM, New York (2016)
4. Gopalsamy, C., Park, S., Rajamanickam, R., Jayaraman, S.: The wearable motherboard: the first generation of adaptive and responsive textile structures (ARTS) for medical applications. Virtual Real. **4**(3), 152–168 (1999)
5. Amft, O., Habetha, J.: Smart medical textiles for monitoring patients with heart conditions. In: Langenhove, L.v. (ed.) Smart Textiles for Medicine and Healthcare, pp. 275–297. Woodhead Publishing Ltd., Cambridge, February 2007. ISBN 1 84569 027 3
6. Schneegass, S., Hassib, M., Zhou, B., Cheng, J., Seoane, F., Amft, O., Lukowicz, P., Schmidt,A.: SimpleSkin: towards multipurpose smart garments. In: Adjunct Proceedings of the 2015 ACM International Joint Conference on Pervasive and Ubiquitous Computing and Proceedings of the 2015 ACM International Symposium on Wearable Computers, UbiComp/ISWC 2015 Adjunct, pp. 241–244. ACM, New York (2015)
7. Choi, S., Jiang, Z.: A wearable cardiorespiratory sensor system for analyzing the sleep condition. Expert Syst. Appl. **35**(12), 317–329 (2008)
8. Scott, R.A. (ed.): Textiles for Protection. Elsevier, Amsterdam (2005)
9. Magenes, G., Curone, D., Secco, E.L., Bonfiglio, A.: The ProeTEX prototype: a wearable integrated system for physiological & environmental monitoring of emergency operators
10. Stoppa, M., Chiolerio, A.: Wearable electronics and smart textiles: a critical review. Sensors **14**(7), 11957–11992 (2014)
11. Stephanidis, C.: Human factors in ambient intelligence environments. Handbook of Human Factors and Ergonomics, pp. 1354–1373 (2012)
12. Podgorski, D., Majchrzycka, K., Dąbrowska, A., Gralewicz, G., Okrasa, M.: Towards a conceptual framework of OSH risk management in smart working environments based on smart PPE, ambient intelligence and the internet of things technologies. Int. J. Occup. Saf. Ergon. **23**(1), 1–20 (2017)

13. Bauk, S., Schmeink, A.: RFID and PPE: concerning workers' safety solutions and cloud perspectives a reference to the Port of Bar (Montenegro). In: 2016 5th Mediterranean Conference on Embedded Computing (MECO), pp. 35–40. IEEE June 2016

14. Lee, J.Y., Park, J., Park, H., Coca, A., Kim, J.H., Taylor, N.A., Tochihara, Y.: What do firefighters desire from the next generation of personal protective equipment? Outcomes from an international survey. Ind. Health **53**(5), 434–444 (2015)

15. Chen, D., Lawo, M.: Smart textiles and smart personnel protective equipment. In: Smart Textiles, pp. 333-357. Springer, Cham (2017)

16. González-Briones, A., Chamoso, P., Yoe, H., Corchado, J.M.: GreenVMAS: virtual organization based platform for heating greenhouses using waste energy from power plants. Sensors **18**(3), 861 (2018)

17. González-Briones, A., Rivas, A., Chamoso, P., Casado-Vara, R., Corchado, J.M.: Case-based reasoning and agent based job offer recommender system. In: The 13th International Conference on Soft Computing Models in Industrial and Environmental Applications, pp. 21–33. Springer, Cham, June 2018

18. Casado-Vara, R., Prieto-Castrillo, F., Corchado, J.M.: A game theory approach for cooperative control to improve data quality and false data detection in WSN. Int. J. Robust Nonlinear Control **28**(16), 5087–5102 (2018)

19. Rodríguez, S., Palomino, C.G., Chamoso, P., Silveira, R.A., Corchado, J.M.: How to create an adaptive learning environment by means of virtual organizations. In: International Workshop on Learning Technology for Education in Cloud, pp. 199–212. Springer, Cham, August 2018

20. Jia, N.: Detecting human falls with a 3-axis digital accelerometer. In: 2009 A Forum for the Exchange of Circuits, Systems, and Software for Real-world Signal Processing, vol. 43(3), p. 3, July 2009

21. Casado-Vara, R., Novais, P., Gil, A.B., Prieto, J., Corchado, J.M.: Distributed continuous-time fault estimation control for multiple devices in IoT networks. IEEE Access **7**, 11972–11984 (2019)

22. Chamoso, P., González-Briones, A., Rivas, A., De La Prieta, F., Corchado, J.M.: Social computing in currency exchange. Knowl. Inf. Syst., (2019)

23. Casado-Vara, R., Prieto-Castrillo, F., Corchado, J.M.: A game theory approach for cooperative control to improve data quality and false data detection in WSN. Int. J. Robust and Nonlinear Control **28**(16), 5087–5102 (2018)

24. Morente-Molinera, J.A., Kou, G., González-Crespo, R., Corchado, J.M., Herrera-Viedma, E.: Solving multi-criteria group decision making problems under environments with a high number of alternatives using fuzzy ontologies and multi-granular linguistic modelling methods. Knowl.-Based Syst. **137**, 54–64 (2017)

25. Li, T., Sun, S., Bolić, M., Corchado, J.M.: Algorithm design for parallel implementation of the SMC-PHD filter. Signal Process. **119**, 115–127 (2016). https://doi.org/10.1016/j.sigpro.2015.07.013

26. Chamoso, P., Rodríguez, S., de la Prieta, F., Bajo, J.: Classification of retinal vessels using a collaborative agent-based architecture. AI Commun. **31**, 1–18 (2018)

Acoustic Anomaly Detection Using Convolutional Autoencoders in Industrial Processes

Taha Berkay Duman$^{(\boxtimes)}$, Barış Bayram, and Gökhan İnce

Faculty of Computer and Informatics Engineering,
Istanbul Technical University, Istanbul, Turkey
{dumanta,baris.bayram,gokhan.ince}@itu.edu.tr

Abstract. In the industrial plants, detection of abnormal events during the processes is a difficult task for human operators who need to monitor the production. In this work, the main aim is to detect anomalies in the industrial processes by an intelligent audio based solution for the new generation of factories. Therefore, this paper presents a Convolutional Autoencoder (CAE) based end-to-end unsupervised Acoustic Anomaly Detection (AAD) system to be used in the context of industrial plants and processes. In this research, a new industrial acoustic dataset has been created by gathering the audio data obtained from a number of videos of industrial processes, recorded in factories involving industrial tools and processes. Due to the fact that the anomalous events in real life are rather rare and the creation of these events is highly costly, anomaly event sounds are superimposed to regular factory soundscape by using different Signal-to-Noise Ratio (SNR) values. To show the effectiveness of the proposed system, the performances of the feature extraction and the AAD are evaluated. The comparison has been made between CAE, One-Class Support Vector Machine (OCSVM), and a hybrid approach of them (CAE-OCSVM) under various SNRs for different anomaly and process sounds. The results showed that CAE with the end-to-end strategy outperforms OCSVM while the respective results are close to the results of hybrid approach.

Keywords: Anomaly detection · Industrial processes ·
Convolutional autoencoders · One-Class Support Vector Machine ·
Signal-to-Noise Ratio · Audio feature extraction

1 Introduction

Anomaly detection aims to determine unusual events by distinguishing them from usual events. It is also frequently applied in the smart factory approaches and tackled in the new generation of industrial manufacturing issues in recent years. From this perspective, anomaly events can decrease the quality of manufactured products, deteriorate the continuity and the reliability of the processes.

© Springer Nature Switzerland AG 2020
F. Martínez Álvarez et al. (Eds.): SOCO 2019, AISC 950, pp. 432–442, 2020.
https://doi.org/10.1007/978-3-030-20055-8_41

Even worse, anomalies can also endanger the safety of people who use industrial machines in the factory. These emergency conditions have to be detected and avoided before they become hazardous. People have to be in safety while they are working in these type of environments. Therefore, monitoring systems have to detect emergency situations immediately when they occur. After they detect, they can work as an alarm system and assist the responsible personnel to prevent possible losses.

Current anomaly detection systems used in the industrial domain are mainly based on sensor data parameters, environment variables and quality metrics of the manufactured products. The detection and monitoring systems can be high-cost and in need of high computational power, especially when used in video and image-based systems. Instead, anomaly detection systems can be designed with low computational power and low-cost with the help of acoustic data. Acoustic Anomaly Detection (AAD) intends to find novel interpretations of acoustic signals by isolating anomalies from normal audio signals the system is trained on.

In this study, a Convolutional Autoencoder (CAE) is developed in the context of AAD. It is able to determine the anomaly patterns in the time and frequency domain when applied to spectrograms as if they are image inputs. Also, convolutional filters are capable of detecting these patterns which represent different sound characteristics [1].

In the industrial domain, there is no study using unsupervised methods for AAD, to our knowledge. In this work, we present the performance evaluations of AAD by CAE with a novel architecture used for feature extraction and as an end-to-end anomaly detector for industrial plants. A CAE is trained with the audio signals gathered from videos recorded in industrial environments. We investigate the anomaly detection performance of AE by comparing it to the performance of a traditional and widely used anomaly detection approach called One-Class Support Vector Machine (OCSVM).

2   Related Work

There are numerous researches based on machine learning with the aim of detecting the anomalies occurring in industrial tools, applications, and processes. Chou et al. used Support Vector Data Description in their novelty detection module designed for wafer -a semiconductor material- quality prediction system in the semiconductor manufacturing process [2]. Also, OCSVM is employed for detecting anomalies in a different types of industrial applications [3–5], respectively kinematic chain, steering system, industrial arm. Another study [6] covers the comparison of the abnormality detection results of k-Nearest Neighbor (kNN), OCSVM, Local Outlier Factor, Principal Component Analysis and Maximum Mean Discrepancy in industrial applications.

In the survey paper [7], various anomaly detection works and fault detection methods used in industrial processes are reviewed. Nonetheless, there is no specific work employing the anomaly detection using audio signals in the industrial domain. However, for detecting anomalies in acoustic data, there are a few works, in which OCSVM with a Radial Basis Function (RBF) kernel is utilized for surveillance applications [8], and One-Class Nearest Neighbor, One-Class Gaussian Mixture and OCSVM are employed for the acoustic fall detection [9].

In the study of Erfani et al. [10], a hybrid approach has been developed by combining AE as a feature extractor and using OCSVM as an anomaly detection approach. We selected OCSVM as a baseline method to compare our proposed AE based approach. In addition to that, a hybrid system is proposed to be used in handling anomaly detection.

With the development of deep learning in recent years, AEs started to be used widely in anomaly detection tasks utilizing acoustic data. In these works [11–13], different types of AEs are proposed for acoustic novelty detection, including basic AE, Compression AE, Denoising AE, Adversarial AE. Droghini et al. presented a CAE based AAD algorithm through the end-to-end strategy for human fall detection in an indoor environment [14]. They compared the performance of the proposed system with OCSVM and showed the improvement with CAE.

Furthermore, Generalized Extreme Value distribution is exploited in their outlier detection algorithm by processing the acoustical signature of the welding process [15]. These works related to industrial AAD only focused on product quality and equipment failure instead of end-to-end industrial processes or applications. In this paper, we also propose to employ unsupervised anomaly detection methods on the acoustic data gathered directly from industrial processes and plants.

3   Acoustic Anomaly Detection System

One of the challenging issues in the new generation factories is to provide intelligent solutions for various tasks such as monitoring and detecting faults and dangers. Visual anomaly detection systems have some drawbacks such as illumination, occlusion by objects, being out of the field of view, etc., which strongly affect the performance of the system. The AAD system, however, is not affected by this kind of problems; thus offers an advantage by using acoustic data features.

3.1   Feature Extraction

The audio feature extraction method proposed by Piczak et al. [16] has been used in this study. In the first step, log-scaled Mel-spectrogram features -a 2D time-frequency representation of a sound- have been extracted from each frame using 32 mel-bands setting. Because of the need for fixed size input in convolutional layers, the features have been divided into the segments consisting of 32 frames (about 500 ms). Segment features with the dimension of 32×32 formed one channel feature input for the network. These spectrograms have been used as an input for CAE and CAE-OCSVM hybrid approach. For the OCSVM approach, a vector is generated for each spectrogram, instead. Then, the vector representations are utilized to generate an OCSVM model.

3.2   Convolutional Autoencoder

In the anomaly detection phase (Fig. 1), the extracted frame-wise features are used as input of AE network. By encoding and decoding this input respectively, the network reconstructs the input as output. Then, the anomaly is detected by calculating the reconstruction error between the input and the output reconstructed by the AE network, [13] using Euclidean distance, as follows;

$$E_R = \frac{1}{N_f} \sum_{i=1}^{N_f} (X_i^a - X_i^r)^T (X_i^a - X_i^r) \tag{1}$$

where E_R is the reconstruction error, N_f denotes the number of frames extracted from a sound, X^a is the actual spectrogram of the frame, X^r is the reconstructed one, and i represents the index of the frame belonging to the sound.

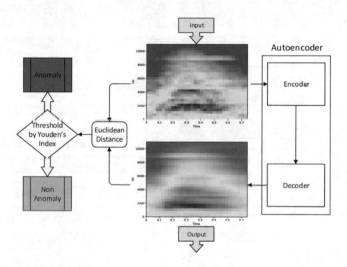

Fig. 1. Acoustic anomaly detection system using autoencoder

The network basically contains 23 convolutional layers, 4 pooling layers, and 4 upsampling layers. For the convolutional layers, we used filters with the receptive field of 3×3. We applied the Rectified Linear Units (ReLU) activation function. Figure 2 shows the CAE architecture we used in our task. Compared to the existing related system in [14], ReLU activation function has been used instead tanh, and Adamax has been used as optimization algorithm instead of Adam. For the loss function, Minimum Squared Error has been applied instead of the logarithmic version. In contrary to [14], a different architecture has been developed and applied. This research is the first study that applies CAE for AAD in the context of the industrial process.

The input consists of $N \times 32 \times 32$ feature arrays where N denotes the number of sound instances. At the end of the network, it reconstructs the input spectrogram.

Youden's index, J, in Eq. (2) is calculated for each threshold exist in the ROC (Receiver Operating Characteristic) curve estimated using the Euclidean distance values [17] and then the threshold which has the maximum J is selected as a threshold, which is estimated as follows;

$$J = Sensitivity + Specificity - 1 \qquad (2)$$

where $Sensitivity$ stands for True Positive Rate (TPR) and $Specificity$ stands for True Negative Rate (TNR). The aim of Youden's index is to maximize the difference between TPR and False Positive Rate (FPR). Thus, the system will set the threshold by maximizing the anomaly detection accuracy and separability rate.

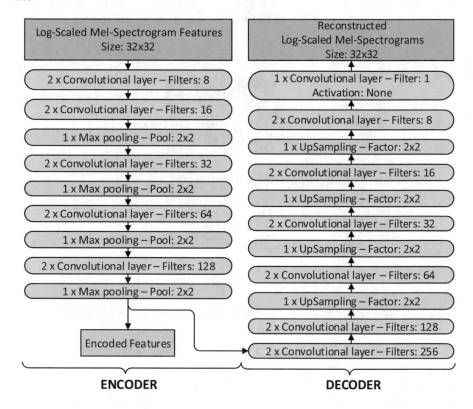

Fig. 2. Convolutional autoencoder architecture

3.3   One-Class Support Vector Machine

For anomaly detection, CAE is compared with OCSVM that is also an unsupervised technique. The normal sounds in a factory setup are used to train the model to estimate the best frontier including the sounds. Then, the sounds outside it are predicted as an anomaly. In this study, we used the RBF kernel for

OCSVM and employed a grid search to estimate the best parameters for kernel coefficient and the bound parameter.

Moreover, the hybrid architecture is implemented by combining CAE and OCSVM in which encoder part of the CAE has been used as a feature extractor. Then, these encoded features of normal sounds are utilized in the training of OCSVM model, and the encoded features including also abnormal ones are predicted via this model.

Table 1. Acoustic datasets

Datasets	Total Duration in Minutes	Number of Recordings
Industrial Painting	26	16
Industrial Cutting	18	15
Industrial Welding	28	14
Industrial Robotic Arm	10	17

4   Experiments and Results

4.1   Experimental Setup

Dataset: We created a new acoustic dataset for the industrial processes and plants by applying the two main steps of dataset creation process [18]:

– Data collection: We retrieved audio data from Youtube videos recorded during industrial processes and anomaly sounds, additionally.
– Data arrangement: We removed the empty (silent), noisy (background noise being high and interfering the target sound) and irrelevant parts from the original audio data.

After generating the dataset, frame-based features are extracted. Due to the nature of unsupervised learning, the frames belonging to the normal factory sounds are directly sent to CAE or OCSVM for training phase. For the testing phase, anomaly sounds are mixed with normal sounds artificially by fixed SNR values, since the data consisting of anomaly sounds is hard to generate due to their rarity in real life. In addition to that, creating anomaly events and recording respective sounds are costly. For generating anomaly sounds, we mixed a number of normal sounds with different types of noises at specific Signal-to-Noise Ratios (SNRs) of $\{10, 0, -10, -20, -30\}$ dB.

We created four acoustic datasets[1] listed in Table 1. All signals in the dataset are sampled at 32000 Hz. Using the audio files from these datasets, we selected three anomaly sounds to be mixed for generating anomaly events; *Explosion*, *Fire*, and *Glass Breaking*.

---

[1] https://kovan.itu.edu.tr/index.php/s/MN2jtWCovolkeGi/download.

Software Specifications: For the extraction of the Mel-spectrogram, Librosa[2] library is utilized. In addition, Keras library is employed for the implementation and application of the CAE network and its experiments. Also, OCSVM technique of scikit-learn library[3] is used in the experiments.

4.2   Evaluation Criteria

We selected the Area Under the ROC Curve (ROC-AUC) score as an evaluation metric to determine the degree of separability. In all possible threshold values, it measures the classification performance of the model. With this metric, we evaluate how the model works while distinguishing between normal and anomaly events. The higher the ROC-AUC score is, the better the model is. A ROC-AUC score which is close to 1.0 means the model can perfectly separate normal and anomaly events.

We used F1-score as a second evaluation metric. It balances precision and recall by giving equal weights to both of them. It indicates the mean of precision and recall for a specific threshold value. This threshold value is specified by Youden's index Eq. (2) in ROC-AUC. At this threshold, we evaluate the correct decision making performance of the model by using F1-score.

4.3   Results

Performance of Acoustic Anomaly Detection: As stated in Tables 2, 3 and 4, for three different industrial data sets, CAE outperforms the other methods and it detects anomalies with more accuracy than OCSVM based methods in various SNRs, including $0, -10,$ and -20 dB. CAE can distinguish normal and anomaly data more clearly compared to the plain OCSVM. Although the hybrid structure improves the performance of pure OCSVM, it can not exceed the CAE performance in the specified SNR values. In general, there is no clear difference between the methods at SNR levels of 10 and -30 dB among the approaches. Thus, it has been shown that CAE can be applied successfully in industrial painting, cutting and welding applications and environments. In addition, it can detect anomalies with a higher rate compared to the standard and traditional methods.

As seen in Table 5 consisting of results belonging to industrial arm data set, the most successful separation is made by the hybrid method except glass breaking in $0, -10,$ and -20 dB SNR values. After the hybrid structure, the best results are obtained by OCSVM and CAE, respectively. For this data set, it is believed that the most important reason for the failure of CAE in relation to other methods is that its data size is small for the context of neural networks. In addition, due to the fact that there are different types of robotic arms and different movement patterns in this data set, the CAE performance is not as good

---

[2] https://librosa.github.io/librosa.

[3] https://scikit-learn.org/stable/modules/generated/sklearn.svm.OneClassSVM.

as other approaches. In such cases, it can be stated that the hybrid approach is more suitable to be applied in a realistic setup.

The results revealed that the audio data can be successfully applied in anomaly detection tasks for industrial processes and the performance of these anomaly detection systems can be boosted by using CAEs.

Performance Comparison Among Anomaly Sounds: According to all Tables 2, 3, 4 and 5, the glass breaking anomaly is more easily detected by all anomaly detection methods. Based on the detection rates, it can be seen that the glass breaking is the most distinguishable anomaly sound. Compared to the spectrograms of explosion and fire anomalies, the glass breakage anomaly has more distinct characteristics and a solid pattern. Also, its power spreads across all frequencies within the spectrogram in contrast to the others. On the other hand, the power of the explosion and fire spectra are concentrated in the lower frequencies, and as the frequency increases, it can be said that the spread pattern gets similar to white noise, and at high frequencies the effect is getting lost. For these reasons, the glass breaking anomaly can be detected with higher accuracy than the other anomalies.

Table 2. Anomaly detection results on industrial painting dataset

Anomaly	Explosion						Fire						Glass Breaking					
Method	CAE		CAE + OCSVM		OCSVM		CAE		CAE + OCSVM		OCSVM		CAE		CAE + OCSVM		OCSVM	
SNR (dB) / Metric	ROC AUC	F1 score	ROC AUC	F1 score	ROC AUC	F1 score	ROC AUC	F1 score	ROC AUC	F1 score	ROC AUC	F1 score	ROC AUC	F1 score	ROC AUC	F1 score	ROC AUC	F1 score
10	0.54	0.50	0.51	0.67	0.50	0.67	0.55	0.50	0.50	0.67	0.50	0.67	**0.82**	**0.76**	0.50	0.67	0.50	0.67
0	**0.84**	**0.78**	0.58	0.71	0.51	0.68	**0.89**	**0.83**	0.56	0.70	0.53	0.68	**0.98**	**0.95**	0.65	0.75	0.58	0.71
-10	**0.98**	**0.89**	0.73	0.79	0.65	0.74	**0.97**	**0.91**	0.77	0.81	0.72	0.78	**0.99**	**0.97**	0.82	0.85	0.78	0.82
-20	**0.99**	**0.96**	0.90	0.91	0.83	0.85	**0.99**	**0.98**	0.91	0.92	0.88	0.89	**0.99**	**0.98**	0.92	0.93	0.92	0.92
-30	0.99	0.99	0.99	0.99	0.97	0.97	0.99	0.99	0.99	0.99	0.99	0.99	0.99	0.99	0.99	0.99	0.99	0.99

Table 3. Anomaly detection results on industrial cutting dataset

Anomaly	Explosion						Fire						Glass Breaking					
Method	CAE		CAE + OCSVM		OCSVM		CAE		CAE + OCSVM		OCSVM		CAE		CAE + OCSVM		OCSVM	
SNR (dB) / Metric	ROC AUC	F1 score	ROC AUC	F1 score	ROC AUC	F1 score	ROC AUC	F1 score	ROC AUC	F1 score	ROC AUC	F1 score	ROC AUC	F1 score	ROC AUC	F1 score	ROC AUC	F1 score
10	**0.73**	**0.67**	0.53	0.67	0.50	0.66	**0.64**	**0.61**	0.52	0.68	0.50	0.66	**0.66**	**0.64**	0.51	0.67	0.50	0.66
0	**0.80**	**0.75**	0.55	0.69	0.50	0.67	**0.80**	**0.77**	0.56	0.69	0.50	0.66	**0.88**	**0.87**	0.54	0.69	0.51	0.66
-10	**0.91**	**0.86**	0.78	0.81	0.52	0.68	**0.88**	**0.84**	0.74	0.81	0.53	0.67	**0.91**	**0.92**	0.72	0.78	0.56	0.68
-20	**0.93**	**0.93**	0.87	0.88	0.63	0.72	**0.93**	**0.92**	0.90	0.91	0.75	0.79	**0.94**	**0.95**	0.89	0.90	0.82	0.85
-30	**0.99**	**0.97**	0.97	0.96	0.84	0.86	0.99	0.96	0.97	0.97	0.92	0.93	0.99	0.99	0.99	0.99	0.92	0.92

Comparison of Datasets: By looking to Tables 2, 3, 4 and 5, it can be seen that the anomalies can be detected better in the industrial painting dataset. This may be due to the fact that the industrial painting sound data has a characteristic which is very close to the white noise and does not have a specific

Table 4. Anomaly detection results on industrial welding dataset

Anomaly Method / SNR (dB)	Explosion						Fire						Glass Breaking					
	CAE		CAE + OCSVM		OCSVM		CAE		CAE + OCSVM		OCSVM		CAE		CAE + OCSVM		OCSVM	
Metric	ROC AUC	F1 score	ROC AUC	F1 score	ROC AUC	F1 score	ROC AUC	F1 score	ROC AUC	F1 score	ROC AUC	F1 score	ROC AUC	F1 score	ROC AUC	F1 score	ROC AUC	F1 score
10	0.64	0.54	0.51	0.67	0.50	0.67	0.60	0.45	0.50	0.67	0.50	0.67	0.61	0.52	0.51	0.67	0.50	0.67
0	**0.81**	**0.73**	0.54	0.69	0.51	0.67	**0.79**	**0.72**	0.58	0.70	0.51	0.67	**0.89**	**0.82**	0.62	0.72	0.51	0.67
-10	**0.95**	**0.89**	0.80	0.83	0.55	0.69	**0.93**	**0.87**	0.82	0.85	0.59	0.71	**0.99**	**0.95**	0.88	0.89	0.66	0.75
-20	**0.99**	**0.98**	0.95	0.96	0.73	0.79	0.99	0.96	0.98	0.98	0.83	0.86	0.99	0.99	0.99	0.99	0.91	0.92
-30	0.99	0.99	0.99	0.99	0.95	0.95	0.99	0.99	0.99	0.99	0.97	0.97	0.99	0.99	0.99	0.99	0.99	0.99

Table 5. Anomaly detection results on industrial arm dataset

Anomaly Method / SNR (dB)	Explosion						Fire						Glass Breaking					
	CAE		CAE + OCSVM		OCSVM		CAE		CAE + OCSVM		OCSVM		CAE		CAE + OCSVM		OCSVM	
Metric	ROC AUC	F1 score	ROC AUC	F1 score	ROC AUC	F1 score	ROC AUC	F1 score	ROC AUC	F1 score	ROC AUC	F1 score	ROC AUC	F1 score	ROC AUC	F1 score	ROC AUC	F1 score
10	0.52	0.52	0.50	0.67	0.50	0.67	0.50	0.61	0.50	0.67	0.50	0.67	0.65	0.62	0.52	0.61	0.51	0.67
0	0.63	0.55	0.51	0.67	0.51	0.67	0.52	0.62	**0.56**	**0.70**	0.52	0.68	**0.82**	**0.75**	0.59	0.71	0.58	0.71
-10	0.66	0.50	**0.69**	**0.77**	0.57	0.70	0.66	0.72	**0.76**	**0.74**	0.66	0.73	**0.90**	**0.82**	0.68	0.76	0.75	0.80
-20	0.86	0.83	**0.89**	**0.90**	0.78	0.82	0.94	0.88	**0.98**	**0.98**	0.90	0.91	0.99	0.99	0.99	0.99	0.95	0.95
-30	0.99	0.99	0.99	0.99	0.97	0.97	0.99	0.99	0.99	0.99	0.99	0.99	0.99	0.99	0.99	0.99	0.99	0.99

formal pattern in the spectrum overall. After adding the corresponding noise to a white noise-like spectrogram, this added noise can be easily detected. For the industrial arm data set, this is exactly the opposite. Based on these quantitative results, we believe that the proposed AAD systems can be more successfully applied in industrial painting facilities.

5 Conclusion

In this study, a CAE based end-to-end acoustic anomaly detection method was proposed which can be used within the scope of industrial plants and applications. The performance of the presented detection system and its ability to distinguish normal and anomaly classes in different SNR values have been evaluated in various industrial datasets created within the framework of this study. As a conclusion of our assessment, it was determined that CAE is more successful than other methods especially in the datasets where industrial painting, cutting and welding processes were included. In addition, in the dataset containing the industrial arm sounds, it was observed that the hybrid architecture which is the combination of the CAE and OCSVM together demonstrates an improved performance. In the end, it has been shown that the CAE-based method can be successfully used for AAD in industrial processes. Thus, with the detection of abnormalities in acoustic audio data, we have shown that we can determine the occurrence of emergencies and created a basic structure for a warning system.

In the future, the dataset will be expanded by collecting sequential normal and anomaly sounds from industrial processes. Using the dataset, sequential AEs will be employed to detect anomalies in the sequential acoustic data.

References

1. Salamon, J., Bello, J.P.: Deep convolutional neural networks and data augmentation for environmental sound classification. IEEE Signal Process. Lett. **24**(3), 279–283 (2017). https://doi.org/10.1109/LSP.2017.2657381
2. Chou, P.H., Wu, M.J., Chen, K.K.: Integrating support vector machine and genetic algorithm to implement dynamic wafer quality prediction system. Expert Syst. Appl. **37**(6), 4413–4424 (2010). https://doi.org/10.1016/j.eswa.2009.11.087
3. Cariño-Corrales, J.A., Saucedo-Dorantes, J.J., Zurita-Millán, D., Delgado-Prieto, M., Ortega-Redondo, J.A., Alfredo Osornio-Rios, R., de Jesus Romero-Troncoso, R.: Vibration-based adaptive novelty detection method for monitoring faults in a kinematic chain. Shock Vibr. **2016**, 1–12 (2016). https://doi.org/10.1155/2016/2417856
4. Carino, J.A., Delgado-Prieto, M., Zurita, D., Millan, M., Redondo, J.A.O., Romero-Troncoso, R.: Enhanced industrial machinery condition monitoring methodology based on novelty detection and multi-modal analysis. IEEE Access **4**, 7594–7604 (2016). https://doi.org/10.1109/ACCESS.2016.2619382
5. Narayanan, V., Bobba, R.B.: Learning based anomaly detection for industrial arm applications. In: Proceedings of the 2018 Workshop on Cyber-Physical Systems Security and PrivaCy, pp. 13–23. ACM (2018). https://doi.org/10.1145/3264888.3264894
6. Jia, X., Zhao, M., Di, Y., Yang, Q., Lee, J.: Assessment of data suitability for machine prognosis using maximum mean discrepancy. IEEE Trans Industr. Electron. **65**(7), 5872–5881 (2018). https://doi.org/10.1109/TIE.2017.2777383
7. Diez-Olivan, A., Del Ser, J., Galar, D., Sierra, B.: Data fusion and machine learning for industrial prognosis: trends and perspectives towards industry 4.0. Inf. Fusion **50**, 92–111 (2019). https://doi.org/10.1016/j.inffus.2018.10.005
8. Aurino, F., Folla, M., Gargiulo, F., Moscato, V., Picariello, A., Sansone, C.: One-class SVM based approach for detecting anomalous audio events. In: 2014 International Conference on Intelligent Networking and Collaborative Systems, pp. 145–151. IEEE (2014). https://doi.org/10.1109/INCoS.2014.59
9. Popescu, M., Mahnot, A.: Acoustic fall detection using one-class classifiers. In: 2009 Annual International Conference of the IEEE Engineering in Medicine and Biology Society, pp. 3505–3508. IEEE (2009)
10. Erfani, S.M., Rajasegarar, S., Karunasekera, S., Leckie, C.: High-dimensional and large-scale anomaly detection using a linear one-class SVM with deep learning. Pattern Recogn. **58**, 121–134 (2016). https://doi.org/10.1016/j.patcog.2016.03.028
11. Marchi, E., Vesperini, F., Eyben, F., Squartini, S., Schuller, B.: A novel approach for automatic acoustic novelty detection using a denoising autoencoder with bidirectional LSTM neural networks. In: 2015 IEEE International Conference on Acoustics, Speech and Signal Processing (ICASSP), pp. 1996–2000. IEEE (2015). https://doi.org/10.1109/ICASSP.2015.7178320
12. Marchi, E., Vesperini, F., Squartini, S., Schuller, B.: Deep recurrent neural network-based autoencoders for acoustic novelty detection. Comput. Intell. Neurosci. vol. 2017, (2017). https://doi.org/10.1155/2017/4694860
13. Principi, E., Vesperini, F., Squartini, S., Piazza, F.: Acoustic novelty detection with adversarial autoencoders. In: 2017 International Joint Conference on Neural Networks (IJCNN), pp. 3324–3330. IEEE (2017). https://doi.org/10.1109/IJCNN.2017.7966273

14. Droghini, D., Ferretti, D., Principi, E., Squartini, S., Piazza, F.: An end-to-end unsupervised approach employing convolutional neural network autoencoders for human fall detection. In: Italian Workshop on Neural Nets, pp. 185–196. Springer (2017). https://doi.org/10.1007/978-3-319-95095-2_18
15. Hartman, D.A.: Real-time detection of processing flaws during inertia friction welding of critical components. In: ASME Turbo Expo 2012: Turbine Technical Conference and Exposition, pp. 1–10. American Society of Mechanical Engineers (2012). https://doi.org/10.1115/GT2012-68014
16. Piczak, K.J.: Environmental sound classification with convolutional neural networks. In: 2015 IEEE 25th International Workshop on Machine Learning for Signal Processing (MLSP), pp. 1–6. IEEE (2015). https://doi.org/10.1109/MLSP.2015.7324337
17. Youden, W.J.: Index for rating diagnostic tests. Cancer **3**(1), 32–35 (1950)
18. Fonseca, E., Plakal, M., Ellis, D.P., Font, F., Favory, X., Serra, X.: Learning sound event classifiers from web audio with noisy labels. arXiv preprint arXiv:1901.01189 (2019)

One-Class Classification to Predict the Success of Private-Participation Infrastructure Projects in Europe

Álvaro Herrero[1]([⊠]) [iD] and Alfredo Jiménez[2] [iD]

[1] Grupo de Inteligencia Computacional Aplicada (GICAP),
Departamento de Ingeniería Civil, Escuela Politécnica Superior,
Universidad de Burgos, Av. Cantabria s/n, 09006 Burgos, Spain
ahcosio@ubu.es
[2] Department of Management, KEDGE Business School, Bordeaux, France
alfredo.jimenez@kedgebs.com

Abstract. Foreign investment has significantly increased in infrastructure projects as governments have realized the potential advantages for the host countries in terms of capital and technical expertise. Private participation projects are a common vehicle for private firms to invest in infrastructures, although these projects are typically subject to pressures from the government, consumers, suppliers, regulatory institutions, and public opinion. Forecasting the success of these projects in advance is a key element to be taken into account when deciding about participation. To support this kind of decisions, present paper proposes the application of some one-class classifiers to check their ability to predict the final success of private participation projects involving infrastructures. To validate the proposed soft-computing models, they are applied to a real-life dataset from the World Bank, comprising information about projects in European countries within the Energy and Telecommunication sectors.

Keywords: Classification · Support Vector Machines · Random trees ·
k nearest neighbors · Private-participation projects · Internationalization

1 Introduction and Previous Work

Participation in infrastructure projects was traditionally not allowed to foreign investors, due to the sensitive and strategic nature of this type of sectors [1]. However, recently, many governments have realized the need for capital and technical expertise that they can only obtain from foreign investors and, as a consequence, the level of foreign participation in infrastructure projects has increased dramatically in the last two decades [2, 3].

Typically nowadays, infrastructure projects in sectors such as the electrical, energy, telecommunication, etc. are undertaken using "private participation projects", in which private domestic or multinational enterprises hold at least some or all equity in the project. This phenomenon has led to the creation of some truly globalized private companies (Grupo ACS, Vinci, FCC, OHL, etc.).

© Springer Nature Switzerland AG 2020
F. Martínez Álvarez et al. (Eds.): SOCO 2019, AISC 950, pp. 443–451, 2020.
https://doi.org/10.1007/978-3-030-20055-8_42

Given the large amount of internal and external stakeholders involved in these projects, which multiples pressures from the government, consumers, suppliers, regulatory institutions, and public opinion, and the intrinsic technical complexity that they usually entail, scholars have recently started to pay increasing attention to the determinants of success in this specific type of investment. As a result, extant literature has demonstrated the critical relevance of the institutional environment in the host country such as political stability [2, 4] and corruption [5]. However, further research is needed to enlarge our understanding of the comprehensive list of factors that significantly affect investments in infrastructure projects. For this reason, in this paper we aim to investigate a sample of private participation projects covering 3892 projects in 18 Central and Eastern European countries. This region is particularly interesting, not only because it is an increasingly attractive destination for international investors [6, 7], but also because it is one of the regions in the world where private participation projects are most frequently employed. We focus on projects in Energy and Telecommunication, two of the most relevant infrastructure sectors for the economy of countries and the ones accounting for the largest number of projects of this type.

A wide variety of classifiers have been used in a wide range of application fields, ranging from cybersecurity [8] to image analysis [9]. On the other hand, enterprise management has been recently enriched by the application of soft-computing techniques [10–13]. However, very few supervised learning models have been applied to problems similar to the one above described. That is the case of [14], where corporate credit rating analysis is conducted based on Support Vector Machine (SVM) and Artificial Neural Networks (ANN). These classifiers are applied to data from for the United States and Taiwan markets trying not only to forecast but also to get a model with better explanatory power. More recently, [15] combined SVM together with fuzzy logic as a real case study in construction management. This hybrid system tried to predict project dispute resolution outcomes (i.e., mediation, arbitration, litigation, negotiation, and administrative appeals) when the dispute category and phase in which a dispute occurs are known during project execution.

In [16] k-Nearest Neighbor (k-NN) is compared to ANN, Discriminant Analysis, Quadratic Discriminant Analysis, and Multinomial Logistic Regression Analysis to provide input to managers who make business decisions. These models were applied to retail department store data, showing that they are most useful when uncertainty is high and a priori classification cannot be made with a high degree of reliability. Additionally, [17] proposed the application of k-NN to multi-criteria inventory classification in order to manage inventory more efficiently. k-NN are compared to SVM, ANN, and Multiple Discriminant Analysis when applied to 4 benchmark datasets. SVM was identified as the most accurate among all of them due to its high generalization capability, as well as its use of kernel functions to increase the learning efficiency.

The rest of this paper is organized as follows: the applied methods are described in Sect. 2, the setup of experiments and the dataset under analysis are described in Sect. 3, together with the results obtained and the conclusions of present study that are stated in Sect. 4.

2   Applied Classifiers

This work proposes the application of two different classifiers to predict the success of infrastructure projects that are described in present section.

2.1   Support Vector Machines

Support Vector Machines (SVMs) [18, 19], based on Statistical Learning Theory, face classification problems from the Structural Risk Minimization perspective as opposed to many other models that are based on the Empirical Risk Minimization paradigm. As a result, they show good generalization performance so they have been applied to wide range of real-life problems [20].

For this problem of finding an hyperplane to separate two classes (one-class classification), SVMs tries to find the optimal hyperplane that not only separates the classes with no error but also maximizes the distance to closest point (for either class).

SVMs can be seen as classifiers where the loss function is the Hinge function, defined as:

$$L[y, f(x)] = max[0, 1 - yf(x)] \tag{1}$$

Being x an observation from input features, y the class x belongs to, and $f(x)$ the output of the classifier.

Once trained, that is the support vectors are identified and the margin is maximized, a SVM can be seen as:

$$f(x) = \sum_{i \in S} \alpha . y_i . \langle x_i, x \rangle + \beta_0 \tag{2}$$

Being S the set of support vectors, α the classifier coefficients, and β the predictor coefficients.

SVMs have been proven as top classifiers and have been also applied to multi-class classification although they were initially designed for single-class classification. In present work, class information is the success of the projects (true or false). Hence, SVMs face a one-class classification in present study.

2.2 k-Nearest Neighbor

k-Nearest Neighbor (k-NN) classifiers [21, 22] are one of the most straightforward approaches to classification as they are based on the idea that a data sample (x) is classified based on the class of its k nearest neighbors. Because they make no underlying assumptions about the statistical structure of the data, they are considered non-parametric classifiers.

In the case of 1-NN, x is assigned the class of its nearest neighbor and all the other data are ignored. In general k-NN, x is assigned the majority class of the nearest k neighbors. In present research, 10 neighbors are used in determining the class ant he Minkowski distance has been applied. This distance metric, which can be considered as a generalization of both the Euclidean distance and the Manhattan distance, is defined as follows:

$$d_{st} = \sqrt[p]{\sum_{j=1}^{n} |x_{sj} - x_{tj}|^p} \tag{3}$$

Where p is a scalar positive value of the exponent, s and t are the indexes of the rows of vector x and j is the index of the column of vector x.

2.3 Random Forest

Classification trees [23] are well-known and inductive learning methods. Within they inner (tree) structure there are two types of nodes; leaf nodes that are those for taking the final decision (prediction) and internal nodes that are those associated to differentiate responses (branches) for a given question regarding the values of a feature from the original training dataset. All internal nodes have at least two child nodes. Labels are assigned to both archs connecting a node to one of its child nodes (their content is related to the responses to the node question) and leaf nodes (their content is one of the classes in the training dataset). They have proved to be valuable tools for many interesting and challenging tasks such as description, classification and generalization of data [24].

Random Forest (RF) algorithm [25] can be seen as an aggregation of a number of classification trees such that each one of them depends on the values of a random vector. This vector is sampled independently and with the same distribution for all trees in the forest. One of the main advantages, when compared to a single classification tree schema, is the reduction of variance. In the case of RF, the prediction is obtained for a new data by aggregating (through majority voting) the predictions made by all the single trees. That is, the new data is assigned to the class that was most often predicted by the individual trees.

In order to select an appropriate number of trees (that is one of the requirements of Random Forests), conclusions from recent work [26] have been taken into account. According to it, for the binary classification problem that is addressed in present paper, no tuning has been carried out and the number of trees has been set to 100. To estimate the accuracy of the RF for each individual (subset of features), the Out-of-Bag (OOB) error rate is measured and analysed. It is calculated by predicting the class for each training sample by using only the trees for which this observation was not included in the bootstrap sample. That is, training data are not used to calculate the OOB error rate of a given tree.

3   Experiments and Results

Present section introduces the analyzed dataset as well as the main obtained results.

3.1   Dataset

We obtained our sample from the World Bank's Private Participation in Infrastructure (PPI) dataset. We follow a previous paper [5] and select a sample of private participation projects from 18 Central and Eastern European countries. As previously mentioned, the interest of these countries is due to a two-fold reason. First, this region is recently attracting increasing flows of foreign investments as a result of their market attractiveness and business-friendly environment. Second, it is a region where the use of private participation projects is frequent. Specifically, the countries covered in our sample are: Armenia, Azerbaijan, Belarus, Bulgaria, Georgia, Kazakhstan, Kyrgyz Republic, Lithuania, Macedonia, Moldova, Romania, Russia, Serbia, Tajikistan, Turkey, Turkmenistan, Ukraine, and Uzbekistan.

We analyze projects in the Energy and Telecommunication sectors, because of their critical impact on the economy and also because they are the two sectors that account for the highest share of private participation projects. 1127 of the projects in the sample are from the Energy sector whereas 2765 are from the Telecommunication one.

Projects in our sample have completed the bidding process, fulfilled the legally binding agreements, and raised the necessary funds. We follow previous studies in the field [2, 5, 27] and code the success or failure of a project using a binary measure. Projects with a status reported in the PPI dataset as "operational", "merged" or "concluded" are considered successful and coded 1. In contrast, projects "cancelled" (when the private sector has exited) or "distressed" (when the government or the operator has either requested termination or is under international arbitration) are considered as failed and coded 0.

As explanatory variables, we replicate the variables employed by [5] and include various host country-level and project-level control variables. At the host-country level, we include the level of corruption as measured by the World Bank Worldwide Governance Indicators (WGI) [28]. For an easier interpretation, we reverse-coded this variable so a higher number indicates higher corruption and a lower number indicates lower corruption [29]. Further, we also included the log of GDP, GDP growth, the log of the unemployment rate and the level of political stability as measured by the Polconv index [30]. At the project level, we included the log of total investment, age, time lag, whether the main investor of the project was a foreign company, whether at least one company from the host country is included in the project consortium, whether the project is publicly traded, whether the host government hold any stake in the ownership of the project, and greenfield versus brownfield projects.

All in all, 13 features are compressed in each one of the datasets for all the project instances (1127 in the case of the Energy sector and 2765 in the case of the Telecommunication sector).

3.2   Results from the Energy Sector

Results obtained when applying SVM, k-NN and RF to the Energy-sector dataset are shown in Table 1.

Table 1. Classification results for the Energy sector.

Classifier	k/trees	Classification loss
SVM	5	0.1171
	10	0.1136
	20	0.1109
k-NN	5	0.0949
	10	0.0932
	20	0.0896
RF	100	0.0648

From these results, it can be said that RF outperforms SVM and k-NN, getting a lowest (41% and 27% respectively) classification loss. As previously explained, 100 trees were built within the RF model, that has obtained an effective prediction of the project success in the Energy sector. On the other hand, the classification loss of both SVM and k-NN decreases when increasing the number of folds (k).

3.3   Results from the Telecommunication Sector

Results obtained when applying SVM, k-NN and RF to the Telecommunication-sector dataset are shown in Table 2.

Table 2. Classification results for the Telecommunication sector.

Classifier	k/trees	Classification loss
SVM	5	0.0253
	10	0.0253
	20	0.0253
k-NN	5	0.0170
	10	0.0156
	20	0.0145
RF	100	0.0116

As for the previous sector, RF outperforms SVM and k-NN, getting a lowest (54% and 20% respectively) classification loss. Similarly, 100 trees were built within the RF model, which has obtained an effective prediction of the project success in the Telecommunication sector.

When comparing results in Tables 1 and 2, it can be concluded that the given features in the dataset let us perform a better forecast in the case of the Telecommunication sector as the lowest classification loss (0.0116) is much lower (44%) than the one obtained for the Energy sector (0.0648). The results therefore suggest that it is easier to forecast the success of the projects in the Telecommunication sector, perhaps as a consequence of the higher exposure of the Energy sector to unpredictable risks that can affect the generation and the consumption of energy (e.g., climate conditions). Also, the results are in line with previous research that has emphasized the particular idiosyncrasy of the Energy sector, in which the role of policy risk has been shown to critically impact investments in energy and where the heterogeneous levels of political capabilities and previous experience dealing with policy risk across firms can greatly affect the investment [31].

4 Conclusions

From the results in Sect. 3, it can be concluded that the proposed classifiers can be effectively applied to predict the success in project participation projects in the Energy and Telecommunication sectors within Europe. RF clearly outperforms SVM and k-NN when trying to classify such projects. These results imply that the level of risk in private participation projects in the Energy sector are subject to greater risk levels as it is more difficult to forecast their success compared to projects in the Telecommunication sector.

While the causes for this higher levels of risk may be due to the higher uncertainty in the generation and consumption of energy due to the higher exposure to uncontrollable factors, the consequences for investors in this type of private participation project can be very relevant in terms of higher premiums in the cost of capital when asking for loans in the financial markets, increased requirements of guarantees from host governments and suppliers, and greater public media surveillance and mistrust from the public media and opinion. In turn, these increase operational costs might reduce the performance of the project (unless they are absorbed by consumer via higher prices), increasing the failure rate compared to other sectors where the lower uncertainty does not translate into higher operating costs.

The lower predictability of success in Energy projects can also have repercussions in terms of the attractiveness of these projects for investors, who might reduce their offers when bidding for the projects in order to buffer themselves from the higher risks. As a result, governments privatizing projects in energy might obtain a lower amount of resources than when privatizing other comparable infrastructure projects in other sectors. Our results are therefore particularly valuable for countries in need of foreign investments as they provide useful recommendations in terms of which infrastructure projects might be better received by international investors and this obtain better offers.

Future work will focus on comparing some additional classifiers as well as extending the present study by considering some other sectors where private participation projects are frequently employed, apart from the Energy and Telecommunication ones, such as Transport or Water Sewerage. Besides, additional explanatory variables from the host country, such as other measures from the institutional environment or from the experience of investors, can be taken into consideration to expand our understanding of the determinants of success for private participation projects.

References

1. Henisz, W.J., Zelner, B.A., Guillén, M.F.: The worldwide diffusion of market-oriented infrastructure reform, 1977–1999. Am. Sociol. Rev. **70**, 871–897 (2005)
2. Jiang, Y., Peng, M.W., Yang, X., Mutlu, C.C.: Privatization, governance, and survival: MNE investments in private participation projects in emerging economies. J. World Bus. **50**, 294–301 (2015)
3. Ramamurti, R., Doh, J.P.: Rethinking foreign infrastructure investment in developing countries. J. World Bus. **39**, 151–167 (2004)
4. Jiménez, A., Salvaj, E., Lee, J.Y.: Policy risk, distance, and private participation projects in Latin America. J. Bus. Res. **88**, 123–131 (2018)
5. Jiménez, A., Russo, M., Kraak, J.M., Jiang, G.F.: Corruption and private participation projects in Central and Eastern Europe. Manage. Int. Rev. **57**, 775–792 (2017)
6. Palmero, J.A., Herrera, J.J.D., de la Fuente Sabaté, J.M.: The role of psychic distance stimuli on the East-West FDI location structure in the EU. Evidence from Spanish MNEs. J. East Eur. Manage. Stud. **18**, 36–65 (2013)
7. Akbar, Y., Balboni, B., Bortoluzzi, G., Dikova, D., Tracogna, A.: Disentangling resource and mode escalation in the context of emerging markets. Evidence from a sample of manufacturing SMEs. J. Int. Manage. **24**, 257–270 (2018)
8. Pinzón, C., Herrero, Á., De Paz, J.F., Corchado, E., Bajo, J.: CBRid4SQL: A CBR Intrusion Detector for SQL Injection Attacks, pp. 510–519. Springer, Heidelberg (2010)
9. Ksieniewicz, P., Krawczyk, B., Woźniak, M.: Ensemble of Extreme Learning Machines with trained classifier combination and statistical features for hyperspectral data. Neurocomputing **271**, 28–37 (2018)
10. Herrero, Á., Jiménez, A.: Improving the management of industrial and environmental enterprises by means of soft computing. Cybernet. Syst. **50**, 1–2 (2019)
11. Jiménez, A., Herrero, Á.: Selecting features that drive internationalization of Spanish firms. Cybernet. Syst. **50**, 25–39 (2019)
12. Simić, D., Svirčević, V., Ilin, V., Simić, S.D., Simić, S.: Particle swarm optimization and pure adaptive search in finish goods' inventory management. Cybernet. Syst. **50**, 58–77 (2019)
13. Herrero, Á., Jiménez, A., Bayraktar, S.: Hybrid unsupervised exploratory plots: a case study of analysing foreign direct investment. Complexity (2019, in press)
14. Huang, Z., Chen, H., Hsu, C.-J., Chen, W.-H., Wu, S.: Credit rating analysis with support vector machines and neural networks: a market comparative study. Decis. Support Syst. **37**, 543–558 (2004)
15. Chou, J.-S., Cheng, M.-Y., Wu, Y.-W.: Improving classification accuracy of project dispute resolution using hybrid artificial intelligence and support vector machine models. Expert Syst. Appl. **40**, 2263–2274 (2013)
16. Malhotra, M.K., Sharma, S., Nair, S.S.: Decision making using multiple models. Eur. J. Oper. Res. **114**, 1–14 (1999)
17. Yu, M.C.: Multi-criteria ABC analysis using artificial-intelligence-based classification techniques. Expert Syst. Appl. **38**, 3416–3421 (2011)
18. Boser, B.E., Guyon, I.M., Vapnik, V.N.: A training algorithm for optimal margin classifiers. In: 5th Annual Workshop on Computational Learning Theory, pp. 144–152. ACM (1992)
19. Cortes, C., Vapnik, V.: Support-vector networks. Mach. Learn. **20**, 273–297 (1995)
20. Byun, H., Lee, S.-W.: Applications of Support Vector Machines for Pattern Recognition: A Survey, pp. 213–236. Springer, Heidelberg (2002)

21. Cunningham, P., Delany, S.J.: k-Nearest neighbour classifiers. Technical report UCD-CSI-2007-4 (2007)
22. Cover, T.M., Hart, P.E.: Nearest neighbor pattern classification. IEEE Trans. Inf. Theory **13**, 21–27 (1967)
23. Safavian, S.R., Landgrebe, D.: A survey of decision tree classifier methodology. IEEE Trans. Syst. Man Cybern. **21**, 660–674 (1991)
24. Sreerama, K.M.: Automatic construction of decision trees from data: a multi-disciplinary survey. Data Min. Knowl. Disc. **2**, 345–389 (1998)
25. Breiman, L.: Random forests. Mach. Learn. **45**, 5–32 (2001)
26. Probst, P., Boulesteix, A.-L.: To tune or not to tune the number of trees in random forest? J. Mach. Learn. Res. **18**, 1–18 (2018)
27. Jiménez, A., Jiang, G.F., Petersen, B., Gammelgaard, J.: Within-country religious diversity and the performance of private participation infrastructure projects. J. Bus. Res. **95**, 13–25 (2019)
28. Keig, D.L., Brouthers, L.E., Marshall, V.B.: Formal and informal corruption environments and multinational enterprise social irresponsibility. J. Manage. Stud. **52**, 89–116 (2015)
29. Cuervo-Cazurra, A.: Better the devil you don't know: types of corruption and FDI in transition economies. J. Int. Manag. **14**, 12–27 (2008)
30. Henisz, W.J.: The institutional environment for infrastructure investment. Ind. Corp. Change **11**, 355–389 (2002)
31. Holburn, G.L., Zelner, B.A.: Political capabilities, policy risk, and international investment strategy: evidence from the global electric power generation industry. Strateg. Manag. J. **31**, 1290–1315 (2010)

Optimizing a Bi-objective Vehicle Routing Problem Appearing in Industrial Enterprises

Ana D. López-Sánchez[1], Alfredo G. Hernández-Díaz[1(✉)], Julián Molina[2], and Manuel Laguna[3]

[1] Pablo de Olavide University, Seville, Spain
agarher@upo.es
[2] Málaga University, Málaga, Spain
[3] University of Colorado, Denver, USA

Abstract. A Muti-Start Multiobjective Local Search algorithm is implemented to solve a bi-objective variant of the Vehicle Routing Problem appearing in industry and environmental enterprises. The problem seeks to design a set of routes for each time on a period and in which the service frequency is a decision variable. The algorithm minimizes total emissions by all vehicles and maximizes service quality measure as the number of times that a customer is visited in order to be served.

1 Introduction

Governments usually impose production suspensions on industrial enterprises regardless of the emission levels in their counties because of inhabitants discontent and the environmental pollution. They apply different measurements regarding the production in order to satisfy environmental requirements since nowadays this is a crucial issue. In this work, we will focus on industrial logistics enterprises. Industrial logistics service implies shipment of raw materials and end products by means of different kind of vehicles (trains or trucks, among others) in order to meet needs of enterprises working different economic sectors.

In this work, we will specifically address the transportation of products delivery by industrial enterprises to their customers representing here different economic sectors. The considered problem is coined as Bi-objective Periodic Vehicle Routing Problem with Service Choice (Bi-PVRP-SC). The model designs vehicle routes for every day of a planning horizon (for example, one week or month) having a fleet of capacitated vehicles to perform routes that begin and end at a single depot. Customers are visited as minimum a preset number of times over the planning horizon. The Bi-PVRP-SC seeks to minimize total emissions because if the quantity is high the industrial enterprises will get production suspensions and to maximize service frequency since different economic sectors considers that the service given by the enterprises will be better if they are served with a high frequency. Note that in our model, vehicles are assigned to customers and days;

© Springer Nature Switzerland AG 2020
F. Martínez Álvarez et al. (Eds.): SOCO 2019, AISC 950, pp. 452–462, 2020.
https://doi.org/10.1007/978-3-030-20055-8_43

customers require a minimum number o visits during the planning period that depends on the demand rate. Furthermore, the service frequency is chosen during the search while considering that each customer requires a minimum number of visits during the planning horizon. Of course, the higher service level is linked to a higher service frequency and hence, a higher quantity of emissions by all vehicles but a better service given by industrial enterprises. Note that total emissions are proportional to the amount of greenhouse gas (GHG) emitted.

The Bi-PVRP-SC can be seen as a generalization of the Periodic Vehicle Routing Problem (PVRP). The PVRP is a single-objective optimization problem and the visits are chosen from a menu of schedule options instead of being another objective function as happen in the Bi-PVRP-SC. The PVRP was proposed by [2], see also, [5,7,12]. For an comprehensive review see [4] and for a recent related problem see [1]. Another variant is the Periodic Vehicle Routing Problem with Service Choice (PVRP-SC) introduced by [6]. The PVRP-SC balances the travel cost and the service level by suggesting efficient single-objective solution methods. To the best of our knowledge the Bi-PVRP-SC has not been tackled in the literature. [9] address a similar problem but they solve a bi-objective waste collection problem in which they minimize the sum of fixed and variable costs throughout the planning horizon (where, fixed costs depend on the class of vehicle used and the day) and service is measured as the waste accumulated throughout the planning horizon.

To solve the Bi-PVRP-SC a dual-phase Multi-Start Multiobjective Local Search (MSMLS) algorithm is proposed, see [3,10]. In the first phase, feasible solutions are generated in order to obtain an initial approximation of the Pareto front. The second phase attempts to improve the approximation of the Pareto front by performing local searches that employ several neighborhoods. We use the classical definition of an efficient solution as one for which no single-objective function value can be improved without deteriorating another objective function value.

The paper is organized as follows. Section 2 describes the proposed algorithm. Computational results are provided in Sect. 3 showing a comparison of our algorithm against the well-known NSGA-II algorithm. Finally, Sect. 4 summarizes the main conclusions.

2   Multi-start Multiobjective Local Search Procedure

We propose a solution method called Multi-Start Multiobjective Local Search (MSMLS), which is a neighborhood based metaheuristic designed to solve multiobjective problems. One of the key aspects of the proposed procedure is its simplicity and robustness. The MSMLS includes two different phases: first, it generates an initial feasible solution and then it applies a multiobjective local search to improve this solution. The phases are repeated until a stopping criterion is satisfied. Note that the goal of multiobjective optimization is to obtain the set of non-dominated solutions that approximates the set of efficient solutions, $\hat{E}$.

2.1   Phase 1. Solution Representation and Construction

We introduce a compact solution representation that is both flexible and realistic, considering the context at hand that is able to use the direct information obtained from the customers (minimum frequency and maximum capacity) without the need to previously generate a set of feasible service profiles. Within this problem, there are three main decisions that lead to the construction of a solution for a given planning horizon, each of them will be represented by using the following variables:

Decision 1. The number of visits to give the service to each customer.

For all customers $i \in N$, $freq_i$ is an integer value among the minimum preset frequency and the planning horizon, i.e., $\gamma_i \leq freq_i \leq |T|$, that indicates the number of days customer i is visited.

Decision 2. The service profile of each customer (i.e., the actual days when each customer will be visited).

For all customers $i \in N, j = \{1, \ldots, freq_i\}$, t_{ij} is an integer value between 1 and $|T|$ that indicates the days that customer i is visited. Note that for a given i all t_{ij} values are different.

Decision 3. The number of routes and the order in which each customer will be visited, on each day.

For all customers $i \in N, j = \{1, \ldots, freq_i\}$, $r_{ij} = 1$ if customer i is the first one visited on a new route in day t_{ij}, otherwise, $r_{ij} = 0$.

For all days $t \in T$ and $i \in N$, O_{it} indicates the position in which customer i is visited on day t.

In order to understand the solution representation clearly, consider an example with $|N| = 5$ and $|T| = 3$, and the following values for the rest of the variables shown in Tables 1 and 2:

Table 1. Assignation of number of visits and determination of services profiles

Customer	Frequency	Days	First visit
Customer 1	$freq_1 = 1$	$t_{11} = 1$	$r_{11} = 1$
Customer 2	$freq_2 = 2$	$t_{21} = 1$	$r_{21} = 0$
		$t_{22} = 3$	$r_{22} = 1$
Customer 3	$freq_3 = 3$	$t_{31} = 3$	$r_{31} = 0$
		$t_{32} = 2$	$r_{32} = 0$
		$t_{33} = 1$	$r_{33} = 0$
Customer 4	$freq_4 = 2$	$t_{41} = 2$	$r_{41} = 0$
		$t_{42} = 3$	$r_{42} = 0$
Customer 5	$freq_5 = 1$	$t_{51} = 3$	$r_{51} = 0$

- $freq_1 = 1$, that is, customer 1 must be visited once during the three days of the planning period; $freq_2 = 2$, customer 2 must be visited twice; $freq_3 = 3$, customer 3 must be visited every day; $freq_4 = 2$, customer 4 must be visited two days; and finally, $freq_5 = 1$, customer 5 must be visited just one day during the three days of the planning period, see Table 1.

Table 2. Ordered list of customers per day

Table 2. Ordered list of customers per day

O_{ij}	Customer 1	Customer 2	Customer 3	Customer 4	Customer 5
Day 1	4	**3**	**2**	**1**	5
Day 2	5	2	**3**	1	**4**
Day 3	1	**4**	**3**	**2**	**5**

- $t_{11} = 1$, i.e., customer 1 is served on day 1; $t_{21} = 1, t_{22} = 3$, customer 2 is served on days 1 and 3; $t_{31} = 3, t_{32} = 2, t_{33} = 1$, customer 3 is served on days 3, 2, and 1; $t_{41} = 2, t_{42} = 3$, customer 4 is served on day 2 and 3; and $t_{51} = 3$, customer 5 is served on day 3, see Table 1.
- $r_{11} = 1, r_{21} = 0, r_{22} = 1, r_{31} = 0, r_{32} = 0, r_{33} = 0, r_{41} = 0, r_{42} = 0, r_{51} = 0$, that is, customer 1 is the first customer visited within a new route on day 1 since $t_{11} = 1$ and $r_{11} = 1$ and customer 2 is the first customer visited within a new route on day 3 since $t_{22} = 3$ and $r_{22} = 1$, see Table 1.
- $O_{11} = 4, O_{12} = 3, O_{13} = 2, O_{14} = 1, O_{15} = 5$, i.e, on day 1 the order to visit customers should be $4 - 3 - 2 - 1 - 5$ but only customers $3 - 2 - 1$ must be visited on such a day, see Table 2 where the customers highlighted in bold are the selected to by visited on the first day.
- $O_{21} = 5, O_{22} = 2, O_{23} = 3, O_{24} = 1, O_{25} = 4$, i.e, on day 2 the order to visit customers should be $5 - 2 - 3 - 1 - 4$ but only customers $3 - 4$ must be visited on such a day, see Table 2 where the customers highlighted in bold are the selected to by visited on the second day.
- $O_{31} = 1, O_{32} = 4, O_{33} = 3, O_{34} = 2, O_{25} = 5$, i.e, on day 3 the order to visit customers should be $1 - 4 - 3 - 2 - 5$ but only customers $4 - 3 - 2 - 5$ must be visited on such a day, see Table 2 where the customers highlighted in bold are the selected to by visited on the third a day.

Taking into account r_{ij}, we obtain the following routes:

- Day 1, route 1: 0-3-2-0
- Day 1, route 2: 0-1-0
- Day 2, route 1: 0-3-4-0
- Day 2, there is no route 2
- Day 3, route 1: 0-4-3-0
- Day 3, route 2: 0-2-5-0

Note that for each day, the first route always starts visiting the first customer in the order list.

One of the best properties of this representation is that solutions are always feasible, except for two cases:

- if a vehicle is full to capacity before reaching the next customer with $r_{ij} = 1$. However, this solution can be easily made feasible. If a customer cannot be visited by a given vehicle due to the capacity constraint of the vehicle, we simply, at that point, set $r_{ij} = 1$ and then a new route starts at that customer.

– if the demand accumulated through the days between any two successive
visits is over the capacity of the customer, D_i for all $i \in N$. In this case, the
solution is not repaired because the algorithm will force visiting the customer
before its accumulated demand overloads.

With this solution representation, we develop an effective construction
method to produce high-quality feasible solutions with a reasonable computational effort. Our construction method employs two strategies to generate the
number of visits during the planning horizon:

1. Set the number of visits to the minimum allowed, in order to generate good
   solutions for the first objective, the minimization of total distance emissions.
2. Select the number of visits randomly between the minimum allowed and the
   total days in the planning horizon, in order to generate good solutions for the
   maximization of the second objective, the service frequency.

In both cases the service profile is design randomly while observing the customers' capacity constraints. Once we have the customers to be visited each
day, with either one of these two procedures, the routes for each day are initially
generated using a nearest neighbor method. Note that more than one route may
be required in those days where the number of visits and/or the accumulated
load is larger than the vehicle capacity.

2.2   Phase 2. Local Search

We use a Multi-objective Local Search procedure because our goal is to approximate the efficient set $\hat{E}$, rather than finding a single optimal point. Therefore,
we designed a local search based on the Proximate Optimality Principle (POP)
that states that efficient solutions are likely to be found close to other efficient
solutions, at an adjacent neighborhood. The POP concept, explained in [11] and
proposed by [8], is considered as a heuristic counterpart of the so-called Principle
of Optimality in dynamic programming. According to [11], the interpretation of
POP within multi-objective optimization is that efficient points are connected
by a curve inside the efficient set, and thus an effective way to obtain a good
approximation of the efficient set is to carry out a series of single-objective linked
local searches (*linked* meaning that the last point of one search becomes the initial point of the next search, and *single-objective* means that just one of the
objective functions is being optimized along the whole search) where each point
visited is checked for inclusion in the efficient set $\hat{E}$. The first local search starts
from the initial point generated in the Construction Phase and attempts to find
the optimal solution to the single-objective problem corresponding to f_1. Let x_1
be the last point visited at the end of this search. A local search is then applied
again to find the best solution to the single-objective problem corresponding to
f_2 using x_1 as the initial solution. Let x_2 be the last point visited at the end of
this search. At this point, we again solve the problem with the first objective f_1
using x_2 as the initial solution, in order to finish a cycle around the efficient set.

In order to complete a final approximation of the efficient set $\hat{E}$, a number NCS of single-objective local searches are implemented where compromise functions, with random weights, are used as the new objective function, as proposed in [11]. These compromise functions try to find efficient intermediate solutions, that is, solutions that balance the total distance and the service frequency.

For each of these local searches, we use a set of nine different neighborhood structures, described below:

$\mathcal{N}_1$: Within a single day, exchange the position of two customers (randomly chosen), but closer than three positions in the same route.

$\mathcal{N}_2$: Exchange the position of two customers (randomly chosen) in the same day, from the same or different routes.

$\mathcal{N}_3$: Exchange the position three customers (randomly chosen) in the same day, from the same or different routes.

$\mathcal{N}_4$: Change a day (randomly chosen) on which a customers (randomly chosen) is visited.

$\mathcal{N}_5$: Move a set of consecutive customers (randomly chosen) to another route (randomly chosen) within a single day.

$\mathcal{N}_6$: Reduce the number of routes in a day (randomly chosen) by trying to link two different routes (randomly chosen).

$\mathcal{N}_7$: Increase the frequency of a customer (randomly chosen) while decreasing the frequency of another one (randomly chosen). This is achieved by adding the first customer to a route (randomly chosen) and removing the second one from that same route.

$\mathcal{N}_8$: Increase the service frequency of a customer (randomly chosen) by one visit.

$\mathcal{N}_9$: Decrease the service frequency of a customer (randomly chosen) by one visit.

Each local search uses these neighborhoods, in the order shown, until $3|N|$ consecutive neighborhood solutions are unable to improve the current solution. That is, instead of limiting the total number of iterations or exploring the whole set of possible neighbors in each case (which may be too time consuming), each neighborhood $\mathcal{N}_j$ is allowed to generate neighbors until $3|N|$ consecutive ones fail to improve the current solution. Once happens neighborhood fails, the local search begins to use the following neighborhood structure, $\mathcal{N}_{j+1}$, until the same stopping condition is achieved. Finally, when the last neighborhood structure, $\mathcal{N}_9$, is used, the local search ends.

3   Computational Results

In this section, the proposed algorithm (MSMLS) and the NSGA-II are tested on a variety of PVRP instances from the literature. Specifically, a set of 32 instances was considered.[1] A description of the instances can be found in Table 3. The first

---

[1] http://www.vrp-rep.org/datasets.html.

Table 3. Instances description

Instance	Nodes	Vehicles	Days	Capacity	f_{min}	f_{max}	f_1	f_2	f_3	f_4	f_5	f_6
p01	50	3	2	160	50	100	50					
p02	50	3	5	160	104	250	17	26			7	
p03	50	1	5	160	50	250	50					
p04	75	2	5	140	75	150	75					
p05	75	6	5	140	153	375	30	34			11	
p06	75	1	10	140	75	750	75					
p07	100	4	2	200	75	200	100					
p08	100	5	5	200	156	500	40	46			14	
p09	100	1	8	200	75	800	100					
p10	100	4	5	200	132	500	40	46	14			
p11	139	4	5	235	192	695	103	22	12	1	1	
p12	163	3	5	140	185	815	148	8	7			
p13	417	9	7	2000	457	2919	337	40				
p14	20	2	4	20	40	80	8	8		4		
p15	38	2	4	30	72	152	16	16		6		
p16	56	2	4	40	104	224	24	24		8		
p17	40	4	4	20	80	160	16	16		8		
p18	76	4	4	30	144	304	32	32		12		
p19	112	4	4	40	208	448	48	48		16		
p20	184	4	4	60	336	736	80	80		24		
p21	60	6	4	20	120	240	24	24		12		
p22	114	6	4	30	216	456	48	48		18		
p23	168	6	4	40	312	672	72	72		24		
p24	51	3	6	20	90	306	36	9				6
p25	51	3	6	20	90	306	36	9				6
p26	51	3	6	20	90	306	36	9				6
p27	102	6	6	20	180	612	72	19				12
p28	102	6	6	20	180	612	72	19				12
p29	102	6	6	20	180	612	72	19				12
p30	153	9	6	20	279	918	108	27				18
p31	153	9	6	20	270	918	108	27				18
p32	153	9	6	20	270	918	108	27				18

column indicates the name of the instance, the second column is the number of nodes (representing the customers) without considering the depot, the third column contains the number of vehicles, the fourth column specifies the number of days in the planning horizon, the fifth column specifies the capacity of the vehicles

Table 4. Indicators

Instance	#Efficient Points		Hypervolume		CPU time	
	MSMLS	NSGA-II	MSMLS	NSGA-II	MSMLS	NSGA-II
p01	48	**52**	**0.6285**	0.5468	36.3	**20.1**
p02	**112**	111	**0.662**	0.1189	**71.6**	119.6
p03	**194**	138	**0.5821**	0.5099	208.0	**136.9**
p04	69	**97**	**0.6790**	0.6036	187.6	**84.0**
p05	**190**	177	**0.6557**	0.4818	**208.3**	625.3
p06	**573**	395	**0.6194**	0.5341	**889.8**	2088.1
p07	**94**	81	0.5751	**0.5848**	**111.5**	298.5
p08	**239**	211	**0.6197**	0.4895	**371.0**	2247.5
p09	**605**	379	**0.5935**	0.5409	**1864.5**	6265.0
p10	**264**	225	**0.6640**	0.5561	**470.0**	1704.3
p11	**395**	306	**0.7238**	0.6892	**2607.7**	10779.0
p12	**481**	371	**0.6735**	0.6219	**3846.0**	41567.2
p13	**1603**	1110	**0.7336**	0.5445	**43963.6**	67157.0
p14	**40**	32	**0.7378**	0.4544	10.3	**7.1**
p15	**81**	72	**0.6386**	0.4779	**46.3**	102.2
p16	118	**122**	**0.6132**	0.5294	**129.6**	450.7
p17	72	**95**	**0.6593**	0.6412	**40.0**	136.2
p18	**150**	146	0.6422	**0.6607**	**108.0**	1879.6
p19	**218**	214	**0.5723**	0.4726	**239.2**	6761.8
p20	**367**	316	0.5950	**0.6665**	**3660.5**	59323.8
p21	**108**	105	**0.6861**	0.5586	**83.9**	597.9
p22	**222**	202	**0.6227**	0.5036	**1230.5**	7514.7
p23	**333**	258	**0.5822**	0.4638	**5969.6**	32290.9
p24	**178**	135	**0.6566**	0.5756	**168.0**	745.5
p25	**187**	143	**0.6749**	0.5987	**178.7**	682.4
p26	**179**	169	**0.6247**	0.5578	**207.6**	682.4
p27	**333**	279	**0.7426**	0.6801	**1568.3**	10778.8
p28	**313**	310	**0.7208**	0.6788	**1568.3**	10614.6
p29	**311**	264	**0.7364**	0.6676	**1330.3**	10315.8
p30	**464**	419	**0.7617**	0.7123	**2223.5**	10315.8
p31	**521**	439	**0.7819**	0.7075	**3944.1**	49801.2
p32	**447**	408	**0.7635**	0.7081	**2554.0**	59062.6

and the remaining columns give the frequency information, where f_i is the number of nodes that must be visited i times. The experiments were conducted on a MacBook Pro 2.7 GHz Intel Core i5 and 16GN 1867 MHz DDR3, and the algorithms are implemented using C++ and executed with Xcode Version 7.3.

460   A. D. López-Sánchez et al.

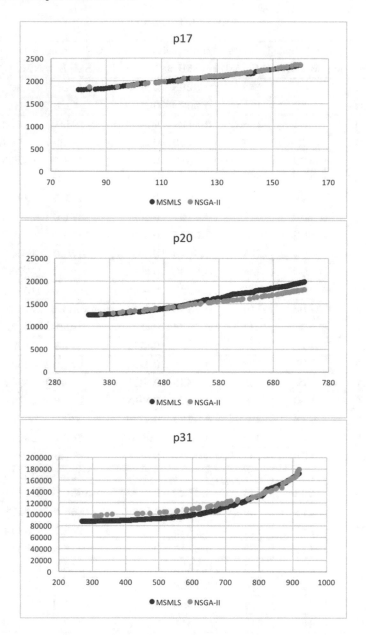

Fig. 1. Trade-off between total emissions and service frequency (MSMLS and NSGA-II)

Table 4 shows that in most of the cases the proposed algorithm (MSMLS) obtains more efficient points than the NSGA-II (except in problems 01, 04, 16, and 17, which are quite similar). Of course, decision makers prefer more rather than fewer efficient points. But furthermore, not only more efficient points are obtained but also a better set of solutions in most of the instances. This is confirmed calculating the hypervolume, that gives the advantage to MSMLS over NSGA-II except in problems 07, 18, and 20. This means that the size of the space covered by the MSMLS algorithm is larger than the space covered by the NSGA-II algorithm. In term of computational effort, it is worth mentioning that the CPU time for the easiest problems is quite similar for both procedures and that when the complexity of the problem increases, NSGA-II is more time consuming.

In order to show graphically some results, Fig. 1 draws the tradeoff of the total emissions and the service frequency associated with the estimated efficient frontier obtained with MSMLS and NSGA-II, for 3 out of 32 instances: p17 shows a similar tradeoff between both algorithms, p20 displays a better tradeoff for the NSGA-II and p31 shows a better tradeoff for the MSMLS. Note that most of instances are similar to instance p31 (as can be seen in Table 4). It can be observed that MSMLS gets denser Pareto fronts than NSGA-II.

4   Conclusions

The bi-objective periodic vehicle routing problem with service choice has been introduced and analyzed. The problem considers two objectives: the minimization of total emissions and the maximization of the service frequency to improve the quality of the service. Both objectives are clearly in conflict because an increase in service (i.e., frequency of visits) typically results in an increase in total emissions.

A solution of the problem consists of a schedule to visit customers during the planning horizon. We propose to represent the requirements of a customer as a minimum number of visits and a capacity. This representation is very flexible and allows the modeling a variety of different practical problems in industrial enterprises. Due to the complexity of the problem, a heuristic procedure is proposed. Specifically, a multi-start multiobjective local search is developed, whose output is an approximation of the efficient frontier that can be used to trade off total emissions and service frequency.

Computational experiments graphically show the performance advantage of using our procedure over an adaptation of NSGA-II. Our results establish the first benchmarks for this problem that can be used for future developments and improvements. Future research will include indicators as the coverage or the ϵ-indicator.

References

1. Archetti, C., Fernández, E., Huerta-Muñoz, D.L.: The flexible periodic vehicle routing problem. Comput. Oper. Res. **85**, 58–70 (2017)
2. Beltrami, E.J., Bodin, L.D.: Networks and vehicle routing for municipal waste collection. Networks **4**(1), 65–94 (1974)
3. Braysy, O., Hasle, G., Dullaert, W.: A multi-start local search algorithm for the vehicle routing problem with time windows. Eur. J. Oper. Res. **159**(3), 586–605 (2004)
4. Campbell, A.M., Wilson, J.H.: Forty years of periodic vehicle routing. Networks **63**(1), 2–15 (2014)
5. Christofides, N., Beasley, J.E.: The period routing problem. Networks **14**(2), 237–256 (1984)
6. Francis, P., Smilowitz, K., Tzur, M.: The period vehicle routing problem with service choice. Transp. Sci. **40**(4), 439–454 (2006)
7. Gaudioso, M., Paletta, G.: A heuristic for the periodic vehicle routing problem. Transp. Sci. **26**(2), 86–92 (1992)
8. Glover, F., Laguna, M.: Tabu Search. Kluwer Academic Publishers, Norwell (1997)
9. Gómez, J.R., Pacheco, J., Gonzalo-Orden, H.: A tabu search method for a bi-objective urban waste collection problem. Comput. Aided Civ. Infrastruct. Eng. **30**(1), 36–53 (2015)
10. Martí, R., Aceves, R., León, M.T., Moreno-Vega, J.M., Duarte, A.: Intelligent Multi-Start Methods, pp. 221–243. Springer International Publishing, Cham (2019)
11. Molina, J., Laguna, M., Marti, R., Caballero, R.: Sspmo: a scatter tabu search procedure for non-linear multiobjective optimization. INFORMS J. Comput. **19**(1), 91–100 (2007)
12. Russell, R., Igo, W.: An assignment routing problem. Networks **9**(1), 1–17 (1979)

An Industrial Application of Soft Computing for the Design of Personalized Call Centers

David Griol$^{(\boxtimes)}$, Jose Manuel Molina, and Araceli Sanchis

Department of Computer Science, Universidad Carlos III de Madrid, Leganés, Spain
{david.griol,josemanuel.molina,araceli.sanchis}@uc3m.es

Abstract. In service industries such as telecommunications, hotels, insurance, banking, retail, or medical services, companies are increasingly paying more attention to human-computer communication systems that are in direct contact with customers, and focused on achieving the desired profit and market share goals. For this reason, chatbots are increasingly used in service industries starting with simple chat conversation up to more complex functionalities based on soft computing methodologies. Evaluation methodologies for chatbots try to provide an efficient means of assessing the quality of the system and/or predicting the user satisfaction. In this paper we present a clustering approach to provide insight on whether user profiles can be automatically detected from the interaction parameters and overall quality predictions, providing a way of corroborating the most representative features for defining user profiles. We have carried out different experiments for a practical dialog system, from which the clustering approach provided an efficient way of easily distinguishing between different user groups and complete a more significant evaluation of the system.

Keywords: Chatbots · Spoken interaction · Soft computing ·
Clustering · User modeling · Evaluation

1 Introduction and Background

Soft computing (SC) is an evolving collection of methodologies, which aims to exploit tolerance for imprecision, uncertainty, and partial truth to achieve robustness, tractability, and low cost. SC provides an attractive opportunity to represent the ambiguity in human thinking with real life uncertainty [8]. It combines machine learning methods (neural networks, support vector machines), fuzzy logic, evolutionary computation (genetic algorithms, differential evolution, metaheuristics and swarm Intelligence), data clustering algorithms (fuzzy clustering, sequential data clustering approaches), and probabilistic methods such as bayesian networks.

This work has been partially supported by Spanish projects TEC2017-88048-C2-2-R and TRA2016-78886-C3-1-R.

© Springer Nature Switzerland AG 2020
F. Martínez Álvarez et al. (Eds.): SOCO 2019, AISC 950, pp. 463–472, 2020.
https://doi.org/10.1007/978-3-030-20055-8_44

During the last years, SC has caused a breakthrough in many engineering and science fields since it can solve problems that have not been able to be solved by traditional analytical methods. In addition, it offers rich knowledge representation, flexible and advanced human-like knowledge acquisition and processing.

The various kinds of cognitive and reactive Artificial Intelligence provided by SC make industrial systems intelligent, adaptive, autonomous, decision supported, and optimized [4]. This innovation potential has been used considerably in human-related fields such as aerospace applications, communication systems, consumer appliances, electric power systems, manufacturing automation and robotics, power electronics and motion control, process engineering, and transportation [9].

Since communication systems involve human beings, soft computing can be effectively applied to such systems to incorporate the previously described human-like and user-friendly adaption of human-computer communication systems to mimic human thinking and behavior. Chatbots and automatic and efficient call centers directly contribute to the success of the whole organizations that use them. These human-computer communication systems are computer programs that engage the user in a dialog using spoken natural language that aims to be similar to that between humans [5,7]. These systems are currently very widespread and employed for an increasing number of applications due to the improvements achieved in their constituent modules (e.g. higher speech recognition accuracy, more fine-grained speech understanding, etc.).

Customer behavior modeling and personalized management are increasingly becoming important in the sales and service oriented industry [8–10]. Within this field, soft computing techniques are employed to identify customers' needs, selecting the prospective customers, understanding the reasons behind customer churn and detect ways to provide better customer service are the key capabilities all the service industries are looking to develop.

Most existing large call centers collect data which is then used to assess and improve the performance of its representatives. Typically, such data includes some form of quality assessment, time management representation, and business processing aspects. While the combination of this data and soft computing and data mining methods have been applied to develop and improve the different modules of chatbots (e.g., Hidden Markov models and Gaussian mixture models for Automatic Speech Recognition; Hidden Vector State model, Stochastic Finite State Transducers, Dynamic Bayesian Networks, Support Vector Machines and Conditional Random Fields for Spoken Language Understanding; Partially observable Markov Decision Processes and Bayesian Networks for Dialog Management; Markov Decision Processes and Reinforcement Learning for Natural Language Generation) [5], there is not much research to apply the data collected and statistical methodologies to detect the different kinds of users interacting with the system and assess real user profiles.

Soft Computing techniques can also be employed in the design of personalized chatbots that deal with information sources related to the users' preferences and requirements and the context of the interaction to [9]: proper handle customer requests, optimize the overall time the chatbot takes to deal with the customers' queries, employ the knowledge about the customer to adapt the services and information provided by the chatbot, predict customers' churn to help companies identifying the prospective customers and preventing them stopping using again the chatbot.

Some chatbots automatically adapt to the users by identifying them and changing their dialog strategies according to what they had previously learned in a corpus of interactions of such users [3]. Such tailored adaptation is efficient mainly in domains in which the system is used very frequently by a reduced number of users. In more open domains in which the system may be used by thousands of not frequent callers, some other strategies must be implemented. Despite the systems that are specially designed for specific population groups such as children or handicapped people, the decision of which groups must be taken into account to adapt the system strategies is not trivial in most application domains and there are few previous studies on the evaluation of the appropriateness of such decision.

In this paper, we present an approach based on clustering to assess whether the user groups considered to implement a system establish meaningful differences in their interaction behavior, which provides industries with a valuable feedback about the appropriateness of considering such user grouping. In order to do so, we propose to carry out a clustering based on interaction parameters and subjective judgments of real users of an interaction corpus and study whether the different user groups are balanced between the clusters or not. We have used this approach with the Let's Go system to obtain feedback about grouping users according to age and technical affinity in its domain.

The rest of the paper is organized as follows. Section 2 describes the CMU Let's Go spoken dialog system. Section 3 presents the experimental set-up describing briefly the corpus employed and the parameters that were computed to carry out the proposed clustering as well as the main experiments carried out. Section 4 presents a discussion of the results obtained. Finally, Sect. 5 presents the conclusions and some guidelines for future work.

2 Let's Go System

Let's Go is a spoken dialog system developed by the Carnegie Mellon University to provide bus schedule information in Pittsburgh. The system has had many users since it was made available for the general public in 2005 - 20,000 calls collected just from March to December of 2005 - [1], so there is a substantial dataset that can be used to train a dialog model. In addition, this large amount of data from spoken interactions has been acquired with real callers, rather than lab testers.

We have chosen the Let's Go task to evaluate our proposal for several reasons. Firstly, the corpus available was gathered from a real task in an operative dialog system that provided its service to real users. This poses a challenge to build realistic user models and find new dialog strategies that are at least as good as the hand-crafted system. Secondly, Let's Go is a common ground for experimentation and evaluation within the dialog system community, which therefore makes our results directly comparable to the alternatives presented by other authors, and this is why it has been intensively used by researchers in the last years.

Figure 1 shows an example of a dialog extracted from the Let's Go corpus. In each dialog, the user needs to provide a place of departure, a destination and a departure time. To model this, the system uses a set of user dialog acts that has been classified into 16 categories. Four of the concepts are used to model where the user is leaving from (monument, pair of road names, neighborhood, or stop). The four concepts used for modeling the place of arrival are similar. Six concepts are used for describing the user's required time of travel (next bus or specific times). The *meth* node describes whether the user is asking for a bus with some constraints, is finished or wants to restart. The concept *disc* models how the user issues "discourse" actions, which relate to only one turn in a dialog.

```
System: Welcome to the Pittsburgh bus information line. You can say start over anytime. Where
are you going to take your bus from?
User: Hi yeah I want to go from McKeesport
System: Sure. Where do you want to go?
User: Duquesne.
System: Did you mean Duquesne?
User: Yes
System: At what time do you want to take it? Say next bus or a time for example 8.40pm.
User: Around 4.15pm
System: I'm going to search for buses around 4.15pm.
Is that OK?
User: Yes.
System: On which day?
User: Today
System: I have found one bus. It is at 4.15pm. There is a bus number 61C from McKeesport. It
will arrive in Duquesne at 4.21pm.
System: Is there anything else I can do for you?
User: No thanks
System: Thanks for calling. Good bye!
```

Fig. 1. Example of dialog extracted from the Let's Go corpus

3   Experimental Set-Up

We have used a corpus of 62 dialogs of real users interacting with the Let's Go system. As shown in Table 1, for our experiments we employed three types of parameters: interaction parameters, user judgments and user profile parameters. Interaction parameters were computed for each dialog so that the total, user and system turn duration (in miliseconds) and the number of words per utterance were averaged along all user or system utterances in the dialog. User judgments correspond to a number between 1 and 5 (1 = bad, 2 = poor, 3 = fair, 4 = good, 5 = excellent) with which the users rated several aspects of the system after interacting with it [2].

Finally, regarding the user profile parameters, we were interested in studying the appropriateness of distinguishing age groups (young or senior) and their self-perceived technical affinity (low or high). In the system there were 32 young and 30 senior users, and 26 users with low and 36 with high technical affinity.

Table 1. Summary of parameters

Summary of parameters	Parameters used
Interaction parameters	Turn duration, user turn duration, system turn duration, number of turns, number of words per user's utterance, number of words per system's utterance, number of help requests in the dialog, task success, concept error rate, number of no matches per dialog, number of repetitions per dialog, number of barge-in per dialog
User judgments	Task rate, overall impression with the interaction, overall impression of the presented system
User profile	Technical affinity, age

For the experiments we employed the X-means clustering algorithm, a variation of K-means clustering that treats cluster allocations by repetitively attempting partition and keeping the optimal resultant splits, until a specific criterion is reached [6]. Using the Weka software, we established a minimum of 2 and maximum of 5 clusters in up to 1,000 interactions, and computed the Euclidean distance between centroids using different features as summarized in Table 2.

Table 2. Summary of experiments

Parameters for computing distance	Parameter studied	Experiment
Interaction parameters	Task rate	1
Interaction parameters	Overall impression with the interaction	1
Interaction parameters	Overall impression of the presented system	1
Interaction parameters	Technical affinity group (low or high)	2
Interaction parameters	Age group (young or senior)	2
Subjective parameters	Technical affinity group (low or high)	3
Subjective parameters	Age group (young or senior)	3
Interaction and subjective parameters	Technical affinity group (low or high)	4
Interaction and subjective parameters	Age group (young or senior)	4

In the first group of experiments, we used interaction parameters to compute the distance measures and studied whether different user judgments were balanced between the clusters or were classified in different clusters and thus are easily distinguishable. The user judgments studied were task rate, overall impression with the interaction, and overall impression of the system. The second group of experiments considered the same features for clustering but studied the balance of user profile features, concretely technical affinity group and age group. The third group of experiments considered only user judgments for clustering and studied the same user profile features. With the fourth group of experiments, we studied the user profile measures by employing both interaction parameters and subjective measures for clustering.

4   Discussion of the Experimental Results

For all the experiments carried out, there were 2 clusters generated. The first group of experiments showed that the overall subjective impressions about the system and the interaction are not distinguished by the clustering algorithm, as shown by the fact that the clusters are composed of a balanced number of dialogs belonging to the different categories (bad, poor, fair, good, and excellent).

However, for the task rate the clustering algorithm clearly differentiated the dialogs in the extremes. As can be observed in Fig. 2, the worst rated dialogs were classified in cluster 1 and the best in cluster 2^1. This might indicate that the judgment of task rate can be somehow derived from the interactions parameters, whereas overall impression might not be so much affected by the real performance of the system during the interaction.

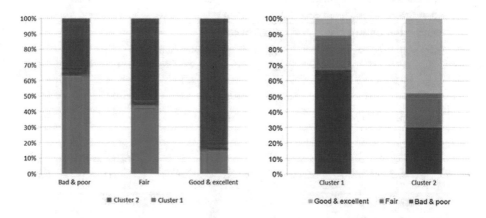

Fig. 2. Results for the task rate when clustering interaction parameters

---

[1] For illustration purposes, we have grouped the five categories into three: *bad&poor*, *fair&good*, and *excellent*.

In the second group of experiments we studied whether the user profile features were classified in different clusters when using only interaction parameters to compute the distances. As shown in Fig. 3, the clusters did not clearly distinguish between users with different technical affinity or age groups, although for age groups a slightly better separation was found.

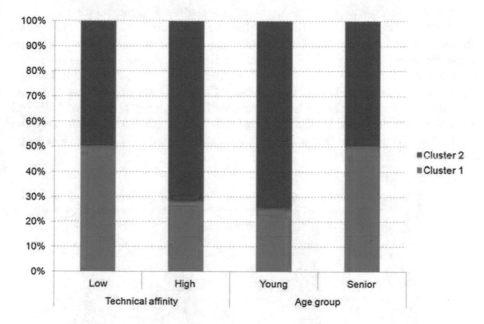

Fig. 3. Distribution of technology affinity and age groups when clustering with interaction parameters

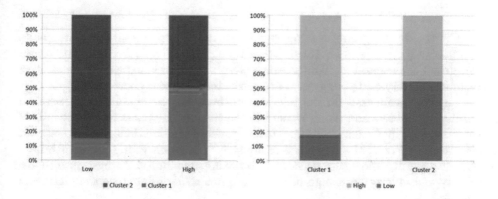

Fig. 4. Distribution of technical affinity when clustering with user judgments

In the third group of experiments we studied whether the user profile features can be determined by user judgments. The experiments revealed that such subjective features distinguish low technical affinity dialogs. As shown in Fig. 4, the 85% of the dialogs with low affinity users were classified in Cluster 2, although the 50% of dialogs is classified in each cluster for high affinity users. The reason for this result may be that users with low affinity systematically evaluate the system with worse rates whereas high affinity users provide more varied judgments.

A close study of the data revealed that the standard deviation for the overall impression with the interaction rating for the low affinity users was 1.1 and the most frequent judgment was 2, whereas for the high affinity users the deviation was 0.6 and the mode value was 3. A similar result was obtained for the age group of the users. As shown in Fig. 5, the 83% of the senior users were classified in cluster 2, whereas the dialogs corresponding to young users were balanced between both clusters. A close study of the data revealed that while only the 19% of the young users had a low technical affinity, the 67% of the senior users had low technical affinity.

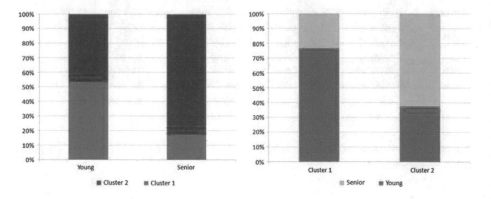

Fig. 5. Distribution of age groups when clustering with user judgments

For the fourth group of experiments we used both interaction parameters and user judgments and obtained that, although low technical affinity and senior age group are separated slightly worse than in the previous experiments, high technical affinity and young users are separated better as shown in Fig. 6.

Finally, we focused on the relationship between the age and the technical affinity parameters and their presence in the clusters, and classified the results of the previous experiments according to the possible combinations of both features. As shown in Fig. 7, the difference really strives between young users with high technical affinity and senior users with low technical affinity. Youngsters with low affinity and elderly with high affinity were not differentiated by the clustering mechanism.

This might be explained by the fact that senior users with low affinity provide a lower assessment of the system despite of how the interaction developed in

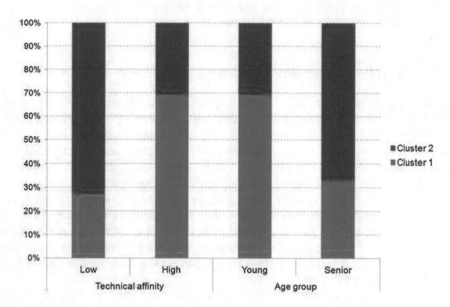

Fig. 6. Distribution of technology affinity and age groups when clustering with all the parameters

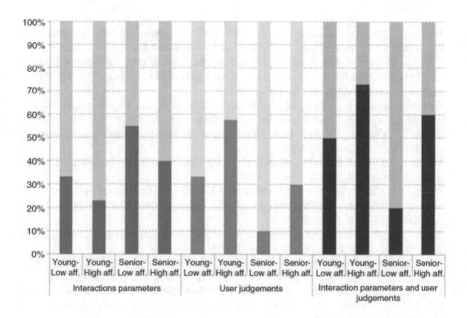

Fig. 7. Combinations of the age technical affinity groups with the different clusterings

terms of the interaction parameters. Moreover, interaction parameters along with the judgments can help to differentiate senior users from young, high-affinity users.

5   Conclusions and Future Work

Successful applications of Soft Computing suggest that it will have increasingly greater impact in the coming years for customer relationship management within the service industry. The development of personalized chatbots is key in the pathway towards this scenario. In this paper, we have used a corpus of real user conversations with the Let's Go dialog system and assessed the appropriateness of grouping the users by a combination of age (senior or young) and self-perceived technical affinity (low or high) by using a clustering approach. From the experiments we can conclude that the profiles of the users elicitated different subjective judgments. The results of the clustering point out that there exists a better grouping that distinguishes between 3 groups: young users with high technical affinity, senior users with low technical affinity, and a third group consider the remaining users.

For future work we plan to quantitative assess the quality of the clustering technique based on different input variables, internal and external measures. Additionally, we intend to replicate the experiments in other application domains considering different user groups.

References

1. Black, A., Burger, S., Langner, B., Parent, G., Eskenazi, M.: Spoken dialog challenge 2010. In: Proceedings of IEEE SLT, pp. 448–453 (2010)
2. Callejas, Z., Griol, D., Engelbrecht, K.: Assessment of user simulators for spoken dialogue systems by means of subspace multidimensional clustering. In: Proceedings of Interspeech'12, pp. 250–253 (2012)
3. Chandramohan, S., Geist, M., Lefevre, F., Pietquin, O.: Clustering behaviors of spoken dialogue systems users. In: Proceedings of ICASSP'12, pp. 306–317 (2012)
4. Dote, Y., Ovaska, S.: Industrial applications of soft computing: a review. Proc. IEEE **89**(9), 1243–1265 (2001)
5. McTear, M.F., Callejas, Z., Griol, D.: The Conversational Interface. Talking to Smart Devices. Springer, Cham (2016)
6. Pelleg, D., Moore, A.: X-means: extending K-means with efficient estimation of the number of clusters. In: Proceedings of 17th International Conference on Machine Learning, pp. 727–734 (2000)
7. Pieraccini, R.: The Voice in the Machine: Building Computers that Understand Speech. The MIT Press, Cambridge (2012)
8. Ravi, K., Ravi, V., Prasad, P.S.R.K.: Fuzzy formal concept analysis based opinion mining for CRM in financial services. Appl. Soft Comput. **60**, 786–807 (2017)
9. Roy, R., Tiwari, A., Shah, S., Hadden, J.: Soft computing in the service industry. In: Proceedings of 6th International Conference on Recent Advances in Soft Computing, pp. 1–9 (2006)
10. Valle, M., Ruz, G., Masías, V.: Using self-organizing maps to model turnover of sales agents in a call center. Appl. Soft Comput. **60**, 763–774 (2017)

A Preliminary Study on Multivariate Time Series Clustering

Iago Váquez[1], José R. Villar[2(✉)], Javier Sedano[1], and Svetlana Simić[3]

[1] Instituto Tecnológico de Castilla y León,
Pol. Ind. Villalonquejar, 09001 Burgos, Spain
{iago.vasquez,javier.sedano}@itcl.es
[2] Computer Science Department, EIMEM, University of Oviedo, Oviedo, Spain
{villarjose,delacal}@uniovi.es
[3] Department of Neurology, Clinical Centre of Vojvodina Novi Sad,
University of Novi Sad, Novi Sad, Republic of Serbia
svetlana.simic@mf.uns.ac.rs

Abstract. Time Series (TS) clustering is one of the most effervescent
research fields due to the Big Data and the IoT explosion. The problem
gets more challenging if we consider the multivariate TS. In the field
of Business and Management, multivariate TS are becoming more and
more interesting as they allow to match events the co-occur in time but
that is hardly noticeable. In this study, Recurrent Neural Networks and
transfer learning have been used to analyze each example, measuring
similarities between variables. All the results are finally aggregated to
create an adjacency matrix that allows extracting the groups. Proof-of-
concept experimentation has been included, showing that the solution
might be valid after several improvements.

1 Introduction

Time Series (TS) clustering is one of the most effervescent research fields due to
the Big Data and the IoT explosion. Until recently, the problem was focused on
univariate TS clustering. For instance, [10] proposed use dynamic time warping
and k-means to cluster the performance of a photovoltaic power plant, so to
predict the meteorological conditions. Similarly, k-means was used to cluster TS
and then predict the weather conditions [8]. Interested readers would read the
review in [1] for a good review on this topic.

However, TS clustering has been moving from univariate to multivariate TS
problems. In these problems, a TS includes more than one variable; i.e., the
pollution measurements in a medium or big city includes several physical and
chemical variables registered in several stations placed all around of a city. Clustering multivariate TS has been found interesting in order to perform complex
event detection or to classify the current scenario. For instance, [4] proposed a
Partitioning around Meriods and Fuzzy C-Meroids clustering for the problem of
detecting high-value pollution records or alarms in the city of Rome.

© Springer Nature Switzerland AG 2020
F. Martínez Álvarez et al. (Eds.): SOCO 2019, AISC 950, pp. 473–480, 2020.
https://doi.org/10.1007/978-3-030-20055-8_45

The similarity among the variables within the TS is one of the most studied topics. PCA similarity factor was combined with the average based Euclidean distance together with a fuzzy clustering scheme in [6]. Discords have been used in multivariate TS to identify anomalies and introduce more efficient search processes [7]. Hash functions have also been used to index and to measure the similarities in multivariate TS searches [15].

Interestingly, models have been also used in measuring the similarity between multivariate TS, i.e., Gaussian Mixture Models [11]. A different approach is based on extracting features and then using these features to group the multivariate TS [5]. Feature extraction together with Self-Organized Maps [14], Hidden Markov Models [9] or Fuzzy Linear [3] are techniques that have been also proposed in solving multivariate TS. Still, this problem cannot be considered solved and a recent study found out that the combination of feature extraction and a classification stage performs better than the current approaches [2].

In this study, a similar idea of that proposed in [11] is revisited for multivariate TS. Recurrent Neural Networks (RNN) are learned to predict a variable from an example and then used to measure the similarity between the different variables. Afterward, the adjacency matrix is found for each example, then aggregated for all the examples and finally binarized to generate the final adjacency matrix. The groups are proposed based on the variables that mutually dependent. To reduce the complexity of the solution transfer learning is proposed.

The organization of this manuscript is as follows. The next section describes the proposal. Section 3 details the dataset and some method's parameters, while Sect. 4 includes the figures and the discussion on the results. The study ends with the conclusions.

2   A Proposal for Multivariate TS Clustering

Let's define multivariate TS dataset as the dataset containing examples, each example is a multivariate TS. A multivariate TS is an arrangement of several TS, each one belonging to a different variable. We assume all the examples having the same variables and, without loss of generality, the same sampling frequency and the same number of samples. Therefore, a multivariate TS example is a matrix of m rows of n variables, where each column represents a univariate TS. However, each example has its own number of samples.

In this approach we propose two stages: (i) to find the similarities between variables in a single example, that is, in a single multivariate TS, and (ii) to aggregate the results among the examples and extract the relationships. To find the similarities between features we propose to model an RNN to predict the test subsequence of a variable, and to use the prediction error over the remaining variables. For this preliminary study, the aggregation of the results was performed with simple thresholding followed by a graph representation. To make the process feasible, we propose to use transfer learning [12].

2.1 Finding Similarities Between Variables from an Example

The procedure is depicted in Algorithm 1. Let's TS^i be the current example, $TS^i = \{X_1^i, \ldots, X_n^i\}$ $\forall i : 1, \ldots, N$, where N is the number of examples, n is the number of variables. Moreover, each variable X_j^i can be written as $X_j^i = (x_{j1}^i, \ldots, x_{jm_i}^i)$ $\forall j : 1, \ldots, n$, with m_i being the number of samples of the TS for each variable in the example i.

Algorithm 1. Computing similarities between features in an example

---

1: **procedure** IN-EXAMPLE-SIMILARITY(TS^i, LoRNN)   ▷ LoRNN list of pre-learnt RNNs, if available
2:     $sim \leftarrow$ zeroes matrix of size $n \times n$
3:     **for** each variable j in TS^i **do**
4:         $X_j^i \leftarrow$ normalize(X_j^i)
5:         $RNN_j^i \leftarrow$ **Train-RNN**(X_j^i, LoRNN[j])
6:         $LoRNN[j] \leftarrow RNN_j^i$
7:         $e_j^i \leftarrow$ RMSE(RNN_j^i, test(X_j^i))
8:         **for** each variable k in TS^i, $k \neq j$ **do**
9:             $X_k^i \leftarrow$ normalize(X_k^i)
10:            $e_{jk}^i \leftarrow$ RMSE(RNN_j^i, test(X_k^i))
11:            $sim[j,k] \leftarrow abs(\frac{e_{jk}^i - e_j^i}{e_j^i})$
12:        **end for**
13:    **end for**
14:    **return** sim
15: **end procedure**
16:
17: **procedure** TRAIN-RNN(X_j^i, RNN)   ▷ RNN is a RNN, if available
18:    **if** is.NULL(RNN) **then**
19:        $RNN \leftarrow$ *full train* RNN for the train part of X_j^i
20:    **else**
21:        $RNN \leftarrow$ *tune RNN* for the train part of X_j^i
22:    **end if**
23:    **return** RNN
24: **end procedure**

Let's also assume that a given percentage (%TRN) of the samples of a TS is kept for training and the remaining for testing. In other words, for any variable X_j^i in example i, $(x_{j1}^i, \ldots, x_{j(\%TRN \times m_i)}^i)$ are kept for training and $(x_{j(\%TRN \times m_i)}^i, x_{jm_i}^i)$ are kept for testing.

It is possible to learn an RNN using the training part of X_j^i to predict its behavior in the testing part, let's call this RNN_j^i. Let us suppose we obtain a good model, and that the aggregation of the prediction error along the test subset for variable X_j^i is e_j^i. This prediction error can be any well-known measurement, as the Root Mean Square Error (RMSE) or similar.

The RNN_j^i is applied to predict each of the remaining variables X_k^i with $k : 1 \cdots n$ and $k \neq j$. The error obtained with RNN_j^i when predicting the test subsequence of the variable X_k^i is denoted as e_{jk}^i. This error is scaled wrt the e_j^i in order to obtain a similarity value: $E_{jk}^i = |(e_{jk}^i - e_j^i)/e_j^i|$. Values close to 0.0 means the TS can be successfully predicted by RNN_j^i. This prediction is also repeated for $-X_k^i$, that is, the normalized test sequence X_k^i is swapped wrt the time axis to consider the case the two TS X_j^i and X_k^i have a negative correlation. Consequently, the minimum of both errors is kept.

Therefore, the similarity between variable j and the remaining variables in the example i is obtained as the vector $sim_j^i = (E_{j1}^i, \cdots, E_{jn}^i)$, with $E_{jj}^i = 0$. Finally, repeating this procedure for each of the variables in the example i, a distance $n \times n$ matrix is obtained, which represents the outcome of this stage.

2.2 RNN and Transfer Learning

As seen in Algorithm 1 and in the previous subsection, an RNN is trained using the train part of variable X_j^i from example TS^i. In this study, the *rnn* R-package [13]; for each training process a simple grid of 12 different learning rate values (from $1/12$ to 1.0), 1 to 12 as the number of epochs and 1 to 12 hidden neurons.

However, training a complete RNN from scratch for each variable and for each example makes this approach unfeasible for even small multivariate TS datasets. A simplification is clearly needed.

To do so, we introduce a simple transfer learning scheme [12]. For the first time, the Train-RNN is call, a NULL value is given as current RNN; thus, full learning of the RNN is performed. However, when it is not NULL, then it is the RNN_j^i trained in the first iteration of the process for variable j and example $i == 1$. We reuse this RNN model, fitting it to the current X_j^i. This adaptation is just a simple weight tuning during a reduced number of iterations (20 in this study).

2.3 Computing the Similarities Within a Multivariate TS Dataset

Once the similarity matrix between the variables from an example is obtained, computing the similarity between the variables for the whole multivariate TS dataset is a matter of choosing the method.

In this preliminary study, we propose to convert each matrix to an adjacency matrix and then to aggregate the adjacency matrices.

The similarity matrix for example TS^i is converted into the SIM adjacency matrix SIM_i as follows. For each pair of variables j and k from the example i, if similarity between j and k is smaller or equal than $th1$ ($sim_i[j,k] \leq th1$), then variable j can predict variable k (denoted as $k \lesssim j$). Thus, $SIM_i[j,k] = 1$; otherwise, $SIM_i[j,k] = 0$.

The adjacency matrices are then aggregated as just the sum of all of them. Therefore, the final aggregated adjacency matrix is $SIM_{ag} = \sum_{\forall i} SIM_i$, each cell contains an integer from 0 to N. Finally, the outcome adjacency matrix

SIM_{final} is obtained by thresholding SIM_{ag} such that whenever $SIM_{ag}[j,k] \geq th2$ then $SIM_{final} = 1$, otherwise $SIM_{final} = 0$.

This binarization produces an adjacency matrix that can be represented in a graph. This visualization can help in deciding what to do with those variables that have not been grouped yet.

3   Experiment and Methods

To evaluate this preliminary study a real-world multivariate TS dataset has been used. This dataset includes up to $n = 11$ variables in each TS example. These multivariate TS have been extracted from a photovoltaic solar power plant, including the following variables:

- Indoor and outdoor temperatures in the weather station (TIN, TOUT)
- Horizontal and Vertical Irradiance reference measurement (HIR and VIR)
- The voltage at the weather station's battery (BV)
- The temperature of 4 photovoltaic panels linked to an inverter (T1 to T4)
- An In-panel Horizontal and Vertical Irradiance measurement (PHI, VHI)

Each example includes data from the evolution of the magnitude of these variables for a period of four days. Although data are available for more than three months, in this preliminary study only the $N = 5$ examples of these data are considered.

The values of the thresholds have been set before any further analysis to $th1 = 0.07$ and $th2 = 3$ (equivalent to require that the 60% of the examples must include that relationship in order to accept a dependence).

4   Results and Discusion

The results are shown from Tables 1, 2 and 3 and in Fig. 1. Table 1 includes the RMSE error measurement obtained when training the RNN_j^i in time series prediction mode. Table 2 depicts the values of E_{jk}^i obtained for the first example in the dataset. Table 3 shows the adjacency matrix obtained after the aggregation of the different examples and pruning with $th2 = 3$. Finally, Fig. 1 shows the graph obtained from that adjacency matrix.

Although this is almost a classroom exercise and the multivariate TS dataset is not complex at all, it seems to fit perfectly in this proof of concept. Firstly, the grouping that was obtained was as expected. Secondly, the several dependency relationships between variables suggest that there could be more variables grouped together but needing further analysis.

Nevertheless, t is clear that the proposal still needs plenty of amendments as well as the structure of the method. However, the obtained results seem to be promising. Items such as different type of TS prediction techniques that might be applied provided transfer learning can be deployed, automatic setting the thresholds to adapt to the problem faced, or the definition of similarity

Table 1. The RMSE error measurement for each of the fully trained RNN.

Variable	Train Error	Variable	Train Error
VIR	0.1194	PHI	0.1471
HIR	0.1481	PVI	0.1346
T1	0.0449	TIN	0.04212
T2	0.0536	TOUT	0.0505
T3 T4	0.0461		

Table 2. The similarity matrix obtained with the first example from the multivariate TS dataset.

	VIR	HIR	BV	T4	T3	T2	T1	PHI	PVI	TIN	TOUT
VIR	0.000	0.011	0.343	0.486	0.487	0.489	0.503	0.006	0.008	0.498	0.438
HIR	0.075	0.000	0.685	0.180	0.155	0.146	0.189	0.080	0.007	0.112	0.271
BV	0.154	0.183	0.000	0.210	0.195	0.194	0.224	0.151	0.176	0.374	0.284
T4	$>10^{14}$	$>10^{14}$	$>10^{14}$	0.000	0.012	0.019	0.058	$>10^{14}$	$>10^{14}$	0.149	0.023
T3	$>10^{14}$	$>10^{14}$	$>10^{14}$	0.009	0.000	0.007	0.046	$>10^{13}$	$>10^{14}$	0.122	0.003
T2	$>10^{14}$	$>10^{14}$	$>10^{14}$	0.006	0.005	0.000	0.038	$>10^{14}$	$>10^{14}$	0.199	0.065
T1	$>10^{14}$	$>10^{13}$	$>10^{14}$	0.056	0.044	0.037	0.000	$>10^{14}$	$>10^{14}$	0.038	0.084
PHI	0.003	0.065	0.601	0.219	0.196	0.188	0.231	0.000	0.059	0.005	0.148
PVI	0.065	0.007	0.648	0.197	0.172	0.163	0.207	0.070	0.000	0.075	0.233
TIN	$>10^{14}$	$>10^{14}$	$>10^{14}$	0.196	0.191	0.184	0.142	$>10^{14}$	$>10^{14}$	0.000	0.151
TOUT	$>10^{14}$	$>10^{14}$	$>10^{14}$	0.078	0.080	0.074	0.036	$>10^{14}$	$>10^{14}$	0.129	0.000

Table 3. Final adjacency matrix obtained using $th2 = 3$.

	VIR	HIR	BV	T4	T3	T2	T1	PVI	PHI	TIN	TOUT
VIR	0	0	0	0	0	0	0	1	0	0	0
HIR	0	0	0	0	0	0	0	0	1	0	0
BV	0	0	0	0	0	0	0	0	0	0	0
T4	0	0	0	0	1	1	1	0	0	0	0
T3	0	0	0	1	0	1	1	0	0	0	0
T2	0	0	0	1	1	0	1	0	0	0	0
T1	0	0	0	1	1	1	0	0	0	0	0
PVI	1	1	0	0	0	0	0	0	1	0	0
PHI	0	1	0	0	0	0	0	0	0	0	0
TIN	0	0	0	0	0	0	0	0	0	0	1
TOUT	0	0	0	0	0	0	0	0	0	1	0

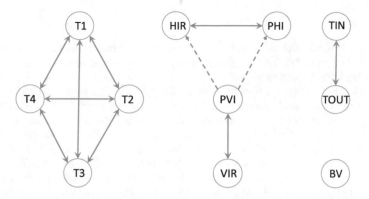

Fig. 1. Final graph: the groups are clearly remarked.

measurements that might be more promising than the scaled RMSE are included among others in the next research to be performed. Furthermore, we do believe that this method could be directly applied in medicine and biology, especially in problems where the experts need support in the analysis of big volumes of multivariate TS.

5   Conclusions

This study proposes a solution for the clustering of multivariate TS datasets. To do so, the prediction error of RNN learned for a variable within an example is used to define a similarity measurement. The aggregation of the obtained similarities among all the examples in the dataset allows developing an adjacency matrix that, finally, is used to group the variables.

A simple proof of concept has been presented, showing that the performance of the method perfectly groups the different variables. Interestingly, this method only contains two parameters (two thresholds) that were easily tuned. We do expect to perform improvements in the algorithm, in the modeling and in the similarity function, so this solution can be applied in data analysis in medicine. Also, the findings in this research would be applied to the identification of interesting groups in areas such as management or to the stock market.

Acknowledgment. This research has been funded by the Spanish Ministry of Science and Innovation, under project MINECO-TIN2017-84804-R.

References

1. Aghabozorgi, S., Shirkhorshidi, A.S., Wah, T.Y.: Time-series clustering - a decade review. Inf. Syst. **53**, 16–38 (2015). http://www.sciencedirect.com/science/article/pii/S0306437915000733
2. Bode, G., Schreiber, T., Baranski, M., Müller, D.: A time series clustering approach for building automation and control systems. Appl. Energy **238**, 1337–1345 (2019). http://www.sciencedirect.com/science/article/pii/S0306261919302089
3. Duan, L., Yu, F., Pedrycz, W., Wang, X., Yang, X.: Time-series clustering based on linear fuzzy information granules. Appl. Soft Comput. **73**, 1053–1067 (2018). http://www.sciencedirect.com/science/article/pii/S1568494618305490
4. D'Urso, P., Giovanni, L.D., Massari, R.: Robust fuzzy clustering of multivariate time trajectories. Int. J. Approx. Reason. **99**, 12–38 (2018). http://www.science direct.com/science/article/pii/S0888613X17306977
5. Ferreira, A.M.S., de Oliveira Fontes, C.H., Cavalcante, C.A.M.T., Marambio, J.E.S.: Pattern recognition as a tool to support decision making in the management of the electric sector. Part II: A new method based on clustering of multivariate time series. Int. J. Electr. Power Energy Syst. **67**, 613–626 (2015). http://www.sciencedirect.com/science/article/pii/S0142061514007285
6. Fontes, C.H., Budman, H.: A hybrid clustering approach for multivariate time series – a case study applied to failure analysis in a gas turbine. ISA Transact. **71**, 513–529 (2017). http://www.sciencedirect.com/science/article/pii/S0019057817305530
7. Hu, M., Feng, X., Ji, Z., Yan, K., Zhou, S.: A novel computational approach for discord search with local recurrence rates in multivariate time series. Inform. Sci. **477**, 220–233 (2019). http://www.sciencedirect.com/science/article/pii/S0020025516320849
8. Lee, Y., Na, J., Lee, W.B.: Robust design of ambient-air vaporizer based on time-series clustering. Comput. Chem. Eng. **118**, 236–247 (2018). http://www.sciencedirect.com/science/article/pii/S0098135418308822
9. Li, J., Pedrycz, W., Jamal, I.: Multivariate time series anomaly detection: a framework of hidden Markov models. Appl. Soft Comput. **60**, 229–240 (2017). http://www.sciencedirect.com/science/article/pii/S1568494617303782
10. Liu, G., Zhu, L., Wu, X., Wang, J.: Time series clustering and physical implication for photovoltaic array systems with unknown working conditions. Sol. Energy **180**, 401–411 (2019). http://www.sciencedirect.com/science/article/pii/S0038092X19300532
11. Øyvind Mikalsen, K., Bianchi, F.M., Soguero-Ruiz, C., Jenssen, R.: Time series cluster kernel for learning similarities between multivariate time series with missing data. Pattern Recogn. **76**, 569–581 (2018). http://www.sciencedirect.com/science/article/pii/S0031320317304843
12. Pan, S.J., Yang, Q.: A survey on transfer learning. IEEE Trans. Knowl. Data Eng. **22**(10), 1345–1359 (2010). https://doi.org/10.1109/TKDE.2009.191
13. Quast, B.: Recurrent neural networks in r February 2019. https://github.com/bquast/rnn
14. Salvo, R.D., Montalto, P., Nunnari, G., Neri, M., Puglisi, G.: Multivariate time series clustering on geophysical data recorded at Mt. Etna from 1996 to 2003. J. Volcanol. Geoth. Res. **251**, 65–74 (2013). http://www.sciencedirect.com/science/article/pii/S0377027312000443, flank instability at Mt. Etna
15. Yu, C., Luo, L., Chan, L.L.H., Rakthanmanon, T., Nutanong, S.: A fast LSH-based similarity search method for multivariate time series. Inf. Sci. **476**, 337–356 (2019). http://www.sciencedirect.com/science/article/pii/S0020025518308430

Adaptive Fault-Tolerant Tracking Control Algorithm for IoT Systems: Smart Building Case Study

Roberto Casado-Vara[1]([✉]), Fernando De la Prieta[1], Sara Rodriguez[1],
Ines Sitton[1], Jose L. Calvo-Rolle[2], G. Kumar Venayagamoorthy[3],
Pastora Vega[4], and Javier Prieto[1]

[1] Bisite Research Group, University of Salamanca,
Calle Espejo 2, 37007 Salamanca, Spain
rober@usal.es
[2] Department of Industrial Engineering, University of A Coruña,
Ferrol, A Coruña, Spain
[3] Holcombe Department of Electrical and Computer Engineering,
Clemson University, Clemson, SC 29634, USA
[4] Department of Automatic and Computer Science, University of Salamanca,
Plaza de los Caídos s/n, 37008 Salamanca, Spain

Abstract. In this paper, the problem of robust adaptive fault-tolerant tracking control with state-prediction performance is analyzed for a class of Iot temperature systems subject to accuracy states uncertainties and external disturbances. In order to ensure the efficiency of our new adaptive temperature control algorithm, we propose a new control strategy which is based on game theory consensus and prediction of accuracy future states to reduce the tracking error and improve the effectiveness of the algorithm. Compared with the existing results, a novel algorithm is developed to improve the operation of the monitoring and control of the Iot networks in order to increase the Energy efficiency of it. According with the simulation information provided by our investigation a new fault tolerant tracking error algorithm guaranteeing the robust tracking of the reference model. It shown that the predicted temperature signal is bounded by a small interval close to the collected temperature data. A case study result is provided to demonstrate the efficacy of the proposed adaptive fault-tolerant tracking control algorithm.

Keywords: Adaptive closed-loop · Control system ·
Algorithm design and analysis · IoT · Non-linear control

1 Introduction

In its Horizon2020 programme, the European Union has energy efficiency in smart buildings as one of its social objectives. Following this research topic many researchers are developing new techniques to increase the energy efficiency of

© Springer Nature Switzerland AG 2020
F. Martínez Álvarez et al. (Eds.): SOCO 2019, AISC 950, pp. 481–490, 2020.
https://doi.org/10.1007/978-3-030-20055-8_46

smart buildings. One of the main problems with optimizing energy consumption in smart buildings is temperature control. Today's smart buildings have the problem that they need a high energy consumption to air-condition them. This may be because the building materials are not the most suitable to retain heat, or on the contrary, because the mechanisms of monitoring and temperature control are not accurate. However, these problems can be left in second place if the Internet of Things (IoT) network that collects smart building data has faulty sensors and actuators. In this way we will improve the accuracy of the sensors and actuators, and if they are inaccurate or failing, we will detect them and ask for a change for a new one. With this algorithm that we have developed, we are going to increase the energy efficiency of smart buildings considerably, since the savings in the air conditioning of the building is going to be very big.

The proposed model in this work is an adaptive control algorithm. This control algorithm optimizes the temperature of the smart building using two modules that improve its performance: (1) State prediction module. This algorithm module uses Markov chains to predict the accuracy states of IoT nodes in future time. In this way, the algorithm can self-correct temperatures and optimize the energy efficiency of the smart building. (2) Data quality. The organization of the rest of this paper is as follows. In Sect. 2, the state of the art is presented. The control algorithm and some preliminaries are presented in Sect. 3. In Sect. 4, a case study is shown and its results is introduced to verify the theoretical model result. Finally, some conclusions end this paper in Sect. 5.

2   Related Work

In real-world network control systems (NCS), failures often occur in system components (e.g. actuators, sensors, controllers and filters) mainly due to complex and difficult working environments as well as limited network resources. It is known that the occurrence of faults can be determined through fault detection techniques that have received substantial research attention, while detailed data on faults can be collected through the fault estimation procedure, which provides a necessary prerequisite for further fault tolerant monitoring [1–3]. With the increasing demands for safety, reliability, economic efficiency and service life of NCCS, the problem of failure detection and estimation has been extensively analyzed so far, and numerous research results have been recorded in recent literature [4–6].

In particular, the issue of flaw identification for nonlinear processes has been investigated in the work of Samuel and Cao using kernel component analysis techniques depth estimation [7]. Furthermore, the parameters of practical NCSs may vary over time as a result of the rapid complexity system and, consequently, the design of fault estimators for time systems has attracted considerable interest in the investigation (see, eg, related work [8–10,28]). For instance, in the work of Dong et al. [8], regarding the merged effects of non-linearity, branched faults and fading channels, time-varying fault estimators have been proposed for stochastic systems over a finite horizon.

In a practical control system, the appearance of system failures is inevitable due to the complexity of the system architecture, long-term work and unexpected changes in the surrounding outside environment. System faults can lead to system degradation or even instability [11,12]. Thus, it is necessary to conduct research on fault-tolerant control (FTC), which has gained more and more attention in recent decades. In general, FTC techniques and methods are divided into passive and active methods. Passive FTC is a method that does not modify the structure and parameters of the controller [13]. Using the robust technique, the fixed controller is designed only for the predefined set of faults. However, the disadvantage is that system performance cannot be guaranteed if faults occur outside the default fault set.

A lot of active control methods have been developed such as sliding model methods [14,27], learning methods [15], multiple-model method [16], observer-based methods [17], and adaptive compensation methods. Among them, adaptive compensation control method have bee widely applied for compensation of unknown failures [18,19]. However, there exists a common restriction in the existing finite-time results that the settling time is relying on initial conditions [20,21]. That is, the convergence time of the system is varying for different initial values. In practice, the initial conditions may be uncertain or even uncertain. In this cases, finite-time control method mentioned above cannot ensure system to achieve a desired performance within an accurate pre-set time [22,25]. On the other hand, some fixed-time control results of nonlinear system have been achieved [23,24,26] based on other methods. However, as far as the authors' knowledge, few results focus on the use of prediction of accuracy states to improve the control signal.

3   Proposed Control Model

This section presents the adaptive control algorithm that we have developed in this paper. The main objective of this algorithm is to control the temperature of the smart building. For this purpose, the control algorithm relies on 3 mechanisms to optimize temperature control: (1) Cooperative control algorithm. This mechanisms improves the quality of data collected by the IoT nodes and finds false data. (2) State prediction. This mechanisms receives the error the IoT nodes are making with respect to the desired temperature and predicts the error they will make in the future. (3) Controller. In this control algorithm a PID controller is used to control the temperature. But the PID input is being optimized by the (1) and (2) mechanisms. This algorithm is shown in Fig. 1.

3.1   Reference Input

The Reference input (RI) in our algorithm is the desired temperature value. In this algorithm the RI is located in a control room where the building manager will set the temperature that the building should have. Then the algorithm controls the actuators (thermostats) so the building reaches that temperature and remains stable.

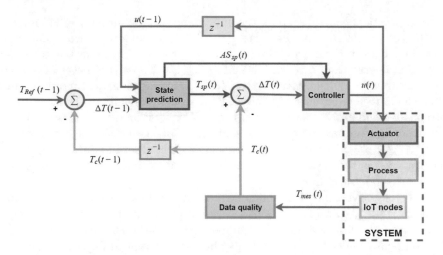

Fig. 1. This algorithm predicts the accuracy state of the sensors via the adaptive control algorithm in the time interval.

3.2   State Predictor

This step of this algorithm is designed to predict that IoT nodes will be in faulty state in time t and then output the predicted temperature $(T_{sp}(t))$. In this subsection we describe in detail the operation of the state predictor.

Initial Accuracy State. It is necessary to design a scale of degradation of precision given in percents. This is according to the data collected by the algorithm we had carried out in previous research [30]. This scale will be the discussion universe of the random variable X_n which determines the current accuracy states of the system in relation to IoT nodes error. Therefore, the IoT nodes' current states are $X_n = \{A = high\ accuracy,\ B = accurate,\ C = low\ accuracy,\ F = failure\}$. Below, Table 1 has the selected for every parameter.

Table 1. Accuracy state of IoT nodes.

X_n	IoT node accuracy state	Error (%)
A	High accuracy	e ≤ 10
B	Accurate	10 < e ≤ 20
C	Low accuracy	20 < e ≤ 35
F	Failure	e ≥ 35

Transition Matrix. Let λ_A be the time the sensor remains in state A (exponential distribution). λ_B and λ_C are defined in a similar way. And let ξ_A be the time the sensor remains in state A. Let μ_A (μ_B, μ_C) be the probability that a sensor in state A (B,C) at time t shifts to state F in the time interval $(t, \Delta t + t)$. Thus, if the IoT nodes was in state A at time $(t - 1)$, the probability of the sensor remaining in state A at time (t) is given by the following equation:

$$P(\xi_A > t + \Delta t | \xi_A > t) = \frac{e^{-\lambda_A (t+\Delta t)}}{e^{-\lambda_A t}} = e^{-\lambda_A \Delta t} = 1 - \lambda_A \Delta t + o(\Delta t) = p_{AA} \tag{1}$$

Similarly, the probability that a sensor in state A at the beginning, will shift to state B, is given by the following equation

$$P(\xi_B > t + \Delta t | \xi_A > t) = 1 - ((1 - \lambda_A \Delta t + o(\Delta t)) - (\mu_A \Delta t + o(\Delta t)))$$
$$= (\lambda_A - \mu_A)\Delta t + o(\Delta t) = p_{AB} \tag{2}$$

In this way, we can build the transition matrix between t and $t + \Delta t$, where the coefficients of the transition matrix are the probabilities of the sensors' switching states (e.g., p_{AF} is the probability that a sensor in state A at the beginning, will eventually shift to state F in the interval $(t, \Delta t + t)$).

In this way, the transition matrix $P(t)$ is built:

$$P(t) = \begin{pmatrix} P(\xi_A > t + \Delta t | \xi_A > t) = p_{AA} & \cdots & p_{AF} \\ \vdots & \ddots & \vdots \\ P(\xi_A > t + \Delta t | \xi_F > t) = p_{FA} & \cdots & p_{FF} \end{pmatrix} \tag{3}$$

Prediction Step. The first action performed by the controller is the prediction step. In this step of the algorithm, the transition matrix of the developed model is used. (see (3)). Let $z^{(t)} : T^{g(t)} \longrightarrow z^{(t)}(T^{g(t)}) = T_{sp}(t)$ be the prediction function of accuracy states (i.e., Prediction step) for each time t and let $t + k$ where $k \in \{1, 2, \cdots\}$ be the predicted time. Given $t^\delta_{i,j} \in T^\delta$, the controller function u is defined as follows:

$$z_{ij}^{(t+k)}(t_{i,j}^g) = max\{\mathbb{P}_{t_{i,j}^{g(t+k)}A}, \mathbb{P}_{t_{i,j}^{g(t+k)}B}, \mathbb{P}_{t_{i,j}^{g(t+k)}C}, \mathbb{P}_{t_{i,j}^{g(t+k)}F}\} \tag{4}$$

Let $z^{(t)}(T^g) = T_{sp}(t)$ be the matrix of the states of accuracy given by the prediction function. The output of this function is the accuracy state of the sensors at time t.

Temperature of the Prediction Step. In the state prediction step the accuracy state that the sensor is going to have is predicted from the measurement error of the sensor with respect to the reference input $(\Delta T(t - 1))$. Using the accuracy error of the sensor, an adjustment factor is used to predict the temperature the sensor will have in time t. Then the state prediction step temperature is calculated as follows:

$$T_{sp}(t) = \begin{pmatrix} 1.05 & 0 & 0 & 0 \\ 0 & 1.15 & 0 & 0 \\ 0 & 0 & 1.25 & 0 \\ 0 & 0 & 0 & 1 \end{pmatrix} \cdot T_c(t-1) \tag{5}$$

3.3   Cooperative Control Algorithm

This algorithm compares the neighborhood temperature of the sensors using a cooperative game based on game theory to detect wrong data and increase the quality of the data collected by the sensors. This algorithm is described in Casado-Vara et al. [30].

4   Results

To test the proposed model, we have chosen a smart building. At the time the IoT nodes measured the temperature, desired temperature in the building is 23 °C. A mesh was used to place the sensors on the surface with the help of laser levels, the IoT nodes were placed vertically one in every section of the building. The type of sensor deployed in the building was a combination of the ESP8266 microcontroller in its commercial version "ESP-01" and a DHT22 temperature and humidity IoT node. The sum of both allows us for greater flexibility when collecting data and adaptability to the case study, since the DHT22 sensor is designed for indoor spaces (it has an operating range of 0 °C to 50 °C) according to its datasheet. The microcontroller obtains data from this sensor through the onewire protocol and communicates it to the environment via Wi-Fi using HTTP standards and GET/POST requests.

The temperature sensor had been collecting data at 5 min intervals, for 6 h in the same day. For the analysis we selected the data collected by the sensors in the following time interval 2018-12-10T08:30:00Z and ended on 2018-12-10T14:30:00Z. To test the efficiency of the control algorithm a disturbance has been introduced in the temperature of smart building (our process) at 1 h intervals to simulate the random behaviour of people's thermostat use (i.e., a group of people could select different temperatures in their office thermostats). These disturbances have been introduced by colleagues in our research group without having a consensus on which temperatures to introduce, so these temperatures can be considered pseudo-random. Below, a statistical summary of the measurements that were made with the sensors is presented in Table 2. In this experiment we have considered the next time interval $(t, t + \Delta t)$: $\frac{1}{365 \cdot 5}$ (i.e., a day). To validate the model we applied the accuracy state prediction model to the data collected by the sensors placed in the building.

In this case study, we are going to consider a system with external disturbances. In the measurement time 4 disturbances have been introduced randomly into the system. In Fig. 2, the temperature of the controlled system is shown. In this figure one can find that the desired temperature son 23 °C. It should be pointed out that, in the smart building, disturbances not only come from our

Table 2. Statistical table of measurements of the IoT nodes.

Timestamp start	Total timestamp	Min temp	Max temp	Mean	Standard deviation
2018-12-10T09:00	06:00:00Z	21.4 °C	24.5 °C	23.11 °C	0.94 °C

case study setup but also arise from a wide range of things (people activities, malfunctioning heating system, etc.). In this case study, we only consider the disturbances introduces by our team.

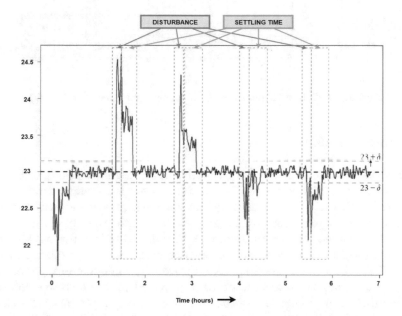

Fig. 2. Temperature collected by the smart building's IoT nodes during the time of the case study. In blue, the 4 disturbances of the office temperature of the smart building are indicated and in green, the settling time is indicated, which takes the temperature to reach the desired temperature after the disturbances.

From Fig. 3, it can be shown that the predicted accuracy states of the IoT nodes of the smart building. In this figure we can see the state prediction performance after the disturbance introduction. During the settling time, the state prediction module output is showing that the IoT nodes are in low accuracy state or even in faulty state.

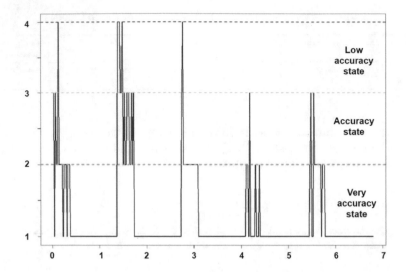

Fig. 3. Prediction trajectories of the accuracy states of IoT nodes.

5   Conclusions

In this paper, the problem of robust adaptive fault-tolerant tracking control with accuracy future states prediction is investigated for IoT temperature networks with external disturbances. By introducing the data quality and state prediction modules, the tracking error can be guaranteed within the apriori given, user-defined commanded temperature bounds. In order to reduce the tracking error, this new algorithm that provides this new two modules can optimize the robustness of the IoT network since it has improved its fault-tolerant performance. Using the accuracy state estimation data provided by the adaptive algorithm, this algorithm can predict the output temperature and compare with the collected temperature from the IoT network. It is shown that the closed-loop signals are bounded by the commanded temperature interval even we introduces some disturbances the settling time is very small. Finally, a case study result is given to demonstrate the efficacy of the proposed algorithm. Future work will be concentrated on robust adaptive algorithm for time delay nonlinear system with external uncertaines disturbances. In future work, multi-agents and artificial intelligence hybrid algorithms will be used to create a recommendation system to optimize the robustness of this algorithm in smart buildings.

Acknowledgment. This paper has been partially supported by the Salamanca Ciudad de Cultura y Saberes Foundation under the Talent Attraction Programme (CHROMOSOME project).

References

1. Basin, M., Li, L., Krueger, M., Ding, S.X.: Finite-time-convergent fault-tolerant control for dynamical systems and its experimental verification for DTS200 three-tank system. IET Control Theory Appl. **9**(11), 1670–1675 (2015)
2. Gao, Z., Cecati, C., Ding, S.X.: A survey of fault diagnosis and fault-tolerant techniques-Part I: fault diagnosis with model-based and signal-based approaches. IEEE Trans. Industr. Electron. **62**(6), 3757–3767 (2015)
3. Gao, Z., Cecati, C., Ding, S.X.: A survey of fault diagnosis and fault-tolerant techniques-Part II: fault diagnosis with knowledge-based and hybrid/active approaches. IEEE Trans. Industr. Electron. **62**, 3768–3774 (2015)
4. Li, L., Ding, S.X., Qiu, J., Peng, K., Yang, Y.: An optimal fault detection approach for piecewise affine systems via diagnostic observers. Automatica **85**, 256–263 (2017)
5. Qiu, A., Gu, J., Wen, C., Zhang, J.: Self-triggered fault estimation and fault tolerant control for networked control systems. Neurocomputing **272**, 629–637 (2018)
6. Shahnazari, H., Mhaskar, P.: Actuator and sensor fault detection and isolation for nonlinear systems subject to uncertainty. Int. J. Robust Nonlinear Control **28**(6), 1996–2013 (2018)
7. Samuel, R.T., Cao, Y.: Nonlinear process fault detection and identification using kernel PCA and kernel density estimation. Syst. Sci. Control Eng. **4**(1), 165–174 (2016)
8. Dong, H., Wang, Z., Ding, S.X., Gao, H.: On H-infinity estimation of randomly occurring faults for a class of nonlinear time-varying systems with fading channels. IEEE Trans. Autom. Control **61**(2), 479–484 (2016)
9. Li, J., Dong, H., Han, F., Hou, N., Li, X.: Filter design, fault estimation and reliable control for networked time-varying systems: a survey. Syst. Sci. Control Eng. **5**(1), 331–341 (2017)
10. Ren, W., Wang, C., Lu, Y.: Fault estimation for time-varying Markovian jump systems with randomly occurring nonlinearities and time delays. J. Franklin Inst. **354**(3), 1388–1402 (2017)
11. Tao, G., Chen, S., Tang, X., Joshi, S.M.: Adaptive Control of Systems with Actuator Failures. Springer, Heidelberg (2013)
12. Fattahi, M., Afshar, A.: Controller-based observer design for distributed consensus of multi-agent systems with fault and delay. J. Control Decis., pp. 1–19 (2018)
13. Dong, J., Yang, G.H.: Reliable state feedback control of T-S fuzzy systems with sensor faults. IEEE Trans. Fuzzy Syst. **23**(2), 421–433 (2015)
14. Alwi, H., Edwards, C.: Fault tolerant control using sliding modes with on-line control allocation. Automatica **44**(7), 1859–1866 (2008)
15. Zhang, X., Parisini, T., Polycarpou, M.M.: Adaptive fault-tolerant control of nonlinear uncertain systems: an information-based diagnostic approach. IEEE Trans. Autom. Control **49**(8), 1259–1274 (2004)
16. Boškovic, J.D., Mehra, R.K.: Multiple-model adaptive flight control scheme for accommodation of actuator failures. J. Guidance Control Dyn. **25**(4), 712–724 (2002)
17. Zhang, K., Jiang, B., Shi, P.: Fast fault estimation and accommodation for dynamical systems. IET Control Theory Appl. **3**(2), 189–199 (2009)
18. Li, X.J., Yang, G.H.: Robust adaptive fault-tolerant control for uncertain linear systems with actuator failures. IET Control Theory Appl. **6**(10), 1544–1551 (2012)

19. Jin, X.: Adaptive fault tolerant control for a class of input and state constrained MIMO nonlinear systems. Int. J. Robust Nonlinear Control **26**(2), 286–302 (2016)
20. Calvo-Rolle, J.L., Quintian-Pardo, H., Corchado, E., del Carmen Meizoso-López, M., García, R.F.: Simplified method based on an intelligent model to obtain the extinction angle of the current for a single-phase half wave controlled rectifier with resistive and inductive load. J. Appl. Log. **13**(1), 37–47 (2015)
21. Manuel Vilar-Martinez, X., Aurelio Montero-Sousa, J., Luis Calvo-Rolle, J., Luis Casteleiro-Roca, J.: Expert system development to assist on the verification of "TACAN" system performance. Dyna **89**(1), 112–121 (2014)
22. Casteleiro-Roca, J.L., Pérez, J.A.M., Piñón-Pazos, A.J., Calvo-Rolle, J.L., Corchado, E.: Modeling the electromyogram (EMG) of patients undergoing anesthesia during surgery. In: 10th International Conference on Soft Computing Models in Industrial and Environmental Applications, pp. 273–283. Springer, Cham (2015)
23. Garcia, R.F., Rolle, J.L.C., Castelo, J.P., Gomez, M.R.: On the monitoring task of solar thermal fluid transfer systems using NN based models and rule based techniques. Eng. Appl. Artif. Intell. **27**, 129–136 (2014)
24. Quintian Pardo, H., Calvo Rolle, J.L., Fontenla Romero, O.: Application of a low cost commercial robot in tasks of tracking of objects. Dyna **79**(175), 24–33 (2012)
25. Sánchez Fernández, A., Francisco Sutil, M., de la Fuente Aparicio, M.J., Vega Cruz, P.: Control predictivo no lineal tolerante a fallos en una planta de tratamiento de aguas residuales (2016)
26. El Bahja, H., Vega, P., Revollar, S.: Economic optimization based on nonlinear parametric GPC for a wastewater treatment plant. In: 2014 IEEE 53rd Annual Conference on Decision and Control (CDC), pp. 3815–3820. IEEE, December 2014
27. Alaayed, I., El Bahja, H., Vega, P.: A sliding mode based on fuzzy logic control for photovoltaic power system using DC-DC boost converter. In: 2013 3rd International Conference on Systems and Control (ICSC), pp. 320–325. IEEE, October 2013
28. Casado-Vara, R., Vale, Z., Prieto, J., Corchado, J.: Fault-tolerant temperature control algorithm for IoT networks in smart buildings. Energies **11**(12), 3430 (2018)
29. Casado-Vara, R., Chamoso, P., De la Prieta, F., Prieto, J., Corchado, J.M.: Nonlinear adaptive closed-loop control system for improved efficiency in IoT-blockchain management. Inf. Fusion **49**, 227–239 (2019)
30. Casado-Vara, R., Prieto-Castrillo, F., Corchado, J.M.: A game theory approach for cooperative control to improve data quality and false data detection in WSN. Int. J. Robust Nonlinear Control **28**, 5087–5102 (2018)

Special Session - Optimization, Modeling and Control by Soft Computing Techniques

Low Voltage Grid Operation Scheduling Considering Forecast Uncertainty

Albert Ferrer, Ferran Torrent-Fontbona[(✉)], Joan Colomer,
and Joaquim Meléndez

University of Girona, 17003 Girona, Catalonia, Spain
ferran.torrent@udg.edu

Abstract. A model for day-ahead scheduling of batteries and branch switches in the low voltage grid, considering forecasts uncertainties, is proposed. The objective is to reduce the energy losses of the distribution lines and avoid critical events such as congestions or over and under-voltages in the local network. Simulations of different day-ahead situations are performed with a modified particle swarm optimisation algorithm. The results show that critical events are avoided and energy self-consumption within the local network is increased.

Keywords: Robust optimisation · Smart grids ·
Particle swarm optimisation · Network reconfiguration ·
Energy storage management

1 Introduction

The quantity of distributed generation (DG) connected to low voltage (LV) grids is rapidly increasing due to the technological advances in DG and policies promoting them. However, LV grids have been designed to be passive elements of the electricity network only used to provide the required energy to costumers. Thus, distribution system operators (DSOs) managed the network under their responsibility at medium voltage (MV) since it was not necessary to actively manage the LV grid. The increase of DG raises the problem of a lack of observability of the LV grid and an absence of systems to actively operate it.

Despite the deficiency of grid operability at LV, there are LV grid actuators such as controllable distributed generators (CGs), battery energy storage systems (BESSs) and branch switches (BSs) that permit to act on the LV grid [1–10]. When these grid assets are available, the DSO can tackle the problem of scheduling them to optimise objectives such as the quality of service and power losses.

This paper formalises the problem of scheduling BESSs and BSs in a grid with renewable energy generation. The scheduling of these grid actuators is performed considering the energy demand and supply forecast with its corresponding uncertainty, in order to prevent or mitigate critical events in the grid and minimise the power losses associated with the transmission.

© Springer Nature Switzerland AG 2020
F. Martínez Álvarez et al. (Eds.): SOCO 2019, AISC 950, pp. 493–502, 2020.
https://doi.org/10.1007/978-3-030-20055-8_47

494   A. Ferrer et al.

Previous studies of BESSs and other storage units have shown that they can reduce power losses, perform peak shaving and solve over-voltages situations [1,2]. Distribution network reconfiguration (DNR) with BSs approaches have also been used to reduce losses, prevent congestions or balance loads [3–5]. The aim of this paper is to use these actuators together to improve their efficiency and reduce their individual costs.

The intrinsic randomness of DGs, such as photovoltaic power generation (PV) or wind power plants, and the uncertainty of load demand are important factors for the decision-making process in the optimisation of the operation scheduling [6,7]. We consider the uncertainty of load demand and renewable generation with a chance constraint formulation [8–10] and we approximate it to robust optimisation.

2   Problem Formulation

Optimisation problems under uncertainties are usually modelled with chance constrained formulations, which consists of setting a confidence level of the conditions of the problem given a set of uncertainties [9,10]. The general expression is as follows,

$$\min\{\mathrm{E}\left[f(\boldsymbol{x},\boldsymbol{\xi})\right]\} \quad \text{s.t.} \quad \Pr\{g_i(\boldsymbol{x},\boldsymbol{\xi}) \leq 0\} \geq \alpha_i \quad i = 1, 2, ..., m, \tag{1}$$

where f is the objective function and g_i are the constraints of the problem, $\boldsymbol{x}$ is a vector with the deterministic variables and $\boldsymbol{\xi}$ the stochastic variables, α_i determines the probability or confidence level of the constraints. $\Pr\{\}$ denotes the probability of the events, considering the probability density function (PDF) of all the random variables. Because the objective function depends on random variables, the trend is to optimize the expected value, but some interesting works also consider the deviations [11].

Although the chance constrained formulation has been proved useful at modelling the optimisation problem with uncertainties, it is known that solving these approaches is very time consuming in computational terms. To reduce the computation time we use a robust optimisation formulation of the problem. The robust optimisation only considers the worst possible case instead of the probability to fulfil the constraints. This is the limit of Eq. (1) when $\alpha_i = 1$,

$$\min\{\mathrm{E}\left[f(\boldsymbol{x},\boldsymbol{\xi})\right]\} \quad \text{s.t.} \quad g_i(\boldsymbol{x},\boldsymbol{\xi}) \leq 0 \quad i = 1, 2, ..., m. \tag{2}$$

With this formulation, we assume that all the conditions of Eq. (1) are fulfilled if the worst situations given $\boldsymbol{\xi}$ meet the conditions in Eq. (2).

The stochastic variables of the study are associated with the energy demand forecast uncertainty and the generation profile of the PVs. The PDFs of both forecast errors are usually modelled as Gaussian distributions [12]. The demand or generation output of each bus in the grid is then the sum of the forecasted value and a random number with a Gaussian distribution $\mathcal{N}(0, \sigma_l^2(t))$ or $\mathcal{N}(0, \sigma_g^2(t))$, being σ_l and σ_g the corresponding standard deviations of energy demand and generation.

2.1   Problem Objectives

The tackled problem consists of scheduling the operation of BESSs and BSs in order to avoid or minimise critical events such as congestion and over/under-voltages, minimise the import/export of energy from/to the MV grid so as to minimise transport losses, and finally minimise the operation costs of BESSs and BSs.

Critical Events. One purpose is to avoid congestions and over/under voltages on the branches, these are considered as constraints of the optimisation. The congestions are avoided as single constraints that assure the loading of each branch to be below a particular threshold, in this case, 90% of the thermal limit. The voltages are secured to be between plus-minus 5% of the nominal value with a joined constrain. In the chance constrained picture

$$\text{Pr}\{\phi_r(\boldsymbol{x}, \boldsymbol{\xi}) < \phi_{\text{thresh}}\} \geq \alpha \quad \forall r, \tag{3}$$

$$\text{Pr}\{V_{\min} < V_r(\boldsymbol{x}, \boldsymbol{\xi}) < V_{\max}\} \geq \beta \quad \forall r, \tag{4}$$

where ϕ_r and V_r are the load and voltage of the grid branch r and ϕ_{thresh}, $V_{\min}$, $V_{\max}$ are the corresponding limits to fulfil with probabilities α and β given all the stochastic possibilities.

Instead, we can consider a robust optimisation approach and reduce this to only looking if the conditions inside $\text{Pr}\{\}$ are met in the three worst scenarios:

1. maximum load demand and maximum variable generation.
2. maximum load demand and minimum variable generation.
3. minimum load demand and maximum variable generation.

The first case corresponds to the worst scenario for congestion, while the second and third to the under- over-voltages respectively.

If the conditions are violated, the corresponding solution is punished for each hour that does not fulfil at least one of these conditions, the term is formulated as

$$f_{\text{critical}} = \sum_{t=t_0}^{t_f} c(t), \tag{5}$$

where c is a binary variable being 0 if the conditions are met for the three scenarios and 1 if any condition is not satisfied, thus f_{critical} is the number of hours that present possible critical events, t_0 is the initial or present time and t_f is the time horizon or final time of the scheduling and time is discretised in n periods of Δt time steps, being $n\Delta t = 24$ h.

Power Losses. On the other hand, we also want the grid to as self-sufficient as possible. To achieve this, we minimise the difference between the total energy given by the variable generators E_{VG} and the BESSs E_{B} with the demanded in the grid E_{L}. Actually, this is the same as minimising the exchange with the external grid E_{EG}, since the energy balance equation between the local and the external grid is

$$E_{\mathrm{EG}}(t) = E_{\mathrm{L}}(t) - E_{\mathrm{B}}(t) - E_{\mathrm{VG}}(t), \tag{6}$$

assuming there are no power losses in the local grid. The self-consumption term for the objective function is formulated as

$$f_{\mathrm{self}} = \sum_{t=t_0}^{t_f} |E_{\mathrm{L}}(t) - E_{\mathrm{B}}(t) - E_{\mathrm{VG}}(t)|. \tag{7}$$

Operational Costs and Restrictions. Critical events and power losses may be avoided or reduced with the scheduling of BESSs, nevertheless its use has associated restrictions and costs. The energy given or stored by a BESS during a time period Δt depends on the change of the state of charge (SoC) as

$$\mathrm{SoC}(t) = \begin{cases} \mathrm{SoC}(t - \Delta t) - \eta_c \frac{E_b(t)}{E_m} \Delta t \text{ if } E_b(t) \leq 0 \\ \\ \mathrm{SoC}(t - \Delta t) - \frac{1}{\eta_d} \frac{E_b(t)}{E_m} \Delta t \text{ if } E_b(t) \geq 0 \end{cases} \tag{8}$$

with $\mathrm{SoC}(t)$ the charge of the battery at time t, η_c and η_d the charging/discharging efficiency of the battery, E_m the energy that the BESS can charge/discharge in an hour, depending on the nominal capacity and the ramping down/up, and $E_b(t)$ the energy exchanged during the period $(t, t + \Delta t)$ between the grid and the BESS b. The BESSs energy exchange depends on the SoC restrictions, which are the initial and final charge and

$$\mathrm{SoC}_{\mathrm{min}} < \mathrm{SoC}(t) < \mathrm{SoC}_{\mathrm{max}}, \tag{9}$$

$$\Delta\mathrm{SoC}_{\mathrm{min}} < |\Delta\mathrm{SoC}(t)| < \Delta\mathrm{SoC}_{\mathrm{max}}, \tag{10}$$

where $\Delta\mathrm{SoC}(t) = \mathrm{SoC}(t) - \mathrm{SoC}(t - \Delta t)$.

In order to maximise the life of the batteries we consider a term to model the use of the battery. Since the wear and tear of the BESS depend on many technicalities of the kind of BESS used, we simply add a term with the total exchanged energy during the time horizon. In this way, the use of all the batteries in the grid is described via

$$f_{\mathrm{batt}} = \sum_{\forall b} \sum_{t=t_0}^{t_f} |E_b(t)|. \tag{11}$$

The presence of BSs can also be very helpful preventing grid critical events, nevertheless changing the state of one switch presents a hard operational cost we want to avoid and it is preferred to keep the BSs inactive. For these reasons we want to reduce the number of BSs changes and the number of time slots the BSs are active. The function associated with these costs is formulated as

$$f_{\mathrm{switches}} = \sum_{\forall s} \left\{ \sum_{t=t_0}^{t_f} a_s(t) + \sum_{t=t_0+\Delta t}^{t_f} |a_s(t) - a_s(t - \Delta t)| \right\}, \tag{12}$$

where $a_s(t)$ is the state of the branch switch s during the period $(t, t + \Delta t)$, with 1 as active and 0 inactive. The first part of Eq. (12) is the number of hours active and the second are the amount of changes.

3 Solution Approach

3.1 Objective Function

The formulation presented in the previous section is a multi-objective optimisation problem of Eqs. (5),(7),(11) and (12). The solution approach proposed for the multiple-criteria decision-making is a hybrid between a hierarchical method and a weighted sum. Through scalarization, the multi-objective problem is converted into a single objective with the following hierarchical preferences.

The first objective to accomplish is to reduce the critical events of the grid Eq. (5), with an assigned coefficient $M_1 \gg 1$. Secondly, minimise the operational cost of the BSs Eq. (12), with a coefficient $M_2 \ll M_1$. Finally, a weighted combination of the self-consumption Eq. (7) and the BESSs use Eq. (11) is also minimised, with a relation weight ν. Altogether, the objective function

$$f = M_1 f_{\text{critical}} + M_2 f_{\text{switches}} + \nu f_{\text{self}} + (1 - \nu) f_{\text{batt}}. \tag{13}$$

3.2 Simulation Algorithm

This paper adopts a particle swarm optimisation (PSO) to solve the model proposed because of its efficiency of computation and adaptable implementation [13–17]. Each particle of the algorithm, represents a solution with all the scheduled BESSs and BSs, these particles move according to simple rules converging to the optimal schedule. All the particles adjust their positions through iterations according to their experience and to the entire community's. The status of the particles is described by its position x_p and velocity v_p which are updated as

$$v_p = \omega v_p + c_1 ran_1(p_p - x_p) + c_2 ran_2(g - x_p), \tag{14}$$

$$x_p = x_p + v_p, \tag{15}$$

where ω is the inertia weight, c_1 and c_2 are the learning factors, ran are random numbers in $(0, 1)$, p_p is the best position of the particle and g the best position of all the particles [14].

If a new velocity moves the particle out of the search space it is changed by the maximum velocity such the particle keeps inside v_{max}.

Each BESS in the grid is a dimension of the particle, while all the BSs act together as one single dimension. At the same time each dimension in the particle consists of an array with a number of values equal to the number of time-series slots we want to schedule. Therefore, we have a position and velocity for every time-slot, for each dimension and for all particles.

We have adopted a dissipative-PSO (DPSO) with a linear time varying (LTV) inertia descending from ω_{max} to ω_{min} to avoid local minima. With the LTV weight we reduce the movements in the search-space progressively, starting with large movements through all the space and increasing the convergence speed when reducing the weight [15].

The dissipative part is introduced changing the positions and velocities randomly after the updates with a given probability d_v for the velocity and d_x for the position. Therefore, we create an open system out from equilibrium and improve the efficiency [16].

To move the BSs in the PSO, the set of binary states of the switches is ordered and converted into a decimal number. After moving the decimal number with the updated velocity it is rounded up and the state of the switches is changed to the corresponding binary state. Other more sophisticated approaches can also be used for the network reconfiguration [17].

After moving the particles, the fitness of each one is evaluated and the best individual and global solution are determined. With the robust approach, we only need to run the three power flow solvers (PFSs) corresponding to the worst possible scenarios described in Sect. 2.1, in order to compute the fitness. Otherwise, if we apply chance constrained optimisation we would have to evaluate the probabilities by sampling, thus, run many PFSs (what takes a lot of computer time). Notice that, the only term with stochastic variables is f_{self}, thus we have to average only this part to compute the mean.

The algorithm parameters used are $\omega_{max} = 0.9$, $\omega_{min} = 0.4$, $c_1 = c_2 = 2$, $d_x = d_v = 0.01$. We have used GridCal as the PFS in our simulations.

4   Experimentation

4.1   Set up

The proposed formulation is simulated with a network topology and historical data of a real LV grid. The pilot grid consists of two radial networks connected to the MV grid. One radial part, network-I, has 9 loads and 2 PVs, while the other, network-II has 7 loads and also 2 PVs. To the grid we add two BESSs, one on each network, and two BSs between buses of the networks. We can see a schematic representation of the grid in Fig. 1.

To determine the constant terms in the objective function, Eq. (13), several simulations were performed, first without critical events and BSs, to get a non-dominated solution of the pareto front depending on ν. Because f_{self} decreases when increasing ν from 0 to 1 but stabilizes around $\nu = 0.65$ and f_{batt} keeps increasing, this weight have been set a priori. To maintain the hierarchy of the other terms described in Sect. 2.1 we have set $M_1 = 10^{10}$ and $M_2 = 10^8$ in the objective function.

The restriction values for the BESSs used in the study are $SoC_{\min} = 0.2$, $SoC_{\max} = 0.8$, $\Delta SoC_{\min} = 0.05$, $\Delta SoC_{\max} = 0.15$, $SoC_{\text{ini}} = SoC_{\text{final}} = 0.5$, $E_m = 100\,\text{kWh}$, $\eta_c = \eta_d = 0.95$.

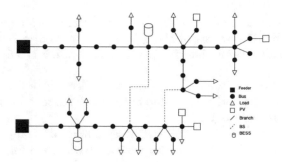

Fig. 1. Pilot grid representation. Network-I in the top and network-II in the bottom.

The standard deviation σ_l determining the error distribution of the energy demand at each bus is set to a corresponding mean absolute percentage error (MAPE) of 10% and a confidence level of 95%. For the deviation σ_g of the PVs generation uncertainty we take the same confidence level but a MAPE of 5%.

The scheduling has been done with a time step $\Delta t = 1\,\text{h}$ (from $00\,\text{h}$ to $23\,\text{h}$), since this is the resolution of the data used. The algorithm parameters used are $N = 60$ particles and 300 iterations.

4.2 Results

Based on the robust optimisation model and the solution approach described above, a day-ahead scheduling of the BESSs and BSs of the grid has been performed. The numerical calculation results of several cases are presented in Table 1. Each case corresponds to a different day with the generation and consumption values increased in order to be close or have critical events in the grid and be coherent with the sizing of the BESSs.

Table 1. Objective function values for different cases with and without the scheduling.

Case	Scheduling	f_{critical}	f_{switches}	f_{self}	f_{batt}	f
1	✗	0	0	894950	0	581717
	✓	0	0	792361	186089	580166
2	✗	2	0	1314848	0	$2.00008 \cdot 10^{10}$
	✓	0	3	1180532	230854	$3.00848 \cdot 10^{8}$
3	✗	1	0	919180	0	$1.00006 \cdot 10^{10}$
	✓	0	3	806909	196834	$3.00593 \cdot 10^{8}$
4	✗	0	0	815202	0	529881
	✓	0	0	673555	220621	515028
5	✗	0	0	1162551	0	755658
	✓	0	0	1174690	204834	835240

Case 1 corresponds to the 1st of March with the values increased 4.5 times, case 2 is 6.5 times the values of the 1st of May, case 3 is 7 times the 1st of July, case 4 is 6.5 times the 1st of September and case 5 is 4.5 times the 1st of December.

Cases 1 and 4 do not present any grid issue. Because of the hierarchy of the objective function, the optimal solution of the BSs is to remain off and the BESS scheduling aims to reduce f_{self}.

We can notice the effect on the energy self-sufficiency comparing the energy exchanged with the MV grid with and without BESSs, see Fig. 2. The peaks of energy surplus (when the generation exceeds the demand) are shaved, and the amount of energy given is also reduced almost at every hour.

In case 5 we do not have a exceed of generation at any time of the day. Thus, there is no BESSs scheduling that can improve the energy self-sufficiency and reduce the fitness of the objective value. Therefore, in this case the scheduling should not be applied since it aggravates the problem.

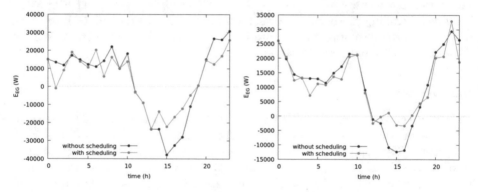

Fig. 2. Energy exchanged with the external grid for the case 1, with and without scheduling. Left for the grid network-I, right for network-II.

Case 2, without scheduling, presents possible congestions at two different hours. One possible critical event is avoided only with the BESSs scheduling, while the other is eliminated by activating one BS during one hour. Similarly, in case 3 one congestion is avoided with the activation of one BS.

Similar results have been found using samplings to evaluate the probabilities of the chance constrained formulation. Since the constraints are less severe, better solutions are possible. Nevertheless, this approach needs much more computer time to find similar results, about 6–7 h compared with 15–25 min of the robust method. Also, we have to take into account the error associated with the random sampling. All this makes unfeasible the chance constrained formulation, since the aim is to propose the scheduling in a few minutes.

With the scheduling, part of the critical events can be avoided. Nevertheless, because of the complexity of the problem, it is not trivial to determine if an event can be solved or not a priori.

5   Conclusions

This paper has presented a novel formulation integrating the scheduling of battery energy storage systems (BESSs) and branch switches (BSs) considering the uncertainties associated with energy generation and demand forecasts. The model has the purposes to avoid possible grid critical events such as congestions and over/under voltages and reduce power losses by improving the local self-sufficiency, while keeping the operational costs of the BESSs and BSs as low as possible.

Simulations of the proposed approach have been done through a dissipative particle swarm optimisation algorithm on a real grid. Results show to successfully avoid possible critical events with the scheduling of BESSs and BSs. The peaks of energy exchange, associated with power losses on the transmission lines, are reduced with the BESSs if the generation in the local grid is greater than the load in at least a period of the scheduled time.

With the approximation of the robust approach versus the chance constrained optimisation, we lose accuracy on the formulation, but the computational implementation is more efficient and makes it possible to obtain a viable solution in less than half an hour.

In future works we expect to adopt a rolling-horizon scheme in order to permit re-scheduling of the assets using more recent forecast. Moreover, the solution approach algorithm could include state-of-the-art techniques such as opposition based learning or improved selective mechanisms to enhance its performance.

Acknowledgments. This work has been developed under the European project RESOLVD of the Horizon 2020 research and innovation program (topic LCE-01-2016-2017) and grant agreement N773715.

References

1. Jayasekara, N., Masoum, M.A., Wolfs, P.J.: Optimal operation of distributed energy storage systems to improve distribution network load and generation hosting capability. IEEE Trans. Sustain. Energy **7**(1), 250–261 (2016)
2. Almeida, M.E., Pires, V.F., Camilo, F.M., Castro, R.: Self-consumption and storage as a way to facilitate the integration of renewable energy in low voltage distribution networks. IET Gener. Transm. Distrib. **10**(7), 1741–1748 (2016)
3. Rao, R.S., Ravindra, K., Satish, K., Narasimham, S.V.: Power loss minimization in distribution system using network reconfiguration in the presence of distributed generation. IEEE Trans. Power Syst. **28**(1), 317–325 (2013)
4. Usman, M., Coppo, M., Bignucolo, F., Turri, R.: Losses management strategies in active distribution networks: a review. Electr. Power Syst. Res. **163**(October), 116–132 (2018)
5. Dorostkar-Ghamsari, M.R., Fotuhi-Firuzabad, M., Lehtonen, M., Safdarian, A.: Value of distribution network reconfiguration in presence of renewable energy resources. IEEE Trans. Power Syst. **31**(3), 1879–1888 (2016)

6. Jia, X., Li, X., Yang, T., Hui, D., Qi, L.: A schedule method of battery energy storage system (BESS) to track day-ahead photovoltaic output power schedule based on short-term photovoltaic power prediction. In: International Conference on Renewable Power Generation (RPG 2015), p. 4 (2015)

7. Gao, H.C., Choi, J.H., Yun, S.Y., Lee, H.J., Ahn, S.J.: Optimal scheduling and real-time control schemes of battery energy storage system for microgrids considering contract demand and forecast uncertainty. Energies 11(6), 1–15 (2018)

8. Cong, P., Tang, W., Zhang, L., Zhang, B., Cai, Y.: Day-ahead active power scheduling in active distribution network considering renewable energy generation forecast errors. Energies 10(9), 1291 (2017)

9. Wu, J., Zhang, B., Jiang, Y., Bie, P., Li, H.: Chance-constrained stochastic congestion management of power systems considering uncertainty of wind power and demand side response. Int. J. Electr. Power Energy Syst. 107, 703–714 (2019)

10. Xu, Y., Zhao, T., Zhao, S., Zhang, J., Wang, Y.: Multi-objective chance-constrained optimal day-ahead scheduling considering BESS degradation. CSEE J. Power Energy Syst. 4(3), 316–325 (2018)

11. Li, P., Arellano-Garcia, H., Wozny, G.: Chance constrained programming approach to process optimization under uncertainty. Comput. Chem. Eng. 32(1–2), 25–45 (2008)

12. Hernandez, L., Baladron, C., Aguiar, J.M., Carro, B., Sanchez-Esguevillas, A.J., Lloret, J., Massana, J.: A survey on electric power demand forecasting: future trends in smart grids, microgrids and smart buildings. IEEE Commun. Surv. Tutor. 16(3), 1460–1495 (2014)

13. Lee, T.Y.: Operating schedule of battery energy storage system in a time-of-use rate industrial user with wind turbine generators: a multipass iteration particle swarm optimization approach. IEEE Trans. Energy Convers. 22(3), 774–782 (2014)

14. Eslami, M., Shareef, H., Khajehzadeh, M., Mohamed, A.: A survey of the state of the art in particle swarm optimization. Res. J. Appl. Sci. Eng. Technol. 4(9), 1181–1197 (2012)

15. Sengupta, S., Basak, S., Peters, R.A.: Particle swarm optimization: a survey of historical and recent developments with hybridization perspectives (2018)

16. Xie, X.F., Zhang, W.J., Yang, Z.L.: Dissipative particle swarm optimization. In: Proceedings of the 2002 Congress on Evolutionary Computation, CEC 2002, vol. 2, pp. 1456–1461 (2002)

17. Pegado, R., Ñaupari, Z., Molina, Y., Castillo, C.: Radial distribution network reconfiguration for power losses reduction based on improved selective BPSO. Electr. Power Syst. Res. 169, 206–213 (2019)

Iterative Learning Control
for a Hydraulic Cushion

Ignacio Trojaola[1]([✉]), Iker Elorza[1], Eloy Irigoyen[2], Aron Pujana[1],
and Carlos Calleja[1]

[1] Ikerlan Technology Research Center, Control and Monitoring Area,
Arrasate-Mondragon, Spain
itrojaola@ikerlan.es
[2] UPV/EHU, Alda. Urquijo S/N, Bizkaia, Spain

Abstract. An Iterative Learning Control algorithm is presented for the
force control circuit of a hydraulic cushion, which improves the existing
control scheme based on classical PI control and feedback linearization.
The circuit contains a proportional valve for regulating the pressure at
the cylinder chamber and, therefore, the force applied by the cushion.
The Iterative Learning Control filter design is based on the rejection of
the valve's dynamics, which are filtered by a fourth-order low-pass filter.
The filter is divided into two second-order filters to carry out forward
and backward filtering and obtain zero-phase filtering. The proposed
methodology improves the performance of the existing control scheme
and reduces considerably the settling time and overshoot of the pressure
signal.

Keywords: Iterative Learning Control · Feed-forward ·
Feedback linearization · Hydraulic press · Force control

1 Introduction

In the context of metal forming presses, the control of the cushion plays a fundamental role in deep drawing processes in which customer-specific force profiles
need to be ensured. The cushion provides a controlled blank holder force necessary for guaranteeing the correct forming of the workpiece.

The cushion force is produced by a hydraulic force control system (see Fig. 1)
which modifies the proportional valve's opening ratio so that the desired pressure in the single-acting cylinder chamber is achieved. There usually exist performance specifications regarding the maximum peak pressure allowed in the
cylinder chamber and the settling time of the pressure signal.

During the drawing process the slide makes contact with the cushion, and
a positive input signal u between 0 and 1 is sent to the valve which moves the
spool to the left, where port A is connected with T, and port P with B. At this
spool position, fluid is released from the cylinder chamber through the valve to
the tank, as the piston is retracted.

© Springer Nature Switzerland AG 2020
F. Martínez Álvarez et al. (Eds.): SOCO 2019, AISC 950, pp. 503–512, 2020.
https://doi.org/10.1007/978-3-030-20055-8_48

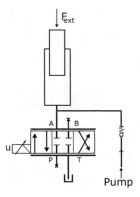

Fig. 1. Hydraulic force control circuit.

When the drawing process is finished and the slide and the cushion are not in contact, a negative input signal between −1 and 0 is sent to the valve, which moves the spool to the right, so that port B is connected with T and port P with A. At this spool position, no fluid flows through the valve as both port P and B are closed, and the fluid from the pump line extends the cylinder to the initial position.

Equation 1 shows the relation between the pressure in the cylinder and the volumetric flow out the valve

$$q = -K_v(y_v)\sqrt{P}, \tag{1}$$

where P is the cylinder pressure and $K_v(y_v)$ is the valve coefficient introduced by [1], here referred to as hydraulic conductance, which is a function of the valve's spool position y_v. The hydraulic conductance function is nonlinear and it is often obtained via empirical tests.

Ideally, the dynamics of the valve would be negligible as they are faster than any other system component dynamics. However, y_v is the valve response to an input command signal, u. For illustrative purposes, the dynamics of the proportional valve from u to y_v can be regarded (see [2]) as a second-order transfer function:

$$G_v(s) = \frac{y_v(s)}{u(s)} = \frac{\omega_n^2}{s^2 + 2\omega_n\zeta s + \omega_n^2}. \tag{2}$$

The following values have been set for the valve parameters: $\omega_n = 400\,\text{rad/s}$ and $\zeta = 1$. These values are used in simulations in order to analyze the performance of the system with the designed control scheme. However, in the design of the control scheme will not be considered as they are approximated values.

PI controllers are used to regulate the valve's opening ratio and, therefore, be able to follow the desired pressure reference in the cylinder. However, the hydraulic force circuit is nonlinear and the PI parameter tuning is confined to a local operating point, resulting in poorer response when the process deviates from said operating point.

Feedback linearization (FL) can be used to minimize the nonlinearities in the system so that the closed-loop results in a linear system. However, as explained later in Sect. 2, to carry out FL it is necessary to have full information of the $K_v(y_v)$ function. The estimation of $K_v(y_v)$ is left out of this paper, as it requires nonlinear regression methods, e.g. neural networks or Gaussian processes, or estimation algorithms such that Kalman filters or state observers. Therefore, an implicit way of estimating $K_v(y_v)$ is proposed, which consists of adding an extra feed-forward (FF) signal based on Iterative Learning Control (ILC).

ILC has been shown to improve the performance of systems that carry out the same operation repeatedly, as is the case of hydraulic presses. By recording the system error at each iteration, a learning algorithm updates the FF signal that is sent to the feedback loop. There exist two ways of combining ILC with a feedback loop [3]. In this paper, the parallel arrangement is employed, and the ILC control input is used instead of the velocity disturbance FF signal. This eliminates the need to know $K_v(y_v)$ explicitly, but restricts the learning to a specific trajectory being required to restart the learning from scratch when the reference trajectory changes.

ILC algorithms have already been used for solving control problems related to hydraulic presses where the controller cannot track the desired position or pressure input accurately. ILC in the position control loop of a hydraulic cylinder was first introduced by [4]. At the first 40 iterations, the iteration error converged towards zero, however it remained constant without converging further for subsequent iterations.

A proportional ILC with a Fourier Domain Filter (FDF) was implemented in [5] for the position control loop of a hydraulic press. The author designed a serial arrangement with the ILC + FDF providing the PI reference, with the objective of accurately tracking a 10 Hz sinusoidal position reference signal. The system was stable as long as no frequencies other than said fundamental frequency were excited, whereas if any disturbance occurred, the system became unstable. For the cushion force controller in this paper, a pressure step reference is used for which stability of the ILC in a specific frequency is not sufficient.

In order to control the cylinder pressure, [6] implemented a proportional ILC and a Butterworth filter that filtered the noise and the model uncertainties at high frequencies. The phase loss of the Butterworth filter was counteracted by introducing a time-shift operator to provide system stability. For a similar hydraulic setup and a higher frequency valve, with a step response of 0.2 bar, the pressure settled after 5 s. In this paper, it is required to have a settling time of less than 0.4 s for a 40 bar step response, so that the cushion force control specifications are fulfilled.

ILC has been implemented either on systems with a limited frequency spectrum or on systems with narrow operating envelopes and without settling time requirements. This paper proposes an ILC algorithm for hydraulic systems with unknown valve dynamics, which considerably reduces the overshoot and settling time of the system to fulfill the cushion force control specifications.

2   Force Controller Design

The relationship between the pressure in the cylinder, the volumetric flow into it and piston motion is as follows:

$$q = A\dot{x} + (V_d + Ax)\beta\dot{P}. \tag{3}$$

Parameters A, V_d and β correspond to the piston area, the dead volume of the cylinder chamber and the hydraulic fluid compressibility, respectively. Variables x and P are the piston position and the cylinder pressure, respectively.

Combining Eqs. 1 and 3, an estimation for the valve spool position can be obtained:

$$\hat{y}_v = -K_v^{-1}\left(\underbrace{\frac{\dot{P}(V_d + Ax)\beta}{\sqrt{P}}}_{FL} + \underbrace{\frac{A\dot{x}}{\sqrt{P}}}_{FF}\right). \tag{4}$$

We may perform FL+FF based on Eq. 4, as indicated. The first term cancels the nonlinearities in the system, while the second is a FF signal that eliminates the velocity disturbance. In the estimation of $\hat{y}_v$, the FL term will not be a significant contributor, as it contains β, which is very small. Therefore, if there exist any uncertainties in the system parameters of the FL, the estimation will still be within an acceptable margin and the deviation will be corrected by the PI controller. However, the velocity disturbance FF term is larger and, if it is not eliminated correctly, it will result in large overshoot and settling time.

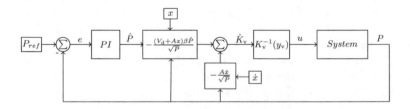

Fig. 2. Force controller block diagram.

Figure 2 shows a block diagram of the PI+FL+FF scheme. PI controller is chosen as it is currently being used in real hydraulic presses, however, other linear controllers could have been chosen, e.g. PID controllers or IP controllers, which could have overshoot improvements with respect to PI controllers [7]. Note that the output of the PI in the block diagram is $\hat{P}$, which in general differs from $\dot{P}$ in the FL+FF equation in 4. If no uncertainty is present in the system, most of the system terms will cancel out except for the valve's dynamics, which are unknown. The following expression will be left

$$\dot{P}(s) = G_v(s)\hat{P}(s), \tag{5}$$

resulting in the following simplified system and controller block diagram (Fig. 3):

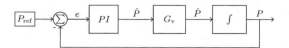

Fig. 3. Simplified force controller block diagram.

Neglecting G_v the closed loop system response is as follows:

$$\frac{P(s)}{P_{\text{ref}}(s)} = \frac{K_{\text{P}}s + K_{\text{I}}}{s^2 + K_{\text{P}}s + K_{\text{I}}}, \tag{6}$$

which is a second-order system with natural frequency $\sqrt{K_{\text{I}}}$. However, this is only true for frequencies low enough for the valve dynamics to remain negligible. This effectively sets an upper limit for K_{I} and K_{P}, and a lower limit for settling time. Therefore, K_{I} and K_{P} are tuned by trial and error to obtain a balance between the overshoot and the settling time in the reference tracking.

3   Iterative Learning Control Design

Given a system that performs the same operation repeatedly under the same operating conditions, ILC is based on the notion that the above mentioned system can improve its performance by learning from previous iterations. ILC was first introduced by Arimoto in [8], who proposed the first learning control scheme

$$U_{j+1}(s) = U_j(s) + L(s)E_j(s), \tag{7}$$

which is a general past-error feed-forward update law. $L(s)$ is the learning function and determines how the tracking error is used to update the control signal from one iteration to the next [9]. $U_j(s)$ is the Laplace transform of the entire input vector at the j-th learning iteration. The iteration error $E_j(s)$ is given by the difference between the reference and the corresponding iteration output, $E_j(s) = R(s) - Y_j(s)$.

In the block diagram shown in Fig. 4, the controller C consists of the PI controller together with FL+FF. The plant G is the hydraulic system, and Q is normally a gain which enlarges the stability region, as shown by Eq. 9.

From the block diagram, the error propagation equation can be obtained:

$$E_{j+1}(s) = Q(s)(1 - G(s)S(s)L(s))E_j(s) \tag{8}$$

where $S = \frac{1}{1+GC}$ is the sensitivity transfer function of the system.

It was shown in [3] that the system will be stable and converge monotonically to $E(s) = 0$ if:

$$|1 - G(jw)S(jw)L(jw)| < \frac{1}{Q(jw)} \qquad \forall \omega \in [-\infty, \infty]. \tag{9}$$

For ease of notation, the term $|1 - G(jw)S(jw)L(jw)|$ will be referred to as the frequency response of the ILC algorithm.

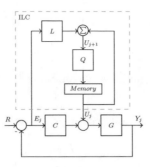

Fig. 4. Parallel ILC scheme.

If L is designed so that $L = S^{-1}G^{-1} = G^{-1} + C$, then the r.h.s in 8 vanishes, resulting in zero error at the second iteration.

The design of the L filter is based on the linearized hydraulic cushion circuit and the existing force controller. The designed L filter will be implemented in a high fidelity nonlinear simulation in Sect. 4.

From Eqs. 1 to 3, a linearized state space system with the pressure of the cylinder and the spool position and velocity as states, and the valve command as input, is derived. The general formulation of the small-signal state equation can be written as follows:

$$\begin{bmatrix} \delta\dot{P} \\ \delta\dot{y}_{\mathrm{v}} \\ \delta\ddot{y}_{\mathrm{v}} \end{bmatrix} = \begin{bmatrix} A_{11} & A_{12} & 0 \\ 0 & 0 & 1 \\ 0 & -\omega_{\mathrm{n}}^2 & -2\omega_{\mathrm{n}}\zeta \end{bmatrix} \begin{bmatrix} \delta P \\ \delta y_{\mathrm{v}} \\ \delta\dot{y}_{\mathrm{v}} \end{bmatrix} + \begin{bmatrix} 0 \\ 0 \\ \omega_{\mathrm{n}}^2 \end{bmatrix} \delta u \qquad \begin{aligned} A_{11} &= -\frac{K_{\mathrm{v}}(\bar{y}_{\mathrm{v}})}{(V_{\mathrm{d}} + Ax)\beta} \frac{1}{2\sqrt{\bar{P}}} \\ A_{12} &= -\frac{\sqrt{\bar{P}}}{(V_{\mathrm{d}} + Ax)\beta} \dot{K}_{\mathrm{v}}(\bar{y}_{\mathrm{v}}) \end{aligned}$$

(10)

where $\delta P = P - \bar{P}$ and $\delta y_{\mathrm{v}} = y_{\mathrm{v}} - \bar{y}_{\mathrm{v}}$, are the small-signal deviation from an operating point obtained in steady-state conditions, $\bar{P}$ and $\bar{y}_{\mathrm{v}}$ respectively.

Equation 10 shows the linearization of the plant depicted in Fig. 1, which would be inverted to carry out the design of $L = G^{-1} + C$. However, the valve's dynamics are generally unknown and, therefore, the states containing valve model parameters in Eq. 10 are not taken into account in the inversion. The designed L is as follows: $L = \frac{s - A_{11}}{A_{12}} + C$.

The L filter is dependent on x, which has been considered constant although the cylinder extends and retracts when the valve is closed and opened respectively. Moreover, the fluid compressibility varies due to pressure and temperature changes taking place in the cylinder chamber, and $K_{\mathrm{v}}(\bar{y}_{\mathrm{v}})$ and its derivative $\dot{K}_{\mathrm{v}}(\bar{y}_{\mathrm{v}})$ are unknown. These system uncertainties will make the magnitude response of L (see Fig. 5) deviate from the linearized model at low frequencies. An investigation of the system response under these parametric uncertainties is left for subsequent communications.

G^{-1} will turn into a non-causal system, which acts as a high-frequency amplifier (see Fig. 5), by which the unknown high frequencies of the valve would not be attenuated. Therefore, a fourth-order low-pass filter has been included in the design of L. The cut-off frequency of said filter has been set lower than the valve's natural frequency, $\omega_{\mathrm{n}} = 400\,\mathrm{rad/s}$, to reduce learning near the latter. The corresponding inverted plant term of the L filter with the addition of the low-pass filter results in $\hat{G}_1^{-1} = \frac{s-A_{11}}{A_{12}} \frac{K}{(s+100)^4}$, where K is a scalar which results in the filter having unit gain.

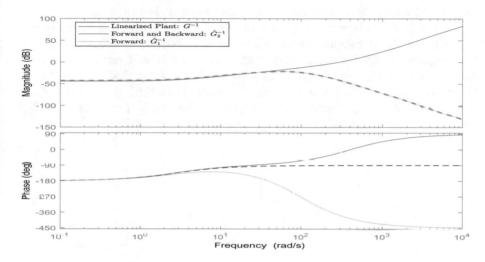

Fig. 5. G^{-1}, $\hat{G}_1^{-1}$ and $\hat{C}_2^{-1}$ Bode diagram.

In Fig. 5, at low frequencies, $\hat{G}_1^{-1}$ has the same response as G^{-1}. However, the low-pass filter attenuates frequencies above 100 rad/s, thus filtering the unknown valve dynamics.

One wishes to design L as close as possible to $G^{-1} + C$ so that the right hand side in 8 vanishes for low frequencies. On a Nyquist plot, this is equivalent to having zero frequency response for all frequencies. In Fig. 6a, the designed L filter has zero frequency response at low frequencies. However, as frequency grows, the frequency response grows apart from zero due to L deviating from $G^{-1} + C$.

The stability criterion 9 may easily be evaluated graphically, as shown by the Nyquist in Fig. 6a, where the frequency response gets out of the unit circle, introducing instability for the designed ILC algorithm, as a result of the phase loss caused by the low-pass filter.

In order to obtain a L design that filters the valve's dynamics and satisfies the stability criterion in 9, zero-phase filtering is carried out. The phase loss introduced by the filter is counteracted by applying a second-order filter to the iteration error vector twice, once in the positive direction of time and another once in the negative direction of time. This is only possible because the ILC's FF input is calculated between iterations, when the entire iteration error vector is available for processing. The result is zero-phase filtering. The new L filter with this *back and forth* design results in: $\hat{G}_2^{-1} = \frac{s-A_{11}}{A_{12}} \frac{K}{(s-100)^2(s+100)^2}$. The zero-phase filtering can be seen in Fig. 5, where a second-order low-pass filter is introduced which attenuates high frequencies without losing phase.

Figure 6b shows how the frequency response, after the *back and forth* filtering, starts at zero and remains close to zero at low frequencies. Near point (1,0) the frequency response signal gets slightly out of the unit circle, resulting in instability for high frequencies. A Q gain of 0.99 is set so that the response of all frequencies remains inside the unit circle and the overall control stability is ensured. Although the chosen Q gain is close to one, the performance of the ILC algorithm will be affected resulting in non-perfect tracking [3].

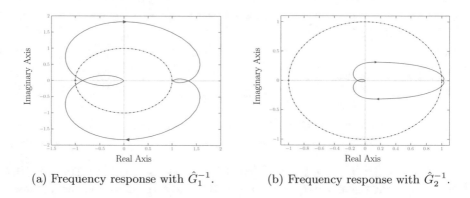

(a) Frequency response with $\hat{G}_1^{-1}$.   (b) Frequency response with $\hat{G}_2^{-1}$.

Fig. 6. ILC filter design.

4   Simulation Results

The designed control scheme has been implemented on a nonlinear high fidelity model of a force hydraulic circuit implemented with Simscape (see Fig. 7a). No velocity disturbance FF signal has been included in the simulations, as it is desired to analyze the ILC performance in the absence of velocity disturbance FF.

At iteration number eight (see Fig. 7b) the pressure references are tracked accurately without overshoot. The ILC learns implicitly the exact spool position required for each instant, which results in no overshoot in the pressure.

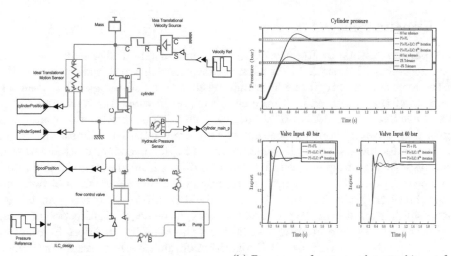

(a) Simscape implementation of the force circuit.

(b) Pressure reference and control input for 40 bar and 60 bar step response.

Fig. 7. Force circuit and ILC implementation results.

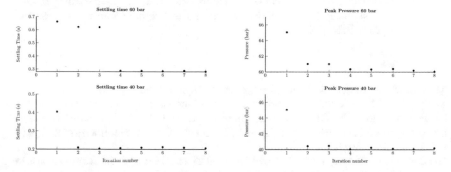

(a) Settling time over iterations for 60 bar and 40 bar reference.

(b) Convergence to 60 bar and 40 bar reference.

Fig. 8. Settling time and peak pressure of the high fidelity model.

Figure 8a shows the settling time to within 2% of the final value (marked by a black dotted line in Fig. 7b) and the peak pressure over eight iterations for 40 bar set-point. Convergence of the overshoot to 40 bar and 60 bar is reached at iteration number four and continues at the same peak level as iterations go by. The settling time is reduced by more than half in both cases compared to the first iteration where only PI and FL are used. Note that it is physically impossible to reach the desired pressure in less than 0.2 ms and 0.3 ms for 40 bar and 60 bar step reference respectively.

5   Conclusions

This paper presents an Iterative Learning Control (ILC) algorithm combined with a classical PI controller, feedback linearization (FL) and feed-forward (FF), for the force control circuit of a hydraulic cushion. The ILC design is based on the inversion of the linearized hydraulic circuit model in which the dynamics of the valve have been discarded as they are unknown. The inverted plant results in a high-frequency amplifier that does not attenuate the valve's dynamics. Therefore, a fourth order low-pass filter has been included in the inversion to attenuate said dynamics. At high frequencies the designed L differs from the inverted plant and introduces a phase loss that causes instability in the ILC algorithm, not fulfilling the stability criterion. The filter has been divided into two second-order filters, one of which is applied in the positive, the other in the negative, direction of time, resulting in zero-phase filtering and fulfilling the stability criterion. Simulation results show that the designed ILC algorithm eliminates the pressure overshoot and significantly reduces the settling time that exists with a PI controller and FL. Therefore, this methodology appears to be suitable to be implemented in a real hydraulic force circuit in the future.

Acknowledgment. This work has been partially funded by the Department of Development and Infrastructures of the Government of the Basque Country, via Industrial Doctorate Program BIKAINTEK.

References

1. Merritt, H.E.: Hydraulic Control Systems. Wiley, Hoboken (1967)
2. Sirouspour, M., Salcudean, S.: On the nonlinear control of hydraulic servo-systems. In: Proceedings 2000 ICRA, Millennium Conference, IEEE International Conference on Robotics and Automation, Symposia Proceedings (Cat. No. 00CH37065), vol. 2, pp. 1276–1282, April 2000
3. Bristow, D.A., Tharayil, M.: A learning-based method for high-performance tracking control. IEEE Control Syst. Mag. **26**, 96–114 (2006)
4. Chen, C.-K., Zeng, W.-C.: The iterative learning control for the position tracking of the hydraulic cylinder (2003)
5. Li, L., Poms, U., Thurner, T.: Accurate position control of a servo-hydraulic test cylinder by iterative learning control technique. In: Proceedings - UKSim-AMSS 8th European Modelling Symposium on Computer Modelling and Simulation, EMS 2014, pp. 297–302 (2014)
6. Gøytil, P.H., Hansen, M.R., Hovland, G.: Iterative learning applied to hydraulic pressure control. Ident. Control **39**(1), 1–14 (2018)
7. Sreekumar, T., Jiji, K.S.: Comparison of proportional-integral (P-I) and integral-proportional (I-P) controllers for speed control in vector controlled induction motor drive. In: ICPCES 2012 - 2012 2nd International Conference on Power, Control and Embedded Systems (2012)
8. Arimoto, S., Kawamura, S., Miyazaki, F.: Bettering operation of robots by learning. J. Robot. Syst. **1**(2), 123–140 (1984)
9. McCranie, K.D., Faulkner, M., French, D., Daddis, G.A., Gow, J., Long, A.: Enhanced iterative learning control with applications to a wafer scanner system. J. Strateg. Stud. **34**(2), 281–293 (2011)

Opinion Mining to Detect Irony
in Twitter Messages in Spanish

Daniela E. Sanjinés, Vivian F. López$^{(\boxtimes)}$, Ana B. Gil, and María N. Moreno

Departamento de Informática y Automática, Facultad de Ciencias,
Universidad de Salamanca, Plaza de los Caídos s/n, 37008 Salamanca, Spain
{idu001885,vivian,abg,mmg}@usal.es

Abstract. Companies, among other sectors, require that the opinions
generated on the web be extracted automatically, obtaining their polarity
on products or services, to achieve their objectives. Since the opinions are
subjective and unstructured, there are still many problems within this
field that must be solved. To mention a few, the problem of ambiguity
and the support of languages, directly affect in the time to make the
right classification of opinions, because most of the tools used in the
processing of texts, they only work well with data in English. With the
aim of contributing to the solution of both problems and evaluating the
real behavior of sentiment analysis for the Spanish language, a system is
proposed that allows determining the positive or negative polarity, trying
to detect the irony as a problem of ambiguity. For the classification,
a supervised learning method was implemented, with the Naive Bayes
algorithm. The evaluation of the results of the classification shows that
the problem of detecting ironies in Spanish, using the classical techniques
of opinion mining, is not completely resolved. However, we believe that
these results can be improved by applying some strategies.

Keywords: Opinion mining · Sentiment analysis ·
Social networks · Natural language processing · Text mining · Twitter ·
Classification algorithms

1 Introduction

With the emergence of web 2.0 and its rapid growth, users went from being
passive users to being active and collaborative on the web, generating different
types of content, such as text messages. The opinions of users are generated daily
in large quantities and in different ways: by rating a purchased product or service,
by expressing their opinion on a topic on social networks, making comments
in forums, criticizing current news, expressing opinions in blogs, among others.
Because this information is very abundant, it was necessary to look for techniques
and mechanisms to its recovery, processing and analysis. It is here that Artificial
Intelligence intervenes by responding to this need through automatic learning or
machine learning and natural language processing (NLP). The opinion mining or

© Springer Nature Switzerland AG 2020
F. Martínez Álvarez et al. (Eds.): SOCO 2019, AISC 950, pp. 513–522, 2020.
https://doi.org/10.1007/978-3-030-20055-8_49

also called sentiment analysis, allows the processing and the automatic extraction of information, classifying it as a positive, negative or neutral, considering that the information on many occasions it is unstructured and is subjective, because it handles natural language.

Despite the fact that in the last five years a lot of work has been done in this field, most of them oriented to the English language, since there are currently several commercial tools to determine the sentiment analysis, the truth is that there is still a long way to go since several obstacles prevail when determining the correct polarity of an opinion [11]. Among the pending challenges are problems of ambiguity, linguistic variability, falsehood [1,2]. All of them affect the sentiment classification, sometimes giving a negative classification being positive or vice versa.

With the aim of contributing to the solution of said problems and evaluating the real behaviour of the sentiment analysis for the Spanish language. It is proposed to elaborate a system of opinion mining, which tries to solve problems of ambiguity, such as the detection of irony. Applying the techniques used with better results currently in sentiment analysis, a two-focus model was designed. The first one using the tool *Natural Language Toolkit*[1] (NLTK) for the pre-processing of data and using unigrams and bigrams to draw characteristics. In the second approach we use *TreeTagger*[2] tool and Part-of-Speech Tagging (PoS Tagging) for the same purposes. For the classification, a supervised learning method was implemented, with the Naive Bayes algorithm. Finally, the results of the sentiment classification were evaluated with the *accuracy, recall, precision and F1 score* metrics [6] to determine the results regarding the detection of the irony. This article is structured as follows: In Sect. 2 we have the state of the art of opinion mining. In Sect. 3, the proposed methodology and experimental work are described. Section 4 we presents the results and their discussion, and finally the conclusions and future work lines.

2   Motivation and Related Work

Opinion mining is a branch of the NLP [8]. It is also classified as a subset of web mining within of data mining. The main methods that have been used in opinion mining correspond to supervised learning [3] and based on Lexicon[3] [4]. These works have been used in both short texts and long texts. Among the current trends, the use of Twitter data [10], stands out, due to its ease of obtaining. The most used methods for its extraction are through *hashtags*, as a supervised form, automatic labelling, and manually with annotators. The semi-supervised method is also used, which requires less amount of tagged data [5].

On the data pre-processing techniques that correspond to the NLP for opinion mining, the *tokenization* and the *PoS Tagging analysis* stand out. Most of

---

[1] http://www.nltk.org/.

[2] http://www.cis.uni-muenchen.de/~schmid/tools/TreeTagger/.

[3] List of words where each one has a numerical value that corresponds to the degree of sentiment whether positive or negative.

the works use English language data, so among the most used tools is Python's NLTK and the tool of Stanford NLP[4] [11]. More adaptive tools for the Spanish language such as TreeTagger[5]. Regarding of feature extraction methods, bag of words is very widely used through the features frequency, unigrams and bigram [5], according to the objectives of the investigation. Some authors try to analyse in a general way the polarity by documents, others through phrases and in a conceptual way. Naive Bayes and Support Vector Machine (SVM) [7] are the most used and best-performing classification techniques. Due to its ease of use and efficiency. On the other hand, there is an increase in the use of Neural Networks. The evaluation metrics of the classification vary, but generally the ones that are used are: accuracy, precision, recall, and F1 score [4].

3   Methodology and Experimentation

The methodology applied for the present work is summarized in Fig. 1. For the experimentation, supervised learning was implemented with two approaches, the first performing the tasks of pre-processing with NLTK and constructing a matrix with unigrams and bigrams. The second one doing the pre-processing with TreeTagger and building the matrix with unigrams, where the unigrams correspond to the *lemma/PoS Tagging* of the word that is extracted previously. The experimental work is detailed below.

3.1   Building of Training Corpus

The data used for the experimentation were obtained from Twitter. The extraction of tweets was done in a supervised way, to obtain an automatic labelling. Three groups of tweets were obtained: positive, negative and ironic through Python 3 using the Tweepy[6], library, which allows connection to the Twitter *REST API*. The *search method* was used retrieving tweets from the last 7 days. After obtaining the tweets of each group, a data cleaning was performed, certain characters were eliminated and some tweets were excluded. The different training sets were created and labelled:

- **Positive tweets:** 15020. It was followed as a search criterion, the tweets that contained in their text, the emoticon happy face ':D', assuming that these tweets correspond to positive polarity.
- **Negative tweets:** 10064. The search criterion was based on the tweets which contained in their text the sad face emoticon ':(', assuming that these tweets correspond to negative polarity.
- **Ironic tweets:** 1125. The search criteria was made with the *hashtags #sarcasmo, #ironía* assuming the tweets are ironic.

---

[4] https://nlp.stanford.edu/software/tokenizer.html.
[5] http://treetaggerwrapper.readthedocs.io/en/latest.
[6] http://tweepy.readthedocs.io/en/v3.5.0/getting_started.html.

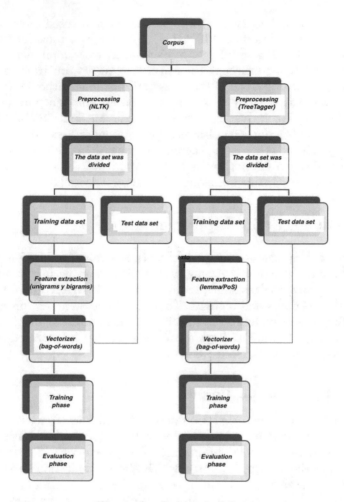

Fig. 1. Applied methodology

Because the number of retrieved ironic tweets was small (1125 tweets), the amount of data for the group of positives and negatives was reduced, taking the same amount. The three groups of tweets were united in a single file called *tuits.csv*. We used the training data set corresponding to 3375 tweets (75% of training and 25% of test).

3.2 Pre-processing

The pre-processing of the text or also called standardization, is the second step of the proposed model. The first approach performs pre-processing with NLTK, and the second approach uses TreeTagger. The treatment carried out for each approach is explained below.

Approach 1 - Pre-processing with NLTK: The tasks performed in the pre-processing stage with NLTK are summarized in Fig. 2.

Fig. 2. Pre-processing with NLTK.

For the analysis and manipulation of the data *Pandas*[7] was used through its *DataFrame*. As we see in Table 1, a data structure was created that contains the *Tag-Num column*, to indicate the type of tweet (positive = 0, negative = 1, ironic = 2). In addition, the pre-processed tweets were stored, eliminating empty words for Spanish with the NLTK method *stopwords.words ('spanish')* and applying the *Snowball Stemmer*[8] ('spanish') to obtain the root of words. The result was stored in the *Pre-processed column*.

Table 1. Pre-processing example with NLTK

Tag	Tweet	Tag-Num	Preprocessing
Positive	'Suena bien! La vida es muy corta para no celebrar el amor'	0	'suen bien ! la vid cort celebr amor'
Negative	'Estoy cansadiiisima de todo ya x favor'	1	'estoy cansadiiisim x favor'
Ironic	'Gracias a todos por respetar a la gente con pareja'	2	'graci respet gent parej'

Approach 2 - Preprocessing with TreeTagger: The tasks performed in the preprocessing with TreeTagger are summarized in Fig. 3.

In the same way, a column was added to the *DataFrame*, a numeric tag equivalent to type of tweet. These were processed with TreeTagger through Python. The tokenization of each tweet was performed. The word, the lemma and the *PoS Tagging* of each word corresponding to the tweets were obtained. In order to use the lemmas related to the *PoS Tagging* as a feature, a column called Preprocessing was created in which only the lemma and PoS Tagging *(lemma/PoS)* of each word were saved (Table 2 that represents an example of the *DataFrame*).

---

[7] https://pandas.pydata.org/.
[8] http://www.snowball.tartarus.org/.

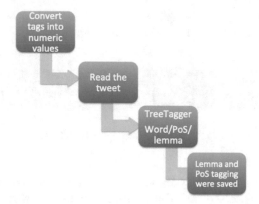

Fig. 3. Pre-processing with TreeTagger.

Table 2. Pre-processing example with TreeTagger

Tag	Tweet	Tag-Num	Preprocessing
Positive	'Suena bien! La vida es muy corta para no celebrar el amor'	0	'Sonar\VLfin bien\ADV !\FS el\ART vida\NC ser\VSF muy\ADV corto\ADJ para\CSUBI no\NEG celebrar\VLinf el\ART amor\NC'
Negative	'Estoy cansadiiisima de todo ya x favor'	1	'Estar\VEfin cansadiiisima\ADJ de\PREP todo\QU ya\ADV x\CARD favor\NC'
Ironic	'Gracias a todos por respetar a la gente con pareja'	2	'Gracia\NC a\PREP todo\QU por\PREP respetar\VLinf a\PREP el\ART gente\NC con\PREP pareja\NC'

3.3 Data Set

The division of the data was done in the same way for both approaches. The pre-processed corpus contains a total of 3375 tweets, this corpus was divided into two parts: the training set, which contained 2531 tweets (75% tagged corpus), and one set for testing, which contained 844 tweets (25% of the data set).

The division of the data set was done with the method *train_test_split* of the Scikit-learn[9] library of Python.

3.4 Feature Extraction

The extraction of feature was made from the data in the Pre-processed column of the *DataFrame* in each of the approaches. For the extraction of features, *CountVectorizer* was also used from Scikit-learn, with which a vocabulary was created from unigrams and bigrams for the first approach. For the second approach the lemma/POS Tagging obtained from the pre-processing was defined. In both approaches, the most frequent words (50% of the corpus) and the least frequent (below 2%) were eliminated.

---

[9] http://scikit-learn.org/stable/index.html.

3.5   Vectorization or Bag-of-Words

Once the features were extracted, the vectorization process was carried out, which is applied in both approaches. The vectorization or also called bag of words, performs the transformation from the text retrieved into a matrix, in which the frequency is obtained by the features present in each tweet. To perform this task we used *vect.transform (train)*. This method uses the adjusted vocabulary to create the matrix. If we visualized it as a *DataFrame* with Pandas, it would have an aspect as in Table 3, in which the first row corresponds to the characteristics obtained in the extraction and then for each tweet there is the frequency in which appear it should be noted that the features do not contain a specific order.

Table 3. Bag of words Example.

Tweet	Tag -Num	Preprocessing	con(PREP)	el(ART)	todo(QU) ...
'Suena bien! La vida es muy corta para no celebrar el amor'	0	'Sonar\VLfin bien\ADV !\FS el\ART vida\NC ser\VSF muy\ADV corto\ADJ para\CSUBI no\NEG celebrar\VLinf el\ART amor\NC'	0	1	0
'Estoy cansadiiisima de todo ya x favor'	1	'Estar\VEfin cansadiiisima\ADJ de\PREP todo\QU ya\ADV x\CARD favor\NC'	0	0	1
'Gracias a todos por respetar a la gente con pareja'	2	'Gracia\NC a\PREP todo\QU por\PREP respetar\VLinf a\PREP el\ART gente\NC con\PREP pareja\NC'	1	1	1
...					

3.6   Classification Algorithm

The Naive Bayes Multinomial probabilistic supervised algorithm was used, because in the experimental work the bag of words technique is used, with the frequency of occurrence of each feature found in the tweet. It has been shown that Naive Bayes classifies better this type of data, which also supports the multiple classification. The algorithm was applied in both approaches in order to predict the classification (positive, negative or ironic). Once the model was trained, the classification of the test data set was carried out through the *nb.predict method of Scikit-learn*.

4   Results Evaluation and Discussion

For the evaluation of the results, the confusion matrix was obtained and the accuracy, precision, recall and F1 score metrics were calculated. Because that the best result obtained was with approach 2, we show the results evaluation using Confusion matrix *Lemma/POS Tagging* with word reduction frequent and less frequent. In Table 4 we can observe the confusion matrix in which we have 293 samples that are positive, of which 144 were predicted correctly and 149 incorrectly. A total of 289 negative tweets were obtained, of which 176 were predicted correctly and 113 incorrectly. A total of 262 ironic tweets were obtained, of which 256 were predicted correctly and 6 were predicted incorrectly.

Table 4. Lemma/PoS matrix and word reduction

Class	Positive	Negative	Ironic
Positive	**144**	110	39
Negative	80	**176**	33
Ironic	3	3	**256**

Evaluation of the Two Approaches in Terms of the Obtained Metrics: In Table 5 it can be seen that the best result obtained was with approach 2, with the application of TreeTagger and reduction of more and less frequent words. On the other hand, the worst results were the approach 1 with its variant bigrams.

4.1   Discussion

Although the best results have been obtained with the approach 2, applying TreeTagger, which has better performance for the Spanish language, these are not good enough for the sentiment analysis. We have observed, in general values not very high in most of the metrics analyzed in both approaches and especially if we compare these results, in relation to those obtained in other researches for the English language. Considering the objectives of this research, we analyze the factors that could have influenced the results in some way. One of the first difficulties is the collection of data to build the corpus, taking into account the quantity and quality of the data obtained for automatic labeling. The data was scarce for the set of ironic tweets. Only tweets could be extracted in the Spanish language *'es'*, labeled as #ironia and #sarcasmo. This greatly limited the amount of tweets obtained, considering that many users do not have the option that specifies their language enabled. On the other hand, although many authors use the *hashtag* as a reliable method of tweet collection. In the case

Table 5. Results of the Metrics

Proposed approach	Accuracy	Precision	Recall	F1 score
Unigrams	0.67	0.66	0.67	0.66
Bigrams	0.36	0.47	0.36	0.28
Lemma/PoS	0.67	0.66	0.67	0.66
Lemma/PoS + reduction of features	**0.68**	**0.67**	**0.68**	**0.67**

of ironic tweets, the analyzed experiment indicates otherwise. Doing a simple data review, we realize that there are phrases that really do not correspond to the irony just because they are tagged on Twitter as #irony or #sarcasmo. This shows that the use of the *hashtag* as a classificatory method for Spanish tweets is not entirely reliable. We consider that the results in the accuracy will be considerably improved if one of the proposals of the revised literature is followed [9,12], which consists in determining first whether the tweet is objective or subjective before proceeding to the classification. As the classification of irony is not a simple job even for the human eye, it would be convenient to carry out manual labeling, even if it requires more time and human reviewers.

5   Conclusion and Future Work

As we have already mentioned, detecting irony is sometimes difficult even for humans. There are different classic techniques that are frequently used in opinion mining in social networks for the English language. In our research, we have tried to evaluate its functioning in order to classify the polarity of sentiment and detect irony. Although the system has worked acceptably, we can conclude that the results are not good enough. Most of the classification evaluation metrics are below 0.7. However, considering the factors that have been able to influence such results, such as the amount of data, the quality of the labeling and the improvement of the support of tools for the Spanish language, we do not rule out the possibility that by applying other approaches and improving these factors, it is possible to predict, with a better percentage of reliability, the ironic texts in Spanish language. As future lines of work, we'll propose to obtain a more reliable corpus, through manual classification or semi-supervised labeling. It would be necessary to work on the improvement of NLP tools for the Spanish language that are used in the sentiment analysis, including specific data preprocessing tools. We also intend to put into practice the proposal that determines first if the tweet is objective or subjective, designing a binary classification model that detects if the opinion is ironic or not, before predict the polarity. For this purpose other types of approaches can be used as a combination of supervised learning with lexicons, or based on rules.

Acknowledgments. This work has been supported by project "IOTEC: Development of Technological Capacities around the Industrial Application of Internet of Things (IoT)". 0123_IOTEC_3_E. FEDER Funds. Interreg Spain-Portugal.

References

1. Appel, O., Chiclana, F., Carter, J., Fujita, H.: A hybrid approach to the sentiment analysis problem at the sentence level. Knowl.-Based Syst. **108**, 110–124 (2016). New Avenues in Knowledge Bases for Natural Language Processing
2. Bouazizi, M., Otsuki, T.: A pattern-based approach for sarcasm detection on Twitter. IEEE Access **4**, 5477–5488 (2016). Cited by 7

3. Charalampakis, B., Spathis, D., Kouslis, E., Kermanidis, K.: A comparison between semi-supervised and supervised text mining techniques on detecting irony in Greek political tweets. Eng. Appl. Artif. Intell. **51**, 50–57 (2016). Mining the Humanities: Technologies and Applications

4. Hemmatian, F., Sohrabi, M.K.: A survey on classification techniques for opinion mining and sentiment analysis. Artif. Intell. Rev. (2017)

5. Joshi, A., Bhattacharyya, P., Carman, M.J.: Automatic sarcasm detection: a survey. CoRR, abs/1602.03426 (2016)

6. Joshi, P.: Artificial Intelligence with Python. Packt Publishing, Birmingham (2017)

7. Kotsiantis, S.B.: Supervised machine learning: a review of classification techniques. Informatika **160**, 3–24 (2007)

8. Mantyla, M.V., Graziotin, D., Kuutila, M.: The evolution of sentiment analysis a review of research topics, venues, and top cited papers. Comput. Sci. Rev. **27**, 16–32 (2018)

9. Mukherjee, S., Bala, P.K.: Detecting sarcasm in customer tweets: an NLP based approach. Ind. Manag. Data Syst. **117**(6), 1109–1126 (2017). Cited by 0

10. Madhavi, D., Piryani, R., Singh, V.: Analytical mapping of opinion mining and sentiment analysis research during 2000 to 2015. Inf. Process. Manag. **53**(1), 122–150 (2017)

11. Sun, S., Luo, C., Chen, J.: A review of natural language processing techniques for opinion mining systems. Inf. Fusion **36**, 10–25 (2017)

12. Wicana, S.G., Ibisoglu, T.Y., Yavanoglu, U.: A review on Sarcasm detection from machine-learning perspective. In: 11th IEEE International Conference on Semantic Computing (ICSC), San Diego, CA, 30 January–01 February 2017, pp. 469–476. IEEE Computer Society (2017)

An Efficient Soft Computing Approach for Solving the Two-Stage Transportation Problem with Fixed Costs

Ovidiu Cosma, Petrica Pop$^{(\boxtimes)}$, and Ioana Zelina

Department of Mathematics and Computer Science,
Technical University of Cluj-Napoca,
North University Center at Baia Mare, Baia Mare, Romania
{ovidiu.cosma,petrica.pop}@cunbm.utcluj.ro, ioanazelina@yahoo.com

Abstract. Transportation problems are nowadays strategic issues which aim at selecting the routes to be opened between different facilities in order to achieve an efficient distribution strategy. This paper presents a soft computing approach for solving the two-stage transportation problem with fixed costs associated to the routes. Our developed a heuristic algorithm embeds an optimization problem within the framework of a genetic algorithm. Computational experiments were performed on two sets of benchmark instances available in the literature and the obtained results prove that our proposed solution approach is highly competitive in comparison with the existing approaches from the literature.

1 Introduction

Transportation problems have been investigated for some time. The two-stage transportation problem is an extension of the basic transportation problem and was considered for the first time by Geoffrion and Graves [5]. Since then the research on the two-stage transportation problem has received considerably attention and several variants of the problem have been studied and several methods, based on exact and heuristic algorithms, have been developed for solving them. In one of these variants, there exists only one manufacturer who has to fulfill the requests of the customers. For more information on this version of the problem and proposed solution approaches for solving it, we refer to [3,9–12]. Another version of the problem takes into consideration the opening costs of the distribution centers in addition to unit transportation costs. For more information on this variant of the problem, we refer to [1,14].

This work deals with a transportation problem with fixed costs associated to the routes in a two-stage supply chain network. In this case, our goal is to determine and select the manufacturers and the distribution centers fulfilling the demands of the customers under minimal distribution costs. The main feature of the considered transportation problem is that a fixed cost is associated with each route that may be opened in addition to the variable transportation cost which is proportional to the amount of goods shipped.

© Springer Nature Switzerland AG 2020
F. Martínez Álvarez et al. (Eds.): SOCO 2019, AISC 950, pp. 523–532, 2020.
https://doi.org/10.1007/978-3-030-20055-8_50

In the form considered in our paper, the problem was introduced by Jawahar and Balaji [8]. They presented a genetic algorithm (GA) with a specific coding scheme suitable for two-stage transportation problems. The same authors proposed a set of 20 benchmark instances and their computational results have been compared to lower bounds and approximate solutions obtained from a relaxation of the problem. Raj and Rajendran [14] called this variant Scenario-1 and developed a two-stage genetic algorithm (TSGA) in order to solve the problem. They also considered a solution representation that allows a single-stage genetic algorithm (SSGA) [15] to solve it. The major characteristic of these methods is a compact representation of a chromosome based on a permutation. Pop et al. [13] described a hybrid algorithm that combines a steady-state genetic algorithm with a local search procedure and recently, Calvet et al. [2] developed a heuristic algorithm which embeds an optimization problem in the framework of an evolutionary algorithm and Cosma et al. [4] described an efficient multi-start Iterated Local Search (ILS) procedure, which constructs an initial solution, makes use of a local search procedure to increase the exploration, a perturbation mechanism and a neighborhood operator in order to diversify the search.

Our novel solution approach is different from the existing ones from the literature: the use of a new efficient representation in which the chromosomes are generated in two stages, the use of an elitist selection in which one of the parents is selected randomly and the other from the fittest 20% individuals and the use of a mutation operator that accepts 5% of the best resulted offspring.

The rest of our paper is organized as follows: in Sect. 2, the two-stage transportation problem with fixed costs associated the routes is defined and a mathematical model of the problem is presented. The developed soft computing approach is presented in Sect. 3 and the computational experiments and the achieved results are presented, analyzed and discussed in Sect. 4. Finally, in the last section, the obtained results in this paper are pointed out and some future research directions are presented.

2   Definition of the Problem

In order to define the two-stage fixed-cost transportation problem with fixed costs associated to the routes, denoted by TS-TP-FC, some notations of the related sets and parameters used in our considered problem are presented:

p is the number of manufacturers and i is the manufacturer identifier;
q is the number of distribution centers (DCs) and j is the DC identifier;
r is the number of customers and k is the customer identifier;
S_i is the capacity of manufacturer i;
D_k is the demand of customer k;
f_{ij} is the fixed transportation cost for the link from manufacturer i to DC j
g_{jk} is the fixed transportation cost for the link from DC j to customer k;
b_{ij} is unit cost of transportation from manufacturer i to DC j;
c_{jk} is unit cost of transportation from DC j to customer k.

Given a set of p manufacturers, a set of q DCs and a set of r customers with the following properties:

1. Each manufacturer may ship to any of the q DCs at a transportation cost b_{ij} per unit from manufacturer i, where $i \in \{1, ..., p\}$, to DC j, where $j \in \{1, ..., q\}$, plus a fixed cost f_{ij} for operating corresponding the route.
2. Each DC may ship to any of the r customers at a transportation cost c_{jk} per unit from DC j, where $j \in \{1, ..., q\}$, to customer k, where $k \in \{1, ..., r\}$, plus a fixed cost g_{jk} for operating the corresponding route.
3. Each manufacturer $i \in \{1, ..., p\}$ has S_i units of supply and each customer $k \in \{1, ..., r\}$ has a given D_k.

then the scope of the two-stage transportation problem with fixed costs associated on the routes is to find the routes to be opened and corresponding shipment quantities on these routes, such that the customer demands and all the shipment constraints are satisfied, and the total distribution costs are minimized.

By introducing the linear variables: x_{ij} representing the number of units transported from manufacturer i to DC j, y_{jk} representing the number of units transported from DC j to customer k, and the binary variables: z_{ij} is 1 if the route from manufacturer i to DC j is used and 0 otherwise, w_{jk} is 1 if the route from DC j to customer k is used and 0 otherwise, then the two-stage transportation problem with fixed costs associated to the routes can be modeled as the following mixed integer problem described by Raj and Rajendran [14]:

$$\min \sum_{i=1}^{p}\sum_{j=1}^{q}(b_{ij}x_{ij} + f_{ij}z_{ij}) + \sum_{j=1}^{q}\sum_{k=1}^{r}(c_{jk}y_{jk} + g_{jk}w_{jk})$$

$$s.t. \sum_{j=1}^{q}x_{ij} \leq S_i, \ \forall\, i \in \{1, ..., p\} \tag{1}$$

$$\sum_{j=1}^{q}y_{jk} = D_k, \ \forall\, k \in \{1, ..., r\} \tag{2}$$

$$\sum_{i=1}^{p}x_{ij} = \sum_{k=1}^{r}y_{jk}, \ \forall\, j \in \{1, ..., q\} \tag{3}$$

$$x_{ij} \geq 0, \ \forall\, i \in \{1, ..., p\}, \ \forall\, j \in \{1, ..., q\} \tag{4}$$

$$y_{jk} \geq 0, \ \forall\, j \in \{1, ..., q\}, \forall\, k \in \{1, ..., r\} \tag{5}$$

$$z_{ij} \in \{0, 1\}, \ \forall\, i \in \{1, ..., p\}, \ \forall\, j \in \{1, ..., q\} \tag{6}$$

$$w_{jk} \in \{0, 1\}, \ \forall\, j \in \{1, ..., q\}, \forall\, k \in \{1, ..., r\} \tag{7}$$

The objective function minimizes the total distribution cost: the fixed costs and transportation per-unit costs. Constraints (1) guarantee that the quantity shipped out from each manufacturer does not exceed the available capacity, constraints (2) guarantee that the total shipment received from DCs by each customer is equal to its demand and constraints (3) are the flow conservation conditions and they guarantee that the units received by a DC from manufacturers are equal to the units shipped from the distribution centers to the customers. The last four constraints ensure the integrality and non-negativity of the decision variables.

An illustration of the investigated two-stage transportation problem with fixed costs is presented in the next Fig. 1.

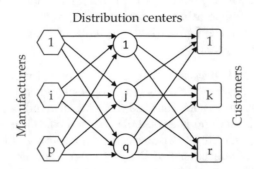

Fig. 1. Illustration of the two-stage transportation problem with fixed costs

The presence of the fixed costs associated to the routes makes the problem more difficult. Hirsch and Dantzig [6] showed that even the fixed-charge transportation problem (FCTP) is an NP-hard problem and therefore our considered two-stage transportation problem with fixed costs associated to the routes, which is an extension of the FCTP, is as well NP-hard.

3   Description of the Proposed Soft Computing Approach

Our developed soft computing approach embeds an optimization problem within the framework of a genetic algorithm. Genetic algorithms (GAs) were introduced by Holland [7] and are search heuristic methods inspired from the theory of natural evolution developed by Charles Darwin based on the "survival of the fittest". GAs have the ability to deliver a "good-enough" solution "fast-enough", making them very attractive in solving optimization problems.

In our GA, each chromosome consists of two parts: the first one is a $p \times q$ matrix associated to the links from manufacturers to DCs and contains the estimated number of units transported, denoted by Ex_{ij} and the second one is a $q \times r$ matrix associated to the links from DCs to customers and contains the estimated number of units transported, denoted by Ey_{jk}.

Taking into account the estimated number of units transported (Ex_{ij}, Ey_{jk}), the following optimization problem is defined:

$$(P) \min \sum_{i=1}^{p}\sum_{j=1}^{q} \tilde{b}_{ij}x_{ij} + \sum_{j=1}^{q}\sum_{k=1}^{r} \tilde{c}_{jk}y_{jk}$$

s.t. (1), (2), (4), (5) and

$$\tilde{b}_{ij} = \begin{cases} b_{ij} + \dfrac{f_{ij}}{Ex_{ij}}, & \text{if } Ex_{ij} > 0 \\ b_{ij} + f_{ij}, & \text{if } Ex_{ij} = 0 \end{cases} \tag{8}$$

$$\tilde{c}_{jk} = \begin{cases} c_{jk} + \dfrac{g_{jk}}{Ey_{jk}}, & \text{if } Ey_{jk} > 0 \\ c_{jk} + g_{jk}, & \text{if } Ey_{jk} = 0 \end{cases} \tag{9}$$

This is a minimum cost flow problem. There are a number of well-known algorithms, which determine the optimal solution of this problem effectively. The experiments presented in this paper were performed using the Network Simplex algorithm. Thus, the amounts x and y that correspond to the optimal solution of this problem are determined.

The chromosome of an individual does not introduce any constraints to the mathematical model of the TS-TP-FC. The only effect is limited by the costs $\tilde{b}_{ij}$ and $\tilde{c}_{jk}$ based on which the optimization problem P is defined. As a consequence, if the constraints (1)–(3) are satisfied, then each chromosome, even a randomly created one will lead to a feasible solution of the TS-TP-FC. However, it is unlikely that an aleatory chromosome will lead to a good solution. In order to improve the quality of the chromosomes (feasible solutions), an algorithm, called Chromosome Optimization is developed.

Algorithm Chromosome Optimization
input: chromosome(Ex, Ey)
output: feasible solution
    1. $EZ \leftarrow \infty$
    2.   Solve the optimization problem P.
    3.   Calculate Z, based on the mathematical model of the problem.
    4.   If $Z > EZ$ or the solution is a duplicate, **then**
        a.   Result $\leftarrow$ saved solution
        b.   **STOP**
    5.   **Else**
        a. $EZ \leftarrow Z$
        b.   Save the solution.
        c.   Adjust EX and EY estimates.
        d.   Continue with step 2.

The estimate of the total cost of the distribution solution (EZ) is initialized in step 1. Step 2 solves optimally in polynomial time the optimization problem (P). Step 3 calculates the cost of the TS-TP-FC solution for the amounts x and y determined at step 2. Steps 2–5 can be repeated several times during the algorithm. However, there is no guarantee that the generated solutions will be better and better. Step 4 takes the decision to continue or to stop the algorithm. The procedure ends if the last solution is weaker than the previous one, or if a duplicate has been reached, which has been generated in previous iterations.

Several different estimates Ex and Ey can lead to the same amounts x, y. Two solutions are considered identical if the amounts x and y are identical, without considering the estimates Ex and Ey. If the algorithm ends, then the last saved solution represents the result. All other solutions generated along the way are destroyed. If step 5 is reached, it means that a better result was achieved at the last iteration than the previous ones. In this case, the solution is saved, the Ex and Ey estimates are adjusted according to: $Ex = x$ and $Ey = y$, after which the solution is re-evaluated.

The initial population in our GA consists of N individuals and it was generated randomly as follows: $\forall\ i \in \{1, ..., p\}$ and $\forall\ j \in \{1, ..., q\}$, Ex_{ij} takes a random value from the interval $[0, S_i]$ and $\forall\ j \in \{1, ..., q\}$ and $\forall\ k \in \{1, ..., r\}$, Ey_{jk} takes a random value from the interval $[0, D_k]$. After generating the chromosomes they are optimized and evaluated according to the previously described algorithm.

Solutions from one population are taken and used to form a new population, motivated by a hope, that the new population will be better than the old one. Solutions which are chosen to form new solutions (offspring) are selected according to the elitist selection strategy based on their fitness values. The better the chromosomes are, the more chances to be selected they have.

Crossover is one of the most significant phases in a GA. In our case, two parents P_1 and P_2 are randomly chosen from the current population, with the restriction that parent P_1 is part of the best 20% individuals of the current population. The genes that make up the chromosome of the offspring are taken either from P_1 or P_2 chromosomes with equal probabilities. This results in an offspring carrying some genetic information from both parents. For each new generation, at least $3N$ crossover operations are performed, resulting at least N new individuals, or the number of crossover operations exceeds $10N$. After being generated, each offspring is optimized and evaluated according to the described Optimization algorithm.

In certain new offspring formed, one of their genes can be subjected to a mutation with a low probability that depends on the number of performed crossover operations. This implies that the value of a gene is altered.

In our GA, the following termination condition was considered: when there has been no improvement in the population for a given number of iterations then the algorithm stops.

4   Computational Results

In order to analyze the performance of our proposed soft computing approach, it was tested on two sets of instances. The first one was generated by Jawahar and Balaji [8] and it consists of 20 test problems of small sizes. The second set of problems contains 10 new randomly generated instances of larger sizes and was generated by Calvete et al. [2].

Our algorithm was coded in Java 8 and 5 independent runs were performed, as there were performed by Calvete et al. [3], for each instance on a PC with

Intel Core i5-4590 3.3 GHz, 4 GB RAM, Windows 10 Education 64 bit operating system.

Genetic parameters are very important for the success of a GA. In our proposed algorithm, based on preliminary computational experiments, the genetic parameters were set as follows: the initial population contains $N = \dfrac{pq + qr}{5}$ individuals and it was generated randomly, the size of the population makes use of the same number of chromosomes, the intermediary size of the population is maximum $2N$, the mutation probability, denoted by p_m depends on the number of performed crossover operations, denoted by CN and is defined as follows:
$p_m = 0$ if $CN < 3N$ and $p_m = \dfrac{1}{11 - \frac{CN}{N}}$ if $3N \leq CN < 10N$.

Table 1 presents the computational results of achieved by our soft computing approach in comparison with the genetic algorithm described by Jawahar and Balaji [8], called JRGA, the two genetic algorithms introduced by Raj and Rajendran [14], denoted by TSGA and SSGA, the hybrid genetic algorithm (HGA) described by Pop et al. [13] and the matheuristic approach proposed by Calvete et al. [2].

The first column in Table 1 gives the size of the instance and the next columns display the solution achieved by the genetic algorithm described by Jawahar and Balaji [8], the two genetic algorithms introduced by Raj and Rajendran [14], the hybrid genetic algorithm described by Pop et al. [13], the matheuristic approach proposed by Calvete et al. [2] and our proposed soft computing approach. The results written in bold represent cases for which the obtained solution is the best existing from the literature.

Analyzing the computational results reported in Table 1, one can observe that our approach has a better computational performance compared to JRGA [8] and TSGA [14]. Compared to the SSGA [14], the hybrid genetic algorithm described by Pop et al. [13], the matheuristic approach proposed by Calvete et al. [2], our algorithm delivered the same optimal solution in all the 20 considered instances.

Regarding the computational times, Jawahar and Balaji [8] provided no information, Raj and Rajendran [14] mentioned only that their algorithms are executed for 10000 generations or 3600 s, Pop et al. [13] mentioned that the average running time is less then 3 s, Calvete et al. [2] reported a time between 0.01 and 0.03 s and in our case each achieved solution was obtained within 1 ms.

Table 2 presents the computational results of our soft computing approach in comparison to the matheuristic approach proposed by Calvete et al. [2] on a set of 10 benchmark instances proposed by the same authors. The first column in Table 2 gives the number of the instance and the second one provides its size. The next two columns contain the optimal solution obtained by CPLEX and the corresponding time. The last columns provide the results reported by Calvete et al. [2] and our achieved results and contain the minimum and the maximum objective function values obtained in the five runs of each instance and the average time of the five runs in seconds spent by the mentioned algorithms.

Analyzing the computational results reported in Table 2, one can observe that our approach provided for all the instances the optimal solution. For the

Table 1. Computational results achieved by our proposed soft computing approach compared to existing methods

Instance	CPLEX	Jawahar [8]	Raj [14]		Pop [13]	Calvete [2]	Our approach
	Z_{opt}	Z_{best}	Z_{best}	Z_{best}	Z_{best}	Z_{best}	Z_{best}
$2 \times 2 \times 3$	**112,600**	**112,600**	**112,600**	**112,600**	**112,600**	**112,600**	**112,600**
$2 \times 2 \times 4$	**237,750**	**237,750**	**237,750**	**237,750**	**237,750**	**237,750**	**237,750**
$2 \times 2 \times 5$	**180,450**	**180,450**	**180,450**	**180,450**	**180,450**	**180,450**	**180,450**
$2 \times 2 \times 6$	**165,650**	**165,650**	**165,650**	**165,650**	**165,650**	**165,650**	**165,650**
$2 \times 2 \times 7$	**162,490**	**162,490**	**162,490**	**162,490**	**162,490**	**162,490**	**162,490**
$2 \times 3 \times 3$	**59,500**	**59,500**	**59,500**	**59,500**	**59,500**	**59,500**	**59,500**
$2 \times 3 \times 4$	**32,150**	**32,150**	**32,150**	**32,150**	**32,150**	**32,150**	**32,150**
$2 \times 3 \times 6$	**65,945**	69,970	67,380	**65,945**	**65,945**	**65,945**	**65,945**
$2 \times 3 \times 8$	**258,730**	263,000	**258,730**	**258,730**	**258,730**	**258,730**	**258,730**
$2 \times 4 \times 8$	**77,400**	80,400	84,600	**77,400**	**77,400**	**77,400**	**77,400**
$2 \times 5 \times 6$	**75,065**	94,565	80,865	**75,065**	**75,065**	**75,065**	**75,065**
$3 \times 2 \times 4$	**47,140**	**47,140**	**47,140**	**47,140**	**47,140**	**47,140**	**47,140**
$3 \times 2 \times 5$	**175,350**	178,950	178,950	**175,350**	**175,350**	**175,350**	**175,350**
$3 \times 3 \times 4$	**57,100**	**57,100**	61,000	**57,100**	**57,100**	**57,100**	**57,100**
$3 \times 3 \times 5$	**152,800**	**152,800**	156,900	**152,800**	**152,800**	**152,800**	**152,800**
$3 \times 3 \times 6$	**132,890**	**132,890**	**132,890**	**132,890**	**132,890**	**132,890**	**132,890**
$3 \times 3 \times 7$ (a)	**99,095**	104,115	106,745	**99,095**	**99,095**	**99,095**	**99,095**
$3 \times 3 \times 7$ (b)	**281,100**	287,360	295,060	**281,100**	**281,100**	**281,100**	**281,100**
$3 \times 4 \times 6$	**76,900**	77,250	81,700	**76,900**	**76,900**	**76,900**	**76,900**
$4 \times 3 \times 5$	**118,450**	**118,450**	**118,450**	**118,450**	**118,450**	**118,450**	**118,450**

Table 2. Computational results achived by our proposed soft computing approach compared to existing methods

Instance	Size of the instance	CPLEX		Calvete et al. [2]			Our approach		
		Z_{opt}	Time	Z_{min}	Z_{max}	Time	Z_{min}	Z_{max}	Time
1	$2 \times 4 \times 6$	71,484	0.2	71,484	71,484	0.0	71,484	71,484	0.016
2	$2 \times 4 \times 8$	102,674	0.4	102,674	102,674	0.0	102,674	102,674	0.012
3	$4 \times 8 \times 12$	124,253	0.3	124,253	124,253	0.2	124,253	124,253	0.039
4	$4 \times 8 \times 16$	136,779	0.3	136,779	136,779	0.2	136,779	136,779	0.076
5	$6 \times 12 \times 14$	150,932	0.3	150,932	150,932	0.2	150,932	150,932	0.247
6	$6 \times 12 \times 18$	200,998	0.9	200,998	200,998	0.2	200,998	200,998	0.244
7	$8 \times 16 \times 24$	147,741	0.5	147,741	147,741	0.4	147,741	147,741	0.557
8	$8 \times 16 \times 32$	196,187	2.4	196,187	196,187	1.8	196,187	196,187	0.685
9	$10 \times 20 \times 30$	162,660	1.2	162,660	162,660	2.2	162,660	162,660	1.209
10	$10 \times 20 \times 40$	216,758	23.1	216,758	216,758	2.6	216,758	216,758	0.976

first instance in Table 2, the optimum solution occurred, in each run, at the initial population generation stage. This situation occurs also at one of the runs in the case of instances 2 and 3. For the other instances in Table 2, the optimum solution occurred at the earliest in generation 1 and at the latest in generation 7. Regarding the efficiency of the algorithms compared in Table 2, as expected,

CPLEX is the slowest, with one exception (instance 9). For instances 3, 4, 8 and 10, our algorithm was the fastest. An efficiency comparison between the algorithms is presented in Fig. 2.

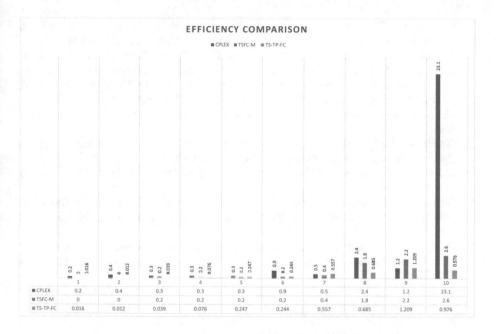

Fig. 2. Efficiency comparison between CPLEX, TSFC-M and TS-TP-FC

5   Conclusions

In this paper an efficient approach was developed in order to the two-stage transportation problem with fixed costs associated to the routes. Our developed a heuristic algorithm embeds an optimization problem within the framework of a genetic algorithm.

The results obtained through the use of our proposed approach are very promising, thus providing a reason to apply this kind of approach to other two-stage transportation problems, with the aim of assessing the real practicality of the described method.

Future research will focus on defining, detailing and adapting some other genetic operators (crossover, mutation and selection) for our GA. In addition, our developed approach is going to be tested in the case of larger size instances of the TS-TP-FC.

References

1. Calvete, H., Gale, C., Iranzo, J.: An improved evolutionary algorithm for the two-stage transportation problem with fixed charge at depots. OR Spectr. **38**, 189–206 (2016)
2. Calvete, H., Gale, C., Iranzo, J., Toth, P.: A matheuristic for the two-stage fixed-charge transportation problem. Comput. Oper. Res. **95**, 113–122 (2018)
3. Cosma, O., Pop, P.C., Matei, O., Zelina, I.: A hybrid iterated local search for solving a particular two-stage fixed-charge transportation problem. In: Proceedings of HAIS 2018, Lecture Notes in Computer Science, vol. 10870, pp. 684-693 (2018)
4. Cosma, O., Pop, P.C., Pop Sitar, C.: An efficient iterated local search heuristic algorithm for the two-stage fixed-charge transportation problem. Carpath. J. Math. (2019, in press)
5. Geoffrion, A.M., Graves, G.W.: Multicommodity distribution system design by Benders decomposition. Manag. Sci. **20**, 822–844 (1974)
6. Hirsch, W.M., Dantzig, G.B.: The fixed charge problem. Naval Res. Log. Q. **15**, 413–424 (1968)
7. Holland, J.H.: Adaptation in Natural and Artificial Systems: An Introductory Analysis with Applications to Biology, Control and Artificial Intelligence. MIT Press, Cambridge (1992)
8. Jawahar, N., Balaji, A.N.: A genetic algorithm for the two-stage supply chain distribution problem associated with a fixed charge. Eur. J. Oper. Res. **194**, 496–537 (2009)
9. Pintea, C.-M., Pop Sitar, C., Hajdu-Macelaru, M., Pop, P.C.: A hybrid classical approach to a fixed-charge transportation problem. In: Corchado, E., et al. (eds.) Proceedings of HAIS 2012, Part I. Lecture Notes in Computer Science, vol. 7208, pp. 557–566 . Springer, Cham (2012)
10. Pintea, C.M., Pop, P.C.: An improved hybrid algorithm for capacitated fixed-charge transportation problem. Log. J. IJPL **23**(3), 369–378 (2015)
11. Pop, P.C., Pintea, C.-M., Pop Sitar, C., Hajdu-Macelaru, M.: An efficient reverse distribution system for solving sustainable supply chain network design problem. J. Appl. Log. **13**(2), 105–113 (2015)
12. Pop, P.C., Matei, O., Pop Sitar, C., Zelina, I.: A hybrid based genetic algorithm for solving a capacitated fixed-charge transportation problem. Carpath. J. Math. **32**(2), 225–232 (2016)
13. Pop, P.C., Sabo, C., Biesinger, B., Hu, B., Raidl, G.: Solving the two-stage fixed-charge transportation problem with a hybrid genetic algorithm. Carpath. J. Math. **33**(3), 365–371 (2017)
14. Raj, K.A.A.D., Rajendran, C.: A genetic algorithm for solving the fixed-charge transportation model: two-stage problem. Comput. Oper. Res. **39**(9), 2016–2032 (2012)
15. Raj, K.A.A.D., Rajendran, C.: A hybrid genetic algorithm for solving single-stage fixed-charge transportation problems. Technol. Oper. Manag. **2**(1), 1–15 (2011)

Takagi-Sugeno Fuzzy Incremental State Model for Optimal Control of a Ball and Beam Nonlinear Model

Basil Mohammed Al-Hadithi[1,2], José Miguel Adánez[1(✉)], and Agustín Jiménez[1]

[1] Intelligent Control Group, Centre for Automation and Robotics UPM - CSIC, Universidad Politécnica de Madrid, C/ J. Gutiérrez Abascal, 2, 28006 Madrid, Spain
jm.adanez@alumnos.upm.es
[2] Department of Electrical, Electronics, Control Engineering and Applied Physics, Higher Technical School of Industrial Design and Engineering, Universidad Politécnica de Madrid, C/ Ronda de Valencia, 3, 28012 Madrid, Spain

Abstract. The optimal control of a ball and beam by an approach based on Takagi-Sugeno incremental state model is proposed. The advantages of incremental state model in comparison with the non incremental one are that the control action cancels steady state errors, the affine terms disappear and incremental state solves the problem of computing the target state, choosing zero as an objective. A generalized version of Takagi-Sugeno identification method is applied. For an optimal control, Linear Quadratic Regulator and optimal state observer are used in each fuzzy rule. Simulation results over the ball and beam nonlinear model show a stable closed loop in the full range, zero steady state error and good transient response.

Keywords: Ball and beam · Takagi-Sugeno ·
Incremental state model · Optimal control

1 Introduction

The ball and beam system [1] is a classical mechanical system with two degrees of freedom. The beam rotates, driven by a torque at the center of rotation. The ball rolls freely in contact with the beam. In spite of its mechanical simplicity, the ball and beam system is nonlinear and unstable which presents significant challenges from the control point of view.

The ball and beam is a common didactic plant in many control laboratories [2], as it is nonlinear, unstable, which means that it is difficult to control, and can be a benchmark for testing several advanced control techniques [3].

Fuzzy logic has become one of the most popular tools for modelling and control of nonlinear systems, especially the Takagi-Sugeno (T-S) fuzzy model [4], which allows the global identification of a nonlinear system as a set of linearized systems connected by a fuzzy bending.

© Springer Nature Switzerland AG 2020
F. Martínez Álvarez et al. (Eds.): SOCO 2019, AISC 950, pp. 533–543, 2020.
https://doi.org/10.1007/978-3-030-20055-8_51

In [5], an approach developed by the authors is presented to improve the estimation of T-S models. The problem is that the original T-S identification method cannot be applied when the triangular membership functions are overlapped by pairs. This restricts the use of this type of membership functions which have been widely used in the controllers design and are popular in industrial applications. The approach, to search for an exact optimal solution, uses the minimum norm method although it increases complexity and computational cost. Another approach was developed in [6], which can be considered as a generalized version of T-S method. This simple method with not much computational cost is based on weighting of parameters. In [7], this T-S identification method was extended to the multivariable case. These methods are characterized by the high accuracy obtained for modelling nonlinear systems in comparison with the original T-S method [4].

Incremental state model is presented in [8] to model multivariable nonlinear delayed systems expressed by a generalized version of T-S fuzzy model. The advantages of incremental state model compared with non incremental one has been defined. First, it solves the problem of computing the target state, choosing zero incremental state as an objective. Second, the control action in an incremental form is equivalent to introduce an integral action. Third, incremental state model makes the affine terms disappear. A comparative analysis has been developed in [9] between incremental state model and traditional one, the control based on these state models have been developed by a linearized model and applied over a nonlinear model of a wind turbine.

The rest of the work is organized as follows. Section 2 describes ball and beam nonlinear model. The fuzzy T-S model and the fuzzy identification method are described in Sect. 3. In Sect. 4, incremental state model is presented. Optimal state controller and optimal state observer designs are described in Sect. 5. In Sect. 6, the proposed fuzzy optimal controller based on incremental state model is applied to the ball and beam nonlinear model.

2   Ball and Beam Nonlinear Model

In this work, the AMIRA BW500 (Fig. 1) [1] is used as ball and beam model. The ball position p, the objective system output, would be measured by a camera, therefore the discrete sample time is supposed to be large. The beam angle α would be measured by an incremental encoder, thus it is considered as a measurable internal variable. The system input F is a force produced by a DC motor, which causes the beam to rotate around its center.

The nonlinear differential equations of the ball and beam model [1] are:

$$\left(m + \frac{I_b}{r^2}\right)\ddot{p} + \left(mr^2 + I_b\right)\frac{1}{r}\ddot{\alpha} - mp\dot{\alpha}^2 = mg\sin(\alpha) \tag{1}$$

$$\left(mp^2 + I_b + I_W\right)\ddot{\alpha} + \left(2mp\dot{p} + bl^2\right)\dot{\alpha} + Kl^2\alpha + \left(mr^2 + I_b\right)\frac{1}{r}\ddot{p} - mgp\cos(\alpha) = Fl\cos(\alpha) \tag{2}$$

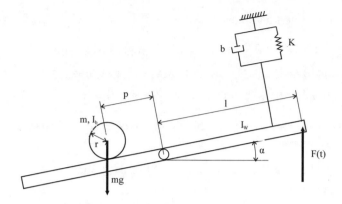

Fig. 1. Ball and beam system.

where p is the ball position, α is the beam angle and F is the drive mechanics force. Table 1 summarizes the model parameters and its values.

Table 1. Ball and beam parameters

Parameter	Meaning	Value
m	Mass of the ball	$0.025\,\mathrm{Kg}$
g	Gravity	$9.81\ \mathrm{m/s^2}$
r	Roll radius of the ball	$0.0167\,\mathrm{m}$
I_b	Inertia moment of the ball	$3.516 \cdot 10^{-6}\,\mathrm{Kgm^2}$
I_W	Inertia moment of the beam	$0.09\ \mathrm{Kgm^2}$
b	Friction coefficient of the drive mechanics	$1.0\,\mathrm{Ns/m}$
K	Stiffness of the drive mechanics	$0.001\,\mathrm{N/m}$
l	Radius of force application	$0.49\,\mathrm{m}$

Since we consider α as a measurable internal variable, the system is modeled as single-input-single-output (SISO), and the relation between output p and input F is described by two second order differential equations, so the global system is fourth order.

3 Fuzzy Takagi-Sugeno Model and System Identification

3.1 Fuzzy T-S Model

Nonlinear systems can be modelled by T-S model, supposing known a set of measurable nonlinear variables $[z_1(k), z_2(k), \ldots, z_m(k)]$ of the system. By choosing $[r_1, r_2, \ldots, r_m]$ number of fuzzy sets for these variables, a monovariable fuzzy system can be defined as follows:

$$S^{(i_1\ldots i_m)}: \text{If } z_1(k) \text{ is } M_1^{i_1} \text{ and } \ldots \text{ and } z_m(k) \text{ is } M_m^{i_m} \text{ then:}$$

$$y(k) = a_0^{(i_1\ldots i_m)} + a_1^{(i_1\ldots i_m)} y(k-1) + \cdots + a_n^{(i_1\ldots i_m)} y(k-n) \quad (3)$$
$$+ b_1^{(i_1\ldots i_m)} u(k-1) + \cdots + b_n^{(i_1\ldots i_m)} u(k-n)$$

In each rule, we can transform the difference equation (3) to state model with affine term as follows:

$$S^{(i_1\ldots i_m)}: \text{If } z_1(k) \text{ is } M_1^{i_1} \text{ and } \ldots \text{ and } z_m(k) \text{ is } M_m^{i_m} \text{ then:}$$
$$x(k) \in \mathfrak{R}^n$$

$$x(k+1) = \begin{bmatrix} a_0^{(i_1\ldots i_m)} \cdot a_1^{(i_1\ldots i_m)} \\ a_0^{(i_1\ldots i_m)} \cdot a_2^{(i_1\ldots i_m)} \\ \vdots \\ a_0^{(i_1\ldots i_m)} \cdot a_n^{(i_1\ldots i_m)} \end{bmatrix} + \begin{bmatrix} a_1^{(i_1\ldots i_m)} & 1 \cdots & 0 \\ a_2^{(i_1\ldots i_m)} & 0 \ddots & 0 \\ \vdots & \vdots \ddots & 1 \\ a_n^{(i_1\ldots i_m)} & 0 \cdots & 0 \end{bmatrix} x(k) + \begin{bmatrix} b_1^{(i_1\ldots i_m)} \\ b_2^{(i_1\ldots i_m)} \\ \vdots \\ b_n^{(i_1\ldots i_m)} \end{bmatrix} u(k)$$

$$y(k) = a_0^{(i_1\ldots i_m)} + \begin{bmatrix} 1 & 0 & \cdots & 0 \end{bmatrix} x(k)$$

$$(4)$$

In matrix form:

$$S^{(i_1\ldots i_m)}: \text{If } z_1(k) \text{ is } M_1^{i_1} \text{ and } \ldots \text{ and } z_m(k) \text{ is } M_m^{i_m} \text{ then:}$$
$$x(k+1) = a_x^{(i_1\ldots i_m)} + A^{(i_1\ldots i_m)} x(k) + B^{(i_1\ldots i_m)} u(k) \quad (5)$$
$$y(k) = a_y^{(i_1\ldots i_m)} + C x(k)$$

3.2   Estimation of T-S Model Parameters

The identification method of T-S fuzzy models [4] is based on the estimation of the fuzzy system parameters minimizing a quadratic performance index. The traditional T-S identification method [4] fails if the triangular membership functions are overlapped by pairs, since the T-S matrix is not of full rank and then it is not invertible [5]. Thus, in [6] a generalized T-S identification was proposed, using a parameters weighting method.

The fuzzy estimation of the output becomes:

$$\hat{y} = \sum_{i_1=1}^{r_1} \cdots \sum_{i_m=1}^{r_m} \beta^{(i_1\ldots i_m)} \left(z_{(i_1\ldots i_m)}(k) \right) \left[a_0^{(i_1\ldots i_m)} + a_1^{(i_1\ldots i_m)} y(k-1) + \right.$$
$$\left. \cdots + a_n^{(i_1\ldots i_m)} y(k-n) + b_1^{(i_1\ldots i_m)} u(k-1) + \cdots + b_n^{(i_1\ldots i_m)} u(k-n) \right] \quad (6)$$

where

$$\beta^{(i_1\ldots i_m)} \left(z_{(i_1\ldots i_m)}(k) \right) = \frac{\mu_{1 i_1}(z_1) \ldots \mu_{m i_m}(z_m)}{\sum_{i_1=1}^{r_1} \cdots \sum_{i_m=1}^{r_m} \left(\mu_{1 i_1}(z_1) \ldots \mu_{m i_m}(z_m) \right)} \quad (7)$$

with $\mu_{ji_j}(z_j)$ being the membership function corresponding to the fuzzy set M_j^{ij}.

It is supposed to have a set of input/output system samples and a first affine linear parameters estimation $p^0 = \begin{bmatrix} a_0^0 & a_1^0 & \dots & a_n^0 & b_1^0 & \dots & b_n^0 \end{bmatrix}$, which could be obtained by a classical input/output identification of the data, for example with least squares method. This first approximation can be utilized as reference parameters for all the subsystems. Then, the fuzzy model parameters can be obtained minimizing:

$$
J = \sum_{k=1}^{s} (y(k) - \hat{y}(k))^2 + \gamma^2 \sum_{i_1=1}^{r_1} \cdots \sum_{i_m=1}^{r_m} \sum_{j=0}^{n} \left(p_j^0 - p_j^{(i_1 \dots i_m)} \right)^2 =
$$

$$
\|Y - XP\|^2 + \gamma^2 \|P_0 - P\|^2 = \left\| \begin{bmatrix} Y \\ \gamma P_0 \end{bmatrix} - \begin{bmatrix} X \\ \gamma I \end{bmatrix} P \right\|^2 = \|Y_a - X_a P\|^2
$$

(8)

where Y are the output data, X are the input/output fuzzy data, P_0 are the linear estimated parameters repeated as many times as the number of fuzzy rules $(P_0 = [p_0, p_0, \dots, p_0])$, and P are the fuzzy T-S model parameters. The γ factor represents the degree of confidence of the linear estimated parameters, and it must be adjusted. It should be noted that the matrix X_a is of full rank, even if the membership functions are triangular overlapped by pairs, which solves the problem where the traditional T-S identification method fails. Thus, the vector P can be computed as:

$$
P = \left(X_a^t X_a \right)^{-1} X_a^t Y_a
$$

(9)

4   Incremental State Model

The control method based on traditional state model, can produce steady state errors in presence of disturbances or modelling errors. This problem can be solved by incremental state model [8,9].

Applying the discrete state model described in (5), at the previous sample $(k-1)$:

$$
\begin{aligned}
x(k) &= a_x + Ax(k-1) + Bu(k-1) \\
y(k-1) &= a_y + Cx(k-1)
\end{aligned}
$$

(10)

Subtracting (10) from (5):

$$
\begin{aligned}
x(k+1) - x(k) &= A\left(x(k) - x(k-1) \right) + B\left(u(k) - u(k-1) \right) \\
y(k) - y(k-1) &= C\left(x(k) - x(k-1) \right)
\end{aligned}
$$

(11)

where the affine terms are cancelled. Defining the incremental state Δx and the incremental input Δu as follows:

$$
\Delta x(k) = x(k) - x(k-1) \qquad \Delta u(k) = u(k) - u(k-1)
$$

(12)

Substituting (12) into (11), it is obtained:

$$
\begin{aligned}
\Delta x(k+1) &= A\Delta x(k) + B\Delta u(k) \\
y(k) &= y(k-1) + C\Delta x(k)
\end{aligned}
$$

(13)

A new state is introduced to complete the formulation, verifying that:

$$y(k+1) = y(k) + C\Delta x(k+1) = y(k) + C\left(A\Delta x(k) + B\Delta u(k)\right) \qquad (14)$$

A new expanded incremental state vector $x_a \in \mathfrak{R}^{(1+n)}$ is defined, obtaining the new state model as follows:

$$\begin{bmatrix} y(k+1) \\ \Delta x(k+1) \end{bmatrix} = \begin{bmatrix} 1 & CA \\ 0 & A \end{bmatrix} \begin{bmatrix} y(k) \\ \Delta x(k) \end{bmatrix} + \begin{bmatrix} CB \\ B \end{bmatrix} \Delta u(k)$$

$$y(k) = \begin{bmatrix} 1 & 0 \end{bmatrix} \begin{bmatrix} y(k) \\ \Delta x(k) \end{bmatrix} \qquad x_a(k) = \begin{bmatrix} y(k) \\ \Delta x(k) \end{bmatrix} \qquad (15)$$

In matrix notation, the expanded state model becomes:

$$x_a(k+1) = A_a x_a(k) + B_a \Delta u(k)$$
$$y(k) = C_a x_a(k) \qquad (16)$$

Using the method described in Eq. (15) in each fuzzy rule defined by T-S state model (5), an incremental T-S state model can be obtained as follows:

$$S^{(i_1 \cdots i_m)}\colon \text{If } z_1(k) \text{ is } M_1^{i_1} \text{ and } \ldots \text{ and } z_m(k) \text{ is } M_m^{i_m} \text{ then:}$$
$$x_a(k+1) = A_a^{(i_1 \cdots i_m)} x_a(k) + B_a^{(i_1 \cdots i_m)} \Delta u(k) \qquad (17)$$
$$y(k) = C_a x_a(k)$$

5   Fuzzy Controller and Observer Design Based on Incremental State Model

5.1   Fuzzy Controller for Zero Steady-State Error

Incremental state feedback control can be designed by any method. In order to calculate the coefficients of the state feedback controller, discrete LQR method is chosen, which allows optimal control weighting the dynamic response and the control action.

The goal is to minimize the cost index J:

$$J = \sum_{k=0}^{\infty} \left[(x_r - x_a(k))^t\, Q\, (x_r - x_a(k)) + \Delta u(k)^t R \Delta u(k) \right] \qquad (18)$$

LQR method is optimal for linear systems, however, in the case of nonlinear systems, it is complex to propose the minimization of any objective function for the global system. In order to solve this problem, it is suggested to minimize the cost of each fuzzy rule instead of the global cost. The solution will be a suboptimal one but with the great advantage of being easy to calculate. With this method, the global stability is not guaranteed, which needs to be analyzed a posteriori, although gaining in return a balance between static and dynamic behavior of the system with admissible control actions.

If the control objective is to approach a steady state output y_r, this is equivalent to achieve the reference expanded state in incremental model [8,9]:

$$x_{ar} = \begin{bmatrix} y_r \\ \Delta x_r \end{bmatrix} = \begin{bmatrix} y_r \\ 0 \end{bmatrix}$$

Then the control action is described as:

$S^{(i_1 \dots i_m)}$: If $z_1(k)$ is $M_1^{i_1}$ and ... and $z_m(k)$ is $M_m^{i_m}$ then:

$$\Delta u(k) = K\left(x_{ar} - x_a(k)\right) = K^{(i_1 \dots i_m)} \begin{bmatrix} y_r - y(k) \\ -\Delta x(k) \end{bmatrix} \tag{19}$$

$$u(k) = u(k-1) + \Delta u(k) \tag{20}$$

It should be noted that the reference expanded state comes naturally without any calculations. Moreover, control action (20) is calculated in incremental form (19), which is equivalent to apply an integral control action. If the feedback system is stable and the steady state is approached, the controlled system has zero steady state error [8,9].

5.2 Fuzzy Observer for Incremental State Model

In the controller algorithm it is supposed that the output $y(k)$ is measurable, but the incremental state $\Delta x(k)$ is not directly accessible, then a state observer is required. The state observer is a parallel dynamic system with a correction term that approximates the estimated state to the real one. The fuzzy incremental state observer [8,9], is formulated as follows:

$S^{(i_1 \dots i_m)}$: If $z_1(k)$ is $M_1^{i_1}$ and ... and $z_m(k)$ is $M_m^{i_m}$ then:

$$\Delta x_e(k+1) = A^{(i_1 \dots i_m)} \Delta x_e(k) + B^{(i_1 \dots i_m)} \Delta u(k) +$$

$$H^{(i_1 \dots i_m)} \left(y(k) - y(k-1) - C^{(i_1 \dots i_m)} \Delta x_e(k) \right) \tag{21}$$

Thus, the expanded estimated state is obtained as follows:

$$x_{ae}(k) = \begin{bmatrix} y(k) \\ \Delta x_e(k) \end{bmatrix} \tag{22}$$

Matrix $H^{(i_1 \dots i_m)} \in \Re^{(n \times 1)}$ coefficients are obtained by optimal state observer design [8,9] in each fuzzy rule.

The optimal observer solves the problem of calculating a matrix H which minimizes the cost index J_o:

$$J_o(H) = \alpha^T (A - HC)(A - HC)^T \alpha \qquad \forall \alpha \in \Re^n \tag{23}$$

It is verified in [8] that for any value of α, it fulfills:

$$H = AC^T(CC^T)^{-1} \tag{24}$$

In the proposed state representation C matrix is defined in observable canonical form, therefore it holds that $CC^T = I$, so that:

$$H = AC^T \tag{25}$$

The separation principle holds between optimal observer and control design.

6   Results

The ball and beam model is defined in continuous time, but the controller has been developed in discrete time, therefore, a sampler and zero order holding device have been added to the model, with the proposed sampling time $T = 0.05\,\mathrm{s}$. All system variables are supposed to be ideally measured.

For a generalized T-S identification of the system, the following weighting factor $\gamma = 10^{-11}$ and the following membership functions defined in Fig. 2 has been chosen, obtaining an identification error of $1.6169 \cdot 10^{-11}$.

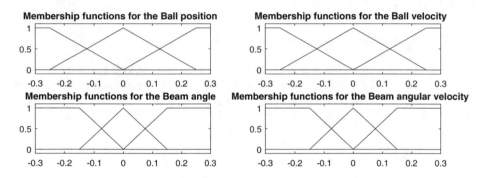

Fig. 2. Membership functions of the fuzzy sets.

With the generalized T-S identification method and Eq. (15), a fuzzy T-S incremental state model has been obtained:

$S^{(1,1,1,1)}$: If $p(k)$ is M_1^1 and $\dot{p}(k)$ is M_2^1 and $\alpha(k)$ is M_3^1 and $\dot{\alpha}(k)$ is M_4^1 then:

$$\begin{bmatrix} y(k+1) \\ \Delta x(k+1) \end{bmatrix} = \begin{bmatrix} 1 & 3.9181 & 1 & 0 & 0 \\ 0 & 3.9181 & 1 & 0 & 0 \\ 0 & -5.7542 & 0 & 1 & 0 \\ 0 & 3.7543 & 0 & 0 & 1 \\ 0 & -0.9181 & 0 & 0 & 0 \end{bmatrix} \begin{bmatrix} y(k) \\ \Delta x(k) \end{bmatrix} + 10^{-3} \begin{bmatrix} -0.0502 \\ -0.0502 \\ 0.1562 \\ 0.1262 \\ -0.0571 \end{bmatrix} \Delta u(k)$$

$$y(k) = \begin{bmatrix} 1 & 0 & 0 & 0 & 0 \end{bmatrix} \begin{bmatrix} y(k) \\ \Delta x(k) \end{bmatrix}$$

Thus, the controller matrix K is designed in each rule by discrete LQR algorithm, using the incremental state matrices A_a and B_a, and the positive definite weighting matrices $Q = I$ and $R = 1$. The observer matrix is defined by the optimal state observer method [8,9], using Eq. (25). Obtaining:

$S^{(1,1,1,1)}$: If $p(k)$ is M_1^1 and $\dot{p}(k)$ is M_2^1 and $\alpha(k)$ is M_3^1 and $\dot{\alpha}(k)$ is M_4^1 then:

$$K^{(1,1,1,1)} = 10^3 \left[0.0008\ 4.0174\ 3.2390\ 2.5627\ 1.9811 \right]$$

$$H^{(1,1,1,1)} = \left[3.9181\ -5.7542\ 3.7543\ -0.9181 \right]^t$$

With this fuzzy controller and observer design method, the global stability cannot be theoretically proved, so it has to be analyzed a posteriori with the simulation results.

Figure 3, shows the ball position and the observation error of the ball position, when the controlled system is subjected to changes in the ball position reference. Moreover, Fig. 4 shows the beam angle and the input force.

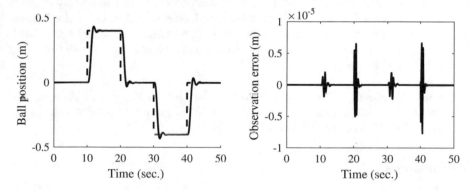

Fig. 3. Ball and beam variables: (left) Ball position. (right) Observation error of ball position.

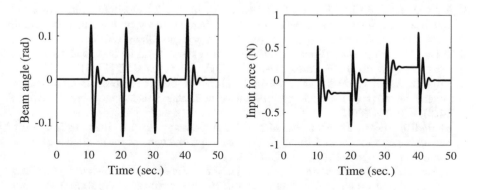

Fig. 4. Ball and beam variables: (left) Beam angle. (right) Input force.

In Fig. 3 (left) it can be seen that the system present stable and good transient response and zero steady state error. In Fig. 3 (right) it is shown that the observation error is of small range and tends to zero. In Fig. 4 it can be seen that, the system variables present smooth and stable transient responses. Thus, the controlled ball and beam model has a stable response in the full range of the system, presents a good transient response and zero steady state error.

7   Conclusion

In this work, we have shown the obtained results of T-S incremental state model and optimal state controller and observer, applied in a ball and beam nonlinear model. The advantages of incremental state model over traditional one comes in a natural way and they are that the affine terms disappear, the problem of computing the target state is solved and the control action cancels steady state errors. The optimal controller and observer has been designed in each fuzzy rule, so a suboptimal solution have been found, but easy of calculate and compute. The results show that the ball and beam controlled nonlinear model has a stable behavior, good transient response and zero steady state error on the full range of the ball and beam system.

Acknowledgements. This work is funded by the Spanish Ministry of Economy and Competitiveness through the project: COGDRIVE: Cognition inspired navigation for autonomous driving (grant DPI2017-86915-C3-3-R).

References

1. AMIRA GmbH: BW500 Laboratory Setup Ball and Beam. Deutschland, Germany (1999)
2. Ballesteros, J., Rodriguez, D., Sanchez, L., Castillo, F., Payo, I.: Teaching-learning model based on the design of didactic equipment for laboratory practices in engineering schools. In: Proceedings of EDULEARN16 Conference, 4th–6th July, Barcelona, Spain (2016)
3. Adánez, J.M., Al-Hadithi, B.M., Jiménez, A., Matía, F.: Optimal control of a ball and beam nonlinear model based on Takagi-Sugeno fuzzy model. In: Advances in Fuzzy Logic and Technology 2017, pp. 1–11. Springer, Cham (2017)
4. Takagi, T., Sugeno, M.: Fuzzy identification of systems and its applications to modeling and control. IEEE Trans. Syst. Man Cybern. **1**, 116–132 (1985)
5. Jiménez, A., Al-Hadithi, B.M., Matía, F., Haber-Haber, R.: Improvement of Takagi-Sugeno fuzzy model for the estimation of nonlinear functions. Asian J. Control **14**(2), 320–334 (2012)
6. Al-Hadithi, B.M., Jiménez, A., Matía, F.: A new approach to fuzzy estimation of Takagi-Sugeno model and its applications to optimal control for nonlinear systems. Appl. Soft Comput. **12**(1), 280–290 (2012)
7. Al-Hadithi, B.M., Barragán, A.J., Andújar, J.M., Jiménez, A.: Chattering-free fuzzy variable structure control for multivariable nonlinear systems. Appl. Soft Comput. **39**, 165–187 (2016)

8. Al-Hadithi, B.M., Jiménez, A., Perez-Oria, J.: New incremental Takagi-Sugeno state model for optimal control of multivariable nonlinear time delay systems. Eng. Appl. Artif. Intell. **45**, 259–268 (2015)
9. Adánez, J.M., Al-Hadithi, B.M., Jiménez, A.: Wind turbine multivariable optimal control based on incremental state model. Asian J. Control **20**(6), 2075–2087 (2018)

Time-Oriented System to Control Critical Medications

Cristina Puente[1]([✉]), Alejandro Sobrino[2], Augusto Villa-Monte[3],
and Jose Angel Olivas[4]

[1] Advanced Technical Faculty of Engineering ICAI,
Comillas Pontifical University, Madrid, Spain
cristina.puente@icai.comillas.edu
[2] Faculty of Philosophy,
University of Santiago de Compostela, La Coruña, Spain
alejandro.sobrino@usc.es
[3] Institute of Research in Computer Science LIDI, Faculty of Computer Science,
National University of La Plata, La Plata, Buenos Aires, Argentina
avillamonte@lidi.info.unlp.edu.ar
[4] Department of Information Technologies and Systems,
University of Castilla-La Mancha, Ciudad Real, Spain
joseangel.olivas@uclm.es

Abstract. It is a truism that time has a high relevance in medicine. The prescription of several medicines to a patient usually must satisfy timing intervals according to their composition. An incorrect administration can lead to severe consequences to the patient or in the best of the cases lower the effect of other drugs. In this paper we want to provide a solution for this problem introducing a prototype that is able to detect anomalies between doses administration in critical medicines. In this vein, we develop a software in three steps: the first one is in charge of filling a database with the commercial name of medicines, taken from the Mayo Clinic website. Once the database is complete, we design an interface to request the information about the medication of a patient and verifying the time restrictions. In a third step, if the patient selects a medicine with time restrictions, the system will search using a web scrapper and a parser for the components that may interact with the first medication, and will set a time restriction alert, facilitating the controls to provide a correct administration.

Keywords: Timely medication administration · Time constraints · Time-oriented systems · Time-decision support in medicine

1 Introduction

Time is a natural property, but also a social construct. As a natural fact, Einstein placed it as the fifth dimension of the universe. As a social fact, humans have invented calendars and clocks to measure and control the time and exist an

© Springer Nature Switzerland AG 2020
F. Martínez Álvarez et al. (Eds.): SOCO 2019, AISC 950, pp. 544–553, 2020.
https://doi.org/10.1007/978-3-030-20055-8_52

objective or chronological time, that measured by devices, and a subjective time, or how humans experiment in different ways the same duration (eternal hours or hours like a breath) [4].

Being a natural property or a social construct, time has a long influence in human activities, ordering agriculture practices, sequencing the food customs or longing for our eternity. In more circumscribed field, such as medicine, time is also relevant. There are several medicine areas time-sensitive. These are some of them involved in the treatment; i.e., the administration of drugs during certain periods of time or under certain temporary restrictions:

(i) *Time-critically medications.* There are some drugs very sensitive to time administration. The Institute for Safe Medication Practices (Canada) published a "List of High-Alert Medications in Acute Care Settings" [3], reporting that some drugs must be provided in a specific temporal period in order to achieve the intended results. Failure to do so may result in serious harm to the patient, either because they are ineffective or because they provoke an additional damage. Troubles caused by the ineffectiveness of the treatments can be extended to public institutions, as claims demanding compensations or extra hospital expenses. Guo et al. reported that the estimated costs of adverse drug events amounts 8,75 dollars per hospital stay (3,5 billion for 400,000 cases) in USA hospitals [6]. So, mistakes in the time sensitive medication are not only a health problem, but also a financial one.

(ii) *Chronotherapy.* The toxicity and effectiveness of some medicines is dependent on the time they are administered, because it is verified that 24-h circadian rhythm influences behavioral and psychological aspects involved in the drugs reception. For instance, the administration of corticosteroids once daily in the morning reduces the adrenocortical suppression, but if the dose is divided into four administrations, a hypothalamus-pituitary-adrenal axis suppression is produced. Time administration recommendations on chronotherapy for regular-prescribed drugs are a relevant topic for health professionals, contributing to the therapeutics efficiency [12].

(iii) *Time dependent drug-drug interactions.* The administration of two or more drugs being one of them incompatible with the other may provoke a decrease in the absorption, affecting its metabolism or producing serious damages (including death) in case of absolute incompatibility, as paradigmatically show certain psychotropic medicines. Administering drugs separately by an adequate time interval (frequently, 2–4 h) can avoid those inconvenient effects [13]. Because time dependent drug-drug administration is in charge of nurses, alert systems warning misuses may be useful to avoid mistakes.

The non-observance of time causes errors in medication with consequences for patients and extra expense for institutions. The *European Medicines Agency* (https://www.ema.europa.eu/en) manifest that "mistakes in the prescribing, dispensing, storing, preparation and administration of a medicine are the most common preventable cause of undesired adverse events in medication practice and present a major public health burden" and *The Office of Diseases Prevention and Health Promotion* (https://health.gov/) defines Adverse Drugs Events

(ADEs) as injuries consequence from the use of a drug, distinguishing between damages provoked by a drug and damages from the use of a drug. According to [9], the 28% of the ADES comes from the medication administration step.

Technological approaches, such as radio frequency tagging, has long used to improve the association between patients and medication. So, [8] proposed a smart RFID (Radio Frequency Identification) device for medicine administration. Other papers approached bar-coded administration and the extent to which ADEs are prevented [11], [5] showed how computerized provided order entries (CPOE) improving safety patients reducing ADS episodes. [13] approached an alert software for inhibiting medication errors in time dependent drug-drug interactions showing a low influence in the ADEs practice.

Professionals associations in charge of the safeguard of the nurses and patients' rights promote the development of the bill of rights to safe medication administration, in order to limit the medication errors and so, contributing to the safety of the patient and the good management of hospitals. Several studies show that a high percentage of all prescriptions in the acute care settings are supplied at an incorrectly time [10]. In this paper we will contribute to solve this deficiency providing a system that locates patients with critical drug administration providing an alert system to control the doses and avoid human errors.

The rest of the paper is organized as follows: In Sect. 2 we will explain how to locate the name of critical medicines [2] in base to their active principles. So that we have used the Mayo Clinic website [1] to get this information by means of a web crawler and a parser. In Sect. 3, we will explain a user interface to introduce the patient's information and establish time restrictions with an example of use. In Sect. 4 we will conclude with future works and conclusions.

2   System for Controlling Time Restrictions of Medicines

In this work we propose a control system of temporary restrictions for the medical staff in charge of supplying medication to patients. There are treatments that involve medicines with active principles that can not be supplied together. They need to wait for time intervals and our system helps in this task. This system contributes to a correct time management in treatments which involves several medicines with time restrictions. For this, the system must consider temporary information extracted from medical documents and websites.

Figure 1 shows a full schema of the system proposed. It has three perfectly differentiated steps: (1) the active principles, substances or components with critical time intervals are catalogued into a database according to a reference medical document, "IHS BCMA – Timely Admin Policy Sample (Canada)" [2,3], (2) to control those restrictions, the time-sensitive medicines composed of these elements are detected using the website of a recognized medical research group [1] filtering and processing the information to be stored into the database, and (3) patient treatment will be controlled by medical staff using an desktop application to indicate the right time to supply each medicine.

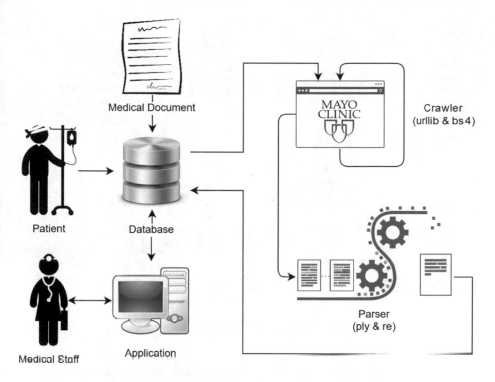

Fig. 1. Schema of the time administration of medicines control process.

The system is basically composed of the time restrictions database, a web crawler, a content parser and the user application. These four important components will be described below.

2.1 Design to Store Time Restrictions in Medical Treatments

As we have exposed, medicines that require an exact administration timing are of interest. Figure 1 shows the database design approached for storing the guidelines of medication scheduling and the treatments of diseases in hospitalized patients. The model consists of nine tables that store all the information required to timely administration of a scheduled drug.

The 'patient' table stores all personal data of a patient in a hospital. For this, the 'bed' and 'hospitalization' tables must be verified. In the first table, all the beds of a clinic are numbered. In the second table, the periods of hospitalization are stored. Once the availability of beds has been checked and the patient is hospitalized, the medical staff begins to perform treatments whose history will be recorded in the 'treatment' table. This table will record with date and time the medicines of a patient. At the same time, each medicine will have associated a set of components whose application time restrictions will be stored

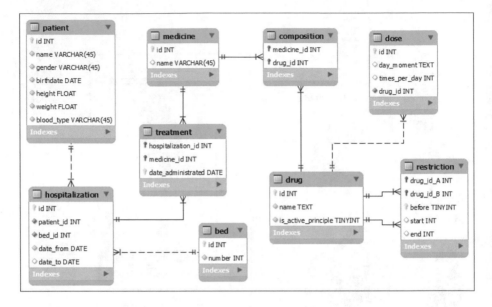

Fig. 2. Timely administration of medicines database.

in the 'restriction' table. In addition, each drug will have its suggested dosage considering the period of the day and the waiting time (in minutes).

The Data Base Management System used was MySQL and to perform the modeling of the database, Visual Database Schema Design from MySQL Workbench was used.

2.2 Crawling the Web

As basis of the software process, we have used the document edited by the Institute for Safe Medication Practices (ISMP) based on nurse's experiences in 2010 regarding the CMS "30-minute rule" [2,3]. From this reference medical document, the list of time-critical scheduled medications was analyzed in detail and loaded into the database. This involved modifying the 'drug', 'dosage' and 'restriction' tables. This task was carefully carried out to avoid altering the document information and ensure a safe control of the times.

Once the active principles were loaded into the database, a script that automatically performed the search of each one on the website of the Mayo Clinic was developed. The script was programmed in Python programming language using the *urllib* and *bs4* libraries. So, with this crawler, we were able to access the 'Drugs&Supplements' site of the Mayo Clinic [1], and automatically search for the active principles that we had stored into our database with a simple query like this: '*https://www.mayoclinic.org/search/search-results?q=*' added to the active principle. For example, if the active is '*fluoroquinolone*', the full query would be '*https://www.mayoclinic.org/search/search-results?q=fluoroquinolone*'.

For each case, the html was decoded to *UTF-8* format to be able to manipulate it correctly as a string. Then, for each results page that returned a search, the content was filtered using the *BeautifulSoup* library of *bs4*. In this way, it was obtained each search snippet from the Mayo Clinic as seen in Fig. 2, being the output of filtering the results returned by the website. These snippets will be the input for a parser developed in the following subsection.

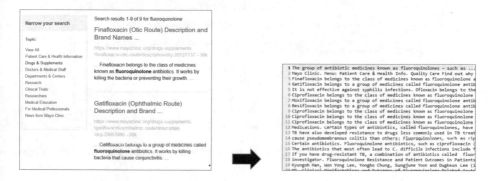

Fig. 3. Example of snippet obtained from the Mayo Clinic website.

2.3 Parsing the Web Content

With all the information related to an active principle stored in a text file, the next step is to extract from here the commercial names of the medicines containing the principle and save them into de database. We have used the *ply* library from Python to parse text, and constructed a two-state automaton that is able to filter a sentence and locate the commercial name of the medicine with a given active principle. Besides, some of the names may appear repeated, as seen in Fig. 3 that shows the web search of the Mayo Clinic page. If the search engine displays more than one result for the same medicine, we will filter those repeated entries to include just one in our data base.

In the text file obtained in the previous step, we have sentences like this: "*Ciprofloxacin belongs to the class of medicines known as fluoroquinolone antibiotics.*", so what we have done in the parse is to isolate each sentence and introduce a bag of words that may precede the commercial name of the medicine, as in this case indicates the word 'belongs'. So, we store the full sentence into a vector, and in the case that any of the words match with the active principle, the program will retrieve the previous word to *belongs*, will check that is not repeated, and if not, will store it into a temporal vector, as seen in Fig. 4.

3 Example of Use

Once the database with all the critical information is completed, a user-friendly app interface to establish time restrictions when administrating doses is pro-

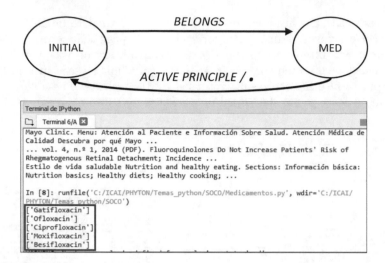

Fig. 4. Schema and output of the process to isolate commercial names of the drugs associated to an active principle.

duced, following the next sequence: First, the interface asks the user for personal information about a patient, disease, room, etc. Then a set of simple questions like the medicine name (in the database we have just introduced only those critical medicaments according to [2,3]). Considering the type of medicine, the system is going to ask for a series of questions corresponding to its administration. This setup interface has to control as well if the user has to take more than one medicine, and its possible conflicts along time with it. If so, the system internally looks if there are conflicts between the two medications sending a web query with the restrictions. If the result is positive, a new time restriction will be created for this patient.

Once the setup information has been introduced, the system will notify when a patient has to take the medicine according to the restrictions introduced. We have created a full example with screenshots to explain how it works.

3.1   Case of Use with the Active Principle *Fluoroquinolone*

The group of antibiotic medicines known as *fluoroquinolones* is used with severe infections. Is one of the indicated groups with time conflicts with other drugs. In particular with *iron, calcium, zinc* or *antacids*, having to take it 2 h before or six after the administration of any of these components. With these indications and the system previously explained, we have set up a working example.

First, the data base has to be filled with the commercial names of the medicines containing *Fluoroquinolone*, using the web crawler and the parser explained in Sects. 2.2 and 2.3. We can repeat that process if we update the software to look up for new medicines, but not every time that we launch the application as it could take its time.

Next, we will have to fill a form like the one displayed in Fig. 5. Additional to the patient's personal information, we need to know the medicines that is taken in order to establish an schedule. In the example, the patient has an urinary infection and is being treated with *Ciprofloxacin*, medicine that contains *Fluoroquinolone*. In this case, the patient is taking a supplement named *Calcium Acetate*, so the system is going to look up into de Mayo Clinic website if this second medicine has any conflicting component such as *iron, calcium, zinc* or is an *antacid*. In this particular case, it will find *calcium*, so will establish a restriction of two hours in the administration of these two medicines displaying a warning alert.

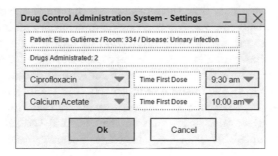

Fig. 5. Patient's medicines form.

In this case, if we want to establish a margin of less than two hours in the dose administration, the system will display a warning message and will change the time of the first dose in the second medication, leaving a margin of two hours.

Once the information of all patients has been introduced in the application, another screen will display only those patients with time restrictions on its medication. In this case, every patient will have a checkbox button enabled in case that is time to take the medicine, and to be checked by the nurse once administrated. If there is a conflicting medicine as in the case of the first and third patients, this check button will appear disabled until the time for no-administration is up, and the drug will turn from red to grey, as seen in Fig. 6.

Drug Control Administration System - 10:36 am				_ □ ×
Patient: Elisa Gutiérrez / Room: 334 / Disease: Urinary infection	Ciprofloxacin 9:30 am	■	Calcium Acetate 11:30 am	
Patient: Elena Arrieta / Room: 267 / Disease: Otitis	Finafloxacin 9:00 am	■		
Patient: Mª Luz García / Room: 306 / Disease: Eye infection	Besifloxacin 8:30 am	■	Iron Supplement 10:30 am	
Patient: Manuel Bravo / Room: 108 / Disease: Urinary infection	Ciprofloxacin 9:00 am	■		

Done   Cancel

Fig. 6. List of all patients with drug conflicts and time scheduling.

4   Conclusions and Future Work

In this paper we consider critical time in the administration of medicines, associated to certain active principles. We have proposed a system to avoid administration errors by controlling timing intervals of medicines, developing a complete tool that contains a web crawler, a parser, and a user-friendly interface.

As consequence of this study, we have observed that time in the administration of drugs is critical in some cases, but in others, it is only partial. It is at this point where soft computing techniques should be used.

As future work, our aim is to analyze how a drug that is prescribed before a second drug can vary, and to what extent, the therapeutic effect of the latter. The objective is to evaluate how the causal effect of a medicine can swing and to what degree the causal effect of a second medicine that is administered after the first, increasing, decreasing, or even nullifying, its curative impact. Helgason & Jobe approached that problem aiming a perception-based representation model of the physiological interaction of drugs in an individual patient, monitoring the evolution over time from the initial conditions [7]. The evolution was represented as a trajectory in a hypercube of relevant variables and the causal influence of one medicine on another is found from the Formal Causal Ground (FCG) and the Clinical Causal Effect (CCE), based on necessary and sufficient causal conditions.

Our aim is to introduce that kind of measures in our software, enabling and effective control over the influence of one drug over another administered after the first.

Another objective for future work is to increase the number of case studies. In this case, we studied the medicines associated to thyroids with an expert supervising, but we would like to include more diseases.

Acknowledgments. This work has been partially supported by the European Regional Development Fund (ERDF/FEDER, UE) and the State Research Agency (AEI) of the Spanish Ministry of Economy, Industry and Competitiveness (MINECO) under grants TIN2016-76843-C4-2-R, TIN2014-56633-C3-1-R, and TIN2017-84796-C2-1-R. A. Villa Monte thanks both the National University of La Plata (Argentina) and the University of Castilla-La Mancha (Spain) for supporting his co-tutelary PhD in Computer Science and Advanced Information Technologies respectively.

References

1. Mayo Clinic Website. https://www.mayoclinic.org/
2. IHS BCMA–Timely Admin Policy Sample (2012). https://www.ihs.gov/bcma/includes/themes/responsive2017/display_objects/documents/resources/BCMA_TimelyAdminPolicySample.pdf
3. ISMP List of High-Alert Medications in Acute Care Settings (2014). https://www.ismp.org/tools/highalertmedications.pdf
4. Bardon, A.: A Brief History of the Philosophy of Time. OUP USA (2013)

5. Devine, E.B., Hansen, R.N., Wilson-Norton, J.L., Lawless, N.M., Fisk, A.W., Blough, D.K., Martin, D.P., Sullivan, S.D.: The impact of computerized provider order entry on medication errors in a multispecialty group practice. J. Am. Med. Inform. Assoc.: JAMIA **17**(1), 78–84 (2010)
6. Guo, J.W., Iribarren, S., Kapsandoy, S., Perri, S., Staggers, N.: Usability evaluation of an electronic medication administration record (eMAR) application. Appl. Clin. Inform. **2**(2), 202–224 (2011)
7. Helgason, C.M., Jobe, T.H.: Perception-based reasoning and fuzzy cardinality provide direct measures of causality sensitive to initial conditions in the individual patient (invited paper) (2002)
8. Iadanza, E.: A smart RFID device for drugs administration. Pharmaceutica Analytica Acta s15 (2012)
9. Institute of Medicine: Preventing Medication Errors: Quality Chasm Series. National Academy Press, Washington (2006)
10. Keers, R., Williams, S.D., Cooke, J., Ashcroft, D.M.: Causes of medication administration errors in hospitals: a systematic review of quantitative and qualitative evidence. Drug Saf. **36**(11), 1045–1067 (2013)
11. Koppel, R., Metlay, J., Cohen, A.: Role of computerized physician order entry systems in facilitating medication errors. ACC Curr. J. Rev. **14**, 8 (2005)
12. Martiny, K., Refsgaard, E., Lund, V., Lunde, M., Thougaard, B., Lindberg, L., Bech, P.: Maintained superiority of chronotherapeutics vs. exercise in a 20-week randomized follow-up trial in major depression. Acta Psychiatr. Scand. **131**(6), 446–457 (2015)
13. Van der Sijs, H., Lammers, L., van den Tweel, A., Aarts, J., Berg, M., Vulto, A., van Gelder, T.: Time-dependent drug-drug interaction alerts in care provider order entry: software may inhibit medication error reductions. J. Am. Med. Inform. Assoc. **16**(6), 864–868 (2009)

Special Session - Soft Computing in Aerospace, Mechanical and Civil Engineering: New Methods and Industrial Applications

An Introduction to Some Methods for Soft Computing in Fluid Dynamics

Soledad Le Clainche(✉)

School of Aerospace Engineering, Universidad Politécnica de Madrid,
28040 Madrid, Spain
soledad.leclainche@upm.es

Abstract. In recent years the terms *big data*, *soft computing* and *machine learning* have been widely extended in several research fields, where the analysis of data (mainly time series) provides useful information to solve different types of engineering problems. *Turbulent flows* and *reduced order models* (ROMs) are two terms related to the field of fluid dynamics that can be compared to the previous expressions, which are generally used in data science. This paper presents a short introduction to some of the methodologies generally used in fluid dynamics for the analysis of complex flows to extract spatio-temporal features. These methods are also used to construct ROMs to predict spatio-temporal events using a small number of variables. The paper intends to connect the field of computer and data science with the classical terms and methodologies used in fluid dynamics for soft computing and machine learning, known as ROMs.

Keywords: Soft computing · Machine learning ·
Reduced order model · Fluid dynamics · Big data · CFD ·
Data science · POD · DMD

1 Introduction to Turbulent Flows and Big Data

Turbulent flows are found in nature and in a wide range of engineering and industrial applications. For example, it is possible to find turbulence in the flow inside engines, in combustion systems or power plants, in heat exchangers, in the flow over transport vehicles such as cars, aircraft, space aircraft, in man-made devices such as micro- aerial or unmanned aerial vehicles (micro-AVs or UAVs), or even in biological flows such as the flow of blood in the heart or in the wake of flying insects, birds and fish. Turbulence produce undesirable effects increasing drag and diminishing efficiency in engineering systems or transport vehicles, which is mainly reflected in the rise of fuel, pollution and cost [17]. Moreover, the effects of turbulence in biological flows could be devastating, for example, in the circulatory system turbulence occurs in pathological situations (i.e.: due to medical implants or medical devices) triggering negative and exclusive biological

© Springer Nature Switzerland AG 2020
F. Martínez Álvarez et al. (Eds.): SOCO 2019, AISC 950, pp. 557–566, 2020.
https://doi.org/10.1007/978-3-030-20055-8_53

responses, such as coronary artery diseases [5]. For all these reasons, studying in depth turbulent flows is a research topic of high interest.

In last 20 years, researchers have paid special attention in understanding the flow physics behind turbulent flows. Global instabilities and flow structures are two key words linked to the bifurcation process describing the transition from laminar to turbulent flow. If the main flow instabilities are identified, then it is possible to predict and control them, making possible to delay or even attenuate turbulence effects [24]. The main drawback in the analysis of turbulence lies in the large amount of spatial and temporal scales describing the flow, which is fairly disparate. Thus, carrying out a proper description of the flow physics entails storing and analysing large amounts of data, requiring powerful facilities managing optimal algorithms post-processing ten of even hundred of Terabytes of information. The link between the expression *big data* and fluid dynamics is naturally established by the need of studying turbulent flows itself, or the flow physics triggering the transition to turbulence [15]. Volume, velocity and variety are the three main features describing big data in general and in particular in the analysis of turbulent flows.

Experiments and numerical simulations are the main tools required to generate large amounts of data that will be analysed with the aim at solving several types of problems. But the *reality* needs to be described using models: either physical models representing prototypes that will be used in the experiments, or mathematical models defining equations that are solved using powerful computers. In computational fluid dynamics (CFD), Navier-Stokes equations model the evolution in time of velocity and pressure fields in a flow. Solving Navier-Stokes equations using high fidelity numerical simulations, called as direct numerical simulations (DNS), is highly expensive in terms of computational time and memory, especially when the flow studied is three-dimensional and fully turbulent [31]. For example, solving the turbulent flow moving around the wing of an aircraft, a typical problem generally solved in the aerospace engineering industry, is modelled by computational domains considering large amounts of grid points (10–100 millions degrees of freedom), which are essential to properly solve the large amount of spatio-temporal scales describing the flow [31]. Only powerful machines are able to solve this type of problems, which are computationally expensive, making that solving and analysing some types of flows is not always feasible.

It is possible to reduce the computational cost solving Navier-Stokes equations extrapolating intrinsic flow mechanisms, by means of using model order reduction strategies for the construction of reduced order models (ROM). Similarly to *machine learning* strategies, it is possible to predict different flow states through available data using ROMs. Thus, ROMs are suitable to approximate solutions using a smaller number of variables and reducing the complexity of the computational problem. This definition, which is similar to *soft computing*, links these two key words for two different highly powerful research communities, fluid dynamics and computer sciences, but with similar interests. This paper focuses on introducing some relevant tools generally used in fluid dynamics as ROMs.

The clear connection between soft computing and ROMs could motivate scientist to exchange information with the aim at improving their research in both fields, fluid dynamics and computer science. Paying attention to the similarities and differences between the algorithms generally used in soft computing or as ROMs, and their multiple applications, will bring new benefits to these two communities.

The paper is organized as follows. Section 2 introduces some of the methodologies used in fluid dynamics for model reduction. Section 3 illustrates the applications of such methodologies to machine learning strategies. Finally, Sect. 4 presents the main conclusions of the work.

2   Soft Computing in Fluid Dynamics

The main benefit of using ROMs in the field of fluid dynamics is that they can be used in both, experimental and numerical data, contributing with several advantages: (i) reducing the computational cost in numerical simulations or even in experiments, (ii) providing efficient tools for several applications such as optimization or flow control [21], and (iii) extracting spatio-temporal information useful to understand the flow physics. Depending on the expert knowledge and the type of data available, it is possible to identify two types of ROMs. The first type of ROMs is based on the modification of the flow governing equations, some examples include: (i) using Galerking projections into sub-spaces of smaller dimension (i.e. projection over a proper orthogonal basis, optimal in terms of energy, called as proper orthogonal decomposition, POD [19]), or (ii) using the linearized Navier-Stokes equations (linear instability analysis [29]). These approaches have been extensively used by several authors in the literature, mainly due to their good performance [22]. However, their main inconvenient is that they require a priory knowledge of the flow equations. The second type of ROMs solve this drawback, since they do not need to know the underlying equations, but they are related to the system identification from only input-output data. This type of ROM is generally used in flight dynamics and aeroelasticity [3], neural networks [21], et cetera. These models provide an accurate description of the temporal evolution of the main dynamics, but the description of the spatial features is more limited. This issue has recently solved combining these ideas with data-driven techniques generally used for the analysis of flow structures in fluid dynamics, such as Dynamic Mode Decomposition (DMD) [7,10,14].

The following sections will briefly introduce the techniques previously mentioned, POD and DMD. These two approaches, generally used to construct ROMs that are suitable to data forecasting in fluid dynamics (and some other fields), are also extensively used to extract flow features, global instabilities and study the flow physics in complex flows (i.e.: turbulent flows - big data analysis). Using these techniques in some other research fields, will bring the users new applications for soft computing, combining machine learning methodologies for data forecasting in time with relevant information extracted from spatial features. This combination will shed light on solving new problems and extending machine learning methodologies to predictions providing relevant spatial information, merging the knowledge acquired by researchers in computed sciences and fluid dynamics.

2.1 Singular Value Decomposition: POD and PCA

Singular value decomposition (SVD) is a type of factorization that captures the directions of a matrix in which vectors can shrink or grow. Such directions are given by the eigenvectors and eigenvalues of a rectangular matrix. SVD approach has been extensively used for different applications, especially for low-rank matrix approximations. This dimensionality reduction is very relevant in fluid dynamics in general, and in the study of turbulent flows in particular. Applying SVD allows reducing the size of the large amount of data analysed, making possible to reduce the computational resources required to carry out this task [4].

POD was introduced by Lumley [16] as a mathematical approach suitable to extract coherent structures from turbulent flows. The main goal of POD is to decompose data as an expansion of modes which are based on optimizing the mean square of a field variable that is analysed. SVD and POD are two terms that are generally used interchangeably in the literature. However, the main difference lies in that SVD is one of the techniques that can be applied to obtain a POD decomposition, which is known as the snapshot method, introduced by Sirovich [25]. The spatial classical POD method [4] is based on the covarianze of a state vector changing in time, whose size is based on the spatial degrees of freedom of the data. For three-dimensional problems, or even large size two-dimensional problems, this method is extremely expensive in terms of computational cost, thus in those cases SVD is used instead, to obtain the POD modes.

The main goal of POD is to decompose a set of spatio-temporal data $v(x,y,z,t)$ into a group of proper orthogonal spatial modes $\boldsymbol{\Phi}_j(x,y,z)$ (x, y and z correspond to the spatial coordinates), weighted by the temporal coefficients $c_j(t)$, as

$$v(x,y,z,t) \simeq \sum_j c_j(t)\boldsymbol{\Phi}_j(x,y,z) \tag{1}$$

To calculate POD modes the data are first organized into a *snapshot matrix*, considering K snapshots evolving in time as

$$X = V_1^K = [v_1, v_2, \ldots, v_k, v_{k+1}, \ldots, v_{K-1}, v_K], \tag{2}$$

where v_k corresponds to the field variable evaluated at time instant t_k, for convenience defined as $v_k = v(t_k)$. It is remarkable, that these snapshots not necessarily require to be equi-distant in time to calculate POD modes, but as it will be explained in the following section, these snapshots must be equi-distant in time for DMD analyses.

Then SVD is applied to such snapshot matrix, thus spatio-temporal data are decomposed into spatial orthogonal modes W, which are the POD modes, temporal modes T and singular values Σ as

$$\boldsymbol{V}_1^K \simeq \boldsymbol{W}\,\boldsymbol{\Sigma}\,\boldsymbol{T}^\top. \tag{3}$$

where $^\top$ denotes the matrix transpose, $\boldsymbol{W}^\top\boldsymbol{W} = \boldsymbol{T}^\top\boldsymbol{T} =$ the $N \times N-$unit matrix and the diagonal of matrix $\boldsymbol{\Sigma}$ contains the singular values $\sigma_1, \cdots, \sigma_K$. POD modes are ranked in decreasing order by their singular values. The modes with highest singular values (highest amplitude) usually represent the general dynamics of the system (i.e.: large size coherent structures), while the modes with smallest singular values could be removed from the approximation, assuming a certain error, as it will be clarified next. Sometimes these small amplitude modes are simply representing the noise contained in the data analysed, as in the case of experimental measurements. Thus, in the analysis of complex flows usually only the POD modes with largest singular values are retained to construct POD expansion (1) or the SVD approximation (3). The number of retained SVD modes, N, can be calculated considering different criteria. Several methodologies are described in [20], where POD or SVD approaches are also called as *Principal Component Analysis* (PCA).

2.2 Dynamic Mode Decomposition

Dynamic mode decomposition (DMD) [26] is a technique recently introduced for the analysis of flow structures. The method decomposes spatio-temporal data as an expansion of DMD modes $\boldsymbol{v}_m(x, y, z, t) = \boldsymbol{v}_m$, weighted with some amplitudes a_m, which oscillate in time with frequency ω_m and either grow, decay or remain permanent in time with growth rate δ_m as

$$\boldsymbol{v}(x, y, z, t_k) \simeq \sum_{m=1}^{M} a_m \boldsymbol{u}_m(x, y, z) e^{(\delta_m + i\omega_m)t_k}, \tag{4}$$

for $k = 1, \cdots, K$, representing the temporal interval of the snapshot matrix (2). DMD decomposition considers data equi-distant in time. Thus, DMD algorithm starts collecting data into the snapshot matrix (2) separated with time interval Δt. Then, the algorithm introduced by Schmid [26] is very simple and only considers two main steps (this algorithm is conveniently defined in this paper as in [13] to connect the different methods presented below). As first step, SVD is applied to the snapshot matrix (2). Thus DMD could be considered as a step further to POD (or SVD), to calculate modes separated in time. Before proceeding with the algorithm description, for convenience Eq. (3) is rewritten as $\boldsymbol{V}_1^K = \boldsymbol{W}\hat{\boldsymbol{V}}_1^K$, with $\hat{\boldsymbol{V}}_1^K = \boldsymbol{\Sigma}\boldsymbol{T}^\top$. Thus, the original snapshot matrix $\boldsymbol{V}_1^K$ is written as product of spatial and temporal terms. As second step, DMD algorithm considers the linear relationship between subsequent snapshots (defined in matrix form) as

$$\hat{\boldsymbol{V}}_2^K \simeq \boldsymbol{R}\hat{\boldsymbol{V}}_1^{K-1}. \tag{5}$$

The Koopman matrix $\boldsymbol{R}$ contains the dynamics of the system, so its eigenvalues represent the DMD frequencies ω_m and growth rates δ_m, while its eigenvectors are used to construct DMD modes $\boldsymbol{u}_m$.

More recently, Le Clainche and Vega [13] introduced an extension of DMD, called as higher order dynamic mode decomposition (HODMD), suitable for the analysis of complex flows, noisy experimental data or flows in transitional regime, where standard DMD may find some limitations [14,15]. HODMD replaces the Koopman assumption (5) by the following *high order Koopman assumption*

$$\hat{\boldsymbol{V}}_{d+1}^K \simeq \boldsymbol{R}_1 \hat{\boldsymbol{V}}_1^{K-d} + \boldsymbol{R}_2 \hat{\boldsymbol{V}}_2^{K-(d-1)} + \ldots + \boldsymbol{R}_d \hat{\boldsymbol{V}}_d^{K-1}, \tag{6}$$

combining DMD algorithm with the Taknes' delayed embedded theorem [28]. So, this algorithm considers d time delayed snapshots (sub-matrices), similar to the window-shift process carried out in Power Spectral Density (PSD). Similarly to DMD, the Koopman operators $\boldsymbol{R}_1, \cdots, \boldsymbol{R}_n$ contain the dynamics of the system, so they are firstly collected into a single matrix to consequently solve its eigenvalue problem and calculate DMD modes, frequencies and growth rates.

2.3   Spectral POD, Multi-scale Analysis and EMD

The analysis of turbulent flows motivate researchers to introduce several variants of POD and DMD analyses [2]. Some interesting examples include the methods based on the extraction of modes related to the fairly disparate number of frequencies representing the flow. Among these methods it is interesting to remark: spectral POD [27], multi-scale analyses [8,18], and an effective method also based on the expansion of modes, which have been recently introduced in this field, empirical mode decomposition (EMD) [1].

Spectral POD [27] is a technique suitable for the analysis of turbulent flows, which is able to provide low-rank approximations of the solution, free of noise, small flow scales or even spurious results. The method simply considers a POD algorithm calculated in the frequency domain. These modes are calculated from a set of realizations of the temporal Fourier transform of the flow field, thus the snapshot matrix (2) is segmented in several blocks and FFT is performed in each one of them. Then SVD is applied over these blocks. In other words, the method could be somehow understood as a variant of HODMD, where FFT is carried out in each one of the sub-snapshot matrices (i.e.: $\hat{\boldsymbol{V}}_1^{K-d}$, $\hat{\boldsymbol{V}}_2^{K-(d+1)}$, et cetera) presented in Eq. (6). Instead of solving the eigenvalue problem of DMD, Fourier modes are promediated four each group. The main benefit of this methodology is that it provides an orthogonal basis of modes at discrete frequencies that are optimally ranked in terms of energy. However, the computational cost is more expensive than HODMD, whose algorithm is optimized, providing solutions without need of manual interaction.

Some variants of POD and DMD algorithms are based on multi-scale analyses. The main goal is to widen the range of applications of these techniques to complex turbulent flows, where the solutions are described by a large number of temporal and spatial modes with different flow scales. The main idea behind these techniques is to filter out, or to separate, the wide variety of high and low frequencies. For example, in muti-resolution POD [18] a first step is carried out before proceeding with the SVD analysis, were correlation matrix $\boldsymbol{S}$ (as previously mentioned, implicit in SVD algorithm) is split into the contribution of different flow scales. This step is carried out via Multi-Resolution Analysis (MRA) using a filter (preserving symmetry of the correlation) or a Wavelet Transform. A similar idea is behind the Multiresolution DMD (mrDMD) [8]. This algorithm separates slow and fast modes (low and high frequency modes) recursively from the collection of snapshots, initially represented in (2). Thus this matrix is divided in several segments, each one related to groups of high and low frequency modes. DMD algorithm is applied into each sub-group of the matrix. Finally, DMD expansion (4) is then represented by the several sub-expansions of DMD modes.

Finally, EMD [1] is a method whose main goal is to extract frequencies in any type of signals. The method is a statistical technique that considers the local maximum of a signal, connecting them by a cubic spline line as the upper envelope. The same procedure is repeated for the minimum and lower envelope. EMD extracts frequencies one by one to approximate a function, filtering the smallest amplitude frequencies. EMD can be compared with Fourier analysis, DMD and Wavelet transform. When the method is applied to turbulent flows, it can provide good results. However, it is interesting to take into account that (i) the method is not based into any physical principle, but it makes a fitting of the solution and (ii) the method does not establish any criterion related to the frequency amplitudes. So, it is more difficult to detect the dominant flow patterns, especially in complex flows.

3   Machine Learning: POD and DMD

Machine learning is one of the theories of soft computing, which is specific to give the machine the ability of thinking (data forecasting), while soft computing is the complete computational intelligence. In CFD, machine learning strategies are based on the construction of ROMs, either using projection strategies, when the flow equations are known, or using matrix-free strategies.

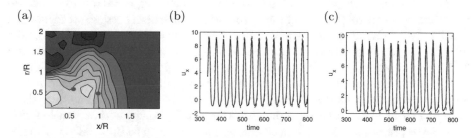

Fig. 1. (a) Instantaneous streamwise velocity at time 700 in a synthetic jet. (b) and (c) compares the original solution (black line) with the approximation using the DMD-based ROM (red line) at two representative points (red dots in (a)).

Generally, the first approach is based on the Galerking projection of the flow equations, Navier-Stokes equations, over the POD modes. The degrees of freedom, represented by the number of grid points of the computational domain modelling the problem solved (i.e.: the wing or an aircraft), are reduced to the number of POD modes retained in expansion (1). This is the ROM. Then, instead of integrating in time the fairly complex equations, which are computationally expensive, the ROM is integrated in time, which is fast and requires low memory capacity [19]. Nevertheless, it is remarkable that the computational cost required to generate the POD modes can be notorious in some cases. To calculate POD modes it is necessary to carry out a initial training to collect information to create the snapshot matrix (2). It is possible to use several approaches to generate this initial information [23]. But, the most extended one is based on the temporal integration of the full (Navier-Stokes) equations, in order to collect relevant information related with different flow states [22]. A different, highly efficient approach, is alternate the integration of the full equations for a short period of time, which is useful to collect information for the POD modes, with the ROM integration for long periods of times. This technique is known as POD on the fly [12,22]. The second approach for machine learning in fluid dynamics is based on data-driven, equation-free methods. A good example is using DMD to construct ROMs. For proper applications of DMD algorithm, it is possible to construct the DMD expansion (4) with small error. Such expansion describes the reconstruction of the original data analysed, collected at time instant t_k, in a time interval defining the snapshot matrix (2) as $t \in [t_1, t_K]$. Nevertheless, for accurate solutions of (4), it is possible to predict temporal events just by adjusting the temporal term $t = t_r \gg t_k$. This simple method defines the application of DMD as a data-driven, equation-free ROM, for simplicity called as DMD-based ROM. DMD has been successfully applied as a ROM for both, temporal and spatial predictions in a wide range of applications [10,14]. A new example is presented in this paper, where DMD has been applied to predict the the the attractor in a numerical simulation modelling a synthetic jet. This type of jet is formed by a piston oscillating into a cavity making the flow to leave and re-enter into the cavity periodically. Studying in depth the flow physics behind this flow is a research topic of high interest due to its multiple industrial applications, for instance, synthetic jet devices are used for fluid mixing, heat transfer enhancement and flow control [15]. Numerical simulations have been carried out to model a synthetic jet in laminar regime and axi-symetric flow conditions (see more details in [9,15]). A group of snapshots have been collected during the first 100 time units: 3 piston oscillation cycles defining the transitory region of the numerical simulation, which is described by a large number of transient modes ($\delta < 0$) interacting with the permanent modes ($\delta \simeq 0$). HODMD has been applied to analyse these data and to construct a DMD expansion as in Eq. (2). Using such expansion, it is possible to predict the solution of the numerical solver up to time $\simeq 800$ with relative error smaller than 10%. Figure 1 compares the real solution of the numerical simulations with the prediction in two representative points of the computational domain. As seen, the method perfectly captures the main flow dynamics driven by periodic flow oscillations, although some small differences are found in the amplitude of the signal, which is related to the small flow scales. Nevertheless, the main flow structures and global instabilities are found in such large size flow scales, thus this DMD-based ROM provides low-rank approximations of the original flow field.

This fact is confirmed (not shown for the sake of brevity) after applying SVD to the same data and reconstructing the original solution using a small number POD modes (see Eq. (4)), suggesting that this DMD-based ROM is a suitable tool to predict flow patterns in complex flows reducing the computational time of the numerical simulations. More details of this application are presented in [9].

4   Concluding Remarks

POD and DMD are the two methodologies selected in this paper as example for constructing ROMs, identified as machine learning and soft computing strategies. The reason is that in the field of fluid dynamics, they also provide useful information about the spatial flow features, named as flow structures and global instabilities. This information is generally used with different ends, but a common extended application is flow control. As it was mentioned in the Introduction of this paper: if flow physics is understood, then it is possible to control it to satisfy our needs.

Acknowledgements. The author is grateful to Prof. J. M. Vega for many interesting discussions and his continuous support. This work was supported by 'Programa Propio UPM'.

References

1. Chen, Y., Ma, J.: Random noise attenuation by f-x empirical-mode decomposition predictive filtering. Geophysics **79**(3), 81–91 (2014)
2. Chen, K.K., Tu, J.H., Rowley, C.W.: Variants of dynamic mode decomposition: boundary condition, Koopman and Fourier analyses. J. Nonlinear Sci. **22**, 8871–8875 (2012)
3. Dowell, H., Hall, K.H.: Modeling of fluid-structure interaction. Ann. Rev. Fluid Mech. **33**(1), 445 (2001)
4. Eckart, C., Young, G.: The approximation of one matrix by another of lower rank. Psychometrika **1**(3), 211–218 (1936)
5. Ferrari, M., Werner, G.S., Bahrmann, P., Richartz, B.M., Figulla, H.R.: Turbulent flow as a cause for underestimating coronary flow reserve measured by Doppler guide wire. Cardiovasc. Ultrasound **14**(4) (2006)
6. Koopman, B.: Hamiltonian systems and transformations in Hilbert space. Proc. Natl. Acad. Sci. U.S.A **17**, 315–318 (1931)
7. Kou, J., Le Clainche, S., Zhang, W.: A reduced-order model for compressible flows with buffeting condition using higher order dynamic mode decomposition. Phys. Fluids **30**(1), 016103 (2018)
8. Kutz, J.N., Fu, X., Brunton, S.L.: Multiresolution dynamic mode decomposition. SIAM J. Appl. Dyn. Sys. **15**(2), 713–735 (2016)
9. Le Clainche, S.: Prediction of the optimal vortex in synthetic jets. Energies (under review) (2019)
10. Le Clainche, S., Ferrer, E.: A reduced order model to predict transient flows around straight bladed vertical axis wind turbines. Energies **11**(3), 566–578 (2018)

11. Le Clainche, S., Lorente, L., Vega, J.M.: Wind predictions upstream wind turbines from a LiDAR database. Energies **11**(3), 543–558 (2018)
12. Le Clainche, S., Varas, F., Vega, J.M.: Accelerating oil reservoir simulations using POD on the fly. Int. J. Numer. Methods Eng. **110**(1), 79–100 (2017)
13. Le Clainche, S., Vega, J.M.: Higher order dynamic mode decomposition. SIAM J. Appl. Dyn. Syst. **16**(2), 882–925 (2017)
14. Le Clainche, S., Vega, J.M.: Higher order dynamic mode decomposition to identify and extrapolate flow patterns. Phys. Fluids **29**, 084102 (2017)
15. Le Clainche, S., Vega, J.M., Soria, J.: Higher order dynamic mode decomposition for noisy experimental data: flow structures on a zero-net-mass-flux jet. Exp. Therm. Fluid Sci. **88**, 336–353 (2017)
16. Lumley, J.L.: The structure of inhomogeneous turbulent flows. In: Yaglam, A.M., Tatarsky, V.I. (eds.) Proceedings of the International Colloquium on the Fine Scale Structure of the Atmosphere and Its Influence on Radio Wave Propagation. Doklady Akademii Nauk SSSR, Nauka, Moscow (1967)
17. Marusic, I., Candler, G., Interrante, V., Subbareddy, P.K., Moss, A.: Real time feature extraction for the analysis of turbulent flows. Semantic Scholar, Chap. 13 (2003). https://doi.org/10.1007/978-1-4615-1733-7-13
18. Mendez, M.A., Balabane, M., Buchlin, J.M.: Multi-scale proper orthogonal decomposition of complex fluid flows. J. Fluid Mech. (submitted). arXiv:1804.09646v3 [physics.flu-dyn]
19. Noack, B.R., Morzynski, M., Tadmor, G.: Reduced-Order Modelling for Flow Control. Springer, New York (2011)
20. Parente, A.: Experimental and numerical investigation of advanced systems for hydrogen-based fuel compustion. Ph.D. Thesis, University of Pisa (2008)
21. Park, K.H., Jun, S.O., Baek, S.M., Cho, M.H., Yee, K.J., Lee, D.H.: Reduced-order model with an artificial neural network for aerostructural design optimization. J. Aircr. **50**(4), 1106 (2013)
22. Rapun, M.L., Vega, J.M.: Reduced order models based on local POD plus Galerkin projection. J. Comput. Phys. **229**(8), 3046–3063 (2010)
23. Rowley, C.: Model reduction for fluids, using balanced proper orthogonal decomposition. Int. J. Bifurcat. Chaos **15**(03), 997 (2005)
24. Rowley, C., Dawson, S.T.M.: Model reduction for flow analysis and control. Ann. Rev. Fluid Mech. **49**, 387–417 (2017)
25. Sirovich, L.: Turbulence and the dynamics of coherent structures. Parts I–III. Q. Appl. Math. **45**(3), 561–571 (1987)
26. Schmid, P.: Dynamic mode decomposition of numerical and experimental data. J. Fluid Mech. **656**, 5–28 (2010)
27. Schmidt, O., Towne, A., Colonius, T., Cavalieri, A., Jordan, P., Bres, G.: Wavepackets and trapped acoustic modes in a Mach 0.9 turbulent jet: a global stability analysis. J. Fluid Mech. **825**, 1153–1181 (2017)
28. Takens, F.: Detecting strange attractors in turbulence. In: Rand, D.A., Young, L.-S. (eds.) Lecture Notes in Mathematics, pp. 366–381. Springer (1981)
29. Theofilis, V.: Advances in global linear instability analysis of nonparallel and three-dimensional flows. Prog. Aerosp. Sci. **39**(4), 249–315 (2003)
30. Tucker, L.R.: Some mathematical notes on three-mode factor analysis. Psychometrika **31**(3), 279–311 (1966)
31. Vinuesa, R., Schlatter, P., Malm, J., Mavriplis, C., Henningson, D.S.: Turbulent boundary layers around wing sections up to Rec = 1,000,000. Int. J. Heat Fluid Flow **72**, 86–99 (2018)

A Data-Driven ROM Based on HODMD

Víctor Beltrán⬤, Soledad Le Clainche⬤, and José M. Vega⁽⊠⁾⬤

School of Aerospace Engineering, Universidad Politécnica de Madrid,
28040 Madrid, Spain
josemanuel.vega@upm.es

Abstract. A preliminary version of a data driven reduced order model
(ROM) for dynamical systems is presented in this Chapter. This ROM
synergically and adaptively combines a black-box full model (FM) of the
system and extrapolate conveniently using a recent extension of standard
dynamic mode decomposition called higher order dynamic mode decomposition (HODMD). These two are applied in interspersed time intervals, called the FM-intervals and the HODMD-intervals, respectively.
The data for the each HODMD-interval is obtained from the application of the FM in the previous FM-interval. The main question is when
extrapolation from HODMD is no longer valid and switching to a new
FM-interval is necessary. This is made attending to two criteria, ensuring
that an estimate of the extrapolation error and a measure of consistency
are both conveniently small. In this sense, the present method is similar
to a previous method called POD on the Fly, which was not a purely
data driven method. Instead, POD on the Fly was based on a Galerkin
projection of the governing equations that thus should be known. The
new method presented in this paper is illustrated with several transient
dynamics for the complex Ginzburg-Landau equation that converges to
either periodic or quasi-periodic attractors. The resulting CPU acceler-
ations factors (compared to the full model) are quite large.

Keywords: Higher order dynamic mode decomposition ·
Adaptive reduced order model · Data driven reduced order model ·
Complex dynamics

1 Introduction

Reduced order models (ROMs) are increasingly used to accelerate computations
in dynamical systems of both scientific and industrial interests. Some of these
ROMs [1,2] include two stages, (i) an off-line stage that is usually very computationally expensive, in which the main ingredients of the ROM are prepared,
and (ii) a very fast on-line stage in which specific computations are performed.
Roughly speaking, these ROMs are appropriate when many specific computations are to be performed, thus compensating the computational cost of the

Supported by the Spanish Ministry of Economy and Competitiveness, under Grant
TRA-2016-75075-R.

F. Martínez Álvarez et al. (Eds.): SOCO 2019, AISC 950, pp. 567–576, 2020.
https://doi.org/10.1007/978-3-030-20055-8_54

off-line stage. When the number of computations is not so large, the off-line stage should be eliminated to obtain a computationally reasonable cost.

The off-line stage is eliminated in the so-called POD on the Fly models (see [3] and references therein) that, roughly speaking use a full model (FM) and a Galerkin projection (GP) of the FM in interspersed time intervals, called FM-intervals and GP-intervals, respectively. For each GP-interval, the Galerkin system is constructed using proper orthogonal decomposition (POD) modes calculated from the data obtained in the previous FM-interval. The main difficulty to construct these models is to decide when, in each GP model, the approximation is no longer valid and switching to the next FM interval is necessary. This is done using two estimates, namely an *error estimate* based on the magnitude of the amplitudes of the higher order POD modes and a *consistency estimate* based on the residual of the Galerkin system. When these two estimates are not conveniently small, the GP-interval is terminated and the computation switches to the next FM-interval.

By its own nature, the POD on the Fly method is not data-driven because constructing the Galerkin system requires knowledge of the equations governing the FM. Here, we construct a data-driven model that is essentially the counterpart of the POD on the Fly method that does not require knowledge of the governing equations, namely the FM can be a black-box model. This is done by using higher order dynamic mode decomposition (HODMD), a purely data driven method, instead of Galerkin projection in the POD on the Fly method. This ROM will be based on the same ideas as the POD on the Fly method. Namely, the FM and the HODMD tools will be used in interspersed FM-intervals and HODMD-intervals, respectively. And switching from each HODMD-interval to the next FM-interval will be based on two estimates, which account for the error and consistency of the HODMD approximation. With these preliminary ideas in mind, the remaining of the paper is organized as follows. Section 2 contains a brief description of the HODMD method and a full description of the ROM that will be used in the paper. Section 3 gives some preliminary results on the application of the constructed ROM to the complex Ginzburg-Landau equation, and Sect. 4, the main conclusions of the paper.

2   Construction of the Data-Driven ROM Based on HODMD

HODMD is a recent extension [4] of standard dynamic mode decomposition (DMD) [5] that has been proved to efficiently deal with numerical and experimental data resulting from a variety of problems ([6] and references therein). Given a set of temporally equispaced snapshots of a discrete dynamical system, $\boldsymbol{u}(t_1) \ldots, \boldsymbol{u}(t_K)$, the HODMD gives a Fourier expansion of the form

$$\boldsymbol{u}(t_k) \simeq \sum_{n=1}^{N} a_n \boldsymbol{q}_n \mathrm{e}^{(\delta_n + \mathrm{i}\omega_n)t_k} \quad \text{for } k = 1 \ldots, K. \tag{1}$$

where the modes $\boldsymbol{q}_n$ are conveniently normalized (which defines the amplitudes $a_n > 0$), the growth rates and frequencies, δ_n and ω_n, respectively, are computed by the method. Here, the vector $\boldsymbol{u}$ may give the state vector in genuinely finite dimensional dynamical systems, but it may account for the vector state variable, such as the velocity vector in fluid flows, in a convenient spatial mesh.

Let us mention here that standard DMD applied to the set of snapshots relays on the following linear relation

$$\boldsymbol{u}(t_{k+1}) \simeq \boldsymbol{R}\,\boldsymbol{u}(t_k) \quad \text{for } k = 1\ldots,K, \tag{2}$$

where the matrix $\boldsymbol{R}$ is independent of k and can be computed via the pseudo-inverse. Now, when this relation holds, solving the linear system (2) leads to an expansion of the form (1). In fact, for very high-dimensional systems, the computational effort to compute $\boldsymbol{R}$ is highly reduced by first applying truncated singular value decomposition (SVD), with a tunable accuracy ε_1, to the set of snapshots to dimension-reduce this set before computing $\boldsymbol{R}$. The logarithms of the eigenvalues and the eigenvectors of the dimension-reduced version of $\boldsymbol{R}$ readily give the exponents and the modes appearing in (1). When this method gives accurate results, it depends on a second tunable parameter ε_2 that defines the number of terms to be retained in (1). However, this method fails with the spatial complexity of the snapshots (defined as the rank of the set of dimension-reduced snapshots) is strictly smaller than the spectral complexity of the expansion (defined as the number of terms that are present in (1). An obvious case in which this happens is that in which the snapshots are scalars (spatial complexity equal to one) but the signal defining the snapshots is very complex (very high spectral complexity).

When standard DMD fails, Eq. (2) can be replaced by

$$\boldsymbol{u}(t_{k+d}) \simeq \boldsymbol{R}_1\,\boldsymbol{u}(t_{k+d-1}) + \boldsymbol{R}_2\,\boldsymbol{u}(t_{k+d-2}) + \ldots + \boldsymbol{R}_d\,\boldsymbol{u}(t_k) \quad \text{for } k = 1\ldots,K, \tag{3}$$

which is the essence of HODMD [4]; see also [7] for a MATLAB executable of the HODMD algorithm. Here, the index $d \geq 1$ is tunable. Note that this equation reduces to (2) when $d = 1$. For $d > 1$ instead, the method could be seen as the result of applying standard DMD to a set of enlarged snapshots [4], where each enlarged snapshot contains not only the original snapshot but also the former $d - 1$ time-delayed snapshots. Thus, HODMD synergically combines the advantages of standard DMD and standard consequences of the Taken's delay-embedding theorem [8]. Appropriate selection of d gives very good results for very general data [6].

Note two differences of HODMD with more standard methods, such as FFT and PSD [9]. These methods exhibit two disadvantages compared to HODMD. First, their application to high-dimensional data is problematic, while for HODMD it is not. Secondly, and more important, FFT and PSD do not naturally give the growth rates, while HODMD does. Now, the accurate computation of the growth rates is very important in many applications. In particular, zero (or very small) growth rates imply that the data lie in an attractor, positive values of the growth rates are associated with instabilities, and negative values, with transient decaying to attractors. In the latter case, retaining only those modes with small growth rates in (1) (and thus, skipping those modes with negative growth rates) leads to a means for computing attractors using data extracted from transient behavior [10]. This idea involves extrapolation in time and could be seen as a first version of the more elaborate adaptive ROM that is considered now.

The adaptive data-driven ROM is constructed as follows. In each FM-interval, the FM is run during a time interval built by three adjacent subintervals, $I = I_0 \bigcup I_1 \bigcup I_2$, where the interval I_0 is absent in the first FM-interval and I_1 and I_2 exhibit equal length. The initial condition in the first FM-interval is fixed by the dynamics we intend to simulate, while in the remaining FM intervals, the initial condition, at the beginning of I_0 is given by the last state at the end of the former HODMD interval; thus, the interval I_0 is introduced to just smooth the reconstructed state from the last HODMD interval. In each FM-interval, HODMD is applied in both I_1 and I_2 using a convenient (tunable) number of equispaced snapshots, and convenient values of the tunable parameters ε_1, ε_2, and d. Both HODMD reconstructions, R_1 and R_2, are compared, requiring that the relative root mean square (RMS) difference between both be smaller than a tunable parameter ε_3. If this condition does not hold, then the interval I_0 is enlarged until this condition holds. Then, the reconstructions R_1 and R_2 are extrapolated in time until one of the following conditions fails:

- *Consistency condition.* The relative RMS difference between the extrapolations from R_1 and R_2 is required to be smaller than some (tunable) ε_4.
- *Truncation error condition.* Using higher order modes (than those retained) in the extrapolation from R_1, the resulting reconstruction is expected to give a good estimate of the truncation error. The relative RMS value of this estimate is required to be smaller than some (tunable) ε_5.

When either of these condition fails, the HODMD-interval is terminated and a new FM-interval is initiated as indicated above.

It must be noted that the HODMD expansions R_1 and R_2 are of the form (1) and, even if they are close to each other in the sub-interval I_2, they will generally exhibit different (though close to each other) values of the growth rates and frequencies, δ_n and ω_n, respectively. These small differences will necessarily promote a divergence of $R_1 - R_2$ as time increases. Therefore, even if the attractor is well-approximated in phase space by both R_1 and R_2 from some value of time on, the consistency condition will not be satisfied in this semi-infinite time interval, and some (may small) FM intervals will be needed at some values of time.

3   Results for the Complex Ginzburg-Landau Equation

Let us now consider the pattern forming dynamical system formulated by the complex Ginzburg-Landau equation (CGLE) with Neumann boundary conditions

$$\partial_t u = (1 + i\alpha)\partial_{xx}^2 u + \mu u - (1 + i\beta)|u|^2 u, \quad \text{with } \partial_x u = 0 \text{ at } x = 0, 1, \quad (4)$$

which a paradigm of a simple nonlinear equation that exhibits intrinsically complex dynamics [11]. In particular, this equation applies to many dynamical systems, since it can be seen as a normal form that applies to infinite dimensional systems near oscillatory instabilities [12]. Thus, this equation is quite appropriate to test ROMs because (i) it can be easily integrated by the reader and (ii) the associated dynamics are quite complex, including periodic, quasi-periodic, and chaotic attractors. The state variable u is scalar and complex. It must be noted that the CGLE is invariant under the $O(2)$ group generated by arbitrary phase shifts and space reflections, namely

$$u \to u e^{ic} \text{ for all } c \quad \text{and } x \to -x.$$

Also, because of invariance under the first action, following [13], it is seen that attractors exhibit expansions of the form (1), with $\delta_n = 0$ (or extremely small) and $\omega_n = \omega_0 + \tilde{\omega}_n$, where ω_0 is generally incommensurable with $\tilde{\omega}_n$. This means that if $|u|$ is periodic, then u is quasi-periodic, exhibiting two fundamental incommensurable frequencies (namely, a 2-torus in phase space). And, more generally, of $|u|$ densely covers a r-torus in phase space, then u covers a $r + 1$ torus.

In the present paper, the CGLE will be considered with $\alpha = -10$, $\beta = 10$ (which is a very demanding case, see [13]) and appropriate values of μ to obtain increasingly complex attractors. The initial condition will be

$$u_0 = \sqrt{\mu} \cdot (1 + i), \quad u_1 = u_0$$

which is very far from the attractor in all considered cases. For the sake of brevity, we shall only consider three values of μ, for which the dynamics are representative of other cases that have also been considered. Also, we shall concentrate in dynamics converging to attractors, which will be indicated in these cases.

The numerical solver is constructed discretizing the spatial interval with centered second order, finite differences using 1000 points. The resulting set of ordinary differential equations is integrated using a Crank-Nicholson plus Adams-Bashforth scheme with time steeping $\Delta t = 10^{-4}$. This is the FM, which is used in this paper as a black-box FM. The time distance between snapshots collected from the outcomes of this FM will be ten times larger, namely $\Delta t = 10^{-3}$.

In all cases considered below, the length of the subinterval I_0 will be 0.1 (unless the adaptation of the method enlarge it), while the length of the subintervals I_1 and I_2 will always be 0.8. Concerning the tunable parameters of the HODMD methods, these will be $\varepsilon_1 = 10^{-8}$, $\varepsilon_2 = 10^{-4}$ ($\varepsilon_2 = 10^{-5}$ for higher order modes than in R_1), and $d \simeq K/10$, where K is the number of considered snapshots. Concerning the tunable parameters of the data-driven ROM, these will be $\varepsilon_3 = 10^{-4}$, $\varepsilon_4 = 5 \cdot 10^{-4}$, and $\varepsilon_5 = 10^{-3}$. These parameters will be maintained for all considered cases, which emphasizes the robustness of the data-driven ROM.

The first case to be considered is $\mu = 35$, for which the attractor is such that $|u|$ is periodic (thus, as anticipated, u is quasi-periodic, building a 2-torus in phase space). Figure 1 illustrates the attractor for this case, giving in the left plot a restricted phase plane, plotting $|u(t, 0.75)|$ vs. $|u(t, 0.25)|$ and a space-time color map of $|u|$ in the right plot. As can be seen in this figure, $|u|$ is periodic of period $T = 0.02529$ time units. Also, it exhibits the following spatio-temporal reflection symmetry $|u(x,t)| = |u(1-x, t+T/2)|$.

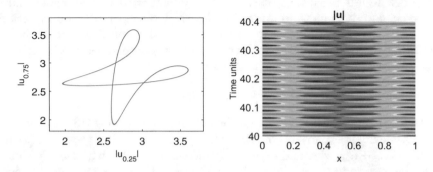

Fig. 1. Restricted phase plane (left) and spatio-temporal color map for $|u|$ (right) in the case $\mu = 35$.

The performance of the data-driven ROM described in the last section for the present value of μ is illustrated in Fig. 2. The left plot gives both the consistency and error estimates. The five FM-intervals that are needed in the considered time interval are guessed because the consistency estimate (which limits the length of the HODMD-intervals in the present case) abruptly decreases in these intervals. Since only four FM-intervals are needed and the computational cost of HODMD is negligible compared to that of the FM, the acceleration factor for the present case is slightly larger than 500.

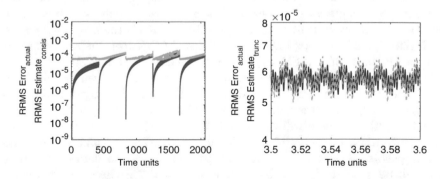

Fig. 2. Performance of the data driven ROM in the case $\mu = 35$. Left: the consistency estimate (purple) and the actual error (green) along the considered time interval. The blue solid line represents ε_4 threshold. Right: blow up of the truncation estimate (red) in the considered shorter time interval and actual error (green).

Now, we consider the case $\mu = 50$, for which the dynamics of $|u|$ are still periodic (thus, as in the previous case, the dynamics of u is quasi-periodic, representing a 2-torus in phase space), but much more complex than in the former case. The counterpart of Fig. 1 for the present case is given in Fig. 3. As can be seen, the dynamics in the attractor are much more complex than in the former case, as anticipated, but it still exhibits the same spatio-temporal reflection symmetry. The counterpart of Fig. 2 for the present case is given in Fig. 4. Note that, for convenience the total timespan in which the dynamics have been simulated has been reduced to $0 \leq t \leq 1000$. However, the number of required FM-intervals is 5, which means that the acceleration factor has been reduced to 200, which is consistent with the higher complexity of the present attractor. Also, the right plot shows that the error estimate remains smaller than 10^{-3} as required.

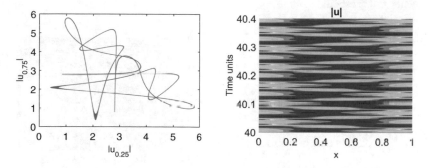

Fig. 3. Restricted phase plane (left) and spatio-temporal color map for $|u|$ (right) in the case $\mu = 50$.

Let us now consider the case $\mu = 40$, in which the attractor exhibits quasi-periodic dynamics (2-torus in phase space for $|u|$ and 3-torus for u). The counterpart of Fig. 1 is given in Fig. 5, where the quasiperiodic nature of the attractor is illustrated. Also, this figure suggest that a spatio-temporal reflection symmetry is still present, though uncovering reflection symmetries in quasi-periodic attractors is a major open problem in the field; see [14] (and references therein), where this problem is addressed using spatio-temporal Koopman decomposition, an extension of HODMD developed in [15].

When using the same tunable parameters as in the previous cases, the performance of the ROM (not illustrated here for the sake of brevity) is not as good. Then, we maintain the remaining parameters, except for the initial length of the subintervals I_1 and I_2, which is just doubled (namely, made equal to 1.6).

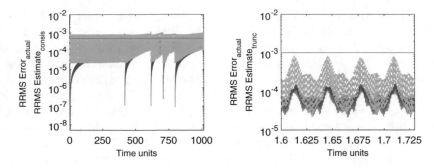

Fig. 4. Performance of the data driven ROM in the case $\mu = 50$. Left: the consistency estimate (purple) and the actual error (green) along the considered time interval. The blue solid line represents ε_4 threshold. Right: blow up of the truncation estimate (red) in the considered shorter time interval and actual error (green). The blue solid line represents ε_5 threshold.

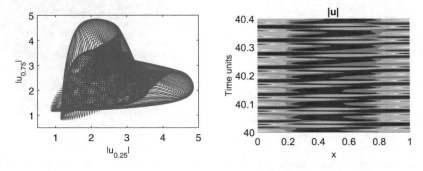

Fig. 5. Restricted phase plane (left) and spatio-temporal color map for $|u|$ (right) in the case $\mu = 40$

As in the former cases, the performance of the data-driven ROM is illustrated in Fig. 6, which is the counterpart of Fig. 2 for the present case. However, for the present value of μ, for convenience, the total timespan in the simulation has been reduced to 180. Note that the data driven ROM produces very good results, namely the truncation error is maintained smaller than 10^{-3} and, in fact, it reproduces quite well the actual error, as clearly seen in the right plot. However, the acceleration factor is now slightly larger than 3, namely much smaller than in the previous cases, which is due to the much higher complexity of the dynamics (see Fig. 5).

This value of the acceleration factor can be improved by a better selection of the tunable parameters, which has not been done here, where emphasizing robustness has been the main focus. This strategy has been followed also in the remaining cases considered above, where the effect of the selected values of the

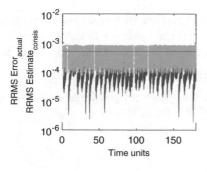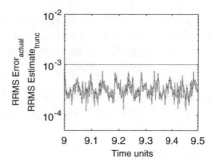

Fig. 6. Performance of the data driven ROM in the case $\mu = 40$. Left: the consistency estimate (purple) and the actual error (green) along the considered time interval. The blue solid line represents ε_4 threshold. Right: blow up of the truncation estimate (red) in the considered shorter time interval and actual error (green). The blue solid line represents ε_5 threshold.

tunable parameters on the performance of the method have not been addressed for the sake of brevity. This topic is beyond the scope of this paper, and left as the object of future research.

4 Conclusions

A purely data driven ROM has been constructed to accelerate the simulation of unsteady, nonlinear dynamical systems that is based on the combined use of a black box FM and extrapolation from HODMD. These two tools are used in interspersed time intervals, whose length is automatically selected by the ROM along the simulation. Thus, the method is adaptive. The method depends on several tunable parameters that must be calibrated for each specific application. However, this calibration can be made in rough terms, namely the method is quite robust in connection with the tunable parameters.

The performance of the data driven ROM has been tested considering the complex Ginzburg Landau equation with Neumann boundary conditions, which represents a very demanding dynamical system in this context. Also, as explained above, this equation has a great interest. The state variable u is complex and, because of the symmetries of the equations, if the orbit of u in phase space is densely covers a r-torus, then $|u|$ fills a $(r-1)$-torus (a closed curve if $r-1 = 1$). Three cases have been considered, in which the dynamics of $|u|$ are associated to transient behavior approaching (i) a fairly simple periodic attractor, (ii) a more complex quasi-periodic attractor, and (iii) a quasi-periodic attractor.

For the three considered cases, the method has been proved to perform as expected, and behave quite robustly in connection with the tunable parameters. Also, the CPU acceleration factors of the data driven ROM, compared to the FM were impressively high for the periodic attractors and fairly high for the quasi-periodic attractor. A better selection of the tunable parameters, omitted in this paper for the sake of brevity, would further improve these acceleration factors. This issue is left as the object of future work, as is the application of the data driven ROM to fluid dynamics problems, such as vortex shedding in the wake of a circular cylinder.

Acknowledgements. This research has been supported by the Spanish Ministry of Economy and Competitiveness, under grant TRA2016-75075-R.

References

1. Lovgren, A.E., Maday, Y., Ronquist, E.M.: A reduced basis element method for complex flow systems. In: Wesseling, P., Oñate, E., Périaux, J. (eds.) European Conference on Computational Fluid Dynamics, ECCOMAS CFD 2006, pp. 1–17. TU Delft, The Netherlands (2006)
2. Chinesta, F., Keunings, R., Leygue, A.: The Proper Generalized Decomposition for Advanced Numerical Simulations. SpringerBriefs in Applied Sciences and Technology. Springer, Berlin (2014)
3. Rapún, M.-L., Terragni, F., Vega, J.M.: Adaptive POD-based low-dimensional modeling supported by residual estimates. Int. J. Numer. Method Eng. **9**, 844–868 (2015)
4. Le Clainche, S., Vega, J.M.: Higher order dynamic mode decomposition. SIAM J. Appl. Dyn. Syst. **16**(2), 882–925 (2017)
5. Schmid, P.J.: Dynamic mode decomposition of numerical and experimental data. J. Fluid Mech. **656**, 5–28 (2010)
6. Le Clainche, S., Vega, J.M.: Analyzing nonlinear dynamics via data-driven dynamic mode decomposition-like methods. Complexity **2018**, 6920783 (2018)
7. https://github.com/LeClaincheVega/HODMD
8. Takens, F.: Detecting strange attractors in turbulence. In: Rand, D.A., Young, L.-S. (eds.) Lecture Notes in Mathematics, pp. 366–381. Springer (1981)
9. Press, W.H., Teukolsky, S.A., Vetterling, W.T., Flannery, B.P.: Numerical Recipes in C. Cambridge University Press, Cambridge (1988)
10. Le Clainche, S., Vega, J.M.: Higher order dynamic mode decomposition to identify and extrapolate flow patterns. Phys. Fluids **29**, 084102 (2017)
11. Aranson, I.S., Kramer, L.: The world of the complex Ginzburg-Landau equation. Rev. Mod. Phys. **74**, 100–142 (2002)
12. Haragus, M., Iooss, G.: Local Bifurcations, Center Manifolds, and Normal Forms in Infinite Dimensional Dynamical Systems. Springer, London (2010)
13. Terragni, F., Vega, J.M.: Construction of bifurcation diagrams using POD on the fly. SIAM J. Appl. Dyn. Syst. **13**, 339–365 (2014)
14. Sanchez Umbria, J., Net, M., Vega, J.M.: Analyzing multidimensional dynamics via spatio-temporal Koopman decomposition. Preprint (2019)
15. Le Clainche, S., Vega, J.M.: Spatio-temporal Koopman decomposition. J. Nonliner Sci. **28**, 1793–1842 (2018)

Soft Computing Techniques to Analyze the Turbulent Wake of a Wall-Mounted Square Cylinder

Christian Amor[1], José M. Pérez[1] (ID), Philipp Schlatter[2] (ID), Ricardo Vinuesa[2], and Soledad Le Clainche[1(✉)] (ID)

[1] School of Aerospace Engineering, Universidad Politécnica de Madrid,
28040 Madrid, Spain
`soledad.leclainche@upm.es`
[2] Linné FLOW Centre, KTH Mechanics, Stockholm, Sweden

Abstract. This paper introduces several methods, generally used in fluid dynamics, to provide low-rank approximations. The algorithm describing these methods are mainly based on singular value decomposition (SVD) and dynamic mode decomposition (DMD) techniques, and are suitable to analyze turbulent flows. The application of these methods will be illustrated in the analysis of the turbulent wake of a wall-mounted cylinder, a geometry modeling a skyscraper. A brief discussion about the large and small size structures of the flow will provide the key ideas to represent the general dynamics of the flow using low-rank approximations. If the flow physics is understood, then it is possible to adapt these techniques, or some other strategies, to solve general complex problems with reduced computational cost. The main goal is to introduce these methods as machine learning strategies that could be potentially used in the field of fluid dynamics, and that can be extended to any other research field.

Keywords: Soft computing · Fluid dynamics · Turbulence flow · CFD · Data science · POD · DMD

1 Introduction

Turbulence is found in the flow modeling within a wide range of industrial applications and in some cases it leads to undesired effects such as increased drag in vehicles, with the consequent raise in cost and pollution. For these reasons, among others, studying the flow physics of turbulent flows has been a research topic of high interest [3,4].

The main problem linked to turbulent flows lies in the large number of fairly disparate spatio-temporal scales describing the flow. To properly describe the physical insight and flow scales, it is necessary to collect data from a large

Supported by Universidad Politécnica de Madrid.

F. Martínez Álvarez et al. (Eds.): SOCO 2019, AISC 950, pp. 577–586, 2020.
https://doi.org/10.1007/978-3-030-20055-8_55

number of grid points, providing a resolution sufficiently high to solve in time and space the evolution of the flow structures. However, this fact implies that analyzing complex turbulent flows is very expensive in terms of computational time and memory, and therefore it is necessary to search for alternatives to overcome such issue.

Nowadays it is possible to find several low-rank approximations, generally extended for soft computing and based on machine learning strategies, also known as reduced order models (ROMs). Some examples include modal expansions based on a set of proper orthogonal modes, optimal in terms of energy [17], modal expansions based on some dynamic modes [18], or some other statistical techniques [1], et cetera. This paper presents some strategies generally used in the field of fluid dynamics to provide low-rank approximations of complex flows. These techniques include proper orthogonal decomposition (POD), singular value decomposition (SVD) [17], dynamic mode decomposition (DMD) [18] and some other variants of these methods, which are more sophisticated, providing better approximations of the original data. The application of these methods is illustrated in the analysis of the turbulent wake of a wall-mounted square cylinder, representing a skyscraper. The general details and conclusions about the performance of these techniques in the flow under consideration, provide an overview of how to use the method for low-rank approximations, in the analysis of flow structures, and the capability of using these methods to some other research field for soft computing or machine learning strategies.

This paper is organized as follows. Section 2 provides a brief description of the algorithm presented. Section 3 presents the main results. Finally, the conclusions are presented in Sect. 4.

2   Methodology

In this section several data-driven and equation-free methods will be introduced. The data analyzed will include spatio-temporal information, described by the velocity vector evaluated at time instant t_k, for convenience described as $v(x, y, z, t_k) = v_k$, where x, y and z represent the streamwise, normal and spanwise components. For simplicity, the data are organized into a *snapshot matrix*, considering K snapshots evolving in time as

$$X = V_1^K = [v_1, v_2, \ldots, v_k, v_{k+1}, \ldots, v_{K-1} v_K]. \tag{1}$$

In the following, these snapshots are equi-distant in time, with time interval Δt. Although it is remarkable that POD or SVD analyses do not need the data to be equally spaced in time.

2.1   Proper Orthogonal Decomposition

Proper orthogonal decomposition (POD) is a techniques that was introduced by Lumley [13] to extract coherent structures from turbulent flows. By means of optimizing the mean square of the field under consideration, POD decomposes data as an expansion of proper orthogonal modes $\boldsymbol{\Phi}_j(x, y, z)$, weighed with some temporal coefficients $c_j(t)$, as

$$\boldsymbol{v}(x, y, z, t) \simeq \sum_j c_j(t)\boldsymbol{\Phi}_j(x, y, z). \tag{2}$$

The classical algorithm behind POD is based on the covarianze of a state vector changing in time. The size of this covarianze matrix is based on the spatial degrees of freedom of the data, making prohibitively expensive the analysis of three-dimensional turbulent flows. To overcome such drawback, Sirovch [17] introduced the snapshot method, also known as singular value decomposition (SVD). Both terms, SVD and POD, are generally used interchangeably in the literature. Although, POD and SVD modes are the same, SVD is only one of the possible technique that can be used to extract POD modes. SVD (or POD) modes capture the directions of a matrix in which vectors can shrink or grow by means of computing the eigenvalues of a rectangular matrix. High dimensional solutions are represented by low-dimensional models by retaining only a few SVD modes, or in other words, the most energetic POD modes.

In order to calculate POD modes in expansion (2), SVD is applied to the snapshot matrix (1) decomposing spatio-temporal data into the POD modes $\boldsymbol{W}$ (orthogonal in space), temporal modes $\boldsymbol{T}$ and singular values $\boldsymbol{\Sigma}$ (representing the kinetic energy of the flow analyzed) as

$$\boldsymbol{V}_1^K \simeq \boldsymbol{W}\,\boldsymbol{\Sigma}\,\boldsymbol{T}^\top, \tag{3}$$

where $^\top$ denotes the matrix transpose, $\boldsymbol{W}^\top\boldsymbol{W} = \boldsymbol{T}^\top\boldsymbol{T}$ are the $N \times N$ unit matrices and the diagonal of matrix $\boldsymbol{\Sigma}$ contains the singular values $\sigma_1, \cdots, \sigma_K$. POD modes are ranked in decreasing order by their singular values, where the most energetic (highest singular value) modes represent the general flow dynamics.

A tolerance ε is set to calculate the standard SVD-error given by the most relevant POD modes retained in the approximation of the original flow field. This error is defined as

$$\sigma_{N+1}/\sigma_1 \le \varepsilon. \tag{4}$$

It is remarkable that this tolerance, set by the user, could represent the level of noise (i.e.: in experimental data) or the relative size of the flow structures that will approximate the general flow field. Finally, applying SVD to each group of data along the temporal component, or each one of the spatial components (organizing snapshot matrix varying in space rather than in time), it is possible to obtain a high order singular value decomposition (HOSVD) [23]. Singular values are calculated with respect to each variable of the flow field, thus it is possible to filter out modes associated to each direction, obtaining further rank reductions to approximate the original data.

580   C. Amor et al.

2.2 Spectral POD

Spectral POD is a method introduced for the analysis of turbulent flows [20,21]. The algorithm calculates POD modes in the frequency domain. To this end, the snapshot matrix (1) is segmented in several blocks and Fast Fourier Transform is performed on each one of the blocks. Thus, spectral POD modes are calculated from a set of realizations of the temporal Fourier transform of the flow field. Finally, SVD is applied over these blocks. The main benefit of spectral POD lies in providing an orthogonal basis of modes at discrete frequencies that are optimally ranked in terms of energy.

2.3 Dynamic Mode Decomposition

Dynamic mode decomposition (DMD) [16,18,19] is a method that decomposes spatio-temporal data as an expansion of DMD modes $\boldsymbol{v}_m(x,y,z,t)$ weighted with some amplitudes a_m, as

$$\boldsymbol{v}(x,y,z,t_k) \simeq \sum_{m=1}^{M} a_m \boldsymbol{u}_m e^{(\delta_m+i\omega_m)t_k}, \tag{5}$$

for $k = 1,\cdots,K$, where ω_m and δ_m are the oscillation frequency and growth rate of the modes. The algorithm to calculate DMD modes was introduced by Schmid [18], although for simplicity in this paper this algorithm is conveniently defined as in Ref. [10], using only two steps, to connect the different methods presented below. As a first step, SVD is applied to the snapshot matrix (1). Thus DMD could be considered as a step further POD (or SVD), to calculate modes separated in time. For convenience Eq. (3) is rewritten as $\boldsymbol{V}_1^K = \boldsymbol{W}\hat{\boldsymbol{V}}_1^K$, with $\hat{\boldsymbol{V}}_1^K = \boldsymbol{\Sigma T}^\top$. As second step, the method considers the Koopman matrix $\boldsymbol{R}$ [5] and the following linear relationship between snapshots

$$\hat{\boldsymbol{V}}_2^K \simeq \boldsymbol{R}\hat{\boldsymbol{V}}_1^{K-1}. \tag{6}$$

The eigenvalues of $\boldsymbol{R}$ represent the DMD frequencies and growth rates, while its eigenvectors are used to construct the DMD modes $\boldsymbol{u}_m$.

Higher order dynamic mode decomposition (HODMD) [10] is a recent extension of DMD introduced for the analysis of complex flows and noisy data [6,8,9,11,12]. HODMD replaces the linear relationship between snapshots (6) by the following *high order Koopman assumption*

$$\hat{\boldsymbol{V}}_{d+1}^K \simeq \boldsymbol{R}_1\hat{\boldsymbol{V}}_1^{K-d} + \boldsymbol{R}_2\hat{\boldsymbol{V}}_2^{K-(d-1)} + \ldots + \boldsymbol{R}_d\hat{\boldsymbol{V}}_d^{K-1}. \tag{7}$$

This algorithm, based on the Taknes' delayed embedded theorem [22], considers d time delayed snapshots (sub-matrices), similar to the window-shift process carried out in Power Spectral Density (PSD). The eigenvectors and eigenvalues of the Koopman operators $\boldsymbol{R}_1,\cdots,\boldsymbol{R}_n$ represent DMD modes, frequencies and growth rates.

The relationship between the spectral POD and the HODMD algorithms is remarkable. Spectral POD can be understood as a variant of HODMD, where FFT is carried out in each one of the sub-snapshot matrices (i.e.: $\hat{V}_1^{K-d}$, $\hat{V}_2^{K-(d+1)}$, et cetera) presented in Eq. (7). Then Fourier modes are averaged four each group, instead of solving the eigenvalue problem of DMD. Note that the main drawback of spectral POD is the higher computational cost when compared with the HODMD algorithm.

2.4 Multi-scale Analyses: POD and DMD

The main idea behind multi-scale analyses is to filter out, or to separate, the wide variety of high and low frequencies. It is possible to find multi-scale analyses combined with the POD and DMD algorithms. Some examples include Multiresolution DMD (mrDMD) [7], which is a method that separates slow and fast modes (low- and high-frequency modes) recursively from the collection of snapshots, initially represented in (1). A second example is muti-resolution POD (mrPOD) [15], where a filter separates low- and high-frequency modes. Then SVD is applied to the data, already filtered in time.

3 Data Analysis

The database modeling the turbulent wake behind a wall-mounted square cylinder is generated containing 306 snapshots equi-distant in time. Figure 1-left shows a three-dimensional view of the computational domain and the flow structures present in the wake of the square cylinder. This flow cases constitutes and excellent platform to analyze complex problems in urban environments, such as effects of gusts and pollutant dispersion [14]. Studying the flow physics in this database is highly complex for several reasons. On the one hand, the turbulent wake of the square cylinder comprises a large number of flow structures with small spatial size, oscillating in time with an almost incommensurable number of frequencies. On the other hand, for each snapshot, the memory size is very large ($\sim Gb$), meaning that to analyze the whole data set it is necessary to use high-performance computing (HPC) resources.

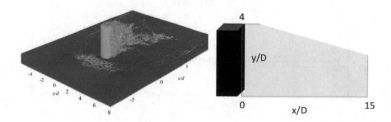

Fig. 1. Left: three-dimensional wake of a wall-mounted square cylinder. Iso-surfaces of instantaneous streamwise velocity $\hat{u} = 0.1U_\infty$. Figure extracted from Ref. [24]. Right: two-dimensional plane extracted in the spanwise symmetry plane.

Using soft computing techniques it is possible to reduce the degrees of freedom of each file, and consequently reducing the computer requirements in this
analysis. The methods previously introduced have been applied to a set of two-
dimensional snapshots, defining the near-wake of the wall-mounted square cylinder. Figure 1-right shows the area of the computational domain corresponding
to the two-dimensional planes, extracted in the symmetry plane of the spanwise
component. The number of spatial points defining streamwise and wall normal
components is 221 × 441. Although the grid points defining this sub-section of
the computational domain is relatively small, this example serves to illustrate
the performance of the method selected for soft computing and the analysis of
turbulent flows.

The first analysis carried out is a high order singular value decomposition.
In other words, SVD or POD modes are calculated orthogonal in space, called
as spatial POD modes, and in time and space, denoted as X-temporal POD
modes and Y-temporal POD modes. Figure 2-top shows the evolution of the singular values. As expected, the total number of singular values calculated for the
spatial, X-temporal and Y-temporal SVD modes correspond to the number of
snapshots (306), and the grid points contained in the spanwise (441) and wall-
normal (221) components, respectively. The amplitude (kinetic energy or singular value) related to these modes decreases very slowly, especially for tolerance
values smaller than 10^{-2}, reflecting the high spatio-temporal complexity of the
turbulent flow analyzed. In other words, obtaining relatively good reconstructions of the original solution implies retaining a large number of POD modes.
For instance, using the tolerance $\varepsilon_1 = \sigma_i/\sigma_1 = 10^{-2}$, the number of X-temporal
and Y-temporal POD modes retained is $\simeq$50 as seen in Fig. 2-bottom. Although,
for retaining similar values of temporal modes it is necessary to rise slightly the
tolerance of the spatial modes, meaning that the temporal complexity is even
larger than the spatial complexity representing the small flow scales.

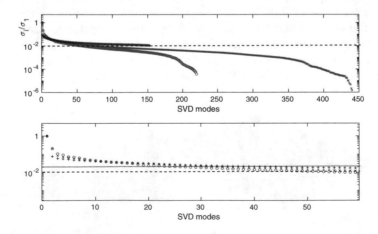

Fig. 2. Top: singular values from the HOSVD analysis. Bottom: zoomed-in view.

Each POD mode is related to a group of frequencies, which will be analyzed in the following section using DMD and spectral POD techniques. Figure 3 shows some representative examples of the spatial POD modes, reflecting the variations of the flow scales. The figure shows the 1st, 5th and 50th most energetic modes. It can be observed that the most energetic mode is composed by a large size flow structures close to the cylinder wall, representing the mean flow. In the remaining modes, it is possible to distinguish some oscillating flow structures related to vortex shedding, mostly focused in the area closest to the cylinder and the horizontal wall. In the 5th mode the size of the flow scales is larger than in the 50th mode, where it is also possible to distinguish a large number of very small flow scales, which could be somehow compared with noise artifacts. This fact shows that the most energetic POD modes represent the large size flow scales driving the flow dynamics related to global instabilities, suitable to describe the flow physics. On the contrary, the smallest flow scales are represented in the less energetic modes. Similar results are found in the X-temporal and Y-temporal POD modes calculated using HOSVD, not shown for the sake of brevity. Limiting the number of POD modes retained to reconstruct the original solution, will provide a good general idea of the data analyzed. SVD, HOSVD and POD are suitable techniques to reduce the degrees of freedom of the solution analyzed, to create machine learning and soft computing strategies. HODMD is applied in the second analysis, using the tolerance previously mentioned for SVD. With the aim at studying the general dynamics of the flow, the method is applied iteratively, only retaining the global instabilities related to the large flow scales. The number of SVD (or POD) modes retained at each iteration decreases, thus, as already shown in Fig. 3, the degrees of freedom of the data analyzed is limited to the modes representing the largest flow scales. Similarly, the number of DMD modes retained in expansion (5) also decreases in each iteration. The shape of these modes is slightly modified at each iteration, since, the number of small size flow scales, related to the POD modes, tends to disappear. When the number of POD modes retained in iterations $(i-1)$ and (i) are the same, the iterative process stops. After 13 iterations, the number of spatial, X-temporal and Y-temporal POD modes retained is 14, 18 and 24, respectively, and the number of DMD modes retained is 12 plus their conjugate complex. Among all these DMD modes it is possible to differentiate a mode with high amplitude and non-dimensional frequency St ~ 0.01 (Strouhal number defined as St $= \frac{fd}{U}$, with f, d and U defining the frequency, diameter of the body and free stream velocity). This value

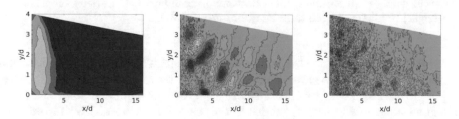

Fig. 3. POD modes. From left to right: 1st, 5th and 50th most energetic modes.

is in good agreement with the predictions carried out by Vinuesa *et al.* [24], who performed FFT analysis in some representative points of the computational domain. The mode related to the mean flow ($\omega = 0$) and the dominant DMD mode with $St \sim 0.01$ are presented in Fig. 4-top. As seen, the shape of these two modes is very similar to the POD modes previously presented. The vortex shedding found in the wake of the cylinder, related to the 5th POD modes is also found in the DMD mode with $St = 0.01$, although the main difference lies in the size of the flow structures. In DMD, the mode is represented by large flow scales and a single frequency. On the contrary, the POD mode, also represented by large flow scales, is composed by a large number of frequencies, making that the aspect of the mode is more noisy than in DMD. Finally, HODMD (non-iterative) has been applied to reconstruct a high fidelity DMD expansion (5) using only 153 snapshots. The method retains 230 modes, representing large and small size flow scales. Then, the temporal term of the DMD expansion (5) has been adjusted to extrapolate the solution to snapshot 306 (prediction). Figure 4-bottom compares the original solution with the extrapolation, where the relative mean square error made in this prediction is $\sim$30%. It can be observed that only the large flow scales are predicted, but this result is sufficiently good to study the general flow dynamics of the flow, reducing the computational requirements by a factor of 2.

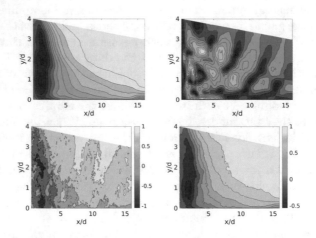

Fig. 4. DMD analyses. Top: DMD modes with frequency $St = 0$ (mean flow) and $St = 0.01$ (highest amplitude mode). Bottom: original flow field in (left) snapshot 306 and (right) prediction of the same snapshot using DMD.

Spectral POD is the third analysis carried out. This algorithm divides the snapshot matrix (1) into several blocks composed by small number of snapshots, in particular 144 blocks of 20 snapshots each for this example. Thus the number of spectral POD (sPOD) modes is 144, each one composed by 11 different modes with different frequencies. The mode with highest amplitude captured by

DMD, St = 0.01, corresponds to the third frequency found in each sPOD mode. Figure 5(a) and (b) show the first and second most energetic sPOD modes corresponding to such frequency. As seen the first sPOD mode is similar to the first DMD mode, while the second sPOD mode is more complex, composed by a larger number of small-scale structures. This technique provides results similar to DMD using modes orthogonal in space, similar to POD, providing good low-rank approximations of the original flow field. Finally, it is remarkable that it is possible to combine this technique with mrPOD or mrDMD, where the high-frequency modes representing the small flow scales could be removed using a low-pass filter before applying these techniques. Figure 5(c) shows the 50th most energetic mrPOD mode. Comparing this solution with standard POD it is possible to see how the filter has removed the small size flow scales from the solution (Fig. 3). Nevertheless, prior knowledge of the dominant frequencies is required in order to avoid filtering high amplitude modes representing the flow.

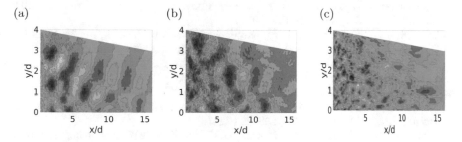

Fig. 5. (a) and (b) Spectral POD analysis: first (a) and second (b) sPOD modes with frequency St = 0.01. (c) mrPOD analysis: 50th mrPOD mode.

4   Conclusions

This paper presents some techniques suitable to analyze turbulent flows and providing low-rank approximations. These techniques, well known in the field of fluid dynamics, are based on SVD and DMD algorithms. The paper includes a short discussion about the relationship between the size of the flow scales and the POD, DMD, sPOD and mrPOD modes.

References

1. Antonia, R.A.: Conditional sampling in turbulence measurements. Ann. Rev. Fluid Mech. **13**, 131–156 (1981)
2. Fischer, P.F., L., J.W. Kerkemeier, S.G.: nek5000 (2008). https://nek5000.mcs.anl.gov
3. Haller, G.: An objective definition of a vortex. J. Fluid Mech. **525**, 1–26 (2005)
4. Hunt, J.C.R., Wray, A., Moin, P.: Eddies, stream, and convergence zones in turbulent flows. Center for Turbulence Research Report, CTR-S88 (1988)

5. Koopman, B.: Hamiltonian systems and transformations in Hilbert space. Proc. Natl. Acad. Sci. U.S.A. **17**, 315–318 (1931)
6. Kou, J., Le Clainche, S., Zhang, W.: A reduced-order model for compressible flows with buffeting condition using higher order dynamic mode decomposition. Phys. Fluids **30**(1), 016103 (2018)
7. Kutz, J.N., Fu, X., Brunton, S.L.: Multiresolution dynamic mode decomposition. SIAM J. Appl. Dyn. Syst. **15**(2), 713–735 (2016)
8. Le Clainche, S., Ferrer, E.: A reduced order model to predict transient flows around straight bladed vertical axis wind turbines. Energies **11**(3), 566–578 (2018)
9. Le Clainche, S., Pérez, J.M., Vega, J.M.: Spatio-temporal flow structures in the three-dimensional wake of a circular cylinder. Fluid Dyn. Res. (2018). https://doi.org/10.1088/1873-7005/aab2f1
10. Le Clainche, S., Vega, J.M.: Higher order dynamic mode decomposition. SIAM J. Appl. Dyn. Syst. **16**(2), 882–925 (2017)
11. Le Clainche, S., Vega, J.M.: Higher order dynamic mode decomposition. Phys. Fluids **16**(2), 882–925 (2017)
12. Le Clainche, S., Vega, J.M., Soria, J.: Higher order dynamic mode decomposition for noisy experimental data: flow structures on a zero-net-mass-flux jet. Exp. Therm. Fluid Sci. **88**, 336–353 (2017)
13. Lumley, J.L.: The structure of inhomogeneous turbulent flows. In: Yaglam, A.M., Tatarsky, V.I. (eds.) Proceedings of the International Colloquium on the Fine Scale Structure of the Atmosphere and Its Influence on Radio Wave Propagation. Doklady Akademii Nauk SSSR, Nauka, Moscow (1967)
14. Monnier, B., Goudarzi, S.A., Vinuesa, R., Wark, C.: Turbulent structure of a simplified urban fluid flow studied through stereoscopic particle image velocimetry. Bound.-Layer Meteorol. **166**, 239–268 (2018)
15. Mendez, M.A., Balabane, M., Buchlin, J.M.: Multi-scale proper orthogonal decomposition of complex fluid flows. J. Fluid Mech. (submitted). arXiv:1804.09646v3 [physics.flu-dyn]
16. Rowley, C.W., Mezić, I., Bagheri, S., Schlatter, P., Henningson, D.S.: Spectral analysis of nonlinear flows. J. Fluid Mech. **641**, 115–127 (2009)
17. Sirovich, L.: Turbulence and the dynamics of coherent structures. Parts I–III. Q. Appl. Math. **45**(3), 561–571 (1987)
18. Schmid, P.: Dynamic mode decomposition of numerical and experimental data. J. Fluid Mech. **656**, 5–28 (2010)
19. Schmid, P.J., Sesterhenn, J.L.: Dynamic mode decomposition of numerical and experimental data. In: Bulletin of American Physical Society, 61st APS Meeting, San Antonio, p. 208 (2008)
20. Schmidt, O., Towne, A., Colonius, T., Cavalieri, A., Jordan, P., Bres, G.: Wavepackets and trapped acoustic modes in a Mach 0.9 turbulent jet: a global stability analysis. J. Fluid Mech. **825**, 1153–1181 (2017)
21. Towne, A., Schmidt, O.T., Colonius, T.: Spectral proper orthogonal decomposition and its relationship to dynamic mode decomposition and resolvent analysis. J. Fluid Mech. **847**, 821–867 (2018)
22. Takens, F.: Detecting strange attractors in turbulence. In: Rand, D.A., Young, L.-S. (eds.) Lecture Notes in Mathematics, pp. 366–381. Springer (1981)
23. Tucker, L.R.: Some mathematical notes on three-mode factor analysis. Psychometrika **31**(3), 279–311 (1966)
24. Vinuesa, R., Schlatter, P., Malm, J., Mavriplis, C., Henningson, D.S.: Direct numerical simulation of the flow around a wall-mounted square cylinder under various inflow conditions. J. Turbulence **16**(6), 555–587 (2015)

Generating Three-Dimensional Fields from Two-Dimensional Soft Computing Strategies

José Miguel Pérez⬤, Soledad Le Clainche⬤, and José Manuel Vega⁽✉⁾⬤

School of Aerospace Engineering, Universidad Politécnica de Madrid,
28040 Madrid, Spain
josemanuel.vega@upm.es

Abstract. The main goal of this work is to provide a new tool to reconstruct three-dimensional data from a few two-dimensional planes. To this purpose, spatio-temporal higher order dynamic mode decomposition (HODMD) is applied to a group of vertical and horizontal planes, located at different positions in space. Firstly, the method extracts the relevant frequencies and wavenumbers related to the field analyzed, providing two-dimensional HODMD expansions. Secondly, a system of equations is constructed in the vertical planes using the previous information. Solving this equation system, it is possible to expand the two-dimensional HODMD solutions to recover a three-dimensional HODMD expansion, representing the complete three-dimensional flow field. This method has been successfully tested to reconstruct a three-dimensional toy problem with great precision (error smaller than 10^{-10}). The method could be potentially used in the analysis of turbulent flows (numerical and experimental data), extracting relevant three-dimensional information from the two-dimensional data with reduced computational cost (soft computing).

Keywords: Three-dimensional reconstruction · HODMD ·
Reduced order models

1 Introduction

Since the past, researchers have paid attention to study in detail complex flows, since they are presented in a wide range of industrial applications and describing events found in nature. Some examples of complex flows include flows transitioning to turbulence or even turbulent flows, i.e.: the flow past cars or aircraft, in heat exchanges, in power plants, the wake of flying insects, etc. A good way to understand the physical phenomena, hitherto unknown, describing these flows, is to find coherence in fluid motions.

Analyzing three-dimensional flow fields provides information suitable for studying the flow physics, with the aim at looking for global instabilities and flow structures finding coherence in fluid motion. However, analyzing three-dimensional data is very expensive in terms of computational time and memory.

© Springer Nature Switzerland AG 2020
F. Martínez Álvarez et al. (Eds.): SOCO 2019, AISC 950, pp. 587–595, 2020.
https://doi.org/10.1007/978-3-030-20055-8_56

The construction of reduced order models (ROMs), also known as soft computing techniques, reduce the number of degrees of freedom in the data analyzed,
providing useful information requiring smaller computational resources. Using
these low-order models is especially necessary in the analysis of turbulent flows,
where databases of ten or even hundred Terabytes are analyzed.

Several techniques have been developed to find structured and well-organized
flow patterns. Some examples include, proper orthogonal decomposition (POD)
[22], representing the flow as an expansion of the most energetic orthogonal
modes, techniques based on statistical analysis [1], or techniques based on the
decomposition of the second invariant of velocity gradient [5,6]. Dynamic mode
decomposition (DMD) [20] is a technique also suitable for the analysis of flow
structures in complex flows. DMD provides an approximation of the Koopman
modes [7,19], providing information about the general dynamics and flow physics
of the field analyzed. This technique have been successfully applied for the analysis of flow structures in data coming from experiments and numerical simulations
[3,4,10,14,21]. However, DMD may present some difficulties when the data analyzed are limited in space (i.e.: computational domain composed by a few spatial
points), or in the analysis of complex or noisy flows [18]. For instance, carrying
out experiments implies collecting data limited to a spatial area. Depending on
the type of experiment performed and the design of the experimental facility,
experimental measurements are carried out either in a few spatial points (i.e.:
Laser Doppler Velocimetry), or in a few planes (i.e.: Particle Image Velocimetry).
In order to overcome the drawback that DMD may find, higher order dynamic
mode decomposition (HODMD), an extension of the algorithm DMD, can be
used instead.

The main goal of this Chapter is to present a new application of HODMD
as a tool for reconstructing three-dimensional data starting from a limited
amount of information. This method, equation-free and data-driven, is generally applied for the analysis of flow structures in complex flows [9,13], although
it can be extended to any field [11,12]. The possibility of extracting three-
dimensional information from two-dimensional data, opens new horizons for
analyzing large databases of turbulent flows or experimental data. The computational resources required for two-dimensional analyses are much smaller than for
three-dimensional tests. Thus, using this new method, it is possible to consider
the two-dimensional data as a ROM, approximating the real three-dimensional
information, calculated with a small computational cost. This new tool sheds
light on new applications for soft computing in fluid dynamics, that could be
potentially extended to any kind of field.

This Chapter is organized as follows. Section 2 introduces the algorithm
for HODMD, while Sect. 3 describes the methodology to reconstruct three-
dimensional fields from two-dimensional data. Finally, Sects. 5 and 6 presents
the main results and conclusions of this work.

2   Spatio-Temporal Higher Order Dynamic Mode Decomposition

Higher order dynamic mode decomposition (HODMD) [15] is an extension of DMD suitable for the analysis of complex flows. Similarly to DMD, this method decomposes spatio-temporal data $\mathbf{v}(x, y, z, t)$ as an expansion of DMD modes as

$$\mathbf{v}(x, y, z, t) \equiv \sum_{m=1}^{M} a_m^t \mathbf{u}_m^t(x, y, z) e^{(\delta_m + i\omega_m)t}. \tag{1}$$

where $\mathbf{u}_m^t$ represents the DMD modes, and a_m^t, ω_m and δ_m are their corresponding amplitudes, temporal frequencies and growth rates. The value M, the number of modes retained in the expansion, is defined as the spectral complexity, while the spatial complexity N is given by the dimensions of the span of the DMD modes. When the spatial complexity N is smaller than the spectral complexity M, DMD may find some difficulties or even fail, allowing HODMD as a good alternative to overcome such drawbacks. It is possible to find a wide variety of cases in which $N < M$, i.e.: in transient flows [18], in the analysis of data coming from real databases (LiDAR experiments, carried out to measure wind velocity [11]), and in the transitory region of a numerical simulation [2,8,9,16].

It is possible obtaining spatio-temporal expansions using HODMD algorithm sequentially in time and space [13], or using a more advanced version of this method, called as Spatio-temporal Koopman Decomposition (STKD) [17] where HODMD is applied in time and space in parallel. Although, for simplicity this Chapter only considers the sequential algorithm.

Applying HODMD to the DMD modes $a_m^t \mathbf{u}_m^t(x, y, z)$, obtained in (1), it is possible to obtain the following spatial DMD expansion (along the spanwise direction)

$$a_m^t \mathbf{u}_m^t(x, y, z) \simeq \sum_{k=1}^{K} a_{mk}^{tz} \mathbf{u}_{mk}^{tz}(x, y) e^{(\alpha_k + i\beta_k)z}, \tag{2}$$

where α_k and β_k are the spatial growth rates and wavenumbers of the spanwise direction, respectively, $\mathbf{u}_{mj}^{tz}(x, y)$ are the *sequential* DMD modes and a_{mn}^{tz} their corresponding amplitudes. In this equation K is the spectral complexity in the spanwise direction.

Introducing Eq. (2) into Eq. (1), leads to the following general spatio-temporal DMD expansion,

$$\mathbf{v}(x, y, z, t) \simeq \sum_{m=1}^{M} \sum_{k=1}^{K} a_{mk}^{tz} \mathbf{u}_{mk}^{tz}(x, y) e^{(\alpha_k + i\beta_k)z + (\delta_m + i\omega_m)t}. \tag{3}$$

Similarly, it is possible to obtain spatio-temporal DMD expansions related to additional spatial components as,

$$\mathbf{v}(x, y, z, t) \simeq \sum_{m=1}^{M} \sum_{k=1}^{K} \sum_{l=1}^{L} a_{mkl}^{tzx} \mathbf{u}_{mkl}^{tzx}(y) e^{(\gamma_l + i\sigma_l)x + (\alpha_k + i\beta_k)z + (\delta_m + i\omega_m)t}. \tag{4}$$

where γ_l and σ_l are the spatial growth rates and wavenumbers of the streamwise direction (x), respectively, $\mathbf{u}_{mkl}^{tzx}(y)$ are the *sequential* DMD modes and a_{mnl}^{tzx} their corresponding amplitudes. In this case L is the spectral complexity in the spanwise direction (z).

2.1   HODMD Algorithm

HODMD algorithm considers the following snapshot matrix

$$\mathbf{V}_1^K = [\mathbf{v}_1 \mathbf{v}_2 \cdots \mathbf{v}_k \cdots \mathbf{v}_K], \tag{5}$$

whose snapshots $\mathbf{v}_k = \mathbf{v}(t_k)$ are equi-distant in time with time interval Δt. Then, the method considers only two steps:

- **Step 1: SVD.** A singular value decomposition (SVD) is applied to the snapshot matrix. So, the spatio-temporal data are decomposed into spatial modes, singular values and temporal modes. The method retains N SVD modes (where N is the spatial complexity), based on a tolerance ε_1 (set by the user), which could be somehow related with the level of noise of the data analyzed or the spatial size of the flow structures.
- **Step 2: DMD-d.** Standard DMD algorithm [20] is combined to the Takens' delayed embedding theorem [23]. This theorem is based on the "sliding-window" process of PSD (power spectral density), considering d time lagged snapshots. This step is applied to the temporal SVD modes, weighted with their corresponding singular values. This is called the DMD-d algorithm and, when d is set to 1, the algorithm is equivalent to standard DMD [20]. A second tolerance ε_2 (also set by the user), determines the number of M DMD modes retained in the expansion (spectral complexity). More details of the algorithm can be found in [15].

3   Methodology for Constructing Three-Dimensional Approximations from a Set of Planes

Given a three-dimensional function, some horizontal and vertical planes are extracted along the computational domain containing relevant information, as shown in Fig. 1. DMD expansion (3) is applied to analyze the data in the vertical planes, while the DMD expansion (4) is applied to analyze the horizontal planes. In the intersection of these planes, these two expansions are identical, making possible to construct the following identity

$$a_{mk}^{tz} \mathbf{u}_{mk}^{tz}(x_i, y_j) = \sum_{l=1}^{L} a_{mkl}^{tzx} \mathbf{u}_{mkl}^{tzx}(y_j) e^{(\gamma_l + i\sigma_l)x_i}, \tag{6}$$

where x_i and y_j are defined in the intersection between planes. This expression can be generalized for any value of the wall-normal component y along the vertical planes (expansion (4)). In that case the values of the modes $\mathbf{u}_{mkl}$ evaluated in every y_j are unknowns. These unknown variables can be calculated from the following linear system of equations,

$$B_{ij}^{mk} = A_{im}^{kl} X_{mj}^{kl}, \tag{7}$$

where $B_{ij}^{mk} = a_{mk}^{tz} \mathbf{u}_{mk}^{tz}(x_i, y_j)$, $A_{im}^{kl} = a_{mkl}^{tzx} e^{(\gamma_l + i\sigma_l)x_i}$ and $X_{mj}^{kl} = \mathbf{u}_{mkl}^{tzx}(y_j)$ $\forall i, j$. The solution of this system of equations provides the DMD modes in the full three-dimensional domain, making possible to reconstruct the full three-dimensional field analyzed.

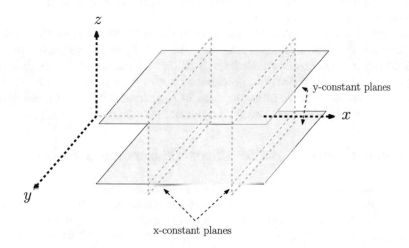

Fig. 1. Two-dimensional planes extracted in a three-dimensional computational domain.

4   Definition of the Three-Dimensional Toy Model

The reconstruction described in the previous section has been applied to a three-dimensional toy problem. Based on Eq. (4), the modes and amplitudes of this toy model are defined as

$$\mathbf{u}_{mkl}(y) = \frac{1}{\sqrt{2\pi(1 + l^2 + k^2)}} e^{-y^2/(2(1+l^2+k^2))}, \tag{8}$$

$$a_{mkl} = \frac{1}{1 + l^4 + k^4 + m^2}.$$

These modes are Gaussian functions and their integrals over the whole domain are normalized as $\int_{-\infty}^{\infty} |\mathbf{u}_{mkl}(y)| dy = 1, \forall m, k, l$.

Fifteen temporal modes were considered in the simulation with temporal frequency equal to $\omega_{\pm 7} = \pm 4.1$, $\omega_{\pm 6} = \pm 2.0$, $\omega_{\pm 5} = \mp 0.1$, $\omega_{\pm 4} = \pm 3.1$, $\omega_{\pm 3} = \mp 1.0$, $\omega_{\pm 2} = \pm 2.1$ and $\omega_0 = 0$. In all cases, $\delta_m = 0 \ \forall m$. Five modes were considered in the spanwise direction, with zero spatial growth rate ($\alpha_k = 0 \ \forall k$) and spatial frequency equal to $\beta_{\pm 2} = \pm 2$, $\beta_{\pm 1} = \pm 1$ and $\beta_0 = 0$. Similarly, five spatial modes were considered in the streamwise direction.

Two types of toy models have been defined. The first one is periodic along the streamwise direction, with $\gamma_l = 0 \ \forall l$, while the problem is non-periodic for the second case, with $\gamma_{\pm 2} = \pm 0.2$, $\gamma_{\pm 1} = \pm 0.1$ and $\gamma_0 = 0$. Regarding the spatial frequencies in the streamwise direction, these are the same in both cases; $\sigma_{\pm 2} = 2$, $\sigma_{\pm 1} = \pm 1$ and $\sigma_0 = 0$.

The size of the computational domain in the streamwise, spanwise and normal direction was equal to $L_x = \pm 8$, $L_z = \pm 8$ and $L_y = \pm 6$, respectively. A set of 200 equi-distant snapshots with $\Delta t = 0.201005$ where considered in all studies. Regarding spatial discretization, equispaced nodes were considered in all directions. The number of these nodes was: 36 in the streamwise (x) direction, 36 in the spanwise(z) direction and 30 in the normal direction (y).

5   Reconstruction of the Three-Dimensional Fields

The three-dimensional reconstruction of the original function has been carried out. The data analyzed consider a group of 5 vertical planes distributed equidistant along the streamwise direction (x) and 7 horizontal planes distributed equidistant along the normal direction (y). The parameters used for the temporal and spatial HODMD analyses are $d = 15$ and $d = 5$, respectively, with tolerances $\varepsilon_1 = 10^{-4}$ and $\varepsilon_2 = 10^{-6}$ in both cases. In the analyses carried out in both test cases, the number of temporal and spatial modes matched the theoretical values previously described in Sect. 4.

Figure 2 shows the comparison between the original and the reconstructed field, in the first test case (periodic function in the streamwise direction). As can be seen in the figure, there is a good agreement between the reconstructed and the original field. The root-mean-square deviation between the original and reconstructed field was $\approx 10^{-11}$, while the module of the imaginary part of the reconstructed solution was smaller than $\approx 10^{-12}$. Similar results were obtained for the second test case, were the error made in the approximation is also smaller than $\approx 10^{-11}$ (Fig. 3).

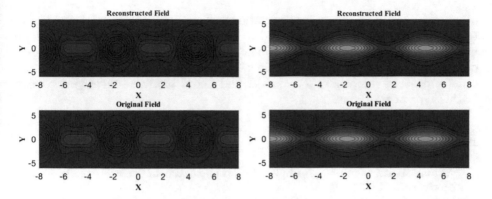

Fig. 2. Toy model 1 (periodic): original and reconstructed solution at $z = -1.6$ at two different time instants. Left: time instant 20. Right: time instant 40. Data analyzed in 5 vertical planes along the streamwise direction and 7 horizontal planes along the normal direction.

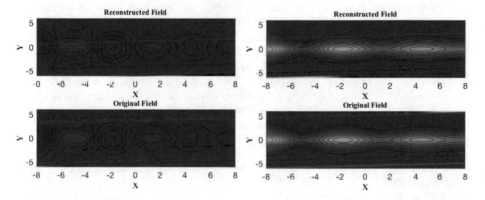

Fig. 3. Counterpart of Fig. 2 for the toy model 2 (non-periodic).

6 Conclusions

A new tool for the reconstruction of three-dimensional field from a few two-dimensional planes of the original field has been presented in this Chapter. To this aim, spatio-temporal HODMD has been applied to a set of planes parallel and perpendicular to the streamwise direction, which allows to obtain the relevant frequencies, wavenumbers and amplitudes. This leads to a linear system of equations, of reduced dimension, whose solution allows to calculate the spatio-temporal modes $\mathbf{u}_{mkl}(y)$ for all available y values, from the data obtained from HODMD. This method has been successfully applied to a three-dimensional toy problem, obtained an error about 10^{-10}.

One of the reason of these good results, is that we considered a field which is well characterized by a small set of well differentiated temporal and spatial modes. As future work, we intend to use this methodology in turbulent flows obtained from direct numerical simulation and/or experiments, with the aim at testing the capability of this method to deal with realistic problems with high number of spatio-temporal modes. Finally, the capability of reconstructing three-dimensional fields from two-dimensional planes can be used to define reduced order models of complex flow dynamics and understand the underlying physics in this type of problems.

References

1. Antonia, R.A.: Conditional sampling in turbulence measurements. Ann. Rev. Fluid Mech. **13**, 131–156 (1981)
2. Beltran, V., Le Clainche, S., Vega, J.M.: Temporal extrapolation of quasi-periodic solutions via DMD-like methods. In: Proceedings of AIAA Fluid Dynamics Conference Atlanta, Georgia. AIAA paper 2018-3092 (2018). https://doi.org/10.2514/6.2018-3092.c1
3. Ferrer, E., De Vicente, J., Valero, E.: Low cost 3D global instability analysis and flow sensitivity based on dynamic mode decomposition and high-order numerical tools. Int. J. Numer. Methods Fluids **76**(3), 169–184 (2014)
4. Ghebali, S., Garicano-Mena, J., Ferrer, E., Valero, E.: Identification of flow structures in fully developed canonical and wavy channels by means of modal decomposition techniques. J. Phys.: Conf. Ser. **1001**(1), 012010 (2018)
5. Haller, G.: An objective definition of a vortex. J. Fluid Mech. **525**, 1–26 (2005)
6. Hunt, J.C.R., Wray, A., Moin, P.: Eddies, stream, and convergence zones in turbulent flows. Center for Turbulence Research Report CTR-S88 (1988)
7. Koopman, B.: Hamiltonian systems and transformations in Hilbert space. Proc. Natl. Acad. Sci. U.S.A. **17**, 315–318 (1931)
8. Kou, J., Le Clainche, S.: A reduced-order model for compressible flows with buffeting condition using higher order dynamic mode decomposition. Phys. Fluids **30**(1), 016103 (2018)
9. Le Clainche, S., Ferrer, E.: A reduced order model to predict transient flows around straight bladed vertical axis wind turbines. Energies **11**(3), 566–578 (2018)
10. Le Clainche, S., Li, I.J., Theofilis, V., Soria, J.: Flow around a hemisphere-cylinder at high angle of attack and low reynolds number. Part I: experimental and numerical investigation. Aerosp. Sci. Technol. **44**, 77–87 (2015)
11. Le Clainche, S., Lorente, L., Vega, J.M.: Wind predictions upstream wind turbines from a LiDAR database. Energies **11**(3), 543–558 (2018)
12. Le Clainche, S., Moreno-Ramos, R., Taylor, P., Vega, J.M.: New robust method to study flight flutter testing. J. Aircr. (2018). https://doi.org/10.2514/1.C034863
13. Le Clainche, S., Pérez, J.M., Vega, J.M.: Spatio-temporal flow structures in the three-dimensional wake of a circular cylinder. Fluid Dyn. Res. (2018). https://doi.org/10.1088/1873-7005/aab2f1
14. Le Clainche, S., Rodríguez, D., Theofilis, V., Soria, J.: Flow around a hemisphere-cylinder at high angle of attack and low reynolds number. Part II: POD and DMD applied to reduced domains. Aerosp. Sci. Technol. **44**, 88–100 (2015)
15. Le Clainche, S., Vega, J.M.: Higher order dynamic mode decomposition. SIAM J. Appl. Dyn. Syst. **16**(2), 882–925 (2017)
16. Le Clainche, S., Vega, J.M.: Higher order dynamic mode decomposition to identify and extrapolate flow patterns. Phys. Fluids **29**(8), 084102 (2017)

17. Le Clainche, S., Vega, J.M.: Spatio-temporal Koopman decomposition. J. Nonlinear Sci. **28**, 1793–1842 (2018)
18. Le Clainche, S., Vega, J.M., Soria, J.: Higher order dynamic mode decomposition for noisy experimental data: flow structures on a Zero-Net-Mass-Flux jet. Exp. Therm. Fluid Sci. **88**, 336–353 (2017)
19. Rowley, C.W., Mezić, I., Bagheri, S., Schlatter, P., Henningson, D.S.: Spectral analysis of nonlinear flows. J. Fluid Mech. **641**, 115–127 (2009)
20. Schmid, P.: Dynamic mode decomposition of numerical and experimental data. J. Fluid Mech. **656**, 5–28 (2010)
21. Schmid, P.: Application of the dynamic mode decomposition to experimental data. Exp. Fluids **50**(4), 1123–1130 (2011)
22. Sirovich, L.: Turbulence and the dynamics of coherent structures. Parts I–III. Q. Appl. Math. **45**(3), 561–571 (1987)
23. Takens, F.: Detecting strange attractors in turbulence. In: Rand, D.A., Young, L.-S. (eds.) Lecture Notes in Mathematics, vol. 525, pp. 366–381. Springer (1981)

Low Cost Methods for Computing Instabilities in Boundary Layer Flows

Juan A. Martin[1]([✉]) and Pedro Paredes[2]

[1] School of Aerospace Engineering, Universidad Politécnica de Madrid,
28040 Madrid, Spain
`juanangel.martin@upm.es`
[2] National Institute of Aerospace, Hampton, VA 23666, USA

Abstract. Delaying laminar to turbulent transition is of crucial importance in aerospace, since friction drag in the turbulent regime is considerably higher than on laminar surfaces. Several procedures have been reported for transition delay, with the objective of reducing fuel and emissions. One of the most promising strategy consists of modulating the boundary layer velocity profile for achieving transition delay. An usual way to perform it is by perturbing the flow with counter-rotating vortices that exhibit transient, non-modal growth and lead to streamwise aligned streaks inside the boundary layer, which have been proved (theoretical and experimentally) to be very robust flow structures.

Hence, we propose a numerical tool to analyze the stability properties of the streaky flow without big computational resources. Since streaks depend on how they are generated, and it is linked to the geometric configuration of the surface, a cheap CPU numerical analysis is required in order to perform large parametric studies for streak optimization. For this end, we use the Boundary Region Equations (BREs) to simulate the nonlinear downstream evolution of the perturbed boundary layer flow. Regarding the stability analysis, the linear three-dimensional plane-marching Parabolized Stability Equations concept constitutes the best candidate for this task.

Herein, we present the results of a thorough parametric study of the instability characteristics of the modified incompressible zero-pressure-gradient flat-plate boundary layer modulated through finite-amplitude linearly optimal disturbances, and by means of steady vortices developing in the streamwise direction. The instability characteristics of this kind of boundary layer flows will be presented and compared with high order computational methods.

Keywords: Stability analysis · Boundary Region Equations ·
Plane-marching Parabolized Stability Equations · Boundary layer

1 Introduction

Streaks are three-dimensional flow structures that appear inside the boundary layer and take form of spanwise thin and streamwise elongated regions alternating high and low speed flow in the spanwise direction. The stability of the

© Springer Nature Switzerland AG 2020
F. Martínez Álvarez et al. (Eds.): SOCO 2019, AISC 950, pp. 596–606, 2020.
https://doi.org/10.1007/978-3-030-20055-8_57

boundary layer with finite amplitude streaky distortions has been theoretically analyzed [4] and experimentally tested [7], detecting that the increase of the amplitude of the streaks can have a stabilizing effect in the Blasius boundary layer, delaying the onset of turbulence.

Steady and stable streaks can be employed then with delaying laminar-turbulent transition purposes. They can be excited by means of different experimental procedures, arising streaks with distinct characteristics and shapes. Some of those mechanism are, for instance, the use of a spanwise array of small cylindrical roughness elements attached to the plate near the leading edge, which can be found in [6–8], where streaks of moderate intensity (around 12% of the free-stream velocity) are reported. Recently, a new mechanism of generating streaks has been proposed in [9], consisting on a spanwise array of small winglet pairs known as miniature vortex generators (MVGs). The streaks created by MVGs are able to reach maximum amplitudes up to 32% of the free-stream velocity, increasing considerably the potential stabilization effect, as detailed in the experiments of [24] and references therein.

Therefore, the development of a cheap numerical method able to reproduce the different configurations of the resulting streaks is of crucial importance to optimize this passive flow control mechanism. For this purpose, in order to describe the base flow, we follow the Boundary Region Equations (BREs) formulation [23]. The asymptotic structure of the streaks when $Re = \sqrt{xU/\nu} \gg 1$ exhibits one long spatial scale in the streamwise direction and two short spatial scales, i.e., the wall normal and spanwise directions, leading to a system of equations fully parabolic in the streamwise direction [17].

Regarding the stability analysis, the 3D plane-marching Parabolized Stability Equations (PSE) can be derived from the three-dimensional stability equations when the base flow can be assumed to experience slow variations along one of the three spatial directions. The 3D plane-marching PSE are valid for convectively unstable flows, such as the present case. The complete procedure can be followed in [19].

In this work, we employ the formulation introduced in previous paragraph, and present the results of a thorough parametric study of the instability characteristic for the flat plate boundary layer flow modulated through finite-amplitude linearly optimal disturbances, and by means of steady streamwise vortices.

2 Theory

The next section is devoted to present the numerical techniques employed for the simulation of the streaky base flow and the posterior instability analysis.

2.1 Streaky Base Flow Formulation

The configuration here considered is a flat plate boundary layer at zero angle of incidence with a spanwise periodic array of stationary streamwise streaks. Streaks can be naturally generated by small perturbances on the free stream that enter the boundary layer producing the so-called Klebanoff modes [15]. They can be also artificially created through (active or passive) devices placed inside the boundary layer, as already pointed on Sect. 1. The asymptotic structure of the streaks when the Reynolds number (defined in the usual way, $Re = U_\infty L/\nu$, where U_∞ is the free-stream velocity, L is the characteristic spatial length and ν is the kinematic viscosity) is $Re \gg 1$ exhibits one long spatial scale in the streamwise direction, and two short spatial scales, in the wall normal and spanwise directions:

$$\bar{x} \sim L, \quad (\bar{y}, \bar{z}) \sim L/\sqrt{Re}, \quad \bar{u} \sim U_\infty, \quad (\bar{v}, \bar{w}) \sim U_\infty/\sqrt{Re}$$

Therefore, the asymptotic downstream evolution of streaks is described by the BREs, which can be written in their non-dimensional form as:

$$
\begin{aligned}
u_x + v_y + w_z &= 0 \\
uu_x + vu_y + wu_z &= u_{yy} + u_{zz} \\
uv_x + vv_y + wv_z &= -p_y + v_{yy} + v_{zz} \\
uw_x + vw_y + ww_z &= -p_z + w_{yy} + w_{zz}.
\end{aligned}
\tag{1}
$$

where the streamwise pressure variation is given by the first order of that system, providing $\dfrac{\partial p}{\partial x} = \dfrac{dU(x)}{dx} = 0$. The appropriate boundary conditions are periodicity in z, together with no slip at the bottom wall and, at the upper edge of the boundary layer, the velocities must behave as

$$(v, w) \to (1, 0) \quad \text{as} \quad y \to \infty. \tag{2}$$

The BREs are derived from the complete incompressible Navier-Stokes formulation using a boundary layer like approximation in the limit of high Reynolds number. The BRE formulation is well known [23], and it has been previously used to describe the nonlinear development of Görtler vortices in curved wall boundary layers [11], and for the analysis of the sustained states that come out of the nonlinear interaction of a slow streamwise vortex and a fast TS wave (see [12] and references therein). The BRE were used to analyze the streak excitation by vortical structures on the free stream, both in steady and unsteady conditions, in [10, 16, 22, 30].

The system of Eq. (1) is fully parabolic in the streamwise direction, and it is integrated using a downstream marching scheme. A detailed description of the derivation of the BREs and of the applied numerical integration procedure can be found in [17].

2.2 Stability Analysis

The equations of fluid motion, the Navier-Stokes equations (NS), consist of an Initial Value Problem (IVP) for $\mathbf{q} = [u, v, w, p]^T$, which is the vector of the primitive flow variables. Stability analysis theory considers the decomposition of all flow quantities into a base flow, $\bar{\mathbf{q}} = [\bar{u}, \bar{v}, \bar{w}, \bar{p}]^T$, upon which small-amplitude perturbations, $\tilde{\mathbf{q}} = [\tilde{u}, \tilde{v}, \tilde{w}, \tilde{p}]^T$, are superposed such that $\mathbf{q} = \bar{\mathbf{q}} + \varepsilon\tilde{\mathbf{q}}$, where $\varepsilon \ll 1$. By introducing this decomposition into the equations of motion, the linearized Navier-Stokes equations (LNS) are recovered.

Spatial Biglobal Analysis. Assuming that the basic flow depends on two out of the three spatial coordinates, the two-dimensional parallel flow is assumed and the biglobal instability theory is applicable (see [28,29] for a review).

In the present spatial context, the disturbances are three-dimensional, but a sinusoidal dependence with the homogeneous x-direction is assumed, with the periodicity length $L_x = 2\pi/\alpha$, as follows

$$\tilde{\mathbf{q}}(x, y, z, t) = \hat{\mathbf{q}}(y, z)exp[i(\alpha x - \omega t)], \tag{3}$$

where $\hat{\mathbf{q}} = [\hat{u}, \hat{v}, \hat{w}, \hat{p}]^T$, is the vector of amplitude functions, and ω is the temporal wavenumber. Substituting the ansatz (3) into the incompressible LNS results in a two-dimensional Generalized Eigenvalue Problem (GEVP) nonlinear on eigenvalue α, which is converted into a linear eigenvalue problem, using the companion matrix method [3,27], in which an auxiliary vector is defined, $\hat{\mathbf{q}}_{ext} = [\hat{u}, \hat{v}, \hat{w}, \hat{p}, \alpha\hat{u}, \alpha\hat{v}, \alpha\hat{w}]^T$, and the resulting GEVP is written as

$$\mathbf{A}\hat{\mathbf{q}}_{ext} = \alpha\mathbf{B}\hat{\mathbf{q}}_{ext}. \tag{4}$$

The entries of the matrices $\mathbf{A}$ and $\mathbf{B}$ and more details about the derivation and the solution procedure are found in [19,21].

Plane-Marching PSE. The plane-marching PSE, also known as PSE-3D [19, 20], can be derived from the three-dimensional stability equations when the basic flow can be assumed to experience slow variations along the streamwise direction, while exhibits a strong dependence on the other two spatial directions.

Following the same reasoning made on the classical (line-marching) PSE (see [13]), the plane-marching PSE are valid for convectively unstable flows. The disturbance quantities are expanded as a fast varying wavy function with a slowly varying amplitude function $\hat{\mathbf{q}}(x, y, z)$, as

$$\tilde{\mathbf{q}}(x, y, z) = \hat{\mathbf{q}}(x, y, z) \exp\left[i\left(\int_x \alpha(x')dx' - \omega t\right)\right]. \tag{5}$$

The linear form of the plane-marching PSE can be written in a compact form as

$$\mathbf{L}\hat{\mathbf{q}} + \mathbf{M}\frac{\partial\hat{\mathbf{q}}}{\partial x} = 0. \tag{6}$$

The entries of $\mathbf{L}$ and $\mathbf{M}$ and more details about the derivation, the numerical properties, and the solution procedure of the plane-marching PSE methodology are found in [19].

3   Results

The next section is dedicated to present the main results obtained that involve, in one hand, the stabilizing effect of the streaks on the Tollmien-Schlichting (TS) waves, and in the other hand the arise of secondary instability (SI) modes beyond a certain streak amplitude, which are amplified as a consequence of the three-dimensional shear layer formed by the streaks.

3.1   Stability of Optimal Streaks

The configuration considered herein is a flat plate boundary layer at zero angle of incidence with a spanwise periodic array of steady streaks developing in the streamwise direction, acting as the inlet conditions. For the simulation of the basic flow, the perturbation profile $(u_p, v_p, w_p,)$ corresponds to the optimal streak obtained in [1] described as

$$\left(u^0, v^0, w^0,\right) = (U_b, 0, 0,) + A_{s0}\left(u_p \cos\left(\beta_0 z\right), v_p \cos\left(\beta_0 z\right), w_p \sin\left(\beta_0 z\right)\right), \quad (7)$$

where U_b is the Blasius profile and $\beta_0 = 0.45$. The starting point is set to $x = 0.4$ and the amplitude of the streak is estimated using the expression

$$A_s\left(x\right) = \frac{1}{2}\left(\max_{y,z}\left(u - U_b\right) - \min_{y,z}\left(u - U_b\right)\right), \quad (8)$$

which was introduced in [2] and basically measures the spanwise departure of u from the Blasius profile. The initial amplitude of the streaks A_{s0} is selected to reach a maximum value of A_s between the range of values 0 (i.e., Blasius profile) and 0.4.

Although generating and analyzing a three-dimensional base flow requires a high computational cost, employing the formulation explained in Sect. 2 allows to perform a complete parametric study without large computational resources.

Figure 1 shows the most distinctive results when the maximum streak amplitude varies. In Fig. 1(a) the neutral stability curves ($\sigma_K = 0$, where σ_K is the growth rate based on the kinetic energy, $\sigma_K = -\alpha_i + (1/2)\, d/dx(\log \hat{K}(x))$, where $\alpha_i = \Im(\alpha)$ and $\hat{K}$ is the kinetic energy of the slow varying amplitude function) are displayed on a $Re - F$ chart ($F = \omega/Re \times 10^6$) corresponding to the different values of maximum streak amplitudes computed $A_{s_{max}} = 0.0(A), 0.14(B), 0.20(C), 0.25(D), 0.30(E), 0.34(F)$ and $0.38(G)$. Two groups of curves can be observed. Packed lines, representing low intensity streaks that have the effect of damping the Tollmien-Schlichting waves (the unstable region is reduced), increasing this result with the streak intensity, which is in agreement with the previous studies performed in [5] for linear optimal streaks. Isolated dash lines depict streaks with a maximum amplitude value above the critical intensity of 26%, streaks E, F and G, thus they are subjected to experience secondary instability [2] and they exhibit a growing shear layer mode. In this case, the unstable region is clearly enlarged for the two higher intensity streaks, compared with the regions corresponding to sub-critical streaks, and at same time the point where the disturbances start to grow is displaced downstream. It is significant to highlight that the neutral curves are plotted on a local level corresponding to a value of $Re_{\delta_0} = 400$.

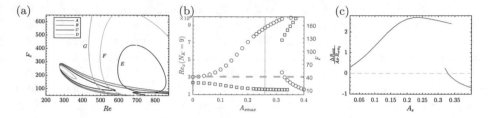

Fig. 1. (a) Neutral stability curves for the streaks with a maximum amplitude of $A_{s_{max}} = 0.00$ (A), 0.14 (B), 0.20 (C), 0.25 (D), 0.30 (E), 0.34 (F) and 0.38 (G). (b) Critical Reynolds number (blue circles) and critical frequency (red squares) as function of maximum streak amplitude. Vertical line depicts the streak intensity from which SI arises. (c) Stabilization efficiency as function of maximum streak amplitude red TS waves, blue SI perturbations.

Figure 1(b) depicts the variation of the critical Reynolds number and the corresponding critical frequency. It is defined as the earliest point were the logarithmic amplification N-factor reaches a value of 9, and transition from laminar to turbulent regime is assumed to occur [14]. The logarithmic amplification N-factor is based on the kinetic energy, and defined as $N_K = \int_{x_i}^{x} \sigma_K \, dx$, where x_i denotes the lower branch of the neutral stability curve. Horizontal dash line marks the critical Reynolds number for the unperturbed boundary layer flow. It can be observed that the critical Reynolds number (upper blue circles branch above dash line) is displaced downstream, as the streak amplitude is increased till a certain point (vertical line) where SI instabilities arises (lower blue circles branch) and it is brought upstream. The maximum streak amplitude for delaying transition can be estimated then when the SI branch encounters the value of $Re(N_K = 9)$ for the unperturbed flow, taking a value of $A_{s_{max}} = 0.33$. Regarding the critical frequency, it exhibits low values when TS waves are damped, while changing to high values as SI arises.

Figure 1(c) compares the variation of the critical Reynolds number, Re_{crit} (defined as pointed in previous paragraph) with the one of the unperturbed boundary layer flow, Re_{crit_0}. If we introduce in the expression the maximum streak intensity (as a way of measuring the amount of modified flow), a streak efficiency can be defined, as $\eta = \dfrac{\Delta Re_{crit}}{A_{s_{max}} Re_{crit_0}}$. Red line represents the efficiency variation for low intensity streaks, where the damping of TS waves is found; while blue line corresponds to high intensity streaks, arising the amplification of SI perturbations. As the streak amplitude is increased, a growth in the efficiency can be observed, until reaching a maximum around streak intensity of 0.23. Afterwards, a clear drop is followed once SI region is entered.

3.2   Stability of Streamwise Vortices

The configuration examined in this section is a flat plate boundary layer at zero angle of incidence with a spanwise periodic array of steady vortices developing in the streamwise direction, being the inlet conditions. The periodic array of vortex pairs is placed at a distance X_v from the leading edge of the flat plate and a distance h_v from the wall. Based on the analysis of [25], the simpler Rankine vortex model is chosen. Therefore, we take the same values for the vortex parameters and we simulate the flow induced by free-stream vortices.

A complete description of the procedure to obtain the following results, including the validation of the method, is reported in [18]. A parametric study is performed by placing the Rankine vortex pairs at several distances h_v from the plate wall, while located with a constant location from the leading edge. The vortex intensity is changed in such a way that the reached maximum streak amplitude, $A_{s_{max}}$, reaches (a) a value of $A_{s_{max}} = 0.20$, (it will yield to stable streaks that are expected to affect the TS instabilities, according to [2]); and (b) a value of $A_{s_{max}} = 0.30$, larger than the threshold for streak instability given by [2] of 26% of the free-stream velocity. Therefore, sinuous, shear layer, secondary instabilities are expected to be amplified.

Firstly, we analyze the stability characteristics of low intensity vortices. Figure 2 displays the neutral stability curves ($\sigma_K = 0$), for the selected values of the location of the vortices, h_v. The utmost neutral curve (dash-dot line) displays the neutral curve of the Blasius boundary layer. Figure 2(a) shows the corresponding curves for wall distance locations $h_v = 0.36$, 0.72, 1.09, and 1.45; uniformly placed into higher Re_δ values. It is significant to highlight that the neutral curves are plotted on a local level corresponding to a value of $Re_{\delta_0} = 726$. For these values of h_v, there is a clear damping effect on the TS waves, being increased as moved away from the wall, i.e., displacing further downstream the instability region. Figure 2(b) depicts the neutral curves for $h_v = 2.18$ and 2.90. The red line stands for $h_v = 2.18$, and, even though the critical Reynolds number is found in an upstream position, the instability region corresponds to the one with the minimum area among all the cases simulated. By increasing further distance from the wall to $h_v = 2.90$, the TS damping effect almost does not occur until high Re_δ values are reached, meaning that the perturbation delays in affecting the flow. And secondly, we examine the stability characteristics of high intensity vortices. Figure 3(a) displays the neutral stability curve ($\sigma_K = 0$) for the vortices situated at wall distances of $h_v = 0.72$, 1.09, 1.45, 2.18, 2.90 and 3.63, along with the corresponding curve for Blasius boundary layer. As can be appreciated, the streaks closer to the wall ($h_v = 0.72$, 1.09, 1.45) involve a wide range of the frequency domain with an unstable region (stripped area), located in an upstream position after the vortex initialization. This can be assumed to be the unstable area that emerges in the near-wake of the MVGs, as pointed by [26]. For the selected cases, it is remarkable to note that these modes are not found to be linked to sinuous modes, because for the range of frequencies studied they monotonically decrease downstream. The sinuous instability mode is unstable for $h_v = 1.45$, 2.18, 2.90, and 3.63. The effect of the distance now is revealed shifting downstream and enlarging the neutral curve region. Consequently, for streaks presenting the same intensity, the N-factor threshold for bypass transition depends on the initial wall normal position of the vortices, being reached in different streamwise locations.

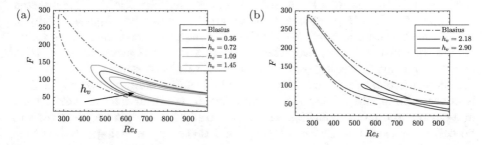

Fig. 2. Neutral stability curves for vortices with maximum streak amplitude of 0.20 and located at a distance (a) $h_v = 0.36, 0.72, 1.09$ and 1.45 and (b) $h_v = 2.18$ and 2.90 from the wall.

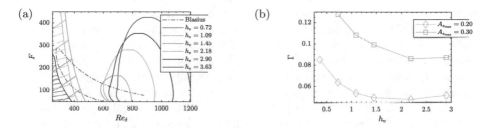

Fig. 3. (a) Neutral stability curves for vortices with maximum streak amplitude of 0.30 and located at a distance $h_v = 0.72, 1.09, 1.45, 2.18, 2.90$ and 3.63. (b) Circulation (Γ) of the initial vortices required to reach a maximum streak amplitude of $A_{s_{max}} = 0.20$ and $A_{s_{max}} = 0.30$ as a function of the vortex pair wall normal location, h_v.

Finally, is worth to mention that the initial vortex energy (Γ) needed to reach a given streak intensity depends on where the vortices are initially located. Figure 3(b) shows the initial circulation of the vortices as a function of the distance to the wall, h_v, required to have $A_{s_{max}} = 0.20$ and $A_{s_{max}} = 0.30$. The circulation presents a minimum for both cases, with the wall-normal location at $h_v = 2.18$. Therefore, this value plays the optimal wall normal location for the vortex pair to induce the streak employing the minimum amount of initial intensity, and achieve the most damping on TS waves as previously analyzed.

4   Conclusions

Results on the stability characteristics of a boundary layer flow perturbed by streaks are presented. The theoretical analysis performed shows that the correct asymptotic behaviour of this type of flows is properly modeled by the use of the Boundary Region Equations (BREs), and the stability analysis is completed by employing the 3D plane-marching Parabolized Stability Equations (PSE). A very small computational effort is therefore required, accomplishing accurate results with just a desktop workstation (Intel Xeon CPU processor with 32Gb of RAM memory), needing only minutes to obtain each streaky flow, hours to complete a neutral curve, and days to fulfil the entire parametric study. Direct Numerical Simulations (DNS) took from days to weeks to achieve same results in a cluster computation facility. Thus, the parametric study here presented can be easily achieved (even with a laptop) and taken as a starting point to tune the streak governing parameters with a minor computational cost, to further increase the accuracy of the results employing a more expensive CPU method, as DNS.

This analysis reports the amplification of Tollmien-Schlichting (TS) waves and secondary instabilities (SI). A parametric study has been executed, comprising the analysis of optimal streaks, typically employed to model the streaky flow after roughness elements placed inside the boundary layer; and of streamwise vortices, usually representing the perturbation induced by miniature vortex generators (MVGs). In the first case, a wide range of different maximum amplitude of the streaks has been simulated, recovering the known attenuating effect on the TS waves. Likewise, the emerging SI for high amplitude streaks has been captured, locating the streak intensity for maximum efficiency, and the stabilizing limit of the streaks. In the second case, the maximum amplitude of the streaks has been kept constant up to two values, varying the wall normal position of the streamwise vortices. The main finding has been the stabilizing effect over the TS waves as the vortices moves away from the plate surface, up to a certain value, from which the effect ceases. Regarding the SI, the unstable region of sinuous shear layer instabilities is shifted downstream and enlarged as the streak moves away from the plate. This study reveals the existence of an optimal wall normal distance for the vortex pair, which requires the minimal vortex circulation to achieve a given streak intensity, matching with the location that achieves the most damping on TS waves.

References

1. Andersson, P., Berggren, M., Henningson, D.S.: Optimal disturbances and bypass transition in boundary layers. Phys. Fluids **11**, 134–150 (1999)
2. Andersson, P., Brandt, L., Bottaro, A., Henningson, D.: On the breakdown of boundary layer streaks. J. Fluid Mech. **428**, 29–60 (2001)
3. Bridges, T., Morris, P.: Differential eigenvalue problems in which the parameter appears nonlinearly. J. Comput. Phys. **55**, 437–460 (1984)

4. Cossu, C., Brandt, L.: Stabilization of Tollmien-Schlichting waves by finite amplitude optimal streaks in the Blasius boundary layer. Phys. Fluids **14**(8), L57–L60 (2002)

5. Cossu, C., Brandt, L.: On Tollmien-Schlichting-like waves in streaky boundary layers. Eur. J. Mech. B. Fluids **23**, 815–833 (2004)

6. Fransson, J.H.M., Brandt, L., Talamelli, A., Cossu, C.: Experimental study of the stabilization of Tollmien-Schlichting waves by finite amplitude streaks. Phys. Fluids **17**(5), 054110 (2005)

7. Fransson, J.H.M., Talamelli, A., Brandt, L., Cossu, C.: Delaying transition to turbulence by a passive mechanism. Phys. Rev. Lett. **96**, 064501 (2006)

8. Fransson, J.H.M., Brandt, L., Talamelli, A., Cossu, C.: Experimental and theoretical investigation of the nonmodal growth of steady streaks in a flat plate boundary layer. Phys. Fluids **16**(10), 3627–3638 (2004)

9. Fransson, J.H.M., Talamelli, A.: On the generation of steady streamwise streaks in flat-plate boundary layers. J. Fluid Mech. **698**, 211–234 (2012)

10. Goldstein, M., Sescu, A.: Boundary-layer transition at high free-stream disturbance levels - beyond Klebanoff modes. J. Fluid Mech. **613**, 95–124 (2008)

11. Hall, P.: The nonlinear development of Görtler vortices in growing boundary-layers. J. Fluid Mech. **193**, 243–266 (1988)

12. Hall, P., Smith, F.T.: On strongly nonlinear vortex wave interactions in boundary-layer-transition. J. Fluid Mech. **227**, 641–666 (1991)

13. Herbert, T.: Parabolized stability equations. Ann. Rev. Fluid Mech. **29**, 245–283 (1997)

14. Ingon, J.L.V.: The e^N method for transition prediction. Historical review of work at TU Delft. In: AIAA 38th Fluid Dynamics Conference and Exhibit, p. 3830 (2008)

15. Klebanoff, P.: Effect of free-stream turbulence on a laminar boundary layer. Bull. Am. Phys. Soc. **10**(11), 1323 (1971)

16. Leib, S., Wundrow, D., Goldstein, M.: Effect of free-stream turbulence and other vortical disturbances on a laminar boundary layer. J. Fluid Mech. **380**, 169–203 (1999)

17. Martin, J.A., Martel, C.: Nonlinear streak computation using boundary region equations. Fluid Dyn. Res. **44**, 045503 (2012)

18. Martín, J.A., Paredes, P.: Three-dimensional instability analysis of boundary layers perturbed by streamwise vortices. Theor. Comput. Fluid Dyn. **31**(5), 505–517 (2017)

19. Paredes, P.: Advances in global instability computations: from incompressible to hypersonic flow. Ph.D. thesis, Technical University of Madrid (2014)

20. Paredes, P., Hanifi, A., Theofilis, V., Henningson, D.: The nonlinear PSE-3D concept for transition prediction in flows with a single slowly-varying spatial direction. Procedia IUTAM **14C**, 35–44 (2015)

21. Paredes, P., Hermanns, M., Le Clainche, S., Theofilis, V.: Order 10^4 speedup in global linear instability analysis using matrix formation. CMAME **253**, 287–304 (2013)

22. Ricco, P., Luo, J., Wu, X.: Evolution and instability of unsteady nonlinear streaks generated by free-stream vortical disturbance. J. Fluid Mech. **677**, 1–38 (2011)

23. Rubin, S.G., Tannehill, J.C.: Parabolized reduced Navier-Stokes computational techniques. Ann. Rev. Fluid Mech. **24**, 117–144 (1992)

24. Shahinfar, S., Sattarzadeh, S.S., Fransson, J.H.M.: Passive boundary layer control of oblique distrubances by finite-amplitude streaks. J. Fluid Mech. **749**, 1–36 (2014)

25. Siconolfi, L., Camarri, S., Fransson, J.H.M.: Boundary layer stabilization using free-stream vortices. J. Fluid Mech. **764**, R2 (2015)

26. Siconolfi, L., Camarri, S., Fransson, J.: Stability analysis of boundary layers controlled by miniature vortex generators. J. Fluid Mech. **784**, 596–618 (2015)
27. Theofilis, V.: Spatial stability of incompressible attachment line flow. Theor. Comput. Fluid Dyn. **7**, 159–171 (1995)
28. Theofilis, V.: Advances in global linear instability of nonparallel and three-dimensional flows. Prog. Aerosp. Sci. **39**(4), 249–315 (2003)
29. Theofilis, V.: Global linear instability. Ann. Rev. Fluid Mech. **43**, 319–352 (2011)
30. Wu, X., Zhao, D., Luo, J.: Excitation of steady and unsteady Görtler vortices by free-stream vortical disturbance. J. Fluid Mech. **682**, 66–100 (2011)

Author Index

© Springer Nature Switzerland AG 2020
F. Martínez Álvarez et al. (Eds.): SOCO 2019, AISC 950, pp. 607–609, 2020.
https://doi.org/10.1007/978-3-030-20055-8

Printed in the United States
By Bookmasters